First Course in the Finite Element Method

Structural Analysis

Daryl L. Logan

CENGAGE Learning™

Australia • Brazil • Japan • Korea • Mexico • Singapore • Spain • United Kingdom • United States

CENGAGE Learning™

First Course in the Finite Element Method
Structural Analysis

Executive Editors:
Maureen Staudt
Michael Stranz

Senior Project Development Manager:
Linda deStefano

Marketing Specialist:
Courtney Sheldon

Senior Production/
Manufacturing Manager:
Donna M. Brown

PreMedia Manager:
Joel Brennecke

Sr. Rights Acquisition Account Manager:
Todd Osborne

Cover Image:
Getty Images*

*Unless otherwise noted, all cover images used by Custom Solutions, a part of Cengage Learning, have been supplied courtesy of Getty Images with the exception of the Earthview cover image, which has been supplied by the National Aeronautics and Space Administration (NASA).

Printed in the United States of America

For product information and technology assistance, contact us at
Cengage Learning Customer & Sales Support, 1-800-354-9706
For permission to use material from this text or product,
submit all requests online at **cengage.com/permissions**
Further permissions questions can be emailed to
permissionrequest@cengage.com

This book contains select works from existing Cengage Learning resources and was produced by Cengage Learning Custom Solutions for collegiate use. As such, those adopting and/or contributing to this work are responsible for editorial content accuracy, continuity and completeness.

ISBN-13: 978-1-111-95291-4

ISBN-10: 1-111-95291-4

Cengage Learning
5191 Natorp Boulevard
Mason, Ohio 45040
USA

Cengage Learning is a leading provider of customized learning solutions with office locations around the globe, including Singapore, the United Kingdom, Australia, Mexico, Brazil, and Japan. Locate your local office at:
international.cengage.com/region

Cengage Learning products are represented in Canada by Nelson Education, Ltd.
For your lifelong learning solutions, visit **www.cengage.com /custom**
Visit our corporate website at **www.cengage.com**

Custom Contents

CHAPTER 1

Introduction

Prologue

The finite element method is a numerical method for solving problems of engineering and mathematical physics. Typical problem areas of interest in engineering and mathematical physics that are solvable by use of the finite element method include structural analysis, heat transfer, fluid flow, mass transport, and electromagnetic potential.

For problems involving complicated geometries, loadings, and material properties, it is generally not possible to obtain analytical mathematical solutions. Analytical solutions are those given by a mathematical expression that yields the values of the desired unknown quantities at any location in a body (here total structure or physical system of interest) and are thus valid for an infinite number of locations in the body. These analytical solutions generally require the solution of ordinary or partial differential equations, which, because of the complicated geometries, loadings, and material properties, are not usually obtainable. Hence we need to rely on numerical methods, such as the finite element method, for acceptable solutions. The finite element formulation of the problem results in a system of simultaneous algebraic equations for solution, rather than requiring the solution of differential equations. These numerical methods yield approximate values of the unknowns at discrete numbers of points in the continuum. Hence this process of modeling a body by dividing it into an equivalent system of smaller bodies or units (finite elements) interconnected at points common to two or more elements (nodal points or nodes) and/or boundary lines and/or surfaces is called *discretization*. In the finite element method, instead of solving the problem for the entire body in one operation, we formulate the equations for each finite element and combine them to obtain the solution of the whole body.

Briefly, the solution for structural problems typically refers to determining the displacements at each node and the stresses within each element making up the structure that is subjected to applied loads. In nonstructural problems, the nodal unknowns may, for instance, be temperatures or fluid pressures due to thermal or fluid fluxes.

This chapter first presents a brief history of the development of the finite element method. You will see from this historical account that the method has become a practical one for solving engineering problems only in the past 50 years (paralleling the developments associated with the modern high-speed electronic digital computer). This historical account is followed by an introduction to matrix notation; then we describe the need for matrix methods (as made practical by the development of the modern digital computer) in formulating the equations for solution. This section discusses both the role of the digital computer in solving the large systems of simultaneous algebraic equations associated with complex problems and the development of numerous computer programs based on the finite element method. Next, a general description of the steps involved in obtaining a solution to a problem is provided. This description includes discussion of the types of elements available for a finite element method solution. Various representative applications are then presented to illustrate the capacity of the method to solve problems, such as those involving complicated geometries, several different materials, and irregular loadings. Chapter 1 also lists some of the advantages of the finite element method in solving problems of engineering and mathematical physics. Finally, we present numerous features of computer programs based on the finite element method.

1.1 Brief History

This section presents a brief history of the finite element method as applied to both structural and nonstructural areas of engineering and to mathematical physics. References cited here are intended to augment this short introduction to the historical background.

The modern development of the finite element method began in the 1940s in the field of structural engineering with the work by Hrennikoff [1] in 1941 and McHenry [2] in 1943, who used a lattice of line (one-dimensional) elements (bars and beams) for the solution of stresses in continuous solids. In a paper published in 1943 but not widely recognized for many years, Courant [3] proposed setting up the solution of stresses in a variational form. Then he introduced piecewise interpolation (or shape) functions over triangular subregions making up the whole region as a method to obtain approximate numerical solutions. In 1947 Levy [4] developed the flexibility or force method, and in 1953 his work [5] suggested that another method (the stiffness or displacement method) could be a promising alternative for use in analyzing statically redundant aircraft structures. However, his equations were cumbersome to solve by hand, and thus the method became popular only with the advent of the high-speed digital computer.

In 1954 Argyris and Kelsey [6, 7] developed matrix structural analysis methods using energy principles. This development illustrated the important role that energy principles would play in the finite element method.

The first treatment of two-dimensional elements was by Turner et al. [8] in 1956. They derived stiffness matrices for truss elements, beam elements, and two-dimensional triangular and rectangular elements in plane stress and outlined the procedure commonly known as the *direct stiffness method* for obtaining the total structure stiffness matrix. Along with the development of the high-speed digital computer in the early 1950s, the work of Turner et al. [8] prompted further development of finite element stiffness equations expressed in matrix notation. The phrase *finite element* was introduced by Clough [9] in 1960 when both triangular and rectangular elements were used for plane stress analysis.

A flat, rectangular-plate bending-element stiffness matrix was developed by Melosh [10] in 1961. This was followed by development of the curved-shell bending-element stiffness matrix for axisymmetric shells and pressure vessels by Grafton and Strome [11] in 1963.

Extension of the finite element method to three-dimensional problems with the development of a tetrahedral stiffness matrix was done by Martin [12] in 1961, by Gallagher et al. [13] in 1962, and by Melosh [14] in 1963. Additional three-dimensional elements were studied by Argyris [15] in 1964. The special case of axisymmetric solids was considered by Clough and Rashid [16] and Wilson [17] in 1965.

Most of the finite element work up to the early 1960s dealt with small strains and small displacements, elastic material behavior, and static loadings. However, large deflection and thermal analysis were considered by Turner et al. [18] in 1960 and material nonlinearities by Gallagher et al. [13] in 1962, whereas buckling problems were initially treated by Gallagher and Padlog [19] in 1963. Zienkiewicz et al. [20] extended the method to visco-elasticity problems in 1968.

In 1965 Archer [21] considered dynamic analysis in the development of the consistent-mass matrix, which is applicable to analysis of distributed-mass systems such as bars and beams in structural analysis.

With Melosh's [14] realization in 1963 that the finite element method could be set up in terms of a variational formulation, it began to be used to solve nonstructural applications. Field problems, such as determination of the torsion of a shaft, fluid flow, and heat conduction, were solved by Zienkiewicz and Cheung [22] in 1965, Martin [23] in 1968, and Wilson and Nickel [24] in 1966.

Further extension of the method was made possible by the adaptation of weighted residual methods, first by Szabo and Lee [25] in 1969 to derive the previously known elasticity equations used in structural analysis and then by Zienkiewicz and Parekh [26] in 1970 for transient field problems. It was then recognized that when direct formulations and variational formulations are difficult or not possible to use, the method of weighted residuals may at times be appropriate. For example, in 1977 Lyness et al. [27] applied the method of weighted residuals to the determination of magnetic field.

In 1976 Belytschko [28, 29] considered problems associated with large-displacement nonlinear dynamic behavior, and improved numerical techniques for solving the resulting systems of equations. For more on these topics, consult the text by Belytschko, Liu, and Moran [58].

A relatively new field of application of the finite element method is that of bioengineering [30, 31]. This field is still troubled by such difficulties as nonlinear materials, geometric nonlinearities, and other complexities still being discovered.

From the early 1950s to the present, enormous advances have been made in the application of the finite element method to solve complicated engineering problems. Engineers, applied mathematicians, and other scientists will undoubtedly continue to develop new applications. For an extensive bibliography on the finite element method, consult the work of Kardestuncer [32], Clough [33], or Noor [57].

1.2 Introduction to Matrix Notation

Matrix methods are a necessary tool used in the finite element method for purposes of simplifying the formulation of the element stiffness equations, for purposes of long-hand solutions of various problems, and, most important, for use in programming the methods for high-speed electronic digital computers. Hence matrix notation represents a simple and easy-to-use notation for writing and solving sets of simultaneous algebraic equations.

Appendix A discusses the significant matrix concepts used throughout the text. We will present here only a brief summary of the notation used in this text.

A **matrix** *is a rectangular array of quantities arranged in rows and columns that is often used as an aid in expressing and solving a system of algebraic equations.* As examples of matrices that will be described in subsequent chapters, the force components (F_{1x}, $F_{1y}, F_{1z}, F_{2x}, F_{2y}, F_{2z}, \ldots, F_{nx}, F_{ny}, F_{nz}$) acting at the various nodes or points $(1, 2, \ldots, n)$ on a structure and the corresponding set of nodal displacements (d_{1x}, d_{1y}, d_{1z}, $d_{2x}, d_{2y}, d_{2z}, \ldots, d_{nx}, d_{ny}, d_{nz}$) can both be expressed as matrices:

$$\{F\} = \underline{F} = \begin{Bmatrix} F_{1x} \\ F_{1y} \\ F_{1z} \\ F_{2x} \\ F_{2y} \\ F_{2z} \\ \vdots \\ F_{nx} \\ F_{ny} \\ F_{nz} \end{Bmatrix} \qquad \{d\} = \underline{d} = \begin{Bmatrix} d_{1x} \\ d_{1y} \\ d_{1z} \\ d_{2x} \\ d_{2y} \\ d_{2z} \\ \vdots \\ d_{nx} \\ d_{ny} \\ d_{nz} \end{Bmatrix} \tag{1.2.1}$$

The subscripts to the right of F and d identify the node and the direction of force or displacement, respectively. For instance, F_{1x} denotes the force at node 1 applied in the x direction. The matrices in Eqs. (1.2.1) are called *column matrices* and have a size of $n \times 1$. The brace notation { } will be used throughout the text to denote a column matrix. The whole set of force or displacement values in the column matrix is simply represented by $\{F\}$ or $\{d\}$. A more compact notation used throughout this text to represent any rectangular array is the underlining of the variable; that is, $\underline{F}$ and $\underline{d}$ denote general matrices (possibly column matrices or rectangular matrices—the type will become clear in the context of the discussion associated with the variable).

The more general case of a known rectangular matrix will be indicated by use of the bracket notation [ ]. For instance, the element and global structure stiffness matrices $[k]$ and $[K]$, respectively, developed throughout the text for various element types (such as those in Figure 1–1 on page 10), are represented by square matrices given as

$$[k] = \underline{k} = \begin{bmatrix} k_{11} & k_{12} & \dots & k_{1n} \\ k_{21} & k_{22} & \dots & k_{2n} \\ \vdots & \vdots & & \vdots \\ k_{n1} & k_{n2} & \dots & k_{nn} \end{bmatrix} \tag{1.2.2}$$

and

$$[K] = \underline{K} = \begin{bmatrix} K_{11} & K_{12} & \dots & K_{1n} \\ K_{21} & K_{22} & \dots & K_{2n} \\ \vdots & \vdots & & \vdots \\ K_{n1} & K_{n2} & \dots & K_{nn} \end{bmatrix} \tag{1.2.3}$$

where, in structural theory, the elements k_{ij} and K_{ij} are often referred to as *stiffness influence coefficients.*

You will learn that the global nodal forces $\underline{F}$ and the global nodal displacements $\underline{d}$ are related through use of the global stiffness matrix $\underline{K}$ by

$$\underline{F} = \underline{K}\underline{d} \tag{1.2.4}$$

Equation (1.2.4) is called the *global stiffness equation* and represents a set of simultaneous equations. It is the basic equation formulated in the stiffness or displacement method of analysis. Using the compact notation of underlining the variables, as in Eq. (1.2.4), should not cause you any difficulties in determining which matrices are column or rectangular matrices.

To obtain a clearer understanding of elements K_{ij} in Eq. (1.2.3), we use Eq. (1.2.1) and write out the expanded form of Eq. (1.2.4) as

$$\begin{Bmatrix} F_{1x} \\ F_{1y} \\ \vdots \\ F_{nz} \end{Bmatrix} = \begin{bmatrix} K_{11} & K_{12} & \dots & K_{1n} \\ K_{21} & K_{22} & \dots & K_{2n} \\ \vdots & & & \\ K_{n1} & K_{n2} & \dots & K_{nn} \end{bmatrix} \begin{Bmatrix} d_{1x} \\ d_{1y} \\ \vdots \\ d_{nz} \end{Bmatrix} \tag{1.2.5}$$

Now assume a structure to be forced into a displaced configuration defined by $d_{1x} = 1, d_{1y} = d_{1z} = \cdots d_{nz} = 0$. Then from Eq. (1.2.5), we have

$$F_{1x} = K_{11} \qquad F_{1y} = K_{21}, \dots, F_{nz} = K_{n1} \tag{1.2.6}$$

Equations (1.2.6) contain all elements in the first column of $\underline{K}$. In addition, they show that these elements, $K_{11}, K_{21}, \dots, K_{n1}$, are the values of the full set of nodal forces required to maintain the imposed displacement state. In a similar manner, the second column in $\underline{K}$ represents the values of forces required to maintain the displaced state $d_{1y} = 1$ and all other nodal displacement components equal to zero. We should now have a better understanding of the meaning of stiffness influence coefficients.

Subsequent chapters will discuss the element stiffness matrices $\underline{k}$ for various element types, such as bars, beams, and plane stress. They will also cover the procedure for obtaining the global stiffness matrices $\underline{K}$ for various structures and for solving Eq. (1.2.4) for the unknown displacements in matrix $\underline{d}$.

Using matrix concepts and operations will become routine with practice; they will be valuable tools for solving small problems longhand. And matrix methods are crucial to the use of the digital computers necessary for solving complicated problems with their associated large number of simultaneous equations.

1.3 Role of the Computer

As we have said, until the early 1950s, matrix methods and the associated finite element method were not readily adaptable for solving complicated problems. Even though the finite element method was being used to describe complicated structures, the resulting large number of algebraic equations associated with the finite element method of structural analysis made the method extremely difficult and impractical to use. However, with the advent of the computer, the solution of thousands of equations in a matter of minutes became possible.

The first modern-day commercial computer appears to have been the Univac, IBM 701 which was developed in the 1950s. This computer was built based on vacuum-tube technology. Along with the UNIVAC came the punch-card technology whereby programs and data were created on punch cards. In the 1960s, transistor-based technology replaced the vacuum-tube technology due to the transistor's reduced cost, weight, and power consumption and its higher reliability. From 1969 to the late 1970s, integrated circuit-based technology was being developed, which greatly enhanced the processing speed of computers, thus making it possible to solve larger finite element problems with increased degrees of freedom. From the late 1970s into the 1980s, large-scale integration as well as workstations that introduced a windows-type graphical interface appeared along with the computer mouse. The first computer mouse received a patent on November 17, 1970. Personal computers had now become mass-market desktop computers. These developments came during the age of networked computing, which brought the Internet and the World Wide Web. In the 1990s the Windows operating system was released, making IBM and IBM-compatible PCs more user friendly by integrating a graphical user interface into the software.

The development of the computer resulted in the writing of computational programs. Numerous special-purpose and general-purpose programs have been written to handle various complicated structural (and nonstructural) problems. Programs such as [46–56] illustrate the elegance of the finite element method and reinforce understanding of it.

In fact, finite element computer programs now can be solved on single-processor machines, such as a single desktop or laptop personal computer (PC) or on a cluster of computer nodes. The powerful memories of the PC and the advances in solver programs have made it possible to solve problems with over a million unknowns.

To use the computer, the analyst, having defined the finite element model, inputs the information into the computer. This information may include the position of the element nodal coordinates, the manner in which elements are connected, the material properties of the elements, the applied loads, boundary conditions, or constraints, and the kind of analysis to be performed. The computer then uses this information to generate and solve the equations necessary to carry out the analysis.

1.4 General Steps of the Finite Element Method

This section presents the general steps included in a finite element method formulation and solution to an engineering problem. We will use these steps as our guide in developing solutions for structural and nonstructural problems in subsequent chapters.

For simplicity's sake, for the presentation of the steps to follow, we will consider only the structural problem. The nonstructural heat-transfer and fluid mechanics problems and their analogies to the structural problem are considered in Chapters 13 and 14.

Typically, for the structural stress-analysis problem, the engineer seeks to determine displacements and stresses throughout the structure, which is in equilibrium and is subjected to applied loads. For many structures, it is difficult to determine the distribution of deformation using conventional methods, and thus the finite element method is necessarily used.

There are two general direct approaches traditionally associated with the finite element method as applied to structural mechanics problems. One approach, called the *force*, or *flexibility, method,* uses internal forces as the unknowns of the problem. To obtain the governing equations, first the equilibrium equations are used. Then necessary additional equations are found by introducing compatibility equations. The result is a set of algebraic equations for determining the redundant or unknown forces.

The second approach, called the *displacement*, or *stiffness, method,* assumes the displacements of the nodes as the unknowns of the problem. For instance, compatibility conditions requiring that elements connected at a common node, along a common edge, or on a common surface before loading remain connected at that node, edge, or surface after deformation takes place are initially satisfied. Then the governing equations are expressed in terms of nodal displacements using the equations of equilibrium and an applicable law relating forces to displacements.

These two direct approaches result in different unknowns (forces or displacements) in the analysis and different matrices associated with their formulations (flexibilities or stiffnesses). It has been shown [34] that, for computational purposes, the displacement (or stiffness) method is more desirable because its formulation is simpler for most structural analysis problems. Furthermore, a vast majority of general-purpose finite element programs have incorporated the displacement formulation for solving structural problems. Consequently, only the displacement method will be used throughout this text.

Another general method that can be used to develop the governing equations for both structural and nonstructural problems is the variational method. The variational method includes a number of principles. One of these principles, used extensively throughout this text because it is relatively easy to comprehend and is often introduced in basic mechanics courses, is the theorem of minimum potential energy that applies to materials behaving in a linear-elastic manner. This theorem is explained and used in various sections of the text, such as Section 2.6 for the spring element, Section 3.10 for the bar element, Section 4.7 for the beam element, Section 6.2 for the constant-strain triangle plane stress and plane strain element, Section 9.1 for the axisymmetric element, Section 11.2 for the three-dimensional solid tetrahedral element, and Section 12.2 for the plate bending element. A functional analogous to that used in the theorem of minimum potential energy is then employed to develop the finite element equations for the nonstructural problem of heat transfer presented in Chapter 13.

Another variational principle often used to derive the governing equations is the principle of virtual work. This principle applies more generally to materials that behave in a linear-elastic fashion, as well as those that behave in a nonlinear fashion. The principle of virtual work is described in Appendix E for those choosing to use it for developing the general governing finite element equations that can be applied specifically to bars, beams, and two- and three-dimensional solids in either static or dynamic systems.

The finite element method involves modeling the structure using small interconnected elements called *finite elements*. A displacement function is associated with each finite element. Every interconnected element is linked, directly or indirectly, to every other element through common (or shared) interfaces, including nodes and/or boundary lines and/or surfaces. By using known stress/strain properties for the material making up the structure, one can determine the behavior of a given node in terms of the properties of every other element in the structure. The total set of equations describing the behavior of each node results in a series of algebraic equations best expressed in matrix notation.

We now present the steps, along with explanations necessary at this time, used in the finite element method formulation and solution of a structural problem. The purpose of setting forth these general steps now is to expose you to the procedure generally followed in a finite element formulation of a problem. You will easily understand these steps when we illustrate them specifically for springs, bars, trusses, beams, plane frames, plane stress, axisymmetric stress, three-dimensional stress, plate bending, heat transfer, and fluid flow in subsequent chapters. We suggest that you review this section periodically as we develop the specific element equations.

Keep in mind that the analyst must make decisions regarding dividing the structure or continuum into finite elements and selecting the element type or types to be used in the analysis (step 1), the kinds of loads to be applied, and the types of boundary conditions or supports to be applied. The other steps, 2–7, are carried out automatically by a computer program.

Step 1 Discretize and Select the Element Types

Step 1 involves dividing the body into an equivalent system of finite elements with associated nodes and choosing the most appropriate element type to model most closely the actual physical behavior. The total number of elements used and their variation in size and type within a given body are primarily matters of engineering judgment. The elements must be made small enough to give usable results and yet large enough to reduce computational effort. Small elements (and possibly higher-order elements) are generally desirable where the results are changing rapidly, such as where changes in geometry occur; large elements can be used where results are relatively constant. We will have more to say about discretization guidelines in later chapters, particularly in Chapter 7, where the concept becomes quite significant. The discretized body or mesh is often created with mesh-generation programs or preprocessor programs available to the user.

The choice of elements used in a finite element analysis depends on the physical makeup of the body under actual loading conditions and on how close to the actual behavior the analyst wants the results to be. Judgment concerning the appropriateness of one-, two-, or three-dimensional idealizations is necessary. Moreover, the choice of the most appropriate element for a particular problem is one of the major tasks that must be carried out by the designer/analyst. Elements that are commonly employed in practice—most of which are considered in this text—are shown in Figure 1–1.

The primary line elements [Figure 1–1(a)] consist of bar (or truss) and beam elements. They have a cross-sectional area but are usually represented by line segments. In general, the cross-sectional area within the element can vary, but throughout this text it will be considered to be constant. These elements are often used to model trusses and frame structures (see Figure 1–2 on page 16, for instance). The simplest line element (called a *linear element*) has two nodes, one at each end, although higher-order elements having three nodes [Figure 1–1(a)] or more (called *quadratic, cubic*, etc. *elements*) also exist. Chapter 10 includes discussion of higher-order line elements. The line elements are the simplest of elements to consider and will be discussed in Chapters 2 through 5 to illustrate many of the basic concepts of the finite element method.

The basic two-dimensional (or plane) elements [Figure 1–1(b)] are loaded by forces in their own plane (plane stress or plane strain conditions). They are triangular or quadrilateral elements. The simplest two-dimensional elements have corner nodes only (linear elements) with straight sides or boundaries (Chapter 6), although there are also higher-order elements, typically with midside nodes [Figure 1–1(b)] (called *quadratic elements*) and curved sides (Chapters 8 and 10). The elements can have variable thicknesses throughout or be constant. They are often used to model a wide range of engineering problems (see Figures 1–3 and 1–4 on pages 17 and 18).

The most common three-dimensional elements [Figure 1–1(c)] are tetrahedral and hexahedral (or brick) elements; they are used when it becomes necessary to perform a three-dimensional stress analysis. The basic three-dimensional elements (Chapter 11) have corner nodes only and straight sides, whereas higher-order elements with midedge nodes (and possible midface nodes) have curved surfaces for their sides [Figure 1–1(c)].

The axisymmetric element [Figure 1–1(d)] is developed by rotating a triangle or quadrilateral about a fixed axis located in the plane of the element through 360°. This element (described in Chapter 9) can be used when the geometry and loading of the problem are axisymmetric.

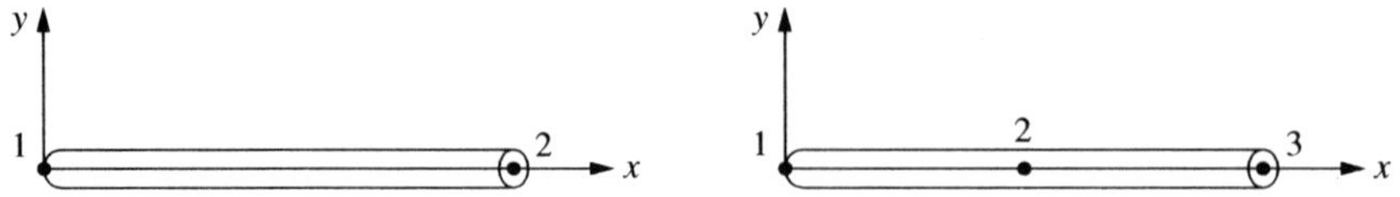

(a) Simple two-noded line element (typically used to represent a bar or beam element) and the higher-order line element

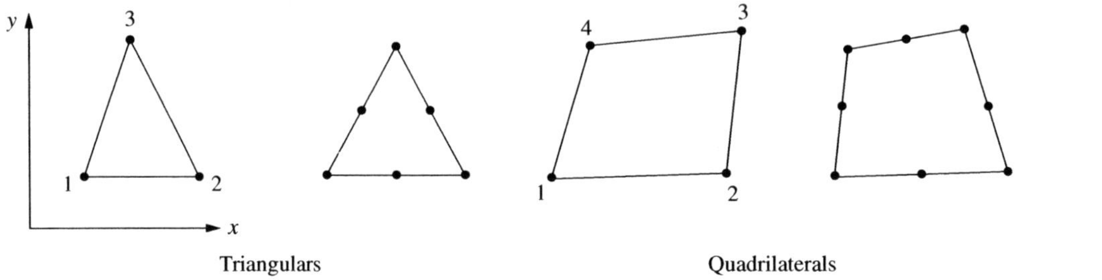

(b) Simple two-dimensional elements with corner nodes (typically used to represent plane stress/strain) and higher-order two-dimensional elements with intermediate nodes along the sides

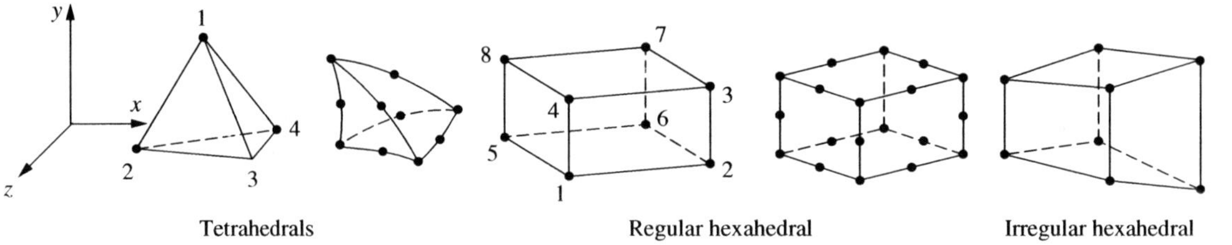

(c) Simple three-dimensional elements (typically used to represent three-dimensional stress state) and higher-order three-dimensional elements with intermediate nodes along edges

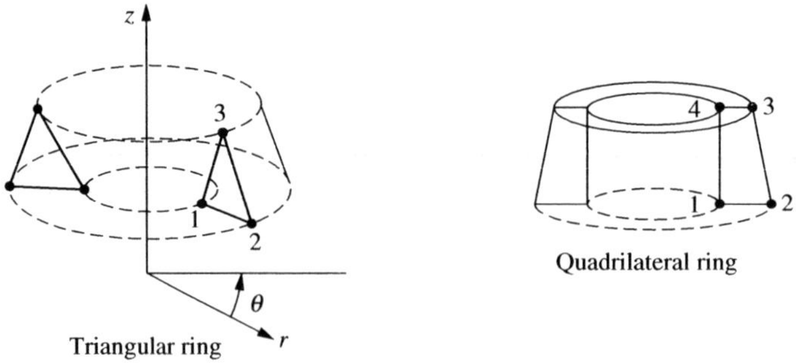

(d) Simple axisymmetric triangular and quadrilateral elements used for axisymmetric problems

Figure 1–1 Various types of simple lowest-order finite elements with corner nodes only and higher-order elements with intermediate nodes

Step 2 Select a Displacement Function

Step 2 involves choosing a displacement function within each element. The function is defined within the element using the nodal values of the element. Linear, quadratic, and cubic polynomials are frequently used functions because they are simple to work with in finite element formulation. However, trigonometric series can also be used. For a two-dimensional element, the displacement function is a function of the coordinates in its plane (say, the x-y plane). The functions are expressed in terms of the nodal unknowns (in the two-dimensional problem, in terms of an x and a y component). The same general displacement function can be used repeatedly for each element. Hence the finite element method is one in which a continuous quantity, such as the displacement throughout the body, is approximated by a discrete model composed of a set of piecewise-continuous functions defined within each finite domain or finite element.

Step 3 Define the Strain/ Displacement and Stress/Strain Relationships

Strain/displacement and stress/strain relationships are necessary for deriving the equations for each finite element. In the case of one-dimensional deformation, say, in the x direction, we have strain ε_x related to displacement u by

$$\varepsilon_x = \frac{du}{dx} \tag{1.4.1}$$

for small strains. In addition, the stresses must be related to the strains through the stress/strain law—generally called the *constitutive law*. The ability to define the material behavior accurately is most important in obtaining acceptable results. The simplest of stress/strain laws, Hooke's law, which is often used in stress analysis, is given by

$$\sigma_x = E\varepsilon_x \tag{1.4.2}$$

where σ_x = stress in the x direction and E = modulus of elasticity.

Step 4 Derive the Element Stiffness Matrix and Equations

Initially, the development of element stiffness matrices and element equations was based on the concept of stiffness influence coefficients, which presupposes a background in structural analysis. We now present alternative methods used in this text that do not require this special background.

Direct Equilibrium Method

According to this method, the stiffness matrix and element equations relating nodal forces to nodal displacements are obtained using force equilibrium conditions for a basic element, along with force/deformation relationships. Because this method is most easily adaptable to line or one-dimensional elements, Chapters 2, 3, and 4 illustrate this method for spring, bar, and beam elements, respectively.

Work or Energy Methods

To develop the stiffness matrix and equations for two- and three-dimensional elements, it is much easier to apply a work or energy method [35]. The principle of virtual work (using virtual displacements), the principle of minimum potential energy, and Castigliano's theorem are methods frequently used for the purpose of derivation of element equations.

The principle of virtual work outlined in Appendix E is applicable for any material behavior, whereas the principle of minimum potential energy and Castigliano's theorem are applicable only to elastic materials. Furthermore, the principle of virtual work can be used even when a potential function does not exist. However, all three principles yield identical element equations for linear-elastic materials; thus which method to use for this kind of material in structural analysis is largely a matter of convenience and personal preference. We will present the principle of minimum potential energy—probably the best known of the three energy methods mentioned here—in detail in Chapters 2 and 3, where it will be used to derive the spring and bar element equations. We will further generalize the principle and apply it to the beam element in Chapter 4 and to the plane stress/strain element in Chapter 6. Thereafter, the principle is routinely referred to as the basis for deriving all other stress-analysis stiffness matrices and element equations given in Chapters 8, 9, 11, and 12.

For the purpose of extending the finite element method outside the structural stress analysis field, a **functional**[1] (a function of another function or a function that takes functions as its argument) analogous to the one to be used with the principle of minimum potential energy is quite useful in deriving the element stiffness matrix and equations (see Chapters 13 and 14 on heat transfer and fluid flow, respectively). For instance, letting π denote the functional and $f(x, y)$ denote a function f of two variables x and y, we then have $\pi = \pi(f(x, y))$, where π is a function of the function f. A more general form of a functional depending on two independent variables $u(x, y)$ and $v(x, y)$, where independent variables are x and y in Cartesian coordinates, is given by:

$$\pi = \iint F(x, y, u, v, u_x, u_y, v_x, v_y, u_{xx}, \ldots, v_{yy})\,dx\,dy \tag{1.4.3}$$

Methods of Weighted Residuals

The methods of weighted residuals are useful for developing the element equations; particularly popular is Galerkin's method. These methods yield the same results as the energy methods wherever the energy methods are applicable. They are especially useful when a functional such as potential energy is not readily available. The weighted residual methods allow the finite element method to be applied directly to any differential equation.

---

[1] Another definition of a functional is as follows: A functional is an integral expression that implicitly contains differential equations that describe the problem. A typical functional is of the form $I(u) = \int F(x, u, u')\,dx$ where $u(x)$, x, and F are real so that $I(u)$ is also a real number.

Galerkin's method, along with the collocation, the least squares, and the subdomain *weighted residual methods* are introduced in Chapter 3. To illustrate each method, they will all be used to solve a one-dimensional bar problem for which a known exact solution exists for comparison. As the more easily adapted residual method, Galerkin's method will also be used to derive the bar element equations in Chapter 3 and the beam element equations in Chapter 4 and to solve the combined heat-conduction/convection/mass transport problem in Chapter 13. For more information on the use of the methods of weighted residuals, see Reference [36]; for additional applications to the finite element method, consult References [37] and [38].

Using any of the methods just outlined will produce the equations to describe the behavior of an element. These equations are written conveniently in matrix form as

$$\begin{Bmatrix} f_1 \\ f_2 \\ f_3 \\ \vdots \\ f_n \end{Bmatrix} = \begin{bmatrix} k_{11} & k_{12} & k_{13} & \dots & k_{1n} \\ k_{21} & k_{22} & k_{23} & \dots & k_{2n} \\ k_{31} & k_{32} & k_{33} & \dots & k_{3n} \\ \vdots & & & & \vdots \\ k_{n1} & & & \dots & k_{nn} \end{bmatrix} \begin{Bmatrix} d_1 \\ d_2 \\ d_3 \\ \vdots \\ d_n \end{Bmatrix} \tag{1.4.4}$$

or in compact matrix form as

$$\{f\} = [k]\{d\} \tag{1.4.5}$$

where $\{f\}$ is the vector of element nodal forces, $[k]$ is the element stiffness matrix (normally square and symmetric), and $\{d\}$ is the vector of unknown element nodal degrees of freedom or generalized displacements, n. Here generalized displacements may include such quantities as actual displacements, slopes, or even curvatures. The matrices in Eq. (1.4.5) will be developed and described in detail in subsequent chapters for specific element types, such as those in Figure 1–1.

Step 5 Assemble the Element Equations to Obtain the Global or Total Equations and Introduce Boundary Conditions

In this step the individual element nodal equilibrium equations generated in step 4 are assembled into the global nodal equilibrium equations. Section 2.3 illustrates this concept for a two-spring assemblage. Another more direct method of superposition (called the *direct stiffness method*), whose basis is nodal force equilibrium, can be used to obtain the global equations for the whole structure. This direct method is illustrated in Section 2.4 for a spring assemblage. Implicit in the direct stiffness method is the concept of continuity, or compatibility, which requires that the structure remain together and that no tears occur anywhere within the structure.

The final assembled or global equation written in matrix form is

$$\{F\} = [K]\{d\} \tag{1.4.6}$$

where $\{F\}$ is the vector of global nodal forces, $[K]$ is the structure global or total stiffness matrix, (for most problems, the global stiffness matrix is square and symmetric) and $\{d\}$ is now the vector of known and unknown structure nodal degrees of freedom or generalized displacements. It can be shown that at this stage, the global stiffness matrix $[K]$ is a singular matrix because its determinant is equal to zero. To remove this singularity problem, we must invoke certain boundary conditions (or constraints or supports) so that the structure remains in place instead of moving as a rigid body. Further details and methods of invoking boundary conditions are given in subsequent chapters. At this time it is sufficient to note that invoking boundary or support conditions results in a modification of the global Eq. (1.4.6). We also emphasize that the applied known loads have been accounted for in the global force matrix $\{F\}$.

Step 6 Solve for the Unknown Degrees of Freedom (or Generalized Displacements)

Equation (1.4.6), modified to account for the boundary conditions, is a set of simultaneous algebraic equations that can be written in expanded matrix form as

$$\begin{Bmatrix} F_1 \\ F_2 \\ \vdots \\ F_n \end{Bmatrix} = \begin{bmatrix} K_{11} & K_{12} & \dots & K_{1n} \\ K_{21} & K_{22} & \dots & K_{2n} \\ \vdots & & & \vdots \\ K_{n1} & K_{n2} & \dots & K_{nn} \end{bmatrix} \begin{Bmatrix} d_1 \\ d_2 \\ \vdots \\ d_n \end{Bmatrix} \tag{1.4.7}$$

where now n is the structure total number of unknown nodal degrees of freedom. These equations can be solved for the ds by using an elimination method (such as Gauss's method) or an iterative method (such as the Gauss–Seidel method). These two methods are discussed in Appendix B. The ds are called the *primary unknowns*, because they are the first quantities determined using the stiffness (or displacement) finite element method.

Step 7 Solve for the Element Strains and Stresses

For the structural stress-analysis problem, important secondary quantities of strain and stress (or moment and shear force) can be obtained because they can be directly expressed in terms of the displacements determined in step 6. Typical relationships between strain and displacement and between stress and strain—such as Eqs. (1.4.1) and (1.4.2) for one-dimensional stress given in step 3—can be used.

Step 8 Interpret the Results

The final goal is to interpret and analyze the results for use in the design/analysis process. Determination of locations in the structure where large deformations and large stresses occur is generally important in making design/analysis decisions. Postprocessor computer programs help the user to interpret the results by displaying them in graphical form.

1.5 Applications of the Finite Element Method

The finite element method can be used to analyze both structural and nonstructural problems. Typical structural areas include

1. Stress analysis, including truss and frame analysis, and stress concentration problems typically associated with holes, fillets, or other changes in geometry in a body
2. Buckling
3. Vibration analysis

Nonstructural problems include

1. Heat transfer
2. Fluid flow, including seepage through porous media
3. Distribution of electric or magnetic potential

Finally, some biomechanical engineering problems (which may include stress analysis) typically include analyses of human spine, skull, hip joints, jaw/gum tooth implants, heart, and eye.

We now present some typical applications of the finite element method. These applications will illustrate the variety, size, and complexity of problems that can be solved using the method and the typical discretization process and kinds of elements used.

Figure 1–2 illustrates a control tower for a railroad. The tower is a three-dimensional frame comprising a series of beam-type elements. The 48 elements are labeled by the circled numbers, whereas the 28 nodes are indicated by the uncircled numbers. Each node has three rotation and three displacement components associated with it. The rotations (θs) and displacements (ds) are called the *degrees of freedom*. Because of the loading conditions to which the tower structure is subjected, we have used a three-dimensional model.

The finite element method used for this frame enables the designer/analyst quickly to obtain displacements and stresses in the tower for typical load cases, as required by design codes. Before the development of the finite element method and the computer, even this relatively simple problem took many hours to solve.

The next illustration of the application of the finite element method to problem solving is the determination of displacements and stresses in an underground box culvert subjected to ground shock loading from a bomb explosion. Figure 1–3 shows the discretized model, which included a total of 369 nodes, 40 one-dimensional bar or truss elements used to model the steel reinforcement in the box culvert, and 333 plane strain two-dimensional triangular and rectangular elements used to model the surrounding soil and concrete box culvert. With an assumption of symmetry, only half of the box culvert need be analyzed. This problem requires the solution of nearly 700 unknown nodal displacements. It illustrates that different kinds of elements (here bar and plane strain) can often be used in one finite element model.

Another problem, that of the hydraulic cylinder rod end shown in Figure 1–4, was modeled by 120 nodes and 297 plane strain triangular elements. Symmetry was also applied to the whole rod end so that only half of the rod end had to be analyzed,

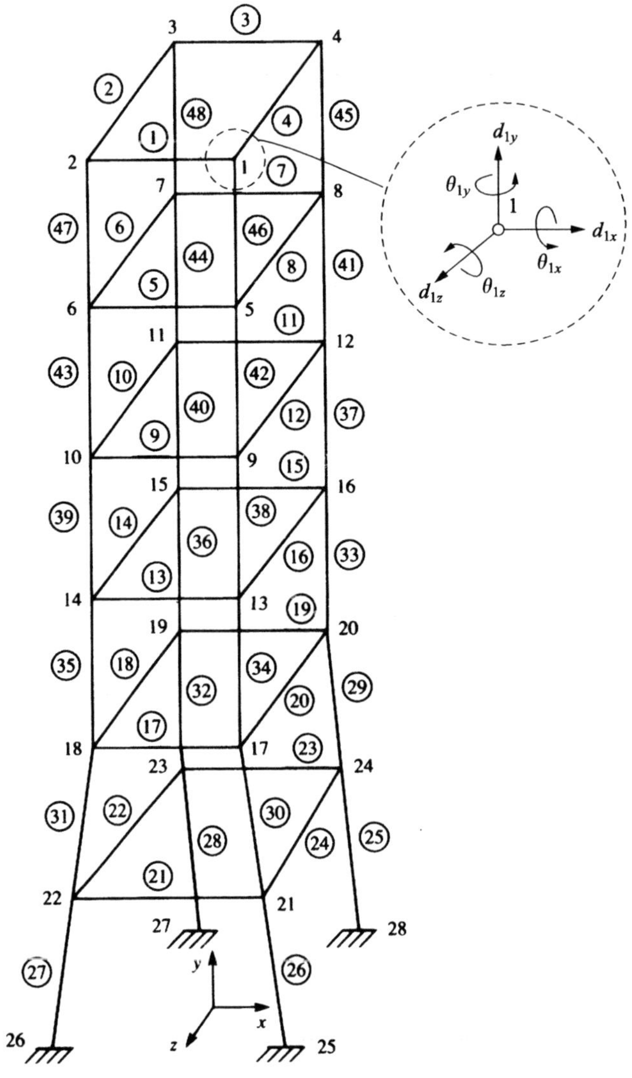

Figure 1–2 Discretized railroad control tower (28 nodes, 48 beam elements) with typical degrees of freedom shown at node 1, for example (By D. L. Logan)

as shown. The purpose of this analysis was to locate areas of high stress concentration in the rod end.

Figure 1–5 shows a chimney stack section that is four form heights high (or a total of 32 ft high). In this illustration, 584 beam elements were used to model the vertical and horizontal stiffeners making up the formwork, and 252 flat-plate elements were used to model the inner wooden form and the concrete shell. Because of the irregular loading pattern on the structure, a three-dimensional model was necessary. Displacements and stresses in the concrete were of prime concern in this problem.

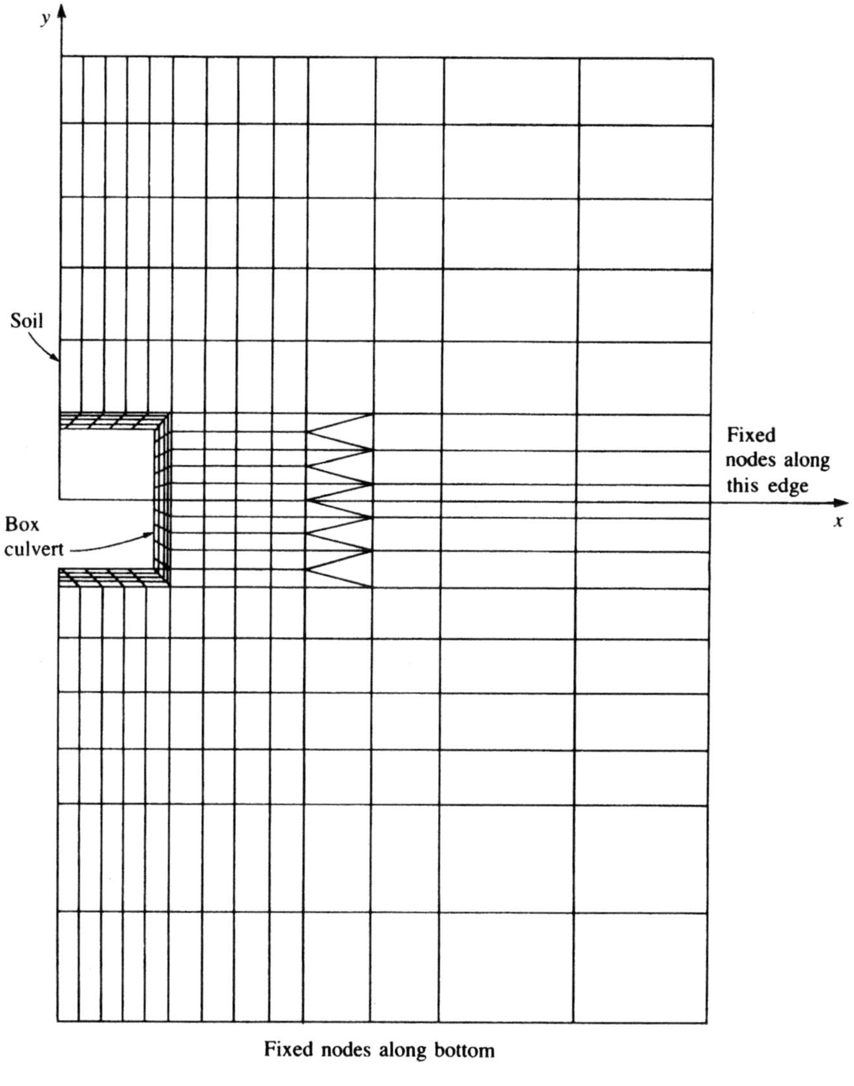

Figure 1–3 Discretized model of an underground box culvert (369 nodes, 40 bar elements, and 333 plane strain elements) [39]

Figure 1–6 shows the finite element discretized model of a proposed steel die used in a plastic film-making process. The irregular geometry and associated potential stress concentrations necessitated use of the finite element method to obtain a reasonable solution. Here 240 axisymmetric elements were used to model the three-dimensional die.

Figure 1–7 illustrates the use of a three-dimensional solid element to model a swing casting for a backhoe frame. The three-dimensional hexahedral elements are

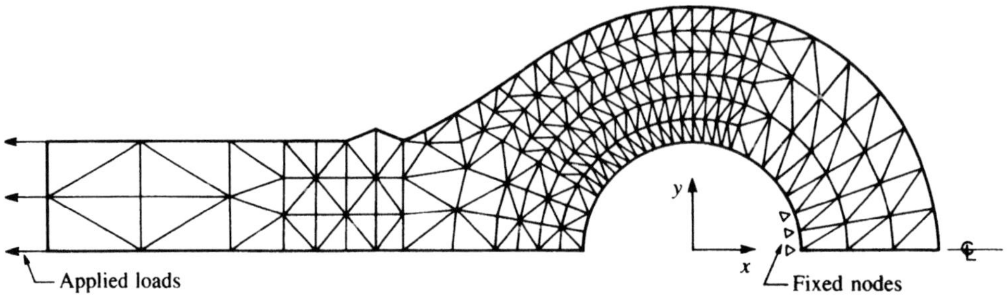

Figure 1–4 Two-dimensional analysis of a hydraulic cylinder rod end (120 nodes, 297 plane strain triangular elements)

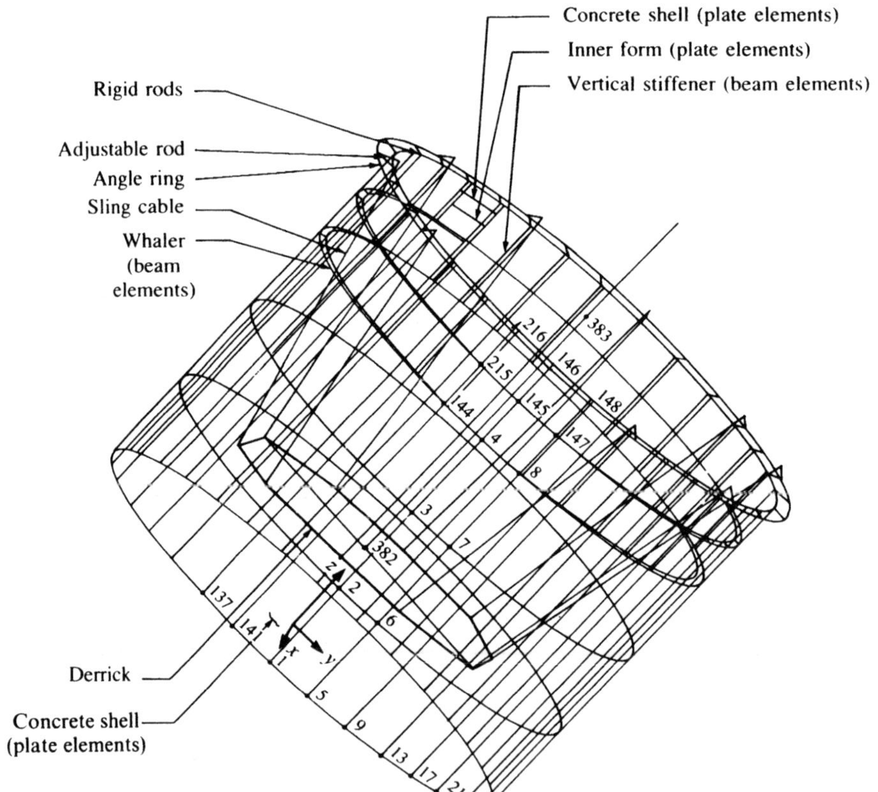

Figure 1–5 Finite element model of a chimney stack section (end view rotated 45°) (584 beam and 252 flat-plate elements) (By D. L. Logan)

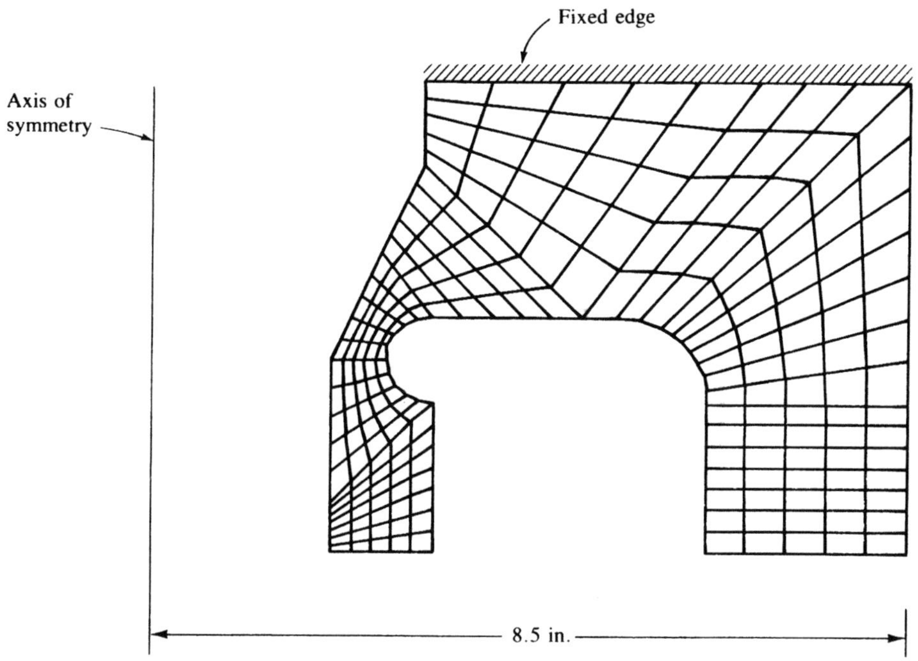

Figure 1–6 Model of a high-strength steel die (240 axisymmetric elements) used in the plastic film industry [40]

necessary to model the irregularly shaped three-dimensional casting. Two-dimensional models certainly would not yield accurate engineering solutions to this problem.

Figure 1–8 illustrates a two-dimensional heat-transfer model used to determine the temperature distribution in earth subjected to a heat source—a buried pipeline transporting a hot gas.

Figure 1–9 shows a three-dimensional finite element model of a pelvis bone with an implant, used to study stresses in the bone and the cement layer between bone and implant.

Finally, Figure 1–10 shows a three-dimensional model of a 710G bucket, used to study stresses throughout the bucket.

These illustrations suggest the kinds of problems that can be solved by the finite element method. Additional guidelines concerning modeling techniques will be provided in Chapter 7.

1.6 Advantages of the Finite Element Method

As previously indicated, the finite element method has been applied to numerous problems, both structural and nonstructural. This method has a number of advantages that have made it very popular. They include the ability to

1. Model irregularly shaped bodies quite easily
2. Handle general load conditions without difficulty

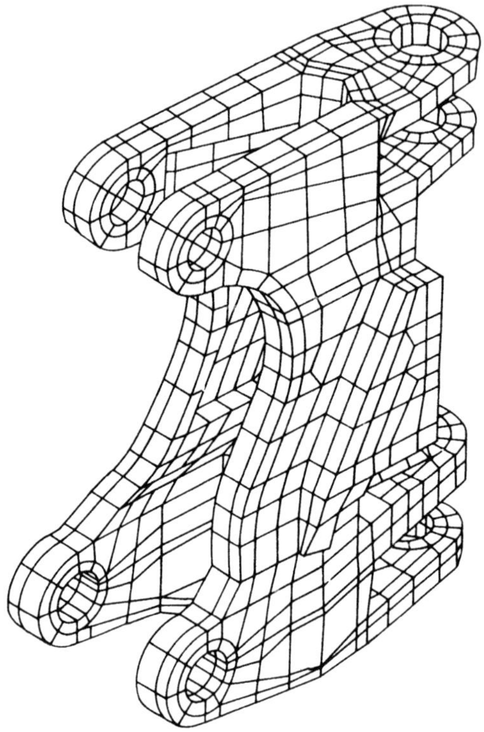

Figure 1–7 Three-dimensional solid element model of a swing casting for a backhoe frame

3. Model bodies composed of several different materials because the element equations are evaluated individually
4. Handle unlimited numbers and kinds of boundary conditions
5. Vary the size of the elements to make it possible to use small elements where necessary
6. Alter the finite element model relatively easily and cheaply
7. Include dynamic effects
8. Handle nonlinear behavior existing with large deformations and nonlinear materials

The finite element method of structural analysis enables the designer to detect stress, vibration, and thermal problems during the design process and to evaluate design changes *before* the construction of a possible prototype. Thus confidence in the acceptability of the prototype is enhanced. Moreover, if used properly, the method can reduce the number of prototypes that need to be built.

Even though the finite element method was initially used for structural analysis, it has since been adapted to many other disciplines in engineering and mathematical physics, such as fluid flow, heat transfer, electromagnetic potentials, soil mechanics, and acoustics [22–24, 27, 42–44].

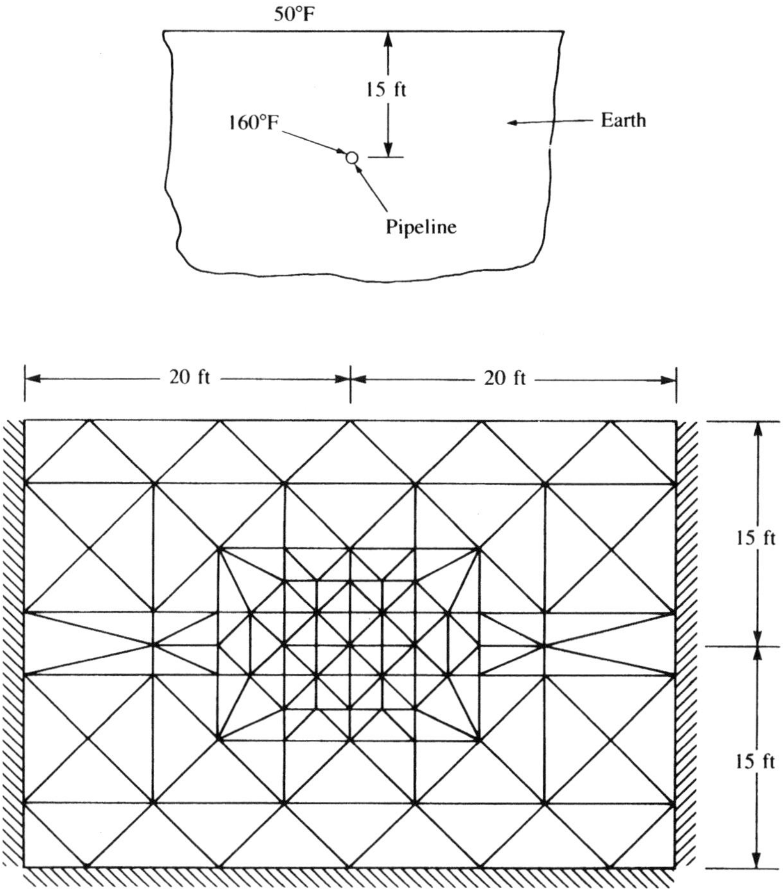

Figure 1–8 Finite element model for a two-dimensional temperature distribution in the earth

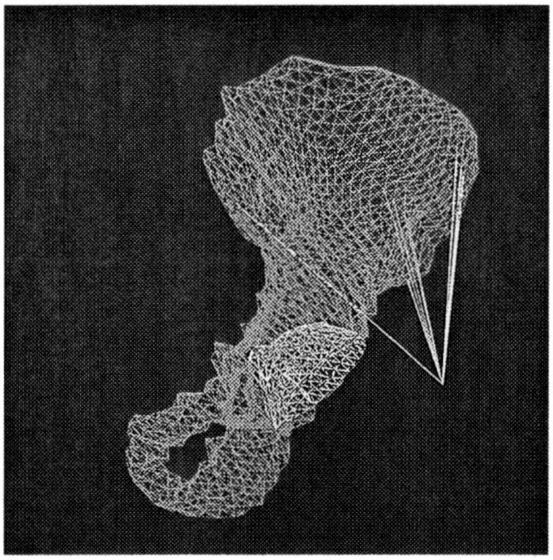

Figure 1–9 Finite element model of a pelvis bone with an implant (over 5000 solid elements were used in the model) (© Thomas Hansen/Courtesy of Harrington Arthritis Research Center, Phoenix, Arizona) [41]

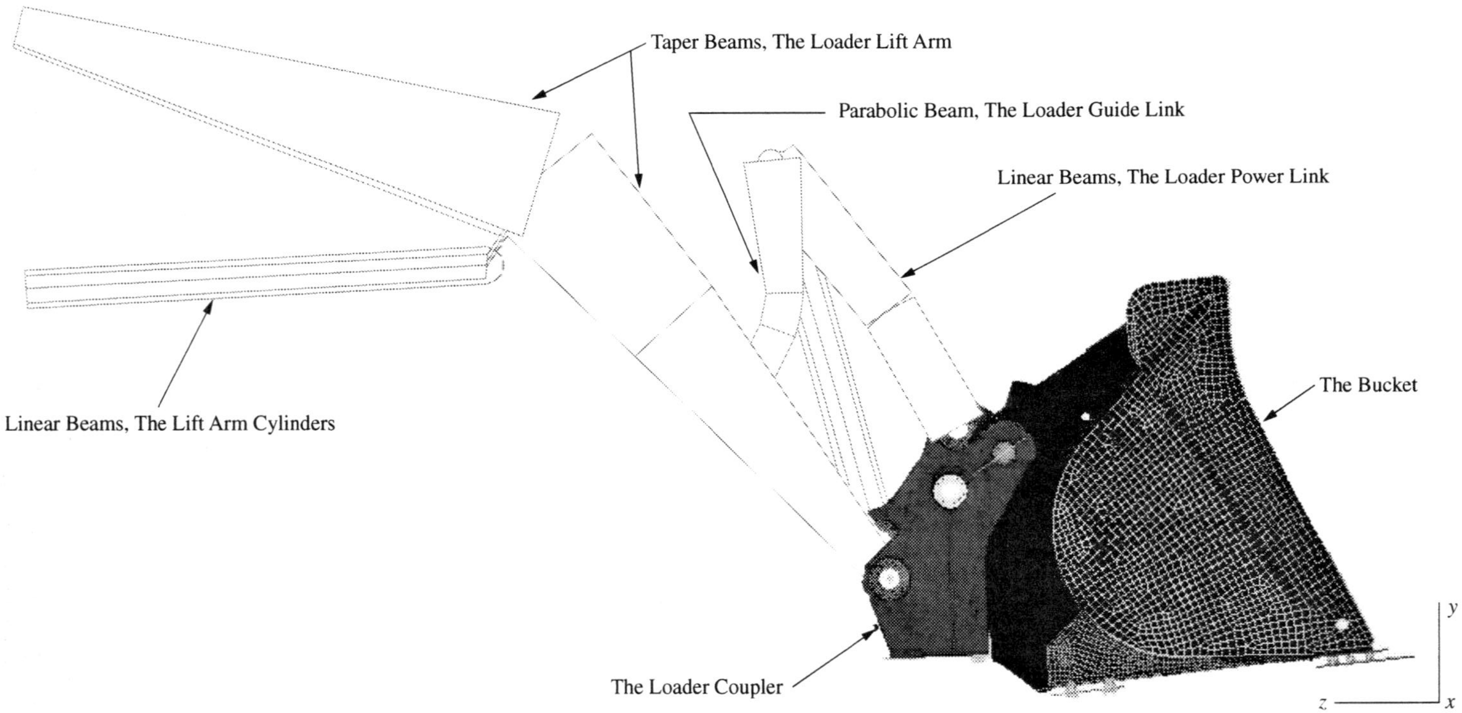

Figure 1–10 Finite element model of a 710G bucket with 169,595 elements and 185,026 nodes used (including 78,566 thin shell linear quadrilateral elements for the bucket and coupler, 83,104 solid linear brick elements to model the bosses, and 212 beam elements to model lift arms, lift arm cylinders, and guide links)(Courtesy of Yousif Omer, Structural Design Engineer, Construction and Forestry Division, John Deere Dubuque Works)

1.7 Computer Programs for the Finite Element Method

There are two general computer methods of approach to the solution of problems by the finite element method. One is to use large commercial programs, many of which have been configured to run on personal computers (PCs); these general-purpose programs are designed to solve many types of problems. The other is to develop many small, special-purpose programs to solve specific problems. In this section, we will discuss the advantages and disadvantages of both methods. We will then list some of the available general-purpose programs and discuss some of their standard capabilities.

Some advantages of general-purpose programs:

1. The input is well organized and is developed with user ease in mind. Users do not need special knowledge of computer software or hardware. Preprocessors are readily available to help create the finite element model.
2. The programs are large systems that often can solve many types of problems of large or small size with the same input format.
3. Many of the programs can be expanded by adding new modules for new kinds of problems or new technology. Thus they may be kept current with a minimum of effort.
4. With the increased storage capacity and computational efficiency of PCs, many general-purpose programs can now be run on PCs.
5. Many of the commercially available programs have become very attractive in price and can solve a wide range of problems [45, 56].

Some disadvantages of general-purpose programs:

1. The initial cost of developing general-purpose programs is high.
2. General-purpose programs are less efficient than special-purpose programs because the computer must make many checks for each problem, some of which would not be necessary if a special-purpose program were used.
3. Many of the programs are proprietary. Hence the user has little access to the logic of the program. If a revision must be made, it often has to be done by the developers.

Some advantages of special-purpose programs:

1. The programs are usually relatively short, with low development costs.
2. Small computers are able to run the programs.
3. Additions can be made to the program quickly and at a low cost.
4. The programs are efficient in solving the problems they were designed to solve.

The major disadvantage of special-purpose programs is their inability to solve different classes of problems. Thus one must have as many programs as there are different classes of problems to be solved.

There are numerous vendors supporting finite element programs, and the interested user should carefully consult the vendor before purchasing any software. However, to give you an idea about the various commercial personal computer programs now available for solving problems by the finite element method, we present a partial list of existing programs.

1. Algor [46]
2. Abaqus [47]
3. ANSYS [48]
4. COSMOS/M [49]
5. GT-STRUDL [50]
6. MARC [51]
7. MSC/NASTRAN [52]
8. NISA [53]
9. Pro/MECHANICA [54]
10. SAP2000 [55]
11. STARDYNE [56]

Standard capabilities of many of the listed programs are provided in the preceding references and in Reference [45]. These capabilities include information on

1. Element types available, such as beam, plane stress, and three-dimensional solid
2. Type of analysis available, such as static and dynamic
3. Material behavior, such as linear-elastic and nonlinear
4. Load types, such as concentrated, distributed, thermal, and displacement (settlement)
5. Data generation, such as automatic generation of nodes, elements, and restraints (most programs have preprocessors to generate the mesh for the model)
6. Plotting, such as original and deformed geometry and stress and temperature contours (most programs have postprocessors to aid in interpreting results in graphical form)
7. Displacement behavior, such as small and large displacement and buckling
8. Selective output, such as at selected nodes, elements, and maximum or minimum values

All programs include at least the bar, beam, plane stress, plate-bending, and three-dimensional solid elements, and most now include heat-transfer analysis capabilities.

Complete capabilities of the programs are best obtained through program reference manuals and websites, such as References [46–56].

References

[1] Hrennikoff, A., "Solution of Problems in Elasticity by the Frame Work Method," *Journal of Applied Mechanics*, Vol. 8, No. 4, pp. 169–175, Dec. 1941.

[2] McHenry, D., "A Lattice Analogy for the Solution of Plane Stress Problems," *Journal of Institution of Civil Engineers*, Vol. 21, pp. 59–82, Dec. 1943.

[3] Courant, R., "Variational Methods for the Solution of Problems of Equilibrium and Vibrations," *Bulletin of the American Mathematical Society*, Vol. 49, pp. 1–23, 1943.

[4] Levy, S., "Computation of Influence Coefficients for Aircraft Structures with Discontinuities and Sweepback," *Journal of Aeronautical Sciences*, Vol. 14, No. 10, pp. 547–560, Oct. 1947.

[5] Levy, S., "Structural Analysis and Influence Coefficients for Delta Wings," *Journal of Aeronautical Sciences*, Vol. 20, No. 7, pp. 449–454, July 1953.

[6] Argyris, J. H., "Energy Theorems and Structural Analysis," *Aircraft Engineering*, Oct., Nov., Dec. 1954 and Feb., Mar., Apr., May 1955.

[7] Argyris, J. H., and Kelsey, S., *Energy Theorems and Structural Analysis*, Butterworths, London, 1960 (collection of papers published in *Aircraft Engineering* in 1954 and 1955).

[8] Turner, M. J., Clough, R. W., Martin, H. C., and Topp, L. J., "Stiffness and Deflection Analysis of Complex Structures," *Journal of Aeronautical Sciences*, Vol. 23, No. 9, pp. 805–824, Sept. 1956.

[9] Clough, R. W., "The Finite Element Method in Plane Stress Analysis," *Proceedings*, American Society of Civil Engineers, 2nd Conference on Electronic Computation, Pittsburgh, PA, pp. 345–378, Sept. 1960.

[10] Melosh, R. J., "A Stiffness Matrix for the Analysis of Thin Plates in Bending," *Journal of the Aerospace Sciences*, Vol. 28, No. 1, pp. 34–42, Jan. 1961.

[11] Grafton, P. E., and Strome, D. R., "Analysis of Axisymmetric Shells by the Direct Stiffness Method," *Journal of the American Institute of Aeronautics and Astronautics*, Vol. 1, No. 10, pp. 2342–2347, 1963.

[12] Martin, H. C., "Plane Elasticity Problems and the Direct Stiffness Method," *The Trend in Engineering*, Vol. 13, pp. 5–19, Jan. 1961.

[13] Gallagher, R. H., Padlog, J., and Bijlaard, P. P., "Stress Analysis of Heated Complex Shapes," *Journal of the American Rocket Society*, Vol. 32, pp. 700–707, May 1962.

[14] Melosh, R. J., "Structural Analysis of Solids," *Journal of the Structural Division*, Proceedings of the American Society of Civil Engineers, pp. 205–223, Aug. 1963.

[15] Argyris, J. H., "Recent Advances in Matrix Methods of Structural Analysis," *Progress in Aeronautical Science*, Vol. 4, Pergamon Press, New York, 1964.

[16] Clough, R. W., and Rashid, Y., "Finite Element Analysis of Axisymmetric Solids," *Journal of the Engineering Mechanics Division*, Proceedings of the American Society of Civil Engineers, Vol. 91, pp. 71–85, Feb. 1965.

[17] Wilson, E. L., "Structural Analysis of Axisymmetric Solids," *Journal of the American Institute of Aeronautics and Astronautics*, Vol. 3, No. 12, pp. 2269–2274, Dec. 1965.

[18] Turner, M. J., Dill, E. H., Martin, H. C., and Melosh, R. J., "Large Deflections of Structures Subjected to Heating and External Loads," *Journal of Aeronautical Sciences*, Vol. 27, No. 2, pp. 97–107, Feb. 1960.

[19] Gallagher, R. H., and Padlog, J., "Discrete Element Approach to Structural Stability Analysis," *Journal of the American Institute of Aeronautics and Astronautics*, Vol. 1, No. 6, pp. 1437–1439, 1963.

[20] Zienkiewicz, O. C., Watson, M., and King, I. P., "A Numerical Method of Visco-Elastic Stress Analysis," *International Journal of Mechanical Sciences*, Vol. 10, pp. 807–827, 1968.

[21] Archer, J. S., "Consistent Matrix Formulations for Structural Analysis Using Finite-Element Techniques," *Journal of the American Institute of Aeronautics and Astronautics*, Vol. 3, No. 10, pp. 1910–1918, 1965.

[22] Zienkiewicz, O. C., and Cheung, Y. K., "Finite Elements in the Solution of Field Problems," *The Engineer*, pp. 507–510, Sept. 24, 1965.

[23] Martin, H. C., "Finite Element Analysis of Fluid Flows," *Proceedings of the Second Conference on Matrix Methods in Structural Mechanics*, Wright-Patterson Air Force Base, Ohio, pp. 517–535, Oct. 1968. (AFFDL-TR-68-150, Dec. 1969; AD-703-685, N.T.I.S.)

[24] Wilson, E. L., and Nickel, R. E., "Application of the Finite Element Method to Heat Conduction Analysis," *Nuclear Engineering and Design*, Vol. 4, pp. 276–286, 1966.
[25] Szabo, B. A., and Lee, G. C., "Derivation of Stiffness Matrices for Problems in Plane Elasticity by Galerkin's Method," *International Journal of Numerical Methods in Engineering*, Vol. 1, pp. 301–310, 1969.
[26] Zienkiewicz, O. C., and Parekh, C. J., "Transient Field Problems: Two-Dimensional and Three-Dimensional Analysis by Isoparametric Finite Elements," *International Journal of Numerical Methods in Engineering*, Vol. 2, No. 1, pp. 61–71, 1970.
[27] Lyness, J. F., Owen, D. R. J., and Zienkiewicz, O. C., "Three-Dimensional Magnetic Field Determination Using a Scalar Potential. A Finite Element Solution," *Transactions on Magnetics*, Institute of Electrical and Electronics Engineers, pp. 1649–1656, 1977.
[28] Belytschko, T., "A Survey of Numerical Methods and Computer Programs for Dynamic Structural Analysis," *Nuclear Engineering and Design*, Vol. 37, No. 1, pp. 23–34, 1976.
[29] Belytschko, T., "Efficient Large-Scale Nonlinear Transient Analysis by Finite Elements," *International Journal of Numerical Methods in Engineering*, Vol. 10, No. 3, pp. 579–596, 1976.
[30] Huiskies, R., and Chao, E. Y. S., "A Survey of Finite Element Analysis in Orthopedic Biomechanics: The First Decade," *Journal of Biomechanics*, Vol. 16, No. 6, pp. 385–409, 1983.
[31] *Journal of Biomechanical Engineering*, Transactions of the American Society of Mechanical Engineers, (published quarterly) (1st issue published 1977).
[32] Kardestuncer, H., ed., *Finite Element Handbook*, McGraw-Hill, New York, 1987.
[33] Clough, R. W., "The Finite Element Method After Twenty-Five Years: A Personal View," *Computers and Structures*, Vol. 12, No. 4, pp. 361–370, 1980.
[34] Kardestuncer, H., *Elementary Matrix Analysis of Structures*, McGraw-Hill, New York, 1974.
[35] Oden, J. T., and Ripperger, E. A., *Mechanics of Elastic Structures*, 2nd ed., McGraw-Hill, New York, 1981.
[36] Finlayson, B. A., *The Method of Weighted Residuals and Variational Principles*, Academic Press, New York, 1972.
[37] Zienkiewicz, O. C., *The Finite Element Method*, 3rd ed., McGraw-Hill, London, 1977.
[38] Cook, R. D., Malkus, D. S., and Plesha, M. E., *Concepts and Applications of Finite Element Analysis*, 3rd ed., Wiley, New York, 1989.
[39] Koswara, H., *A Finite Element Analysis of Underground Shelter Subjected to Ground Shock Load*, M.S. Thesis, Rose-Hulman Institute of Technology, 1983.
[40] Greer, R. D., "The Analysis of a Film Tower Die Utilizing the ANSYS Finite Element Package," M.S. Thesis, Rose-Hulman Institute of Technology, Terre Haute, Indiana, May 1989.
[41] Koeneman, J. B., Hansen, T. M., and Beres, K., "The Effect of Hip Stem Elastic Modulus and Cement/Stem Bond on Cement Stresses," 36th Annual Meeting, Orthopaedic Research Society, Feb. 5–8, 1990, New Orleans, Louisiana.
[42] Girijavallabham, C. V., and Reese, L. C., "Finite-Element Method for Problems in Soil Mechanics," *Journal of the Structural Division*, American Society of Civil Engineers, No. Sm2, pp. 473–497, Mar. 1968.
[43] Young, C., and Crocker, M., "Transmission Loss by Finite-Element Method," *Journal of the Acoustical Society of America*, Vol. 57, No. 1, pp. 144–148, Jan. 1975.
[44] Silvester, P. P., and Ferrari, R. L., *Finite Elements for Electrical Engineers*, Cambridge University Press, Cambridge, England, 1983.
[45] Falk, H., and Beardsley, C. W., "Finite Element Analysis Packages for Personal Computers," *Mechanical Engineering*, pp. 54–71, Jan. 1985.
[46] Algor Interactive Systems, 150 Beta Drive, Pittsburgh, PA 15238.

[47] Web site http://www.abaqus.com.
[48] Swanson, J. A., ANSYS-Engineering Analysis Systems User's Manual, Swanson Analysis Systems, Inc., Johnson Rd., P.O. Box 65, Houston, PA 15342.
[49] COSMOS/M, Structural Research & Analysis Corp., 12121 Wilshire Blvd., Los Angeles, CA 90025.
[50] web site http://ce6000.cegatech.edu.
[51] web site http://www.mscsoftware.com.
[52] MSC/NASTRAN, MacNeal-Schwendler Corp., 600 Suffolk St., Lowell, MA, 01854.
[53] web site http://emrc.com.
[54] Toogood, Roger, Pro/MECHANICA Structure Tutorial, SDC Publications, 2001.
[55] Computers & Structures, Inc., 1995 University Ave., Berkeley, CA 94704.
[56] STARDYNE, Research Engineers, Inc., 22700 Savi Ranch Pkwy, Yorba Linda, CA 92687.
[57] Noor, A. K., "Bibliography of Books and Monographs on Finite Element Technology," *Applied Mechanics Reviews*, Vol. 44, No. 6, pp. 307–317, June 1991.
[57] Belytschko, T., Liu W. K., and Moran, B., *Nonlinear Finite Elements For Continua and Structures*, John Wiley, 1996.

Problems

1.1 Define the term *finite element.*

1.2 What does *discretization* mean in the finite element method?

1.3 In what year did the modern development of the finite element method begin?

1.4 In what year was the direct stiffness method introduced?

1.5 Define the term *matrix.*

1.6 What role did the computer play in the use of the finite element method?

1.7 List and briefly describe the general steps of the finite element method.

1.8 What is the displacement method?

1.9 List four common types of finite elements.

1.10 Name three commonly used methods for deriving the element stiffness matrix and element equations. Briefly describe each method.

1.11 To what does the term *degrees of freedom* refer?

1.12 List five typical areas of engineering where the finite element method is applied.

1.13 List five advantages of the finite element method.

CHAPTER 2

Introduction to the Stiffness (Displacement) Method

Introduction

This chapter introduces some of the basic concepts on which the direct stiffness method is founded. The linear spring is introduced first because it provides a simple yet generally instructive tool to illustrate the basic concepts. We begin with a general definition of the stiffness matrix and then consider the derivation of the stiffness matrix for a linear-elastic spring element. We next illustrate how to assemble the total stiffness matrix for a structure comprising an assemblage of spring elements by using elementary concepts of equilibrium and compatibility. We then show how the total stiffness matrix for an assemblage can be obtained by superimposing the stiffness matrices of the individual elements in a direct manner. The term *direct stiffness method* evolved in reference to this technique.

After establishing the total structure stiffness matrix, we illustrate how to impose boundary conditions—both homogeneous and nonhomogeneous. A complete solution including the nodal displacements and reactions is thus obtained. (The determination of internal forces is discussed in Chapter 3 in connection with the bar element.)

We then introduce the principle of minimum potential energy, apply it to derive the spring element equations, and use it to solve a spring assemblage problem. We will illustrate this principle for the simplest of elements (those with small numbers of degrees of freedom) so that it will be a more readily understood concept when applied, of necessity, to elements with large numbers of degrees of freedom in subsequent chapters.

2.1 Definition of the Stiffness Matrix

Familiarity with the stiffness matrix is essential to understanding the stiffness method. We define the stiffness matrix as follows: *For an element, a* **stiffness matrix** $\hat{\underline{k}}$ *is a matrix such that* $\hat{\underline{f}} = \hat{\underline{k}}\hat{\underline{d}}$, *where* $\hat{\underline{k}}$ *relates local-coordinate* $(\hat{x}, \hat{y}, \hat{z})$ *nodal displacements* $\hat{\underline{d}}$ *to local forces* $\hat{\underline{f}}$ *of a single element.* (Throughout this text, the underline notation denotes

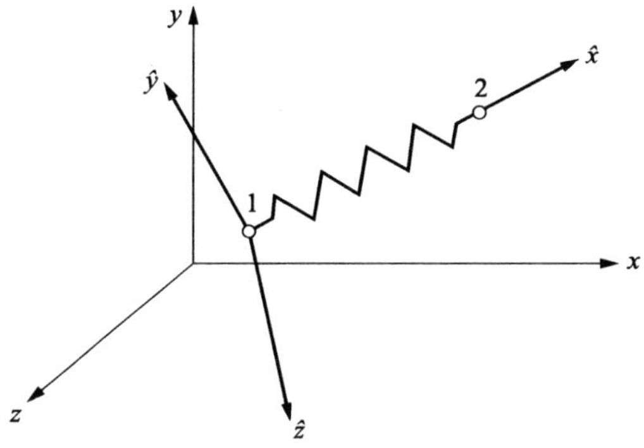

Figure 2–1 Local $(\hat{x}, \hat{y}, \hat{z})$ and global (x, y, z) coordinate systems

a matrix, and the ˆ symbol denotes quantities referred to a local-coordinate system set up to be convenient for the element as shown in Figure 2–1.)

For a continuous medium or structure comprising a series of elements, a stiffness matrix $\underline{K}$ relates global-coordinate (x, y, z) nodal displacements $\underline{d}$ to global forces $\underline{F}$ of the whole medium or structure. (Lowercase letters such as x, y, and z without the ˆ symbol denote global-coordinate variables.)

2.2 Derivation of the Stiffness Matrix for a Spring Element

Using the direct equilibrium approach, we will now derive the stiffness matrix for a one-dimensional linear spring—that is, a spring that obeys Hooke's law and resists forces only in the direction of the spring. Consider the linear spring element shown in Figure 2–2. Reference points 1 and 2 are located at the ends of the element. These reference points are called the *nodes* of the spring element. The local nodal forces are $\hat{f}_{1x}$ and $\hat{f}_{2x}$ for the spring element associated with the local axis $\hat{x}$. The local axis acts in the direction of the spring so that we can directly measure displacements and forces along the spring. The local nodal displacements are $\hat{d}_{1x}$ and $\hat{d}_{2x}$ for the spring element. These nodal displacements are called the *degrees of freedom* at each node. Positive directions for the forces and displacements at each node are taken in the positive $\hat{x}$ direction as shown from node 1 to node 2 in the figure. The symbol k is called the *spring constant* or *stiffness* of the spring.

Analogies to actual spring constants arise in numerous engineering problems. In Chapter 3, we see that a prismatic uniaxial bar has a spring constant $k = AE/L$, where A represents the cross-sectional area of the bar, E is the modulus of elasticity, and L is the bar length. Similarly, in Chapter 5, we show that a prismatic circular-cross-section bar in torsion has a spring constant $k = JG/L$, where J is the polar moment of inertia and G is the shear modulus of the material. For one-dimensional heat conduction (Chapter 13), $k = AK_{xx}/L$, where K_{xx} is the thermal conductivity of

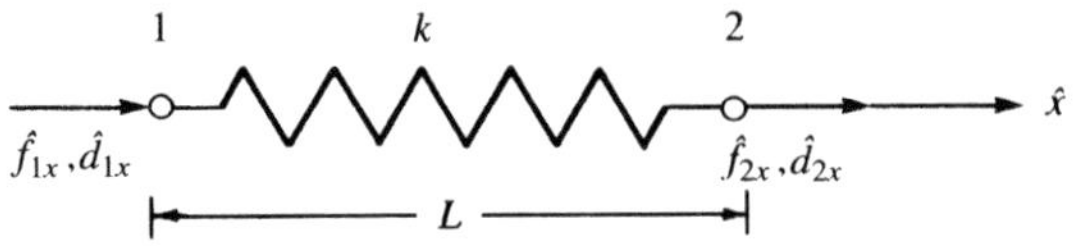

Figure 2–2 Linear spring element with positive nodal displacement and force conventions

the material, and for one-dimensional fluid flow through a porous medium (Chapter 14), $k = AK_{xx}/L$, where K_{xx} is the permeability coefficient of the material.

We will then observe that the stiffness method can be applied to nonstructural problems, such as heat transfer, fluid flow, and electrical networks, as well as structural problems by simply applying the proper constitutive law (such as Hooke's law for structural problems, Fourier's law for heat transfer, Darcy's law for fluid flow and Ohm's law for electrical networks) and a conservation principle such as nodal equilibrium or conservation of energy.

We now want to develop a relationship between nodal forces and nodal displacements for a spring element. This relationship will be the stiffness matrix. Therefore, we want to relate the nodal force matrix to the nodal displacement matrix as follows:

$$\begin{Bmatrix} \hat{f}_{1x} \\ \hat{f}_{2x} \end{Bmatrix} = \begin{bmatrix} k_{11} & k_{12} \\ k_{21} & k_{22} \end{bmatrix} \begin{Bmatrix} \hat{d}_{1x} \\ \hat{d}_{2x} \end{Bmatrix} \tag{2.2.1}$$

where the element stiffness coefficients k_{ij} of the $\underline{\hat{k}}$ matrix in Eq. (2.2.1) are to be determined. Recall from Eqs. (1.2.5) and (1.2.6) that k_{ij} represent the force F_i in the ith degree of freedom due to a unit displacement d_j in the jth degree of freedom while all other displacements are zero. That is, when we let $d_j = 1$ and $d_k = 0$ for $k \neq j$, force $F_i = k_{ij}$.

We now use the general steps outlined in Section 1.4 to derive the stiffness matrix for the spring element in this section (while keeping in mind that these same steps will be applicable later in the derivation of stiffness matrices of more general elements) and then to illustrate a complete solution of a spring assemblage in Section 2.3. Because our approach throughout this text is to derive various element stiffness matrices and then to illustrate how to solve engineering problems with the elements, step 1 now involves only selecting the element type.

Step 1 Select the Element Type

Consider the linear spring element (which can be an element in a system of springs) subjected to resulting nodal tensile forces T (which may result from the action of adjacent springs) directed along the spring axial direction $\hat{x}$ as shown in Figure 2–3, so as to be in equilibrium. The local $\hat{x}$ axis is directed from node 1 to node 2. We represent the spring by labeling nodes at each end and by labeling the element number. The original distance between nodes before deformation is denoted by L. The material property (spring constant) of the element is k.

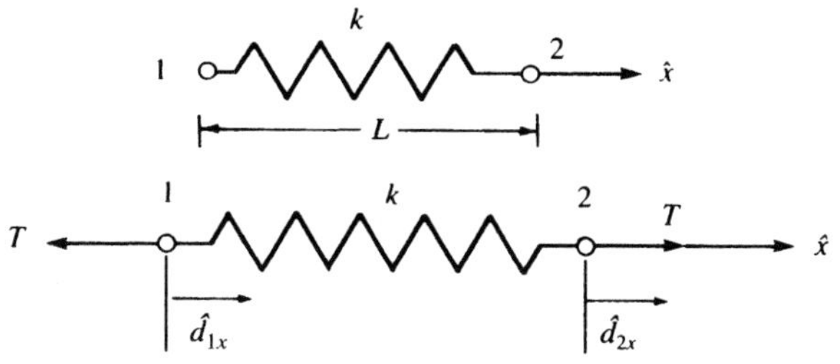

Figure 2–3 Linear spring subjected to tensile forces

Step 2 Select a Displacement Function

We must choose in advance the mathematical function to represent the deformed shape of the spring element under loading. Because it is difficult, if not impossible at times, to obtain a closed form or exact solution, we assume a solution shape or distribution of displacement within the element by using an appropriate mathematical function. The most common functions used are polynomials.

Because the spring element resists axial loading only with the local degrees of freedom for the element being displacements $\hat{d}_{1x}$ and $\hat{d}_{2x}$ along the $\hat{x}$ direction, we choose a displacement function $\hat{u}$ to represent the axial displacement throughout the element. Here a linear displacement variation along the $\hat{x}$ axis of the spring is assumed [Figure 2–4(b)], because a linear function with specified endpoints has a unique path. Therefore,

$$\hat{u} = a_1 + a_2\hat{x} \tag{2.2.2}$$

In general, the total number of coefficients a is equal to the total number of degrees of freedom associated with the element. Here the total number of degrees of freedom is

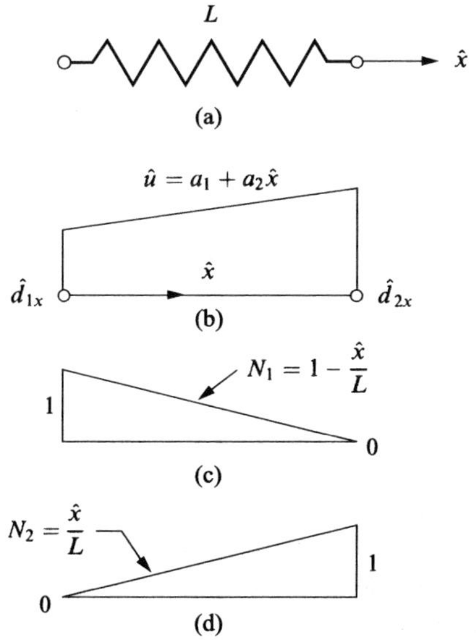

Figure 2–4 (a) Spring element showing plots of (b) displacement function $\hat{u}$ and shape functions (c) N_1 and (d) N_2 over domain of element

two—an axial displacement at each of the two nodes of the element (we present further discussion regarding the choice of displacement functions in Section 3.2). In matrix form, Eq. (2.2.2) becomes

$$\hat{u} = [1 \quad \hat{x}]\begin{Bmatrix} a_1 \\ a_2 \end{Bmatrix} \tag{2.2.3}$$

We now want to express $\hat{u}$ as a function of the nodal displacements $\hat{d}_{1x}$ and $\hat{d}_{2x}$. as this will allow us to apply the physical boundary conditions on nodal displacements directly as indicated in Step 3 and to then relate the nodal displacements to the nodal forces in Step 4. We achieve this by evaluating $\hat{u}$ at each node and solving for a_1 and a_2 from Eq. (2.2.2) as follows:

$$\hat{u}(0) = \hat{d}_{1x} = a_1 \tag{2.2.4}$$

$$\hat{u}(L) = \hat{d}_{2x} = a_2 L + \hat{d}_{1x} \tag{2.2.5}$$

or, solving Eq. (2.2.5) for a_2,

$$a_2 = \frac{\hat{d}_{2x} - \hat{d}_{1x}}{L} \tag{2.2.6}$$

Upon substituting Eqs. (2.2.4) and (2.2.6) into Eq. (2.2.2), we have

$$\hat{u} = \left(\frac{\hat{d}_{2x} - \hat{d}_{1x}}{L}\right)\hat{x} + \hat{d}_{1x} \tag{2.2.7}$$

In matrix form, we express Eq. (2.2.7) as

$$\hat{u} = \left[1 - \frac{\hat{x}}{L} \quad \frac{\hat{x}}{L}\right]\begin{Bmatrix} \hat{d}_{1x} \\ \hat{d}_{2x} \end{Bmatrix} \tag{2.2.8}$$

or

$$\hat{u} = [N_1 \quad N_2]\begin{Bmatrix} \hat{d}_{1x} \\ \hat{d}_{2x} \end{Bmatrix} \tag{2.2.9}$$

Here

$$N_1 = 1 - \frac{\hat{x}}{L} \qquad \text{and} \qquad N_2 = \frac{\hat{x}}{L} \tag{2.2.10}$$

are called the *shape functions* because the N_i's express the shape of the assumed displacement function over the domain ($\hat{x}$ coordinate) of the element when the ith element degree of freedom has unit value and all other degrees of freedom are zero. In this case, N_1 and N_2 are linear functions that have the properties that $N_1 = 1$ at node 1 and $N_1 = 0$ at node 2, whereas $N_2 = 1$ at node 2 and $N_2 = 0$ at node 1. See Figure 2–4(c) and (d) for plots of these shape functions over the domain of the spring element. Also, $N_1 + N_2 = 1$ for any axial coordinate along the bar. (Section 3.2 further explores this important relationship.) In addition, the N_i's are often called *interpolation functions* because we are interpolating to find the value of a function between given nodal values. The interpolation function may be different from the actual function except at the endpoints or nodes, where the interpolation function and actual function must be equal to specified nodal values.

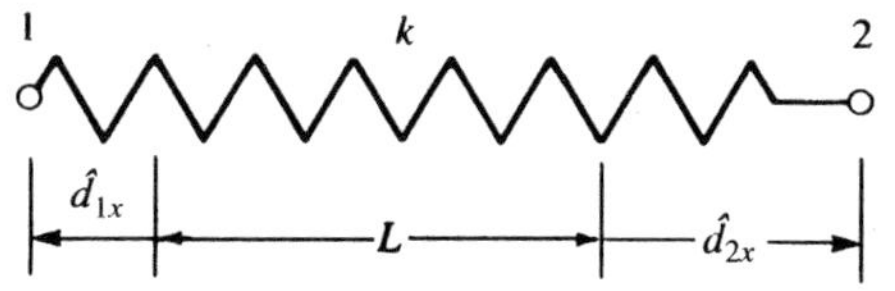

Figure 2–5 Deformed spring

Step 3 Define the Strain/ Displacement and Stress/Strain Relationships

The tensile forces T produce a total elongation (deformation) δ of the spring. The typical total elongation of the spring is shown in Figure 2–5. Here $\hat{d}_{1x}$ is a negative value because the direction of displacement is opposite the positive $\hat{x}$ direction, whereas $\hat{d}_{2x}$ is a positive value.

The deformation of the spring is then represented by

$$\delta = \hat{u}(L) - \hat{u}(0) = \hat{d}_{2x} - \hat{d}_{1x} \tag{2.2.11}$$

From Eq. (2.2.11), we observe that the total deformation is the difference of the nodal displacements in the $\hat{x}$ direction.

For a spring element, we can relate the force in the spring directly to the deformation. Therefore, the strain/displacement relationship is not necessary here.

The stress/strain relationship can be expressed in terms of the force/deformation relationship instead as

$$T = k\delta \tag{2.2.12}$$

Now, using Eq. (2.2.11) in Eq. (2.2.12), we obtain

$$T = k(\hat{d}_{2x} - \hat{d}_{1x}) \tag{2.2.13}$$

Step 4 Derive the Element Stiffness Matrix and Equations

We now derive the spring element stiffness matrix. By the sign convention for nodal forces and equilibrium, we have

$$\hat{f}_{1x} = -T \qquad \hat{f}_{2x} = T \tag{2.2.14}$$

Using Eqs. (2.2.13) and (2.2.14), we have

$$\begin{aligned} T &= -\hat{f}_{1x} = k(\hat{d}_{2x} - \hat{d}_{1x}) \\ T &= \hat{f}_{2x} = k(\hat{d}_{2x} - \hat{d}_{1x}) \end{aligned} \tag{2.2.15}$$

Rewriting Eqs. (2.2.15), we obtain

$$\begin{aligned} \hat{f}_{1x} &= k(\hat{d}_{1x} - \hat{d}_{2x}) \\ \hat{f}_{2x} &= k(\hat{d}_{2x} - \hat{d}_{1x}) \end{aligned} \tag{2.2.16}$$

Now expressing Eqs. (2.2.16) in a single matrix equation yields

$$\begin{Bmatrix} \hat{f}_{1x} \\ \hat{f}_{2x} \end{Bmatrix} = \begin{bmatrix} k & -k \\ -k & k \end{bmatrix} \begin{Bmatrix} \hat{d}_{1x} \\ \hat{d}_{2x} \end{Bmatrix} \tag{2.2.17}$$

This relationship holds for the spring along the $\hat{x}$ axis. From our basic definition of a stiffness matrix and application of Eq. (2.2.1) to Eq. (2.2.17), we obtain

$$\hat{\underline{k}} = \begin{bmatrix} k & -k \\ -k & k \end{bmatrix} \tag{2.2.18}$$

as the stiffness matrix for a linear spring element. Here $\hat{\underline{k}}$ is called the *local stiffness matrix* for the element. We observe from Eq. (2.2.18) that $\hat{\underline{k}}$ is a symmetric (that is, $k_{ij} = k_{ji}$) square matrix (the number of rows equals the number of columns in $\hat{\underline{k}}$). Appendix A gives more description and numerical examples of symmetric and square matrices.

Step 5 Assemble the Element Equations to Obtain the Global Equations and Introduce Boundary Conditions

The global stiffness matrix and global force matrix are assembled using nodal force equilibrium equations, force/deformation and compatibility equations from Section 2.3, and the direct stiffness method described in Section 2.4. This step applies for structures composed of more than one element such that

$$\underline{K} = [K] = \sum_{e=1}^{N} \underline{k}^{(e)} \qquad \text{and} \qquad \underline{F} = \{F\} = \sum_{e=1}^{N} \underline{f}^{(e)} \tag{2.2.19}$$

where $\underline{k}$ and $\underline{f}$ are now element stiffness and force matrices expressed in a global reference frame. (Throughout this text, the $\sum$ sign used in this context does not imply a simple summation of element matrices but rather denotes that these element matrices must be assembled properly according to the direct stiffness method described in Section 2.4.)

Step 6 Solve for the Nodal Displacements

The displacements are then determined by imposing boundary conditions, such as support conditions, and solving a system of equations, $\underline{F} = \underline{K}\underline{d}$, simultaneously.

Step 7 Solve for the Element Forces

Finally, the element forces are determined by back-substitution, applied to each element, into equations similar to Eqs. (2.2.16).

2.3 Example of a Spring Assemblage

Structures such as trusses, building frames, and bridges comprise basic structural components connected together to form the overall structures. To analyze these structures, we must determine the total structure stiffness matrix for an interconnected system of elements. Before considering the truss and frame, we will determine the total structure stiffness matrix for a spring assemblage by using the force/displacement matrix relationships derived in Section 2.2 for the spring element, along with fundamental concepts of nodal equilibrium and compatibility. Step 5 above will then have been illustrated.

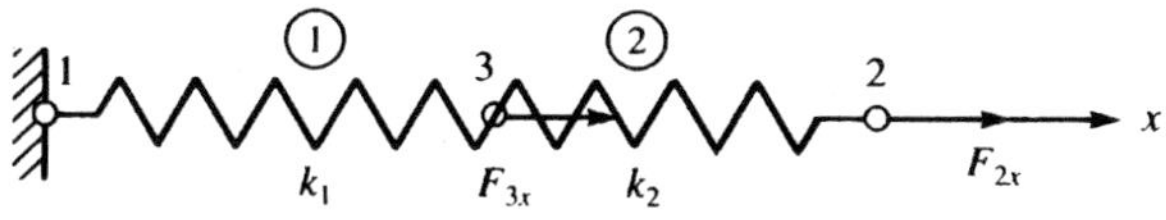

Figure 2–6 Two-spring assemblage

We will consider the specific example of the two-spring assemblage shown in Figure 2–6*. This example is general enough to illustrate the direct equilibrium approach for obtaining the total stiffness matrix of the spring assemblage. Here we fix node 1 and apply axial forces for F_{3x} at node 3 and F_{2x} at node 2. The stiffnesses of spring elements 1 and 2 are k_1 and k_2, respectively. The nodes of the assemblage have been numbered 1, 3, and 2 for further generalization because sequential numbering between elements generally does not occur in large problems.

The x axis is the global axis of the assemblage. The local $\hat{x}$ axis of each element coincides with the global axis of the assemblage.

For element 1, using Eq. (2.2.17), we have

$$\begin{Bmatrix} f_{1x} \\ f_{3x} \end{Bmatrix} = \begin{bmatrix} k_1 & -k_1 \\ -k_1 & k_1 \end{bmatrix} \begin{Bmatrix} d_{1x}^{(1)} \\ d_{3x}^{(1)} \end{Bmatrix} \tag{2.3.1}$$

and for element 2, we have

$$\begin{Bmatrix} f_{3x} \\ f_{2x} \end{Bmatrix} = \begin{bmatrix} k_2 & -k_2 \\ -k_2 & k_2 \end{bmatrix} \begin{Bmatrix} d_{3x}^{(2)} \\ d_{2x}^{(2)} \end{Bmatrix} \tag{2.3.2}$$

Furthermore, elements 1 and 2 must remain connected at common node 3 throughout the displacement. This is called the *continuity* or *compatibility requirement*. The compatibility requirement yields

$$d_{3x}^{(1)} = d_{3x}^{(2)} = d_{3x} \tag{2.3.3}$$

where, throughout this text, the superscript in parentheses above d refers to the element number to which they are related. Recall that the subscripts to the right identify the node and the direction of displacement, respectively, and that d_{3x} is the node 3 displacement of the total or global spring assemblage.

Free-body diagrams of each element and node (using the established sign conventions for element nodal forces in Figure 2–2) are shown in Figure 2–7.

Based on the free-body diagrams of each node shown in Figure 2–7 and the fact that external forces must equal internal forces at each node, we can write nodal equilibrium equations at nodes 3, 2, and 1 as

$$F_{3x} = f_{3x}^{(1)} + f_{3x}^{(2)} \tag{2.3.4}$$

$$F_{2x} = f_{2x}^{(2)} \tag{2.3.5}$$

$$F_{1x} = f_{1x}^{(1)} \tag{2.3.6}$$

* Throughout this text, element numbers in figures are shown with circles around them.

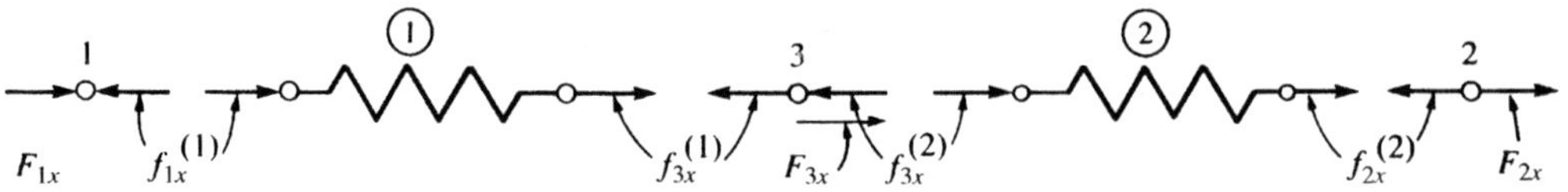

Figure 2–7 Nodal forces consistent with element force sign convention

where F_{1x} results from the external applied reaction at the fixed support.

Here Newton's third law, of equal but opposite forces, is applied in moving from a node to an element associated with the node. Using Eqs. (2.3.1)–(2.3.3) in Eqs. (2.3.4)–(2.3.6), we obtain

$$
\begin{aligned}
F_{3x} &= (-k_1 d_{1x} + k_1 d_{3x}) + (k_2 d_{3x} - k_2 d_{2x}) \\
F_{2x} &= -k_2 d_{3x} + k_2 d_{2x} \\
F_{1x} &= k_1 d_{1x} - k_1 d_{3x}
\end{aligned} \tag{2.3.7}
$$

In matrix form, Eqs. (2.3.7) are expressed by

$$
\begin{Bmatrix} F_{3x} \\ F_{2x} \\ F_{1x} \end{Bmatrix} = \begin{bmatrix} k_1 + k_2 & -k_2 & -k_1 \\ -k_2 & k_2 & 0 \\ -k_1 & 0 & k_1 \end{bmatrix} \begin{Bmatrix} d_{3x} \\ d_{2x} \\ d_{1x} \end{Bmatrix} \tag{2.3.8}
$$

Rearranging Eq. (2.3.8) in numerically increasing order of the nodal degrees of freedom, we have

$$
\begin{Bmatrix} F_{1x} \\ F_{2x} \\ F_{3x} \end{Bmatrix} = \begin{bmatrix} k_1 & 0 & -k_1 \\ 0 & k_2 & -k_2 \\ -k_1 & -k_2 & k_1 + k_2 \end{bmatrix} \begin{Bmatrix} d_{1x} \\ d_{2x} \\ d_{3x} \end{Bmatrix} \tag{2.3.9}
$$

Equation (2.3.9) is now written as the single matrix equation

$$
\underline{F} = \underline{K}\underline{d} \tag{2.3.10}
$$

where $\underline{F} = \begin{Bmatrix} F_{1x} \\ F_{2x} \\ F_{3x} \end{Bmatrix}$ is called the *global nodal force matrix*, $\underline{d} = \begin{Bmatrix} d_{1x} \\ d_{2x} \\ d_{3x} \end{Bmatrix}$ is called the *global nodal displacement matrix*, and

$$
\underline{K} = \begin{bmatrix} k_1 & 0 & -k_1 \\ 0 & k_2 & -k_2 \\ -k_1 & -k_2 & k_1 + k_2 \end{bmatrix} \tag{2.3.11}
$$

is called the *total* or *global* or *system stiffness matrix*.

In summary, to establish the stiffness equations and stiffness matrix, Eqs. (2.3.9) and (2.3.11), for a spring assemblage, we have used force/deformation relationships Eqs. (2.3.1) and (2.3.2), compatibility relationship Eq. (2.3.3), and nodal force equilibrium Eqs. (2.3.4)–(2.3.6). We will consider the complete solution to this example problem after considering a more practical method of assembling the total stiffness matrix in Section 2.4 and discussing the support boundary conditions in Section 2.5.

2.4 Assembling the Total Stiffness Matrix by Superposition (Direct Stiffness Method)

We will now consider a more convenient method for constructing the total stiffness matrix. This method is based on proper superposition of the individual element stiffness matrices making up a structure (also see References [1] and [2]).

Referring to the two-spring assemblage of Section 2.3, the element stiffness matrices are given in Eqs. (2.3.1) and (2.3.2) as

$$\underline{k}^{(1)} = \begin{array}{c} \begin{array}{cc} d_{1x} & d_{3x} \end{array} \\ \begin{bmatrix} k_1 & -k_1 \\ -k_1 & k_1 \end{bmatrix} \end{array} \begin{array}{c} d_{1x} \\ d_{3x} \end{array} \qquad \underline{k}^{(2)} = \begin{array}{c} \begin{array}{cc} d_{3x} & d_{2x} \end{array} \\ \begin{bmatrix} k_2 & -k_2 \\ -k_2 & k_2 \end{bmatrix} \end{array} \begin{array}{c} d_{3x} \\ d_{2x} \end{array} \tag{2.4.1}$$

Here the d_{ix}'s written above the columns and next to the rows in the $\underline{k}$'s indicate the degrees of freedom associated with each element row and column.

The two element stiffness matrices, Eqs. (2.4.1), are not associated with the same degrees of freedom; that is, element 1 is associated with axial displacements at nodes 1 and 3, whereas element 2 is associated with axial displacements at nodes 2 and 3. Therefore, the element stiffness matrices cannot be added together (superimposed) in their present form. To superimpose the element matrices, we must expand them to the order (size) of the total structure (spring assemblage) stiffness matrix so that each element stiffness matrix is associated with all the degrees of freedom of the structure. To expand each element stiffness matrix to the order of the total stiffness matrix, we simply add rows and columns of zeros for those displacements not associated with that particular element.

For element 1, we rewrite the stiffness matrix in expanded form so that Eq. (2.3.1) becomes

$$k_1 \begin{array}{c} \begin{array}{ccc} d_{1x} & d_{2x} & d_{3x} \end{array} \\ \begin{bmatrix} 1 & 0 & -1 \\ 0 & 0 & 0 \\ -1 & 0 & 1 \end{bmatrix} \end{array} \begin{Bmatrix} d_{1x}^{(1)} \\ d_{2x}^{(1)} \\ d_{3x}^{(1)} \end{Bmatrix} = \begin{Bmatrix} f_{1x}^{(1)} \\ f_{2x}^{(1)} \\ f_{3x}^{(1)} \end{Bmatrix} \tag{2.4.2}$$

where, from Eq. (2.4.2), we see that $d_{2x}^{(1)}$ and $f_{2x}^{(1)}$ are not associated with $\underline{k}^{(1)}$. Similarly, for element 2, we have

$$k_2 \begin{array}{c} \begin{array}{ccc} d_{1x} & d_{2x} & d_{3x} \end{array} \\ \begin{bmatrix} 0 & 0 & 0 \\ 0 & 1 & -1 \\ 0 & -1 & 1 \end{bmatrix} \end{array} \begin{Bmatrix} d_{1x}^{(2)} \\ d_{2x}^{(2)} \\ d_{3x}^{(2)} \end{Bmatrix} = \begin{Bmatrix} f_{1x}^{(2)} \\ f_{2x}^{(2)} \\ f_{3x}^{(2)} \end{Bmatrix} \tag{2.4.3}$$

Now, considering force equilibrium at each node results in

$$\begin{Bmatrix} f_{1x}^{(1)} \\ 0 \\ f_{3x}^{(1)} \end{Bmatrix} + \begin{Bmatrix} 0 \\ f_{2x}^{(2)} \\ f_{3x}^{(2)} \end{Bmatrix} = \begin{Bmatrix} F_{1x} \\ F_{2x} \\ F_{3x} \end{Bmatrix} \tag{2.4.4}$$

where Eq. (2.4.4) is really Eqs. (2.3.4)–(2.3.6) expressed in matrix form. Using Eqs. (2.4.2) and (2.4.3) in Eq. (2.4.4), we obtain

$$k_1 \begin{bmatrix} 1 & 0 & -1 \\ 0 & 0 & 0 \\ -1 & 0 & 1 \end{bmatrix} \begin{Bmatrix} d_{1x}^{(1)} \\ d_{2x}^{(1)} \\ d_{3x}^{(1)} \end{Bmatrix} + k_2 \begin{bmatrix} 0 & 0 & 0 \\ 0 & 1 & -1 \\ 0 & -1 & 1 \end{bmatrix} \begin{Bmatrix} d_{1x}^{(2)} \\ d_{2x}^{(2)} \\ d_{3x}^{(2)} \end{Bmatrix} = \begin{Bmatrix} F_{1x} \\ F_{2x} \\ F_{3x} \end{Bmatrix} \tag{2.4.5}$$

where, again, the superscripts on the d's indicate the element numbers. Simplifying Eq. (2.4.5) results in

$$\begin{bmatrix} k_1 & 0 & -k_1 \\ 0 & k_2 & -k_2 \\ -k_1 & -k_2 & k_1 + k_2 \end{bmatrix} \begin{Bmatrix} d_{1x} \\ d_{2x} \\ d_{3x} \end{Bmatrix} = \begin{Bmatrix} F_{1x} \\ F_{2x} \\ F_{3x} \end{Bmatrix} \tag{2.4.6}$$

Here the superscripts indicating the element numbers associated with the nodal displacements have been dropped because $d_{1x}^{(1)}$ is really d_{1x}, $d_{2x}^{(2)}$ is really d_{2x}, and, by Eq. (2.3.3), $d_{3x}^{(1)} = d_{3x}^{(2)} = d_{3x}$, the node 3 displacement of the total assemblage. Equation (2.4.6), obtained through superposition, is identical to Eq. (2.3.9).

The expanded element stiffness matrices in Eqs. (2.4.2) and (2.4.3) could have been added directly to obtain the total stiffness matrix of the structure, given in Eq. (2.4.6). This reliable method of directly assembling individual element stiffness matrices to form the total structure stiffness matrix and the total set of stiffness equations is called the *direct stiffness method.* It is the most important step in the finite element method.

For this simple example, it is easy to expand the element stiffness matrices and then superimpose them to arrive at the total stiffness matrix. However, for problems involving a large number of degrees of freedom, it will become tedious to expand each element stiffness matrix to the order of the total stiffness matrix. To avoid this expansion of each element stiffness matrix, we suggest a direct, or short-cut, form of the direct stiffness method to obtain the total stiffness matrix. For the spring assemblage example, the rows and columns of each element stiffness matrix are labeled according to the degrees of freedom associated with them as follows:

$$\underline{k}^{(1)} = \begin{array}{c} \begin{matrix} d_{1x} & d_{3x} \end{matrix} \\ \begin{bmatrix} k_1 & -k_1 \\ -k_1 & k_1 \end{bmatrix} \end{array} \begin{matrix} d_{1x} \\ d_{3x} \end{matrix} \qquad \underline{k}^{(2)} = \begin{array}{c} \begin{matrix} d_{3x} & d_{2x} \end{matrix} \\ \begin{bmatrix} k_2 & -k_2 \\ -k_2 & k_2 \end{bmatrix} \end{array} \begin{matrix} d_{3x} \\ d_{2x} \end{matrix} \tag{2.4.7}$$

$\underline{K}$ is then constructed simply by directly adding terms associated with degrees of freedom in $\underline{k}^{(1)}$ and $\underline{k}^{(2)}$ into their corresponding identical degree-of-freedom locations in $\underline{K}$ as follows. The d_{1x} row, d_{1x} column term of $\underline{K}$ is contributed only by element 1, as only element 1 has degree of freedom d_{1x} [Eq. (2.4.7)], that is, $k_{11} = k_1$. The d_{3x} row,

d_{3x} column of $\underline{K}$ has contributions from both elements 1 and 2, as the d_{3x} degree of freedom is associated with both elements. Therefore, $k_{33} = k_1 + k_2$. Similar reasoning results in $\underline{K}$ as

$$\underline{K} = \begin{array}{c} \\ \\ \\ \end{array}\begin{matrix} d_{1x} & d_{2x} & d_{3x} \\ \left[\begin{matrix} k_1 & 0 & -k_1 \\ 0 & k_2 & -k_2 \\ -k_1 & -k_2 & k_1+k_2 \end{matrix}\right] \end{matrix} \begin{matrix} \\ d_{1x} \\ d_{2x} \\ d_{3x} \end{matrix} \tag{2.4.8}$$

Here elements in $\underline{K}$ are located on the basis that degrees of freedom are ordered in increasing node numerical order for the total structure. Section 2.5 addresses the complete solution to the two-spring assemblage in conjunction with discussion of the support boundary conditions.

2.5 Boundary Conditions

We must specify boundary (or support) conditions for structure models such as the spring assemblage of Figure 2–6, or $\underline{K}$ will be singular; that is, the determinant of $\underline{K}$ will be zero, and its inverse will not exist. This means the structural system is unstable. Without our specifying adequate kinematic constraints or support conditions, the structure will be free to move as a rigid body and not resist any applied loads. In general, the number of boundary conditions necessary to make $[\underline{K}]$ nonsingular is equal to the number of possible rigid body modes.

Boundary conditions are of two general types. Homogeneous boundary conditions—the more common—occur at locations that are completely prevented from movement; nonhomogeneous boundary conditions occur where finite nonzero values of displacement are specified, such as the settlement of a support.

To illustrate the two general types of boundary conditions, let us consider Eq. (2.4.6), derived for the spring assemblage of Figure 2–6. which has a single rigid body mode in the direction of motion along the spring assemblage. We first consider the case of homogeneous boundary conditions. Hence all boundary conditions are such that the displacements are zero at certain nodes. Here we have $d_{1x} = 0$ because node 1 is fixed. Therefore, Eq. (2.4.6) can be written as

$$\begin{bmatrix} k_1 & 0 & -k_1 \\ 0 & k_2 & -k_2 \\ -k_1 & -k_2 & k_1+k_2 \end{bmatrix} \begin{Bmatrix} 0 \\ d_{2x} \\ d_{3x} \end{Bmatrix} = \begin{Bmatrix} F_{1x} \\ F_{2x} \\ F_{3x} \end{Bmatrix} \tag{2.5.1}$$

Equation (2.5.1), written in expanded form, becomes

$$\begin{aligned} k_1(0) + (0)d_{2x} - k_1 d_{3x} &= F_{1x} \\ 0(0) + k_2 d_{2x} - k_2 d_{3x} &= F_{2x} \\ -k_1(0) - k_2 d_{2x} + (k_1 + k_2)d_{3x} &= F_{3x} \end{aligned} \tag{2.5.2}$$

where F_{1x} is the unknown reaction and F_{2x} and F_{3x} are known applied loads.

Writing the second and third of Eqs. (2.5.2) in matrix form, we have

$$\begin{bmatrix} k_2 & -k_2 \\ -k_2 & k_1 + k_2 \end{bmatrix} \begin{Bmatrix} d_{2x} \\ d_{3x} \end{Bmatrix} = \begin{Bmatrix} F_{2x} \\ F_{3x} \end{Bmatrix} \tag{2.5.3}$$

We have now effectively partitioned off the first column and row of $\underline{K}$ and the first row of $\underline{d}$ and $\underline{F}$ to arrive at Eq. (2.5.3).

For homogeneous boundary conditions, Eq. (2.5.3) could have been obtained directly by deleting the row and column of Eq. (2.5.1) corresponding to the zero-displacement degrees of freedom. Here row 1 and column 1 are deleted because one is really multiplying column 1 of $\underline{K}$ by $d_{1x} = 0$. However, F_{1x} is not necessarily zero and can be determined once d_{2x} and d_{3x} are solved for.

After solving Eq. (2.5.3) for d_{2x} and d_{3x}, we have

$$\begin{Bmatrix} d_{2x} \\ d_{3x} \end{Bmatrix} = \begin{bmatrix} k_2 & -k_2 \\ -k_2 & k_1 + k_2 \end{bmatrix}^{-1} \begin{Bmatrix} F_{2x} \\ F_{3x} \end{Bmatrix} = \begin{bmatrix} \dfrac{1}{k_2} + \dfrac{1}{k_1} & \dfrac{1}{k_1} \\ \dfrac{1}{k_1} & \dfrac{1}{k_1} \end{bmatrix} \begin{Bmatrix} F_{2x} \\ F_{3x} \end{Bmatrix} \tag{2.5.4}$$

Now that d_{2x} and d_{3x} are known from Eq. (2.5.4), we substitute them in the first of Eqs. (2.5.2) to obtain the reaction F_{1x} as

$$F_{1x} = -k_1 d_{3x} \tag{2.5.5}$$

We can express the unknown nodal force at node 1 (also called the *reaction*) in terms of the applied nodal forces F_{2x} and F_{3x} by using Eq. (2.5.4) for d_{3x} substituted into Eq. (2.5.5). The result is

$$F_{1x} = -F_{2x} - F_{3x} \tag{2.5.6}$$

Therefore, for all homogeneous boundary conditions, we can delete the rows and columns corresponding to the zero-displacement degrees of freedom from the original set of equations and then solve for the unknown displacements. This procedure is useful for hand calculations. (However, Appendix B.4 presents a more practical, computer-assisted scheme for solving the system of simultaneous equations.)

We now consider the case of nonhomogeneous boundary conditions. Hence some of the specified displacements are nonzero. For simplicity's sake, let $d_{1x} = \delta$, where δ is a known displacement (Figure 2–8), in Eq. (2.4.6). We now have

$$\begin{bmatrix} k_1 & 0 & -k_1 \\ 0 & k_2 & -k_2 \\ -k_1 & -k_2 & k_1 + k_2 \end{bmatrix} \begin{Bmatrix} \delta \\ d_{2x} \\ d_{3x} \end{Bmatrix} = \begin{Bmatrix} F_{1x} \\ F_{2x} \\ F_{3x} \end{Bmatrix} \tag{2.5.7}$$

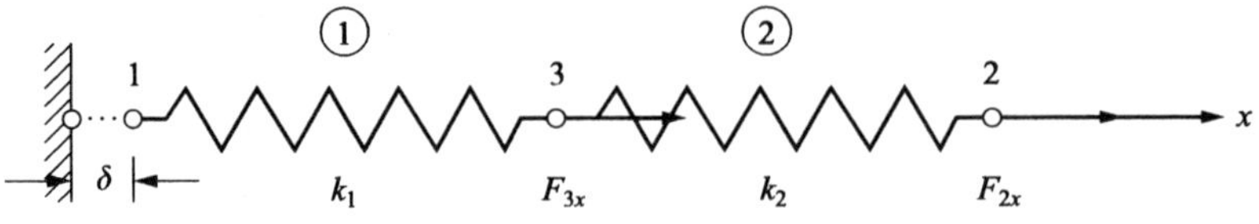

Figure 2–8 Two-spring assemblage with known displacement δ at node 1

Equation (2.5.7) written in expanded form becomes

$$\begin{aligned} k_1\delta + 0d_{2x} - k_1 d_{3x} &= F_{1x} \\ 0\delta + k_2 d_{2x} - k_2 d_{3x} &= F_{2x} \\ -k_1\delta - k_2 d_{2x} + (k_1 + k_2) d_{3x} &= F_{3x} \end{aligned} \tag{2.5.8}$$

where F_{1x} is now a reaction from the support that has moved an amount δ. Considering the second and third of Eqs. (2.5.8) because they have known right-side nodal forces F_{2x} and F_{3x}, we obtain

$$\begin{aligned} 0\delta + k_2 d_{2x} - k_2 d_{3x} &= F_{2x} \\ -k_1\delta - k_2 d_{2x} + (k_1 + k_2) d_{3x} &= F_{3x} \end{aligned} \tag{2.5.9}$$

Transforming the known δ terms to the right side of Eqs. (2.5.9) yields

$$\begin{aligned} k_2 d_{2x} - k_2 d_{3x} &= F_{2x} \\ -k_2 d_{2x} + (k_1 + k_2) d_{3x} &= +k_1\delta + F_{3x} \end{aligned} \tag{2.5.10}$$

Rewriting Eqs. (2.5.10) in matrix form, we have

$$\begin{bmatrix} k_2 & -k_2 \\ -k_2 & k_1 + k_2 \end{bmatrix} \begin{Bmatrix} d_{2x} \\ d_{3x} \end{Bmatrix} = \begin{Bmatrix} F_{2x} \\ k_1\delta + F_{3x} \end{Bmatrix} \tag{2.5.11}$$

Therefore, when dealing with nonhomogeneous boundary conditions, we cannot initially delete row 1 and column 1 of Eq. (2.5.7), corresponding to the nonhomogeneous boundary condition, as indicated by the resulting Eq. (2.5.11) because we are multiplying each element by a nonzero number. Had we done so, the $k_1\delta$ term in Eq. (2.5.11) would have been neglected, resulting in an error in the solution for the displacements. For nonhomogeneous boundary conditions, we must, in general, transform the terms associated with the known displacements to the right-side force matrix before solving for the unknown nodal displacements. This was illustrated by transforming the $k_1\delta$ term of the second of Eqs. (2.5.9) to the right side of the second of Eqs. (2.5.10).

We could now solve for the displacements in Eq. (2.5.11) in a manner similar to that used to solve Eq. (2.5.3). However, we will not further pursue the solution of Eq. (2.5.11) because no new information is to be gained.

However, on substituting the displacement back into Eq. (2.5.7), the reaction now becomes

$$F_{1x} = k_1\delta - k_1 d_{3x} \tag{2.5.12}$$

which is different than Eq. (2.5.5) for F_{1x}.

At this point, we summarize some properties of the stiffness matrix in Eq. (2.5.7) that are also applicable to the generalization of the finite element method.

1. $\underline{K}$ is symmetric, as is each of the element stiffness matrices. If you are familiar with structural mechanics, you will not find this symmetry property surprising. It can be proved by using the reciprocal laws described in such References as [3] and [4].

2. $\underline{K}$ is singular, and thus no inverse exists until sufficient boundary conditions are imposed to remove the singularity and prevent rigid body motion.
3. The main diagonal terms of $\underline{K}$ are always positive. Otherwise, a positive nodal force F_i could produce a negative displacement d_i—a behavior contrary to the physical behavior of any actual structure.

In general, specified support conditions are treated mathematically by partitioning the global equilibrium equations as follows:

$$\begin{bmatrix} \underline{K}_{11} & \underline{K}_{12} \\ \underline{K}_{21} & \underline{K}_{22} \end{bmatrix} \begin{Bmatrix} \underline{d}_1 \\ \underline{d}_2 \end{Bmatrix} = \begin{Bmatrix} \underline{F}_1 \\ \underline{F}_2 \end{Bmatrix} \tag{2.5.13}$$

where we let $\underline{d}_1$ be the unconstrained or free displacements and $\underline{d}_2$ be the specified displacements. From Eq. (2.5.13), we have

$$\underline{K}_{11}\underline{d}_1 = \underline{F}_1 - \underline{K}_{12}\underline{d}_2 \tag{2.5.14}$$

and

$$\underline{F}_2 = \underline{K}_{21}\underline{d}_1 + \underline{K}_{22}\underline{d}_2 \tag{2.5.15}$$

where $\underline{F}_1$ are the known nodal forces and $\underline{F}_2$ are the unknown nodal forces at the specified displacement nodes. $\underline{F}_2$ is found from Eq. (2.5.15) after $\underline{d}_1$ is determined from Eq. (2.5.14). In Eq. (2.5.14), we assume that $\underline{K}_{11}$ is no longer singular, thus allowing for the determination of $\underline{d}_1$.

To illustrate the stiffness method for the solution of spring assemblages we now present the following examples.

Example 2.1

For the spring assemblage with arbitrarily numbered nodes shown in Figure 2–9, obtain (a) the global stiffness matrix, (b) the displacements of nodes 3 and 4, (c) the reaction forces at nodes 1 and 2, and (d) the forces in each spring. A force of 5000 lb is applied at node 4 in the x direction. The spring constants are given in the figure. Nodes 1 and 2 are fixed.

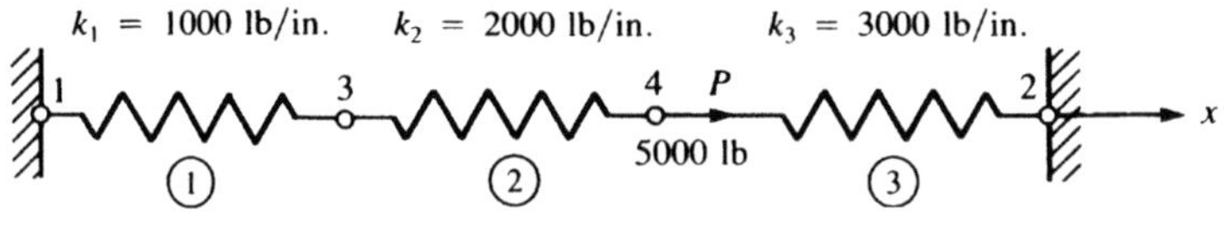

Figure 2–9 Spring assemblage for solution

(a) We begin by making use of Eq. (2.2.18) to express each element stiffness matrix as follows:

$$\underline{k}^{(1)} = \begin{array}{c} \begin{matrix} 1 & 3 \end{matrix} \\ \begin{bmatrix} 1000 & -1000 \\ -1000 & 1000 \end{bmatrix} \end{array} \begin{matrix} 1 \\ 3 \end{matrix} \qquad \underline{k}^{(2)} = \begin{array}{c} \begin{matrix} 3 & 4 \end{matrix} \\ \begin{bmatrix} 2000 & -2000 \\ -2000 & 2000 \end{bmatrix} \end{array} \begin{matrix} 3 \\ 4 \end{matrix}$$

$$\underline{k}^{(3)} = \begin{array}{c} \begin{matrix} 4 & 2 \end{matrix} \\ \begin{bmatrix} 3000 & -3000 \\ -3000 & 3000 \end{bmatrix} \end{array} \begin{matrix} 4 \\ 2 \end{matrix} \tag{2.5.16}$$

where the numbers above the columns and next to each row indicate the nodal degrees of freedom associated with each element. For instance, element 1 is associated with degrees of freedom d_{1x} and d_{3x}. Also, the local $\hat{x}$ axis coincides with the global x axis for each element.

Using the concept of superposition (the direct stiffness method), we obtain the global stiffness matrix as

$$\underline{K} = \underline{k}^{(1)} + \underline{k}^{(2)} + \underline{k}^{(3)}$$

or

$$\underline{K} = \begin{array}{c} \begin{matrix} 1 & 2 & 3 & 4 \end{matrix} \\ \begin{bmatrix} 1000 & 0 & -1000 & 0 \\ 0 & 3000 & 0 & -3000 \\ -1000 & 0 & 1000 + 2000 & -2000 \\ 0 & -3000 & -2000 & 2000 + 3000 \end{bmatrix} \end{array} \begin{matrix} 1 \\ 2 \\ 3 \\ 4 \end{matrix} \tag{2.5.17}$$

(b) The global stiffness matrix, Eq. (2.5.17), relates global forces to global displacements as follows:

$$\begin{Bmatrix} F_{1x} \\ F_{2x} \\ F_{3x} \\ F_{4x} \end{Bmatrix} = \begin{bmatrix} 1000 & 0 & -1000 & 0 \\ 0 & 3000 & 0 & -3000 \\ -1000 & 0 & 3000 & -2000 \\ 0 & -3000 & -2000 & 5000 \end{bmatrix} \begin{Bmatrix} d_{1x} \\ d_{2x} \\ d_{3x} \\ d_{4x} \end{Bmatrix} \tag{2.5.18}$$

Applying the homogeneous boundary conditions $d_{1x} = 0$ and $d_{2x} = 0$ to Eq. (2.5.18), substituting applied nodal forces, and partitioning the first two equations of Eq. (2.5.18) (or deleting the first two rows of $\{F\}$ and $\{d\}$ and the first two rows and columns of $\underline{K}$ corresponding to the zero-displacement boundary conditions), we obtain

$$\begin{Bmatrix} 0 \\ 5000 \end{Bmatrix} = \begin{bmatrix} 3000 & -2000 \\ -2000 & 5000 \end{bmatrix} \begin{Bmatrix} d_{3x} \\ d_{4x} \end{Bmatrix} \tag{2.5.19}$$

Solving Eq. (2.5.19), we obtain the global nodal displacements

$$d_{3x} = \frac{10}{11} \text{ in.} \qquad d_{4x} = \frac{15}{11} \text{ in.} \tag{2.5.20}$$

(c) To obtain the global nodal forces (which include the reactions at nodes 1 and 2), we back-substitute Eqs. (2.5.20) and the boundary conditions $d_{1x} = 0$ and

$d_{2x} = 0$ into Eq. (2.5.18). This substitution yields

$$\begin{Bmatrix} F_{1x} \\ F_{2x} \\ F_{3x} \\ F_{4x} \end{Bmatrix} = \begin{bmatrix} 1000 & 0 & -1000 & 0 \\ 0 & 3000 & 0 & -3000 \\ -1000 & 0 & 3000 & -2000 \\ 0 & -3000 & -2000 & 5000 \end{bmatrix} \begin{Bmatrix} 0 \\ 0 \\ \frac{10}{11} \\ \frac{15}{11} \end{Bmatrix} \tag{2.5.21}$$

Multiplying matrices in Eq. (2.5.21) and simplifying, we obtain the forces at each node

$$F_{1x} = \frac{-10{,}000}{11}\ \text{lb} \qquad F_{2x} = \frac{-45{,}000}{11}\ \text{lb} \qquad F_{3x} = 0$$
$$F_{4x} = \frac{55{,}000}{11}\ \text{lb} \tag{2.5.22}$$

From these results, we observe that the sum of the reactions F_{1x} and F_{2x} is equal in magnitude but opposite in direction to the applied force F_{4x}. This result verifies equilibrium of the whole spring assemblage.

(d) Next we use local element Eq. (2.2.17) to obtain the forces in each element.

Element 1

$$\begin{Bmatrix} \hat{f}_{1x} \\ \hat{f}_{3x} \end{Bmatrix} = \begin{bmatrix} 1000 & -1000 \\ -1000 & 1000 \end{bmatrix} \begin{Bmatrix} 0 \\ \frac{10}{11} \end{Bmatrix} \tag{2.5.23}$$

Simplifying Eq. (2.5.23), we obtain

$$\hat{f}_{1x} = \frac{-10{,}000}{11}\ \text{lb} \qquad \hat{f}_{3x} = \frac{10{,}000}{11}\ \text{lb} \tag{2.5.24}$$

A free-body diagram of spring element 1 is shown in Figure 2–10(a). The spring is subjected to tensile forces given by Eqs. (2.5.24). Also, $\hat{f}_{1x}$ is equal to the reaction force F_{1x} given in Eq. (2.5.22). A free-body diagram of node 1 [Figure 2–10(b)] shows this result.

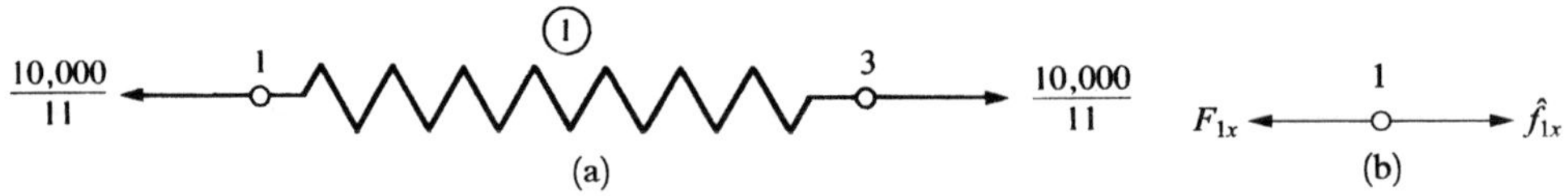

Figure 2–10 (a) Free-body diagram of element 1 and (b) free-body diagram of node 1.

Element 2

$$\begin{Bmatrix} \hat{f}_{3x} \\ \hat{f}_{4x} \end{Bmatrix} = \begin{bmatrix} 2000 & -2000 \\ -2000 & 2000 \end{bmatrix} \begin{Bmatrix} \frac{10}{11} \\ \frac{15}{11} \end{Bmatrix} \tag{2.5.25}$$

Simplifying Eq. (2.5.24), we obtain

$$\hat{f}_{3x} = \frac{-10{,}000}{11}\ \text{lb} \qquad \hat{f}_{4x} = \frac{10{,}000}{11}\ \text{lb} \tag{2.5.26}$$

A free-body diagram of spring element 2 is shown in Figure 2–11. The spring is subjected to tensile forces given by Eqs. (2.5.26).

Figure 2–11 Free-body diagram of element 2

Element 3

$$\begin{Bmatrix} \hat{f}_{4x} \\ \hat{f}_{2x} \end{Bmatrix} = \begin{bmatrix} 3000 & -3000 \\ -3000 & 3000 \end{bmatrix} \begin{Bmatrix} \frac{15}{11} \\ 0 \end{Bmatrix} \tag{2.5.27}$$

Simplifying Eq. (2.5.27) yields

$$\hat{f}_{4x} = \frac{45{,}000}{11}\ \text{lb} \qquad \hat{f}_{2x} = \frac{-45{,}000}{11}\ \text{lb} \tag{2.5.28}$$

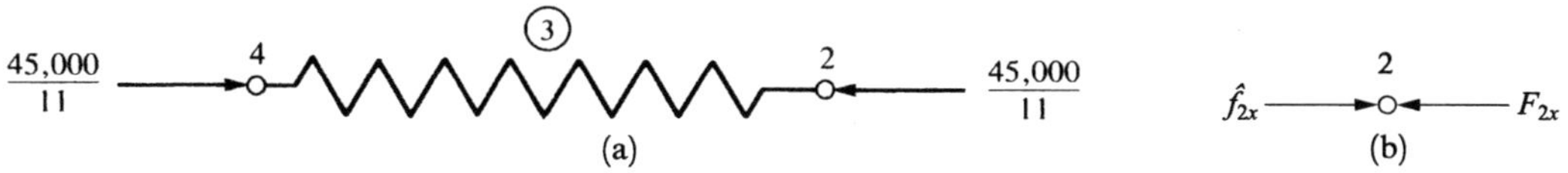

Figure 2–12 (a) Free-body diagram of element 3 and (b) free-body diagram of node 2

A free-body diagram of spring element 3 is shown in Figure 2–12(a). The spring is subjected to compressive forces given by Eqs. (2.5.28). Also, $\hat{f}_{2x}$ is equal to the reaction force F_{2x} given in Eq. (2.5.22). A free-body diagram of node 2 (Figure 2–12b) shows this result. ■

Example 2.2

For the spring assemblage shown in Figure 2–13, obtain (a) the global stiffness matrix, (b) the displacements of nodes 2–4, (c) the global nodal forces, and (d) the local element forces. Node 1 is fixed while node 5 is given a fixed, known displacement $\delta = 20.0$ mm. The spring constants are all equal to $k = 200$ kN/m.

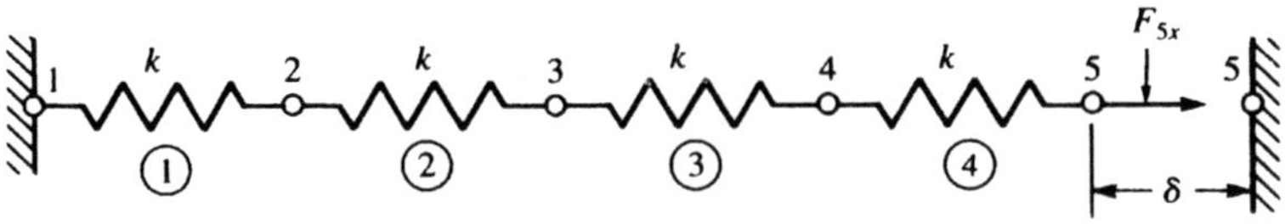

Figure 2–13 Spring assemblage for solution

(a) We use Eq. (2.2.18) to express each element stiffness matrix as

$$\underline{k}^{(1)} = \underline{k}^{(2)} = \underline{k}^{(3)} = \underline{k}^{(4)} = \begin{bmatrix} 200 & -200 \\ -200 & 200 \end{bmatrix} \tag{2.5.29}$$

Again using superposition, we obtain the global stiffness matrix as

$$\underline{K} = \begin{bmatrix} 200 & -200 & 0 & 0 & 0 \\ -200 & 400 & -200 & 0 & 0 \\ 0 & -200 & 400 & -200 & 0 \\ 0 & 0 & -200 & 400 & -200 \\ 0 & 0 & 0 & -200 & 200 \end{bmatrix} \frac{\text{kN}}{\text{m}} \tag{2.5.30}$$

(b) The global stiffness matrix, Eq. (2.5.30), relates the global forces to the global displacements as follows:

$$\begin{Bmatrix} F_{1x} \\ F_{2x} \\ F_{3x} \\ F_{4x} \\ F_{5x} \end{Bmatrix} = \begin{bmatrix} 200 & -200 & 0 & 0 & 0 \\ -200 & 400 & -200 & 0 & 0 \\ 0 & -200 & 400 & -200 & 0 \\ 0 & 0 & -200 & 400 & -200 \\ 0 & 0 & 0 & -200 & 200 \end{bmatrix} \begin{Bmatrix} d_{1x} \\ d_{2x} \\ d_{3x} \\ d_{4x} \\ d_{5x} \end{Bmatrix} \tag{2.5.31}$$

Applying the boundary conditions $d_{1x} = 0$ and $d_{5x} = 20$ mm ($= 0.02$ m), substituting known global forces $F_{2x} = 0$, $F_{3x} = 0$, and $F_{4x} = 0$, and partitioning the first and fifth equations of Eq. (2.5.31) corresponding to these boundary conditions, we obtain

$$\begin{Bmatrix} 0 \\ 0 \\ 0 \end{Bmatrix} = \begin{bmatrix} -200 & 400 & -200 & 0 & 0 \\ 0 & -200 & 400 & -200 & 0 \\ 0 & 0 & -200 & 400 & -200 \end{bmatrix} \begin{Bmatrix} 0 \\ d_{2x} \\ d_{3x} \\ d_{4x} \\ 0.02 \text{ m} \end{Bmatrix} \tag{2.5.32}$$

We now rewrite Eq. (2.5.32), transposing the product of the appropriate stiffness coefficient (-200) multiplied by the known displacement (0.02 m) to the left side.

$$\begin{Bmatrix} 0 \\ 0 \\ 4 \text{ kN} \end{Bmatrix} = \begin{bmatrix} 400 & -200 & 0 \\ -200 & 400 & -200 \\ 0 & -200 & 400 \end{bmatrix} \begin{Bmatrix} d_{2x} \\ d_{3x} \\ d_{4x} \end{Bmatrix} \tag{2.5.33}$$

Solving Eq. (2.5.33), we obtain

$$d_{2x} = 0.005 \text{ m} \qquad d_{3x} = 0.01 \text{ m} \qquad d_{4x} = 0.015 \text{ m} \tag{2.5.34}$$

(c) The global nodal forces are obtained by back-substituting the boundary condition displacements and Eqs. (2.5.34) into Eq. (2.5.31). This substitution yields

$$\begin{aligned}
F_{1x} &= (-200)(0.005) = -1.0 \text{ kN} \\
F_{2x} &= (400)(0.005) - (200)(0.01) = 0 \\
F_{3x} &= (-200)(0.005) + (400)(0.01) - (200)(0.015) = 0 \\
F_{4x} &= (-200)(0.01) + (400)(0.015) - (200)(0.02) = 0 \\
F_{5x} &= (-200)(0.015) + (200)(0.02) = 1.0 \text{ kN}
\end{aligned} \tag{2.5.35}$$

The results of Eqs. (2.5.35) yield the reaction F_{1x} opposite that of the nodal force F_{5x} required to displace node 5 by $\delta = 20.0$ mm. This result verifies equilibrium of the whole spring assemblage.

(d) Next, we make use of local element Eq. (2.2.17) to obtain the forces in each element.

Element 1

$$\begin{Bmatrix} \hat{f}_{1x} \\ \hat{f}_{2x} \end{Bmatrix} = \begin{bmatrix} 200 & -200 \\ -200 & 200 \end{bmatrix} \begin{Bmatrix} 0 \\ 0.005 \end{Bmatrix} \tag{2.5.36}$$

Simplifying Eq. (2.5.36) yields

$$\hat{f}_{1x} = -1.0 \text{ kN} \qquad \hat{f}_{2x} = 1.0 \text{ kN} \tag{2.5.37}$$

Element 2

$$\begin{Bmatrix} \hat{f}_{2x} \\ \hat{f}_{3x} \end{Bmatrix} = \begin{bmatrix} 200 & -200 \\ -200 & 200 \end{bmatrix} \begin{Bmatrix} 0.005 \\ 0.01 \end{Bmatrix} \tag{2.5.38}$$

Simplifying Eq. (2.5.38) yields

$$\hat{f}_{2x} = -1 \text{ kN} \qquad \hat{f}_{3x} = 1 \text{ kN} \tag{2.5.39}$$

Element 3

$$\begin{Bmatrix} \hat{f}_{3x} \\ \hat{f}_{4x} \end{Bmatrix} = \begin{bmatrix} 200 & -200 \\ -200 & 200 \end{bmatrix} \begin{Bmatrix} 0.01 \\ 0.015 \end{Bmatrix} \tag{2.5.40}$$

Simplifying Eq. (2.5.40), we have

$$\hat{f}_{3x} = -1 \text{ kN} \qquad \hat{f}_{4x} = 1 \text{ kN} \tag{2.5.41}$$

Element 4

$$\begin{Bmatrix} \hat{f}_{4x} \\ \hat{f}_{5x} \end{Bmatrix} = \begin{bmatrix} 200 & -200 \\ -200 & 200 \end{bmatrix} \begin{Bmatrix} 0.015 \\ 0.02 \end{Bmatrix} \tag{2.5.42}$$

Simplifying Eq. (2.5.42), we obtain

$$\hat{f}_{4x} = -1 \text{ kN} \qquad \hat{f}_{5x} = 1 \text{ kN} \tag{2.5.43}$$

You should draw free-body diagrams of each node and element and use the results of Eqs. (2.5.35)–(2.5.43) to verify both node and element equilibria. ■

Finally, to review the major concepts presented in this chapter, we solve the following example problem.

Example 2.3

(a) Using the ideas presented in Section 2.3 for the system of linear elastic springs shown in Figure 2–14, express the boundary conditions, the compatibility or continuity condition similar to Eq. (2.3.3), and the nodal equilibrium conditions similar to Eqs. (2.3.4)–(2.3.6). Then formulate the global stiffness matrix and equations for solution of the unknown global displacement and forces. The spring constants for the elements are k_1, k_2, and k_3; P is an applied force at node 2.

(b) Using the direct stiffness method, formulate the same global stiffness matrix and equation as in part (a).

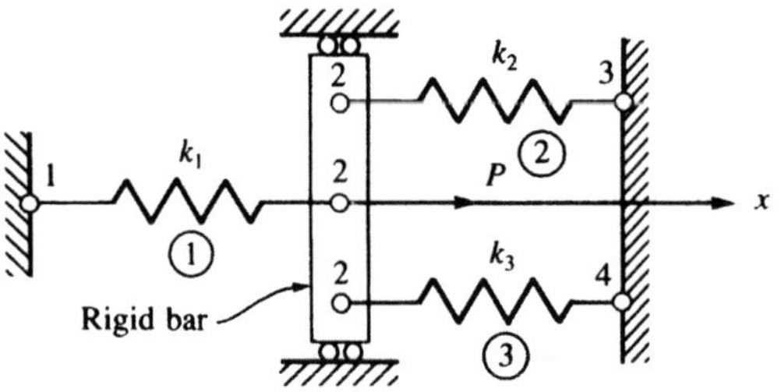

Figure 2–14 Spring assemblage for solution

(a) The boundary conditions are

$$d_{1x} = 0 \qquad d_{3x} = 0 \qquad d_{4x} = 0 \tag{2.5.44}$$

The compatibility condition at node 2 is

$$d_{2x}^{(1)} = d_{2x}^{(2)} = d_{2x}^{(3)} = d_{2x} \tag{2.5.45}$$

The nodal equilibrium conditions are

$$
\begin{aligned}
F_{1x} &= f_{1x}^{(1)} \\
P &= f_{2x}^{(1)} + f_{2x}^{(2)} + f_{2x}^{(3)} \\
F_{3x} &= f_{3x}^{(2)} \\
F_{4x} &= f_{4x}^{(3)}
\end{aligned}
\tag{2.5.46}
$$

where the sign convention for positive element nodal forces given by Figure 2–2 was used in writing Eqs. (2.5.46). Figure 2–15 shows the element and nodal force free-body diagrams.

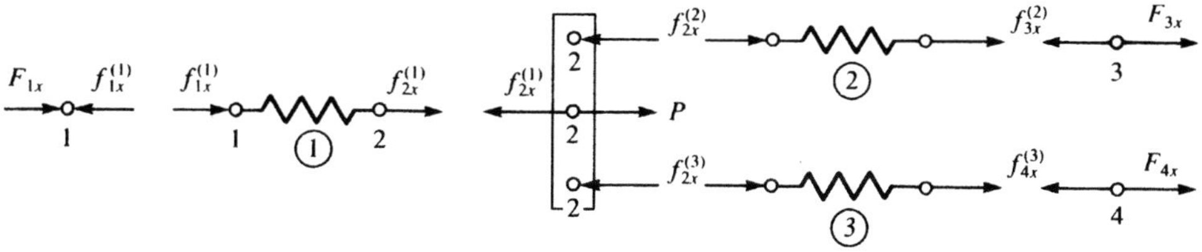

Figure 2–15 Free-body diagrams of elements and nodes of spring assemblage of Figure 2–14

Using the local stiffness matrix Eq. (2.2.17) applied to each element, and compatibility condition Eq. (2.5.45), we obtain the total or global equilibrium equations as

$$
\begin{aligned}
F_{1x} &= k_1 d_{1x} - k_1 d_{2x} \\
P &= -k_1 d_{1x} + k_1 d_{2x} + k_2 d_{2x} - k_2 d_{3x} + k_3 d_{2x} - k_3 d_{4x} \\
F_{3x} &= -k_2 d_{2x} + k_2 d_{3x} \\
F_{4x} &= -k_3 d_{2x} + k_3 d_{4x}
\end{aligned}
\tag{2.5.47}
$$

In matrix form, we express Eqs. (2.5.47) as

$$
\begin{Bmatrix} F_{1x} \\ P \\ F_{3x} \\ F_{4x} \end{Bmatrix} = \begin{bmatrix} k_1 & -k_1 & 0 & 0 \\ -k_1 & k_1 + k_2 + k_3 & -k_2 & -k_3 \\ 0 & -k_2 & k_2 & 0 \\ 0 & -k_3 & 0 & k_3 \end{bmatrix} \begin{Bmatrix} d_{1x} \\ d_{2x} \\ d_{3x} \\ d_{4x} \end{Bmatrix} \tag{2.5.48}
$$

Therefore, the global stiffness matrix is the square, symmetric matrix on the right side of Eq. (2.5.48). Making use of the boundary conditions, Eqs. (2.5.44), and then considering the second equation of Eqs. (2.5.47) or (2.5.48), we solve for d_{2x} as

$$
d_{2x} = \frac{P}{k_1 + k_2 + k_3} \tag{2.5.49}
$$

We could have obtained this same result by deleting rows 1, 3, and 4 in the $\underline{F}$ and $\underline{d}$ matrices and rows and columns 1, 3, and 4 in $\underline{K}$, corresponding to zero displacement, as previously described in Section 2.4, and then solving for d_{2x}.

Using Eqs. (2.5.47), we now solve for the global forces as

$$F_{1x} = -k_1 d_{2x} \qquad F_{3x} = -k_2 d_{2x} \qquad F_{4x} = -k_3 d_{2x} \tag{2.5.50}$$

The forces given by Eqs. (2.5.50) can be interpreted as the global reactions in this example. The negative signs in front of these forces indicate that they are directed to the left (opposite the x axis).

(b) Using the direct stiffness method, we formulate the global stiffness matrix. First, using Eq. (2.2.18), we express each element stiffness matrix as

$$\underline{k}^{(1)} = \overset{\begin{matrix} d_{1x} & \quad d_{2x} \end{matrix}}{\begin{bmatrix} k_1 & -k_1 \\ -k_1 & k_1 \end{bmatrix}} \quad \underline{k}^{(2)} = \overset{\begin{matrix} d_{2x} & \quad d_{3x} \end{matrix}}{\begin{bmatrix} k_2 & -k_2 \\ -k_2 & k_2 \end{bmatrix}} \quad \underline{k}^{(3)} = \overset{\begin{matrix} d_{2x} & \quad d_{4x} \end{matrix}}{\begin{bmatrix} k_3 & -k_3 \\ -k_3 & k_3 \end{bmatrix}} \tag{2.5.51}$$

where the particular degrees of freedom associated with each element are listed in the columns above each matrix. Using the direct stiffness method as outlined in Section 2.4, we add terms from each element stiffness matrix into the appropriate corresponding row and column in the global stiffness matrix to obtain

$$\underline{K} = \overset{\begin{matrix} d_{1x} & \qquad d_{2x} & \qquad d_{3x} & \quad d_{4x} \end{matrix}}{\begin{bmatrix} k_1 & -k_1 & 0 & 0 \\ -k_1 & k_1 + k_2 + k_3 & -k_2 & -k_3 \\ 0 & -k_2 & k_2 & 0 \\ 0 & -k_3 & 0 & k_3 \end{bmatrix}} \tag{2.5.52}$$

We observe that each element stiffness matrix $\underline{k}$ has been added into the location in the global $\underline{K}$ corresponding to the identical degree of freedom associated with the element $\underline{k}$. For instance, element 3 is associated with degrees of freedom d_{2x} and d_{4x}; hence its contributions to $\underline{K}$ are in the 2–2, 2–4, 4–2, and 4–4 locations of $\underline{K}$, as indicated in Eq. (2.5.52) by the k_3 terms.

Having assembled the global $\underline{K}$ by the direct stiffness method, we then formulate the global equations in the usual manner by making use of the general Eq. (2.3.10), $\underline{F} = \underline{K}\underline{d}$. These equations have been previously obtained by Eq. (2.5.48) and therefore are not repeated. ■

Another method for handling imposed boundary conditions that allows for either homogeneous (zero) or nonhomogeneous (nonzero) prescribed degrees of freedom is called the *penalty method.* This method is easy to implement in a computer program.

Consider the simple spring assemblage in Figure 2–16 subjected to applied forces F_{1x} and F_{2x} as shown. Assume the horizontal displacement at node 1 to be forced to be $d_{1x} = \delta$.

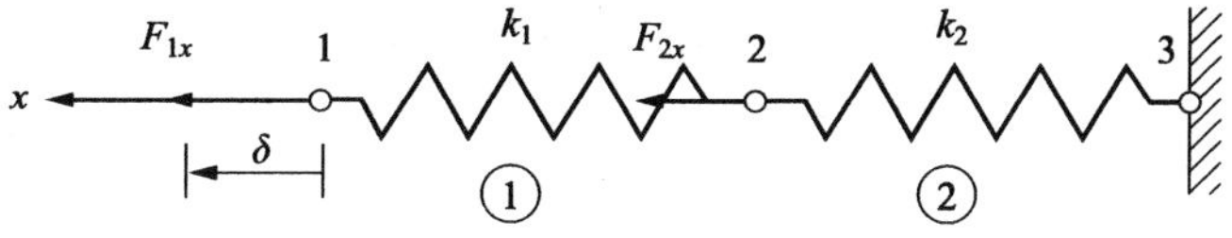

Figure 2–16 Spring assemblage used to illustrate the penalty method

We add another spring (often called a boundary element) with a large stiffness k_b to the assemblage in the direction of the nodal displacement $d_{1x} = \delta$ as shown in Figure 2–17. This spring stiffness should have a magnitude about 10^6 times that of the largest k_{ii} term.

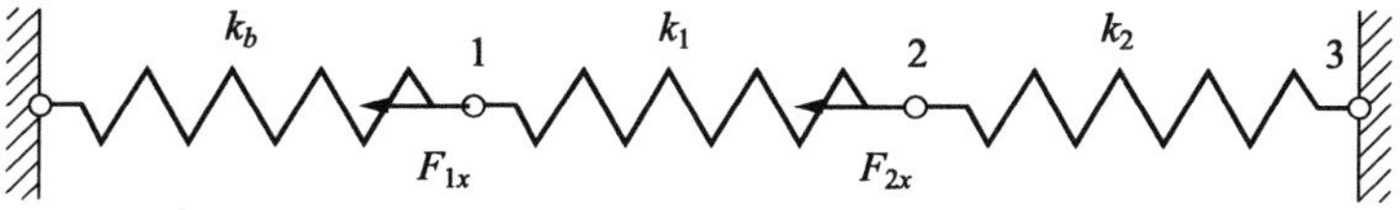

Figure 2–17 Spring assemblage with a boundary spring element added at node 1

Now we add the force $k_b\delta$ in the direction of d_{1x} and solve the problem in the usual manner as follows.

The element stiffness matrices are

$$\underline{k}^{(1)} = \begin{bmatrix} k_1 & -k_1 \\ -k_1 & k_1 \end{bmatrix} \qquad \underline{k}^{(2)} = \begin{bmatrix} k_2 & -k_2 \\ -k_2 & k_2 \end{bmatrix} \tag{2.5.53}$$

Assembling the element stiffness matrices using the direct stiffness method, we obtain the global stiffness matrix as

$$\underline{K} = \begin{bmatrix} k_1 + k_b & -k_1 & 0 \\ -k_1 & k_1 + k_2 & -k_2 \\ 0 & -k_2 & k_2 \end{bmatrix} \tag{2.5.54}$$

Assembling the global $\underline{F} = \underline{K}\underline{d}$ equations and invoking the boundary condition $d_{3x} = 0$, we obtain

$$\begin{Bmatrix} F_{1x} + k_b\delta \\ F_{2x} \\ F_{3x} \end{Bmatrix} = \begin{bmatrix} k_1 + k_b & -k_1 & 0 \\ -k_1 & k_1 + k_2 & -k_2 \\ 0 & -k_2 & k_2 \end{bmatrix} \begin{Bmatrix} d_{1x} \\ d_{2x} \\ d_{3x} = 0 \end{Bmatrix} \tag{2.5.55}$$

Solving the first and second of Eqs. (2.5.55), we obtain

$$d_{1x} = \frac{F_{2x} - (k_1 + k_2)d_{2x}}{-k_1} \tag{2.5.56}$$

and

$$d_{2x} = \frac{(k_1 + k_b)F_{2x} + F_{1x}k_1 + k_b\delta k_1}{k_b k_1 + k_b k_2 + k_1 k_2} \tag{2.5.57}$$

Now as k_b approaches infinity, Eq. (2.5.57) simplifies to

$$d_{2x} = \frac{F_{2x} + \delta k_1}{k_1 + k_2} \tag{2.5.58}$$

and Eq. (2.5.56) simplifies to

$$d_{1x} = \delta \tag{2.5.59}$$

These results match those obtained by setting $d_{1x} = \delta$ initially.

In using the penalty method, a very large element stiffness should be parallel to a degree of freedom as is the case in the preceding example. If k_b were inclined, or were placed within a structure, it would contribute to both diagonal and off-diagonal coefficients in the global stiffness matrix $\underline{K}$. This condition can lead to numerical difficulties in solving the equations $\underline{F} = \underline{K}\underline{d}$. To avoid this condition, we transform the displacements at the inclined support to local ones as described in Section 3.9.

2.6 Potential Energy Approach to Derive Spring Element Equations

One of the alternative methods often used to derive the element equations and the stiffness matrix for an element is based on the principle of *minimum potential energy*. (The use of this principle in structural mechanics is fully described in Reference [4].) This method has the advantage of being more general than the method given in Section 2.2, which involves nodal and element equilibrium equations along with the stress/strain law for the element. Thus the principle of minimum potential energy is more adaptable to the determination of element equations for complicated elements (those with large numbers of degrees of freedom) such as the plane stress/strain element, the axisymmetric stress element, the plate bending element, and the three-dimensional solid stress element.

Again, we state that the principle of virtual work (Appendix E) is applicable for any material behavior, whereas the principle of minimum potential energy is applicable only for elastic materials. However, both principles yield the same element equations for linear-elastic materials, which are the only kind considered in this text. Moreover, the principle of minimum potential energy, being included in the general category of *variational methods* (as is the principle of virtual work), leads to other variational functions (or functionals) similar to potential energy that can be formulated for other classes of problems, primarily of the nonstructural type. These other problems are generally classified as *field problems* and include, among others, torsion of a bar, heat transfer (Chapter 13), fluid flow (Chapter 14), and electric potential.

Still other classes of problems, for which a variational formulation is not clearly definable, can be formulated by *weighted residual methods*. We will describe Galerkin's method in Section 3.12, along with collocation, least squares, and the subdomain weighted residual methods in Section 3.13. In Section 3.13, we will also demonstrate these methods by solving a one-dimensional bar problem using each of the four residual methods and comparing each result to an exact solution. (For more information on weighted residual methods, also consult References [5–7].)

Here we present the principle of minimum potential energy as used to derive the spring element equations. We will illustrate this concept by applying it to the simplest of elements in hopes that the reader will then be more comfortable when applying it to handle more complicated element types in subsequent chapters.

The total potential energy π_p of a structure is expressed in terms of displacements. In the finite element formulation, these will generally be nodal displacements such that $\pi_p = \pi_p(d_1, d_2, \ldots, d_n)$. When π_p is minimized with respect to these displacements, equilibrium equations result. For the spring element, we will show that the same nodal equilibrium equations $\underline{\hat{k}}\underline{\hat{d}} = \underline{\hat{f}}$ result as previously derived in Section 2.2.

We first state the principle of minimum potential energy as follows:

> Of all the geometrically possible shapes that a body can assume, the true one, corresponding to the satisfaction of stable equilibrium of the body, is identified by a minimum value of the total potential energy.

To explain this principle, we must first explain the concepts of potential energy and of a stationary value of a function. We will now discuss these two concepts.

Total potential energy is defined as *the sum of the internal strain energy U and the potential energy of the external forces* Ω; that is,

$$\pi_p = U + \Omega \tag{2.6.1}$$

Strain energy is the capacity of internal forces (or stresses) to do work through deformations (strains) in the structure; Ω is the capacity of forces such as body forces, surface traction forces, and applied nodal forces to do work through deformation of the structure.

Recall that a linear spring has force related to deformation by $F = kx$, where k is the spring constant and x is the deformation of the spring (Figure 2–18).

The differential internal work (or strain energy) dU in the spring for a small change in length of the spring is the internal force multiplied by the change in displacement through which the force moves, given by

$$dU = F\,dx \tag{2.6.2}$$

Now we express F as

$$F = kx \tag{2.6.3}$$

Using Eq. (2.6.3) in Eq. (2.6.2), we find that the differential strain energy becomes

$$dU = kx\,dx \tag{2.6.4}$$

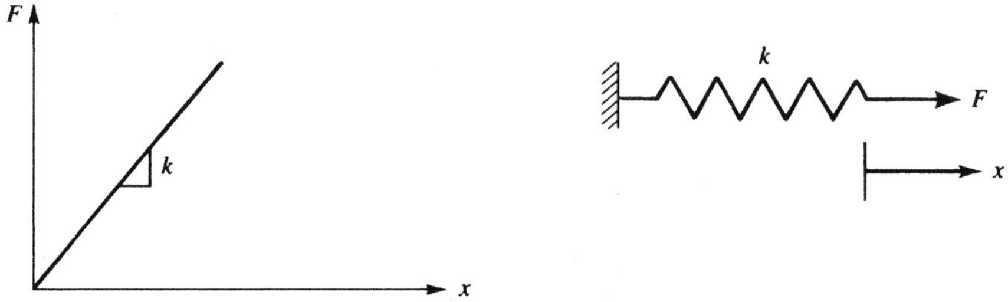

Figure 2–18 Force/deformation curve for linear spring

The total strain energy is then given by

$$U = \int_0^x kx\,dx \tag{2.6.5}$$

Upon explicit integration of Eq. (2.6.5), we obtain

$$U = \tfrac{1}{2}kx^2 \tag{2.6.6}$$

Using Eq. (2.6.3) in Eq. (2.6.6), we have

$$U = \tfrac{1}{2}(kx)x = \tfrac{1}{2}Fx \tag{2.6.7}$$

Equation (2.6.7) indicates that the strain energy is the area under the force/deformation curve.

The potential energy of the external force, being opposite in sign from the external work expression because the potential energy of the external force is lost when the work is done by the external force, is given by

$$\Omega = -Fx \tag{2.6.8}$$

Therefore, substituting Eqs. (2.6.6) and (2.6.8) into (2.6.1), yields the total potential energy as

$$\pi_p = \tfrac{1}{2}kx^2 - Fx \tag{2.6.9}$$

The concept of a *stationary value* of a function G (used in the definition of the principle of minimum potential energy) is shown in Figure 2–19. Here G is expressed as a function of the variable x. The stationary value can be a maximum, a minimum, or a neutral point of $G(x)$. To find a value of x yielding a stationary value of $G(x)$, we use differential calculus to differentiate G with respect to x and set the expression equal to zero, as follows:

$$\frac{dG}{dx} = 0 \tag{2.6.10}$$

An analogous process will subsequently be used to replace G with π_p and x with discrete values (nodal displacements) d_i. With an understanding of variational calculus (see Reference [8]), we could use the first variation of π_p (denoted by $\delta\pi_p$, where δ denotes arbitrary change or variation) to minimize π_p. However, we will avoid the details of variational calculus and show that we can really use the familiar differential calculus to perform the minimization of π_p. To apply the principle of minimum

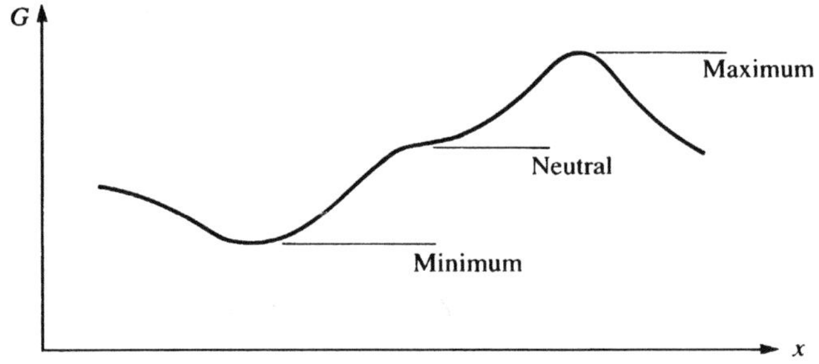

Figure 2–19 Stationary values of a function

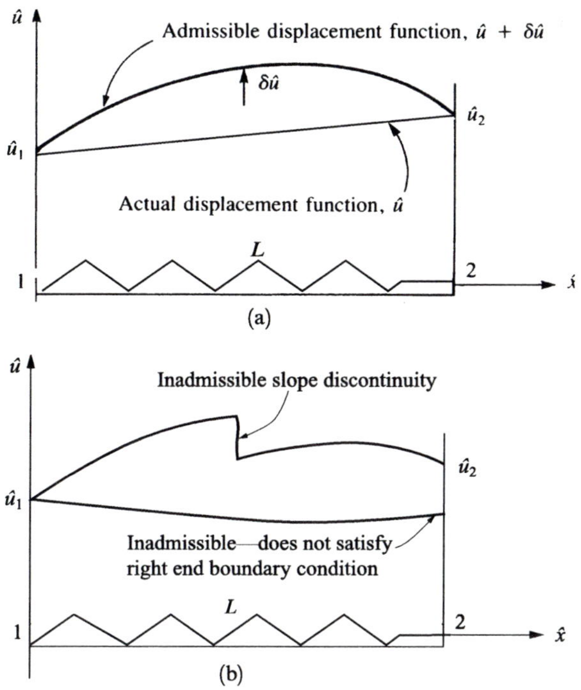

Figure 2–20 (a) Actual and admissible displacement functions and (b) inadmissible displacement functions

potential energy—that is, to minimize π_p—we take the *variation* of π_p, which is a function of nodal displacements d_i defined in general as

$$\delta\pi_p = \frac{\partial \pi_p}{\partial d_1}\delta d_1 + \frac{\partial \pi_p}{\partial d_2}\delta d_2 + \cdots + \frac{\partial \pi_p}{\partial d_n}\delta d_n \tag{2.6.11}$$

The principle states that equilibrium exists when the d_i define a structure state such that $\delta\pi_p = 0$ (change in potential energy $= 0$) for arbitrary admissible variations in displacement δd_i from the equilibrium state. An *admissible variation* is one in which the displacement field still satisfies the boundary conditions and interelement continuity. Figure 2–20(a) shows the hypothetical actual axial displacement and an admissible one for a spring with specified boundary displacements $\hat{u}_1$ and $\hat{u}_2$. Figure 2–20(b) shows inadmissible functions due to slope discontinuity between endpoints 1 and 2 and due to failure to satisfy the right end boundary condition of $\hat{u}(L) = \hat{u}_2$. Here $\delta\hat{u}$ represents the variation in $\hat{u}$. In the general finite element formulation, $\delta\hat{u}$ would be replaced by δd_i. This implies that any of the δd_i might be nonzero. Hence, to satisfy $\delta\pi_p = 0$, all coefficients associated with the δd_i must be zero independently. Thus,

$$\frac{\partial \pi_p}{\partial d_i} = 0 \quad (i = 1, 2, 3, \ldots, n) \qquad \text{or} \qquad \frac{\partial \pi_p}{\partial \{d\}} = 0 \tag{2.6.12}$$

where n equations must be solved for the n values of d_i that define the static equilibrium state of the structure. Equation (2.6.12) shows that for our purposes throughout this text, we can interpret *the variation of* π_p as a compact notation equivalent to differentiation of π_p with respect to the unknown nodal displacements for which π_p is expressed. For linear-elastic materials in equilibrium, the fact that π_p is a minimum is shown, for instance, in Reference [4].

Before discussing the formulation of the spring element equations, we now illustrate the concept of the principle of minimum potential energy by analyzing a single-degree-of-freedom spring subjected to an applied force, as given in Example 2.4. In this example, we will show that the equilibrium position of the spring corresponds to the minimum potential energy.

Example 2.4

For the linear-elastic spring subjected to a force of 1000 lb shown in Figure 2–21, evaluate the potential energy for various displacement values and show that the minimum potential energy also corresponds to the equilibrium position of the spring.

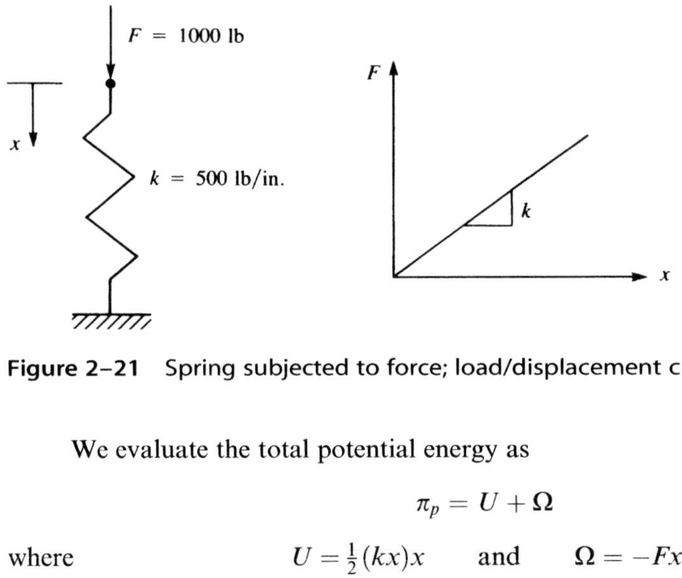

Figure 2–21 Spring subjected to force; load/displacement curve

We evaluate the total potential energy as

$$\pi_p = U + \Omega$$

where $U = \frac{1}{2}(kx)x$ and $\Omega = -Fx$

We now illustrate the minimization of π_p through standard mathematics. Taking the variation of π_p with respect to x, or, equivalently, taking the derivative of π_p with respect to x (as π_p is a function of only one displacement x), as in Eqs. (2.6.11) and (2.6.12), we have

$$\delta\pi_p = \frac{\partial\pi_p}{\partial x}\delta x = 0$$

or, because δx is arbitrary and might not be zero,

$$\frac{\partial\pi_p}{\partial x} = 0$$

Using our previous expression for π_p, we obtain

$$\frac{\partial \pi_p}{\partial x} = 500x - 1000 = 0$$

or

$$x = 2.00 \text{ in.}$$

This value for x is then back-substituted into π_p to yield

$$\pi_p = 250(2)^2 - 1000(2) = -1000 \text{ lb-in.}$$

which corresponds to the minimum potential energy obtained in Table 2–1 by the following searching technique. Here $U = \frac{1}{2}(kx)x$ is the strain energy or the area under the load/displacement curve shown in Figure 2–21, and $\Omega = -Fx$ is the potential energy of load F. For the given values of F and k, we then have

$$\pi_p = \tfrac{1}{2}(500)x^2 - 1000x = 250x^2 - 1000x$$

We now search for the minimum value of π_p for various values of spring deformation x. The results are shown in Table 2–1. A plot of π_p versus x is shown in Figure 2–22, where we observe that π_p has a minimum value at $x = 2.00$ in. This deformed position also corresponds to the equilibrium position because $(\partial \pi_p / \partial x) = 500(2) - 1000 = 0$. ■

We now derive the spring element equations and stiffness matrix using the principle of minimum potential energy. Consider the linear spring subjected to nodal forces shown in Figure 2–23. Using Eq. (2.6.9) reveals that the total potential energy is

$$\pi_p = \tfrac{1}{2}k(\hat{d}_{2x} - \hat{d}_{1x})^2 - \hat{f}_{1x}\hat{d}_{1x} - \hat{f}_{2x}\hat{d}_{2x} \tag{2.6.13}$$

where $\hat{d}_{2x} - \hat{d}_{1x}$ is the deformation of the spring in Eq. (2.6.9). The first term on the right in Eq. (2.6.13) is the strain energy in the spring. Simplifying Eq. (2.6.13), we obtain

$$\pi_p = \tfrac{1}{2}k(\hat{d}_{2x}^2 - 2\hat{d}_{2x}\hat{d}_{1x} + \hat{d}_{1x}^2) - \hat{f}_{1x}\hat{d}_{1x} - \hat{f}_{2x}\hat{d}_{2x} \tag{2.6.14}$$

Table 2–1 Total potential energy for various spring deformations

Deformation x, in.	Total Potential Energy π_p, lb-in.
−4.00	8000
−3.00	5250
−2.00	3000
−1.00	1250
0.00	0
1.00	−750
2.00	−1000
3.00	−750
4.00	0
5.00	1250

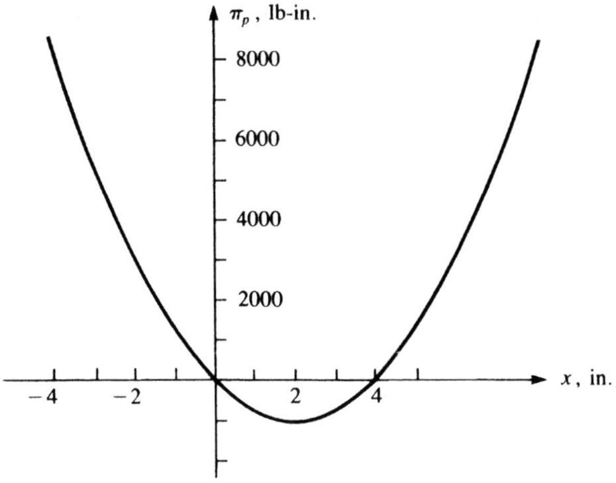

Figure 2–22 Variation of potential energy with spring deformation

Figure 2–23 Linear spring subjected to nodal forces

The minimization of π_p with respect to each nodal displacement requires taking partial derivatives of π_p with respect to each nodal displacement such that

$$\begin{aligned}\frac{\partial \pi_p}{\partial \hat{d}_{1x}} &= \frac{1}{2}k(-2\hat{d}_{2x} + 2\hat{d}_{1x}) - \hat{f}_{1x} = 0 \\ \frac{\partial \pi_p}{\partial \hat{d}_{2x}} &= \frac{1}{2}k(2\hat{d}_{2x} - 2\hat{d}_{1x}) - \hat{f}_{2x} = 0\end{aligned} \tag{2.6.15}$$

Simplifying Eqs. (2.6.15), we have

$$\begin{aligned}k(-\hat{d}_{2x} + \hat{d}_{1x}) &= \hat{f}_{1x} \\ k(\hat{d}_{2x} - \hat{d}_{1x}) &= \hat{f}_{2x}\end{aligned} \tag{2.6.16}$$

In matrix form, we express Eq. (2.6.16) as

$$\begin{bmatrix} k & -k \\ -k & k \end{bmatrix} \begin{Bmatrix} \hat{d}_{1x} \\ \hat{d}_{2x} \end{Bmatrix} = \begin{Bmatrix} \hat{f}_{1x} \\ \hat{f}_{2x} \end{Bmatrix} \tag{2.6.17}$$

Because $\{\hat{f}\} = [\hat{k}]\{\hat{d}\}$, we have the stiffness matrix for the spring element obtained from Eq. (2.6.17):

$$[\hat{k}] = \begin{bmatrix} k & -k \\ -k & k \end{bmatrix} \tag{2.6.18}$$

As expected, Eq. (2.6.18) is identical to the stiffness matrix obtained in Section 2.2, Eq. (2.2.18).

We considered the equilibrium of a single spring element by minimizing the total potential energy with respect to the nodal displacements (see Example 2.4). We also developed the finite element spring element equations by minimizing the total potential energy with respect to the nodal displacements. We now show that the total potential energy of an entire structure (here an assemblage of spring elements) can be minimized with respect to each nodal degree of freedom and that this minimization results in the same finite element equations used for the solution as those obtained by the direct stiffness method.

Example 2.5

Obtain the total potential energy of the spring assemblage (Figure 2–24) for Example 2.1 and find its minimum value. The procedure of assembling element equations can then be seen to be obtained from the minimization of the total potential energy.

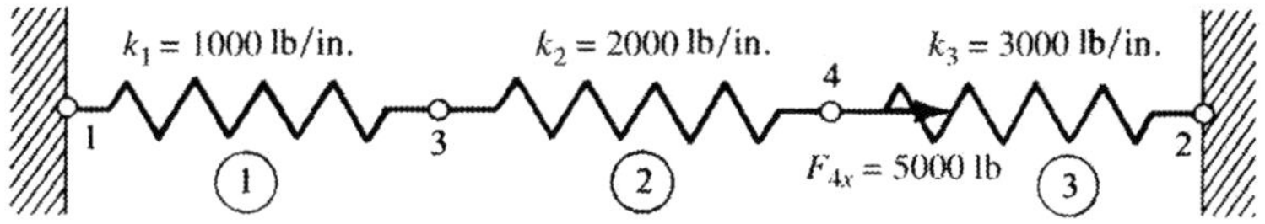

Using Eq. (2.6.10) for each element of the spring assemblage, we find that the total potential energy is given by

$$\begin{aligned}\pi_p = \sum_{e=1}^{3} \pi_p^{(e)} &= \frac{1}{2}k_1(d_{3x} - d_{1x})^2 - f_{1x}^{(1)} d_{1x} - f_{3x}^{(1)} d_{3x} \\ &+ \frac{1}{2}k_2(d_{4x} - d_{3x})^2 - f_{3x}^{(2)} d_{3x} - f_{4x}^{(2)} d_{4x} \\ &+ \frac{1}{2}k_3(d_{2x} - d_{4x})^2 - f_{4x}^{(3)} d_{4x} - f_{2x}^{(3)} d_{2x}\end{aligned} \tag{2.6.19}$$

Upon minimizing π_p with respect to each nodal displacement, we obtain

$$\begin{aligned}\frac{\partial \pi_p}{\partial d_{1x}} &= -k_1 d_{3x} + k_1 d_{1x} - f_{1x}^{(1)} = 0 \\ \frac{\partial \pi_p}{\partial d_{2x}} &= k_3 d_{2x} - k_3 d_{4x} - f_{2x}^{(3)} = 0 \\ \frac{\partial \pi_p}{\partial d_{3x}} &= k_1 d_{3x} - k_1 d_{1x} - k_2 d_{4x} + k_2 d_{3x} - f_{3x}^{(1)} - f_{3x}^{(2)} = 0 \\ \frac{\partial \pi_p}{\partial d_{4x}} &= k_2 d_{4x} - k_2 d_{3x} - k_3 d_{2x} + k_3 d_{4x} - f_{4x}^{(2)} - f_{4x}^{(3)} = 0\end{aligned} \tag{2.6.20}$$

In matrix form, Eqs. (2.6.20) become

$$\begin{bmatrix} k_1 & 0 & -k_1 & 0 \\ 0 & k_3 & 0 & -k_3 \\ -k_1 & 0 & k_1+k_2 & -k_2 \\ 0 & -k_3 & -k_2 & k_2+k_3 \end{bmatrix} \begin{Bmatrix} d_{1x} \\ d_{2x} \\ d_{3x} \\ d_{4x} \end{Bmatrix} = \begin{Bmatrix} f_{1x}^{(1)} \\ f_{2x}^{(3)} \\ f_{3x}^{(1)} + f_{3x}^{(2)} \\ f_{4x}^{(2)} + f_{4x}^{(3)} \end{Bmatrix} \tag{2.6.21}$$

Using nodal force equilibrium similar to Eqs. (2.3.4)–(2.3.6), we have

$$\begin{aligned} f_{1x}^{(1)} &= F_{1x} \\ f_{2x}^{(3)} &= F_{2x} \\ f_{3x}^{(1)} + f_{3x}^{(2)} &= F_{3x} \\ f_{4x}^{(2)} + f_{4x}^{(3)} &= F_{4x} \end{aligned} \tag{2.6.22}$$

Using Eqs. (2.6.22) in (2.6.21) and substituting numerical values for k_1, k_2, and k_3, we obtain

$$\begin{bmatrix} 1000 & 0 & -1000 & 0 \\ 0 & 3000 & 0 & -3000 \\ -1000 & 0 & 3000 & -2000 \\ 0 & -3000 & -2000 & 5000 \end{bmatrix} \begin{Bmatrix} d_{1x} \\ d_{2x} \\ d_{3x} \\ d_{4x} \end{Bmatrix} = \begin{Bmatrix} F_{1x} \\ F_{2x} \\ F_{3x} \\ F_{4x} \end{Bmatrix} \tag{2.6.23}$$

Equation (2.6.23) is identical to Eq. (2.5.18), which was obtained through the direct stiffness method. The assembled Eqs. (2.6.23) are then seen to be obtained from the minimization of the total potential energy. When we apply the boundary conditions and substitute $F_{3x} = 0$ and $F_{4x} = 5000$ lb into Eq. (2.6.23), the solution is identical to that of Example 2.1. ■

References

[1] Turner, M. J., Clough, R. W., Martin, H. C., and Topp, L. J., "Stiffness and Deflection Analysis of Complex Structures," *Journal of the Aeronautical Sciences*, Vol. 23, No. 9, pp. 805–824, Sept. 1956.

[2] Martin, H. C., *Introduction to Matrix Methods of Structural Analysis*, McGraw-Hill, New York, 1966.

[3] Hsieh, Y. Y., *Elementary Theory of Structures*, 2nd ed., Prentice-Hall, Englewood Cliffs, NJ, 1982.

[4] Oden, J. T., and Ripperger, E. A., *Mechanics of Elastic Structures*, 2nd ed., McGraw-Hill, New York, 1981.

[5] Finlayson, B. A., *The Method of Weighted Residuals and Variational Principles*, Academic Press, New York, 1972.

[6] Zienkiewicz, O. C., *The Finite Element Method*, 3rd ed., McGraw-Hill, London, 1977.

[7] Cook, R. D., Malkus, D. S., Plesha, M. E., and Witt, R. J. *Concepts and Applications of Finite Element Analysis*, 4th ed., Wiley, New York, 2002.
[8] Forray, M. J., *Variational Calculus in Science and Engineering*, McGraw-Hill, New York, 1968.

Problems

2.1 **a.** Obtain the global stiffness matrix $\underline{K}$ of the assemblage shown in Figure P2–1 by superimposing the stiffness matrices of the individual springs. Here k_1, k_2, and k_3 are the stiffnesses of the springs as shown.
b. If nodes 1 and 2 are fixed and a force P acts on node 4 in the positive x direction, find an expression for the displacements of nodes 3 and 4.
c. Determine the reaction forces at nodes 1 and 2.
(*Hint:* Do this problem by writing the nodal equilibrium equations and then making use of the force/displacement relationships for each element as done in the first part of Section 2.4. Then solve the problem by the direct stiffness method.)

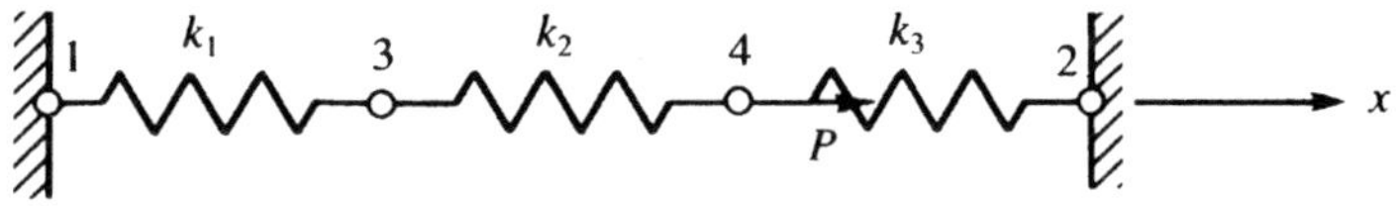

Figure P2–1

2.2 For the spring assemblage shown in Figure P2–2, determine the displacement at node 2 and the forces in each spring element. Also determine the force F_3. Given: Node 3 displaces an amount $\delta = 1$ in. in the positive x direction because of the force F_3 and $k_1 = k_2 = 500$ lb/in.

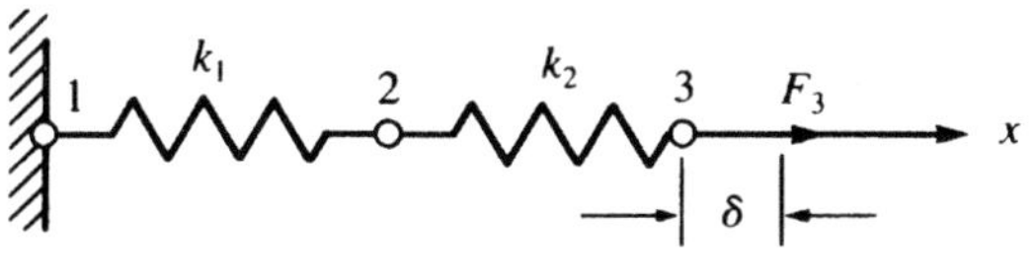

Figure P2–2

2.3 **a.** For the spring assemblage shown in Figure P2–3, obtain the global stiffness matrix by direct superposition.
b. If nodes 1 and 5 are fixed and a force P is applied at node 3, determine the nodal displacements.
c. Determine the reactions at the fixed nodes 1 and 5.

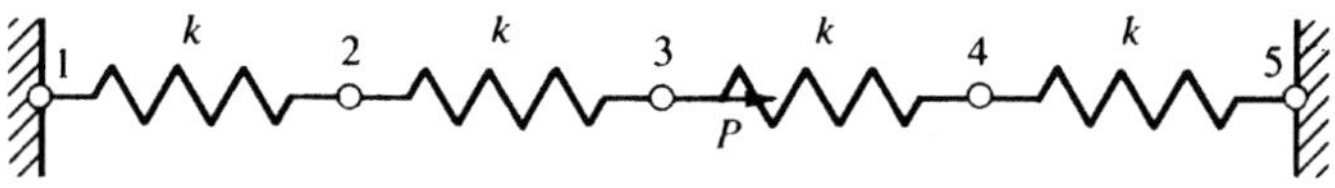

Figure P2–3

2.4 Solve Problem 2.3 with $P = 0$ (no force applied at node 3) and with node 5 given a fixed, known displacement of δ as shown in Figure P2–4.

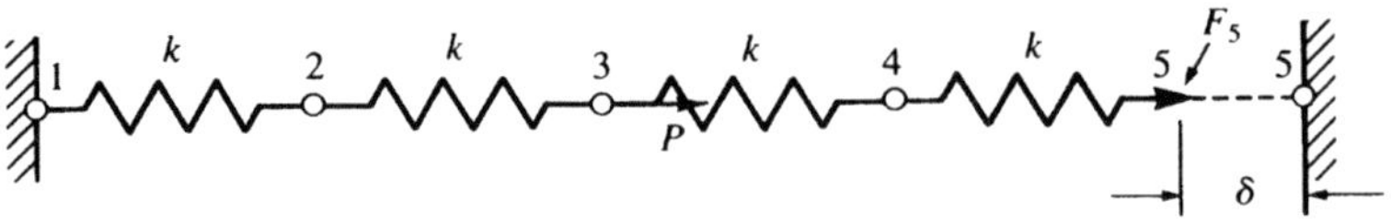

Figure P2–4

2.5 For the spring assemblage shown in Figure P2–5, obtain the global stiffness matrix by the direct stiffness method. Let $k^{(1)} = 1$ kip/in., $k^{(2)} = 2$ kip/in., $k^{(3)} = 3$ kip/in., $k^{(4)} = 4$ kip/in., and $k^{(5)} = 5$ kip/in.

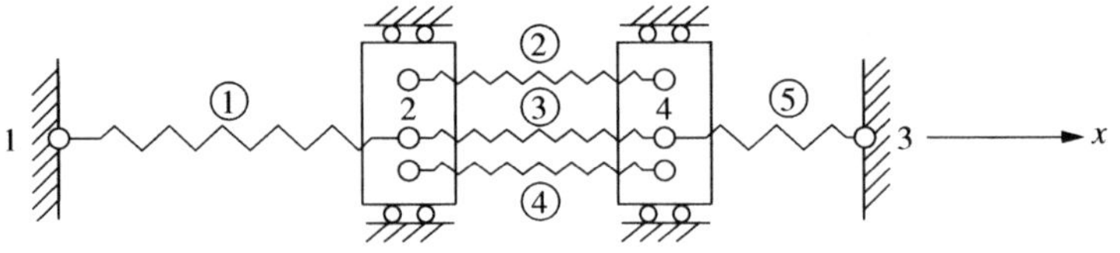

Figure P 2–5

2.6 For the spring assemblage in Figure P2–5, apply a concentrated force of 2 kips at node 2 in the positive x direction and determine the displacements at nodes 2 and 4.

2.7 Instead of assuming a tension element as in Figure P2–3, now assume a compression element. That is, apply compressive forces to the spring element and derive the stiffness matrix.

2.8–2.16 For the spring assemblages shown in Figures P2–8—P2–16, determine the nodal displacements, the forces in each element, and the reactions. Use the direct stiffness method for all problems.

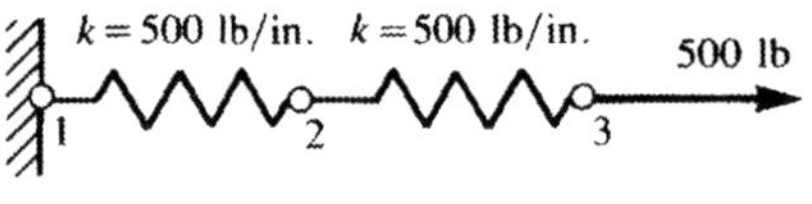

Figure P 2–8

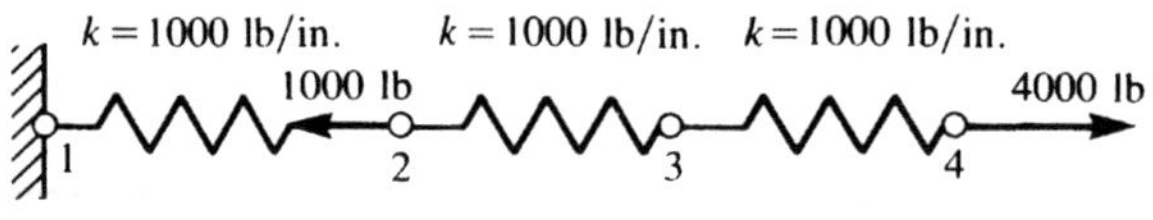

Figure P2–9

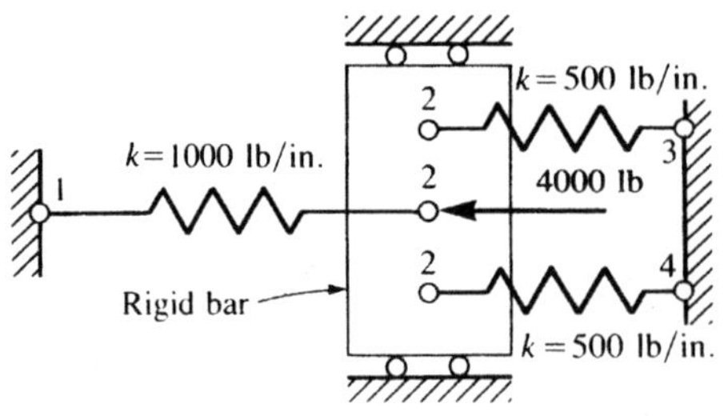

Figure P2–10

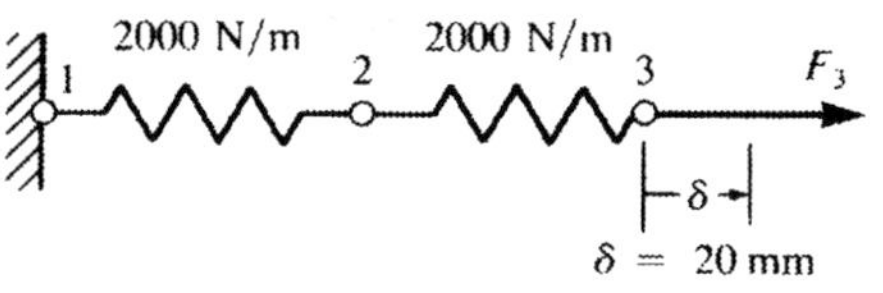

Figure P2–11

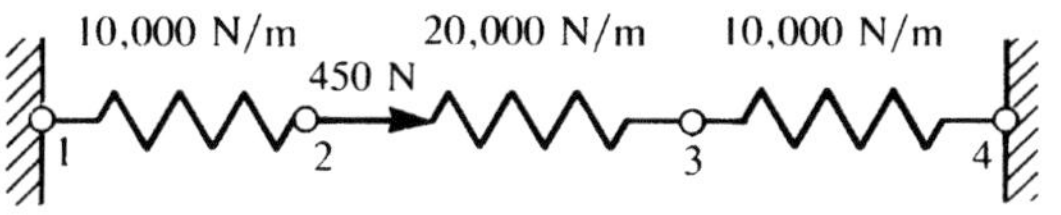

Figure P2–12

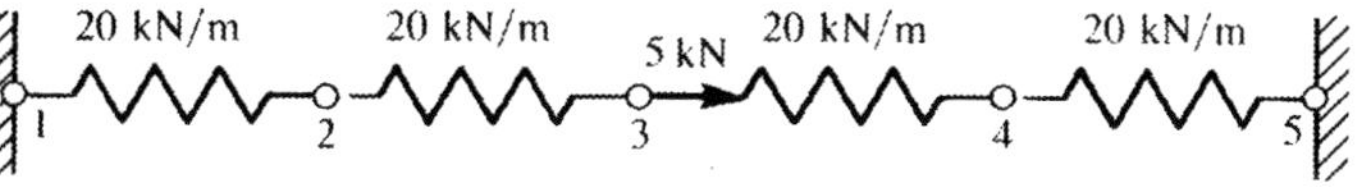

Figure P2–13

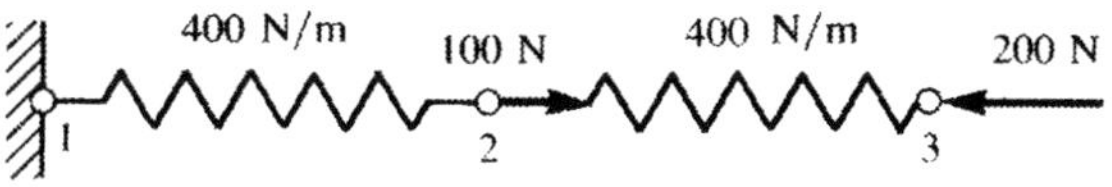

Figure P2–14

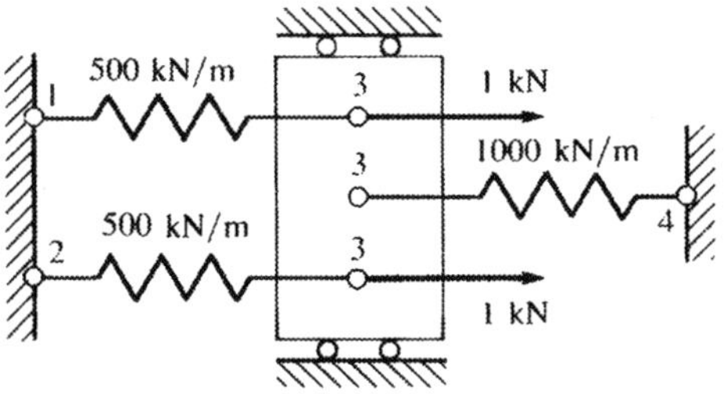

Figure P2–15

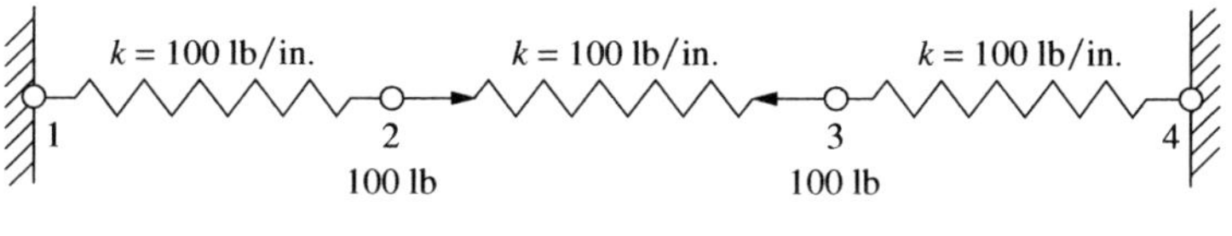

Figure P2–16

2.17 Use the principle of minimum potential energy developed in Section 2.6 to solve the spring problems shown in Figure P2–17. That is, plot the total potential energy for variations in the displacement of the free end of the spring to determine the minimum potential energy. Observe that the displacement that yields the minimum potential energy also yields the stable equilibrium position.

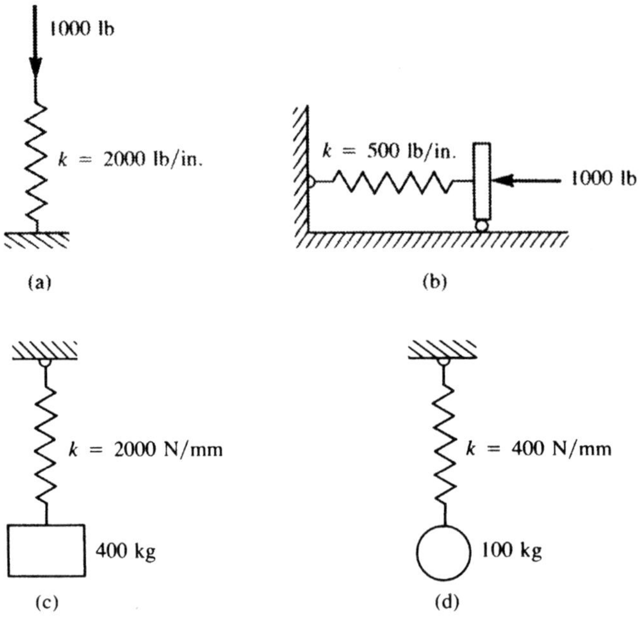

Figure P2–17

2.18 Reverse the direction of the load in Example 2.4 and recalculate the total potential energy. Then use this value to obtain the equilibrium value of displacement.

2.19 The nonlinear spring in Figure P2–19 has the force/deformation relationship $f = k\delta^2$. Express the total potential energy of the spring, and use this potential energy to obtain the equilibrium value of displacement.

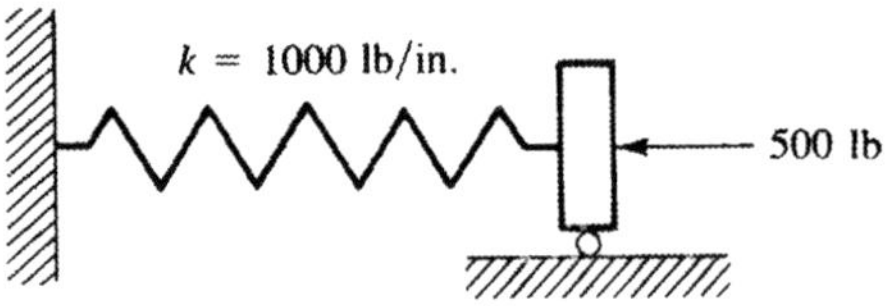

Figure P2–19

2.20–2.21 Solve Problems 2.10 and 2.15 by the potential energy approach (see Example 2.5).

CHAPTER 3

Development of Truss Equations

Introduction

Having set forth the foundation on which the direct stiffness method is based, we will now derive the stiffness matrix for a linear-elastic bar (or truss) element using the general steps outlined in Chapter 1. We will include the introduction of both a local coordinate system, chosen with the element in mind, and a global or reference coordinate system, chosen to be convenient (for numerical purposes) with respect to the overall structure. We will also discuss the transformation of a vector from the local coordinate system to the global coordinate system, using the concept of transformation matrices to express the stiffness matrix of an arbitrarily oriented bar element in terms of the global system. We will solve three example plane truss problems (see Figure 3–1 for a typical railroad trestle plane truss) to illustrate the procedure of establishing the total stiffness matrix and equations for solution of a structure.

Next we extend the stiffness method to include space trusses. We will develop the transformation matrix in three-dimensional space and analyze two space trusses. Then we describe the concept of symmetry and its use to reduce the size of a problem and facilitate its solution. We will use an example truss problem to illustrate the concept and then describe how to handle inclined, or skewed, supports.

We will then use the principle of minimum potential energy and apply it to rederive the bar element equations. We then compare a finite element solution to an exact solution for a bar subjected to a linear varying distributed load. We will introduce Galerkin's residual method and then apply it to derive the bar element equations. Finally, we will introduce other common residual methods—collocation, subdomain, and least squares to merely expose you to these other methods. We illustrate these methods by solving a problem of a bar subjected to a linear varying load.

Figure 3–1 A typical railroad trestle plane truss. (By Daryl L. Logan)

3.1 Derivation of the Stiffness Matrix for a Bar Element in Local Coordinates

We will now consider the derivation of the stiffness matrix for the linear-elastic, constant cross-sectional area (prismatic) bar element shown in Figure 3–2. The derivation here will be directly applicable to the solution of pin-connected trusses. The bar is subjected to tensile forces T directed along the local axis of the bar and applied at nodes 1 and 2.

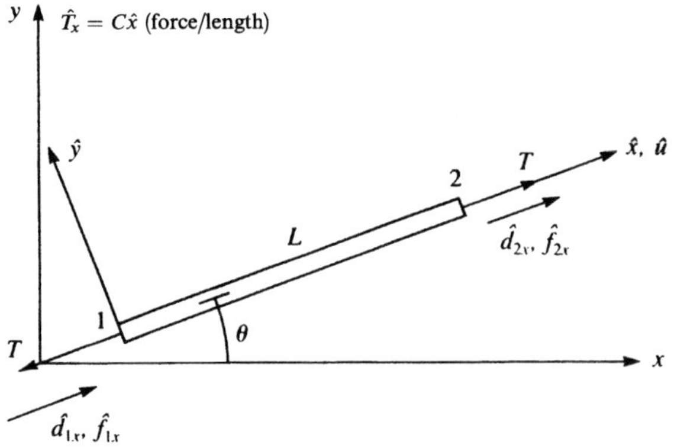

Figure 3–2 Bar subjected to tensile forces T; positive nodal displacements and forces are all in the local $\hat{x}$ direction

Here we have introduced two coordinate systems: a local one $(\hat{x}, \hat{y})$ with $\hat{x}$ directed along the length of the bar and a global one (x, y) assumed here to be best suited with respect to the total structure. Proper selection of global coordinate systems is best demonstrated through solution of two- and three-dimensional truss problems as illustrated in Sections 3.6 and 3.7. Both systems will be used extensively throughout this text.

The bar element is assumed to have constant cross-sectional area A, modulus of elasticity E, and initial length L. The nodal degrees of freedom are local axial displacements (longitudinal displacements directed along the length of the bar) represented by $\hat{d}_{1x}$ and $\hat{d}_{2x}$ at the ends of the element as shown in Figure 3–2.

From Hooke's law [Eq. (a)] and the strain/displacement relationship [Eq. (b) or Eq. (1.4.1)], we write

$$\sigma_x = E\varepsilon_x \tag{a}$$

$$\varepsilon_x = \frac{d\hat{u}}{d\hat{x}} \tag{b}$$

From force equilibrium, we have

$$A\sigma_x = T = \text{constant} \tag{c}$$

for a bar with loads applied only at the ends. (We will consider distributed loading in Section 3.10.) Using Eq. (b) in Eq. (a) and then Eq. (a) in Eq. (c) and differentiating with respect to $\hat{x}$, we obtain the differential equation governing the linear-elastic bar behavior as

$$\frac{d}{d\hat{x}}\left(AE\frac{d\hat{u}}{d\hat{x}}\right) = 0 \tag{d}$$

where $\hat{u}$ is the axial displacement function along the element in the $\hat{x}$ direction and A and E are written as though they were functions of $\hat{x}$ in the general form of the differential equation, even though A and E will be assumed constant over the whole length of the bar in our derivations to follow.

The following assumptions are used in deriving the bar element stiffness matrix:

1. The bar cannot sustain shear force or bending moment, that is, $\hat{f}_{1y} = 0, \hat{f}_{2y} = 0, \hat{m}_1 = 0$ and $\hat{m}_2 = 0$.
2. Any effect of transverse displacement is ignored.
3. Hooke's law applies; that is, axial stress σ_x is related to axial strain ε_x by $\sigma_x = E\varepsilon_x$.
4. No intermediate applied loads.

The steps previously outlined in Chapter 1 are now used to derive the stiffness matrix for the bar element and then to illustrate a complete solution for a bar assemblage.

Step 1 Select the Element Type

Represent the bar by labeling nodes at each end and in general by labeling the element number (Figure 3–2).

Step 2 Select a Displacement Function

Assume a linear displacement variation along the $\hat{x}$ axis of the bar because a linear function with specified endpoints has a unique path. These specified endpoints are the nodal values $\hat{d}_{1x}$ and $\hat{d}_{2x}$. (Further discussion regarding the choice of displacement functions is provided in Section 3.2 and References [1–3].) Then

$$\hat{u} = a_1 + a_2\hat{x} \tag{3.1.1}$$

with the total number of coefficients a_i always equal to the total number of degrees of freedom associated with the element. Here the total number of degrees of freedom is two—axial displacements at each of the two nodes of the element. Using the same procedure as in Section 2.2 for the spring element, we express Eq. (3.1.1) as

$$\hat{u} = \left(\frac{\hat{d}_{2x} - \hat{d}_{1x}}{L}\right)\hat{x} + \hat{d}_{1x} \tag{3.1.2}$$

The reason we convert the displacement function from the form of Eq. (3.1.1) to Eq. (3.1.2) is that it allows us to express the strain in terms of the nodal displacements using the strain/displacement relationship given by Eq. (3.1.5) and to then relate the nodal forces to the nodal displacements in step 4.

In matrix form, Eq. (3.1.2) becomes

$$\hat{u} = [N_1 \quad N_2]\begin{Bmatrix} \hat{d}_{1x} \\ \hat{d}_{2x} \end{Bmatrix} \tag{3.1.3}$$

with shape functions given by

$$N_1 = 1 - \frac{\hat{x}}{L} \qquad N_2 = \frac{\hat{x}}{L} \tag{3.1.4}$$

These shape functions are identical to those obtained for the spring element in Section 2.2. The behavior of and some properties of these shape functions were described in Section 2.2. The linear displacement function $\hat{u}$ (Eq. (3.1.2)), plotted over the length of the bar element, is shown in Figure 3–3. The bar is shown with the same orientation as in Figure 3–2.

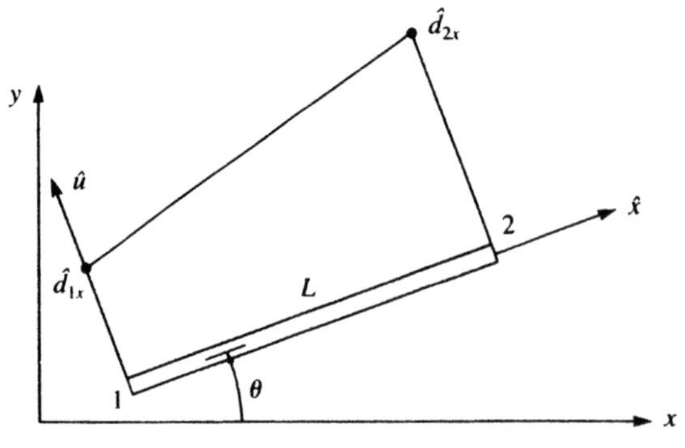

Figure 3–3 Displacement $\hat{u}$ plotted over the length of the element

Step 3 Define the Strain/Displacement and Stress/Strain Relationships

The strain/displacement relationship is

$$\varepsilon_x = \frac{d\hat{u}}{d\hat{x}} = \frac{\hat{d}_{2x} - \hat{d}_{1x}}{L} \tag{3.1.5}$$

where Eqs. (3.1.3) and (3.1.4) have been used to obtain Eq. (3.1.5), and the stress/strain relationship is

$$\sigma_x = E\varepsilon_x \tag{3.1.6}$$

Step 4 Derive the Element Stiffness Matrix and Equations

The element stiffness matrix is derived as follows. From elementary mechanics, we have

$$T = A\sigma_x \tag{3.1.7}$$

Now, using Eqs. (3.1.5) and (3.1.6) in Eq. (3.1.7), we obtain

$$T = AE\left(\frac{\hat{d}_{2x} - \hat{d}_{1x}}{L}\right) \tag{3.1.8}$$

Also, by the nodal force sign convention of Figure 3–2,

$$\hat{f}_{1x} = -T \tag{3.1.9}$$

When we substitute Eq. (3.1.8), Eq. (3.1.9) becomes

$$\hat{f}_{1x} = \frac{AE}{L}(\hat{d}_{1x} - \hat{d}_{2x}) \tag{3.1.10}$$

Similarly, $$\hat{f}_{2x} = T \tag{3.1.11}$$

or, by Eq. (3.1.8), Eq. (3.1.11) becomes

$$\hat{f}_{2x} = \frac{AE}{L}(\hat{d}_{2x} - \hat{d}_{1x}) \tag{3.1.12}$$

Expressing Eqs. (3.1.10) and (3.1.12) together in matrix form, we have

$$\begin{Bmatrix} \hat{f}_{1x} \\ \hat{f}_{2x} \end{Bmatrix} = \frac{AE}{L}\begin{bmatrix} 1 & -1 \\ -1 & 1 \end{bmatrix}\begin{Bmatrix} \hat{d}_{1x} \\ \hat{d}_{2x} \end{Bmatrix} \tag{3.1.13}$$

Now, because $\underline{\hat{f}} = \underline{\hat{k}}\underline{\hat{d}}$, we have, from Eq. (3.1.13),

$$\underline{\hat{k}} = \frac{AE}{L}\begin{bmatrix} 1 & -1 \\ -1 & 1 \end{bmatrix} \tag{3.1.14}$$

Equation (3.1.14) represents the stiffness matrix for a bar element in local coordinates. In Eq. (3.1.14), AE/L for a bar element is analogous to the spring constant k for a spring element.

Step 5 Assemble Element Equations to Obtain Global or Total Equations

Assemble the global stiffness and force matrices and global equations using the direct stiffness method described in Chapter 2 (see Section 3.6 for an example truss). This step applies for structures composed of more than one element such that (again)

$$\underline{K} = [K] = \sum_{e=1}^{N} \underline{k}^{(e)} \quad \text{and} \quad \underline{F} = \{F\} = \sum_{e=1}^{N} \underline{f}^{(e)} \tag{3.1.15}$$

where now all local element stiffness matrices $\hat{\underline{k}}$ must be transformed to global element stiffness matrices $\underline{k}$ (unless the local axes coincide with the global axes) before the direct stiffness method is applied as indicated by Eq. (3.1.15). (This concept of coordinate and stiffness matrix transformations is described in Sections 3.3 and 3.4.)

Step 6 Solve for the Nodal Displacements

Determine the displacements by imposing boundary conditions and simultaneously solving a system of equations, $\underline{F} = \underline{K}\underline{d}$.

Step 7 Solve for the Element Forces

Finally, determine the strains and stresses in each element by back-substitution of the displacements into equations similar to Eqs. (3.1.5) and (3.1.6).

We will now illustrate a solution for a one-dimensional bar problem.

Example 3.1

For the three-bar assemblage shown in Figure 3–4, determine (a) the global stiffness matrix, (b) the displacements of nodes 2 and 3, and (c) the reactions at nodes 1 and 4. A force of 3000 lb is applied in the x direction at node 2. The length of each element is 30 in. Let $E = 30 \times 10^6$ psi and $A = 1$ in^2 for elements 1 and 2, and let $E = 15 \times 10^6$ psi and $A = 2$ in^2 for element 3. Nodes 1 and 4 are fixed.

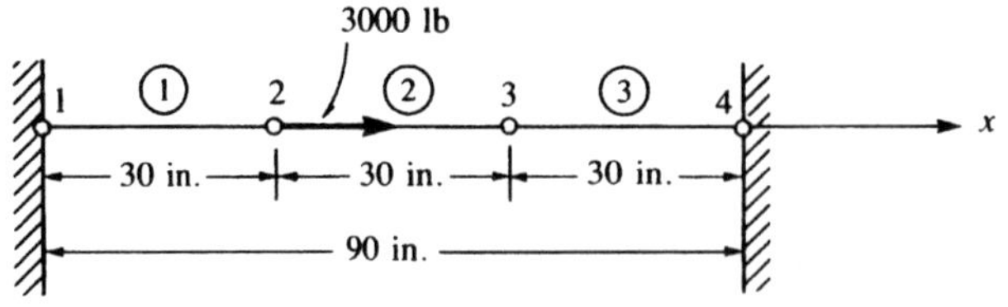

Figure 3–4 Three-bar assemblage

(a) Using Eq. (3.1.14), we find that the element stiffness matrices are

$$\underline{k}^{(1)} = \underline{k}^{(2)} = \frac{(1)(30\times 10^6)}{30}\begin{bmatrix} 1 & -1 \\ -1 & 1 \end{bmatrix} = 10^6 \overset{\begin{matrix} 1 & 2^{(1)} \\ 2 & 3^{(2)} \end{matrix}}{\begin{bmatrix} 1 & -1 \\ -1 & 1 \end{bmatrix}} \frac{\text{lb}}{\text{in.}} \tag{3.1.16}$$

$$\underline{k}^{(3)} = \frac{(2)(15\times 10^6)}{30}\begin{bmatrix} 1 & -1 \\ -1 & 1 \end{bmatrix} = 10^6 \overset{\begin{matrix} 3 & 4 \end{matrix}}{\begin{bmatrix} 1 & -1 \\ -1 & 1 \end{bmatrix}} \frac{\text{lb}}{\text{in.}}$$

where, again, the numbers above the matrices in Eqs. (3.1.16) indicate the displacements associated with each matrix. Assembling the element stiffness matrices by the direct stiffness method, we obtain the global stiffness matrix as

$$\underline{K} = 10^6 \overset{\begin{matrix} d_{1x} & d_{2x} & d_{3x} & d_{4x} \end{matrix}}{\begin{bmatrix} 1 & -1 & 0 & 0 \\ -1 & 1+1 & -1 & 0 \\ 0 & -1 & 1+1 & -1 \\ 0 & 0 & -1 & 1 \end{bmatrix}} \frac{\text{lb}}{\text{in.}} \tag{3.1.17}$$

(b) Equation (3.1.17) relates global nodal forces to global nodal displacements as follows:

$$\begin{Bmatrix} F_{1x} \\ F_{2x} \\ F_{3x} \\ F_{4x} \end{Bmatrix} = 10^6 \begin{bmatrix} 1 & -1 & 0 & 0 \\ -1 & 2 & -1 & 0 \\ 0 & -1 & 2 & -1 \\ 0 & 0 & -1 & 1 \end{bmatrix} \begin{Bmatrix} d_{1x} \\ d_{2x} \\ d_{3x} \\ d_{4x} \end{Bmatrix} \tag{3.1.18}$$

Invoking the boundary conditions, we have

$$d_{1x} = 0 \qquad d_{4x} = 0 \tag{3.1.19}$$

Using the boundary conditions, substituting known applied global forces into Eq. (3.1.18), and partitioning equations 1 and 4 of Eq. (3.1.18), we solve equations 2 and 3 of Eq. (3.1.18) to obtain

$$\begin{Bmatrix} 3000 \\ 0 \end{Bmatrix} = 10^6 \begin{bmatrix} 2 & -1 \\ -1 & 2 \end{bmatrix} \begin{Bmatrix} d_{2x} \\ d_{3x} \end{Bmatrix} \tag{3.1.20}$$

Solving Eq. (3.1.20) simultaneously for the displacements yields

$$d_{2x} = 0.002 \text{ in.} \qquad d_{3x} = 0.001 \text{ in.} \tag{3.1.21}$$

(c) Back-substituting Eqs. (3.1.19) and (3.1.21) into Eq. (3.1.18), we obtain the global nodal forces, which include the reactions at nodes 1 and 4, as follows:

$$\begin{aligned}
F_{1x} &= 10^6(d_{1x} - d_{2x}) = 10^6(0 - 0.002) = -2000 \text{ lb} \\
F_{2x} &= 10^6(-d_{1x} + 2d_{2x} - d_{3x}) = 10^6[0 + 2(0.002) - 0.001] = 3000 \text{ lb} \\
F_{3x} &= 10^6(-d_{2x} + 2d_{3x} - d_{4x}) = 10^6[-0.002 + 2(0.001) - 0] = 0 \\
F_{4x} &= 10^6(-d_{3x} + d_{4x}) = 10^6(-0.001 + 0) = -1000 \text{ lb}
\end{aligned} \tag{3.1.22}$$

The results of Eqs. (3.1.22) show that the sum of the reactions F_{1x} and F_{4x} is equal in magnitude but opposite in direction to the applied nodal force of 3000 lb at node 2. Equilibrium of the bar assemblage is thus verified. Furthermore, Eqs. (3.1.22) show that $F_{2x} = 3000$ lb and $F_{3x} = 0$ are merely the applied nodal forces at nodes 2 and 3, respectively, which further enhances the validity of our solution. ■

3.2 Selecting Approximation Functions for Displacements

Consider the following guidelines, as they relate to the one-dimensional bar element, when selecting a displacement function. (Further discussion regarding selection of displacement functions and other kinds of approximation functions (such as temperature functions) will be provided in Chapter 4 for the beam element, in Chapter 6 for the constant-strain triangular element, in Chapter 8 for the linear-strain triangular element, in Chapter 9 for the axisymmetric element, in Chapter 10 for the three-noded bar element and the rectangular plane element, in Chapter 11 for the three-dimensional stress element, in Chapter 12 for the plate bending element, and in Chapter 13 for the heat transfer problem. More information is also provided in References [1–3].

1. Common approximation functions are usually polynomials such as the simplest one that gives the linear variation of displacement given by Eq. (3.1.1) or equivalently by Eq. (3.1.3), where the function is expressed in terms of the shape functions.
2. The approximation function should be continuous within the bar element. The simple linear function for $\hat{u}$ of Eq. (3.1.1) certainly is continuous within the element. Therefore, the linear function yields continuous values of $\hat{u}$ within the element and prevents openings, overlaps, and jumps because of the continuous and smooth variation in $\hat{u}$ (Figure 3–5).
3. The approximating function should provide interelement continuity for all degrees of freedom at each node for discrete line elements and along common boundary lines and surfaces for two- and three-dimensional elements. For the bar element, we must ensure that nodes

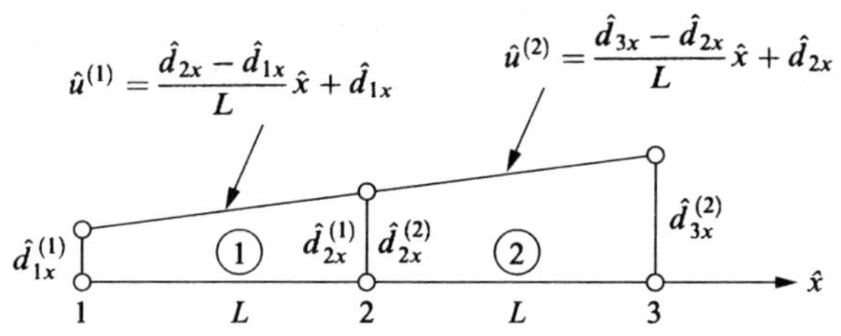

Figure 3–5 Interelement continuity of a two-bar structure

common to two or more elements remain common to these elements upon deformation and thus prevent overlaps or voids between elements. For example, consider the two-bar structure shown in Figure 3–5. For the two-bar structure, the linear function for $\hat{u}$ [Eq. (3.1.2)] within each element will ensure that elements 1 and 2 remain connected; the displacement at node 2 for element 1 will equal the displacement at the same node 2 for element 2; that is, $\hat{d}_{2x}^{(1)} = \hat{d}_{2x}^{(2)}$. This rule was also illustrated by Eq. (2.3.3). The linear function is then called a *conforming*, or *compatible, function* for the bar element because it ensures the satisfaction both of continuity between adjacent elements and of continuity within the element.

In general, the symbol C^m is used to describe the continuity of a piecewise field (such as axial displacement), where the superscript m indicates the degree of derivative that is interelement continuous. A field is then C^0 continuous if the function itself is interelement continuous. For instance, for the field variable being the axial displacement illustrated in Figure 3–5, the displacement is continuous across the common node 2. Hence the displacement field is said to be C^0 continuous. Bar elements, plane elements (see Chapter 7), and solid elements (Chapter 11) are C^0 elements in that they enforce displacement continuity across the common boundaries.

If the function has both its field variable and its first derivative continuous across the common boundary, then the field variable is said to be C^1 continuous. We will later see that the beam and plate elements are C^1 continuous. That is, they enforce both displacement and slope continuity across common boundaries.

4. The approximation function should allow for rigid-body displacement and for a state of constant strain within the element. The one-dimensional displacement function [Eq. (3.1.1)] satisfies these criteria because the a_1 term allows for rigid-body motion (constant motion of the body without straining) and the $a_2\hat{x}$ term allows for constant strain because $\varepsilon_x = d\hat{u}/d\hat{x} = a_2$ is a constant. (This state of constant strain in the element can, in fact, occur if elements are chosen small enough.) The simple polynomial Eq. (3.1.1) satisfying this fourth guideline is then said to be *complete* for the bar element.

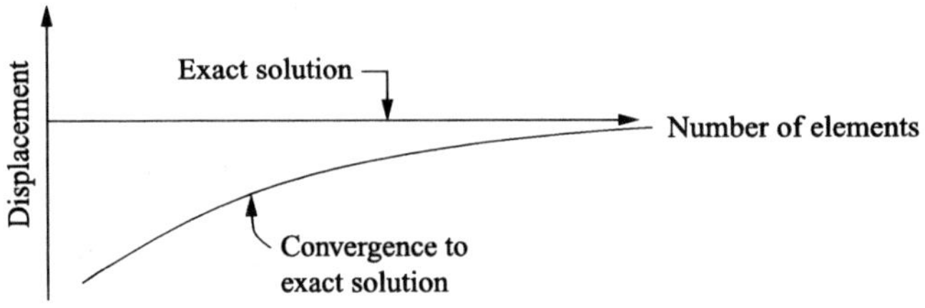

Figure 3–6 Convergence to the exact solution for displacement as the number of elements of a finite element solution is increased

This idea of completeness also means in general that the lower-order term cannot be omitted in favor of the higher-order term. For the simple linear function, this means a_1 cannot be omitted while keeping $a_2\hat{x}$. Completeness of a function is a necessary condition for convergence to the exact answer, for instance, for displacements and stresses (Figure 3–6) (see Reference [3]). Figure 3–6 illustrates monotonic convergence toward an exact solution for displacement as the number of elements in a finite element solution is increased. Monotonic convergence is then the process in which successive approximation solutions (finite element solutions) approach the exact solution consistently without changing sign or direction.

The idea that the interpolation (approximation) function must allow for a rigid-body displacement means that the function must be capable of yielding a constant value (say, a_1), because such a value can, in fact, occur. Therefore, we must consider the case

$$\hat{u} = a_1 \tag{3.2.1}$$

For $\hat{u} = a_1$ requires nodal displacements $\hat{d}_{1x} = \hat{d}_{2x}$ to obtain a rigid-body displacement. Therefore

$$a_1 = \hat{d}_{1x} = \hat{d}_{2x} \tag{3.2.2}$$

Using Eq. (3.2.2) in Eq. (3.1.3), we have

$$\hat{u} = N_1\hat{d}_{1x} + N_2\hat{d}_{2x} = (N_1 + N_2)a_1 \tag{3.2.3}$$

From Eqs. (3.2.1) and (3.2.3), we then have

$$\hat{u} = a_1 = (N_1 + N_2)a_1 \tag{3.2.4}$$

Therefore, by Eq. (3.2.4), we obtain

$$N_1 + N_2 = 1 \tag{3.2.5}$$

Thus Eq. (3.2.5) shows that the displacement interpolation functions must add to unity at every point within the element so that $\hat{u}$ will yield a constant value when a rigid-body displacement occurs.

3.3 Transformation of Vectors in Two Dimensions

In many problems it is convenient to introduce both local and global (or reference) coordinates. Local coordinates are always chosen to represent the individual element conveniently. Global coordinates are chosen to be convenient for the whole structure.

Given the nodal displacement of an element, represented by the vector **d** in Figure 3–7, we want to relate the components of this vector in one coordinate system to components in another. For general purposes, we will assume in this section that **d** is not coincident with either the local or the global axis. In this case, we want to relate global displacement components to local ones. In so doing, we will develop a transformation matrix that will subsequently be used to develop the global stiffness matrix for a bar element. We define the angle θ to be positive when measured counterclockwise from x to $\hat{x}$. We can express vector displacement **d** in both global and local coordinates by

$$\mathbf{d} = d_x\mathbf{i} + d_y\mathbf{j} = \hat{d}_x\hat{\mathbf{i}} + \hat{d}_y\hat{\mathbf{j}} \tag{3.3.1}$$

where **i** and **j** are unit vectors in the x and y global directions and $\hat{\mathbf{i}}$ and $\hat{\mathbf{j}}$ are unit vectors in the $\hat{x}$ and $\hat{y}$ local directions. We will now relate **i** and **j** to $\hat{\mathbf{i}}$ and $\hat{\mathbf{j}}$ through use of Figure 3–8.

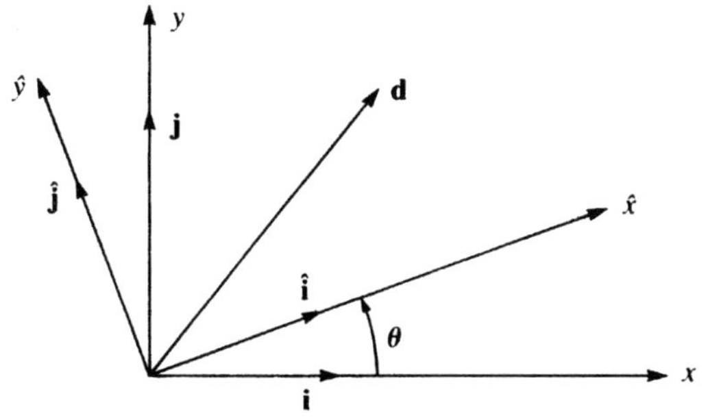

Figure 3–7 General displacement vector **d**

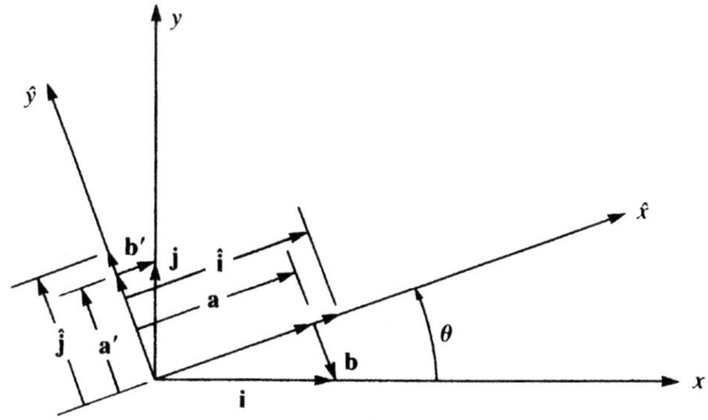

Figure 3–8 Relationship between local and global unit vectors

Using Figure 3–8 and vector addition, we obtain

$$\mathbf{a} + \mathbf{b} = \mathbf{i} \tag{3.3.2}$$

Also, from the law of cosines,

$$|\mathbf{a}| = |\mathbf{i}| \cos\theta \tag{3.3.3}$$

and because **i** is, by definition, a unit vector, its magnitude is given by

$$|\mathbf{i}| = 1 \tag{3.3.4}$$

Therefore, we obtain

$$|\mathbf{a}| = 1 \cos\theta \tag{3.3.5}$$

Similarly,

$$|\mathbf{b}| = 1 \sin\theta \tag{3.3.6}$$

Now **a** is in the $\hat{\mathbf{i}}$ direction and **b** is in the $-\hat{\mathbf{j}}$ direction. Therefore,

$$\mathbf{a} = |\mathbf{a}|\hat{\mathbf{i}} = (\cos\theta)\hat{\mathbf{i}} \tag{3.3.7}$$

and

$$\mathbf{b} = |\mathbf{b}|(-\hat{\mathbf{j}}) = (\sin\theta)(-\hat{\mathbf{j}}) \tag{3.3.8}$$

Using Eqs. (3.3.7) and (3.3.8) in Eq. (3.3.2) yields

$$\mathbf{i} = \cos\theta\hat{\mathbf{i}} - \sin\theta\hat{\mathbf{j}} \tag{3.3.9}$$

Similarly, from Figure 3–8, we obtain

$$\mathbf{a}' + \mathbf{b}' = \mathbf{j} \tag{3.3.10}$$

$$\mathbf{a}' = \cos\theta\hat{\mathbf{j}} \tag{3.3.11}$$

$$\mathbf{b}' = \sin\theta\hat{\mathbf{i}} \tag{3.3.12}$$

Using Eqs. (3.3.11) and (3.3.12) in Eq. (3.3.10), we have

$$\mathbf{j} = \sin\theta\hat{\mathbf{i}} + \cos\theta\hat{\mathbf{j}} \tag{3.3.13}$$

Now, using Eqs. (3.3.9) and (3.3.13) in Eq. (3.3.1), we have

$$d_x(\cos\theta\hat{\mathbf{i}} - \sin\theta\hat{\mathbf{j}}) + d_y(\sin\theta\hat{\mathbf{i}} + \cos\theta\hat{\mathbf{j}}) = \hat{d}_x\hat{\mathbf{i}} + \hat{d}_y\hat{\mathbf{j}} \tag{3.3.14}$$

Combining like coefficients of $\hat{\mathbf{i}}$ and $\hat{\mathbf{j}}$ in Eq. (3.3.14), we obtain

$$\begin{aligned} d_x \cos\theta + d_y \sin\theta &= \hat{d}_x \\ \text{and} \quad -d_x \sin\theta + d_y \cos\theta &= \hat{d}_y \end{aligned} \tag{3.3.15}$$

In matrix form, Eqs. (3.3.15) are written as

$$\begin{Bmatrix} \hat{d}_x \\ \hat{d}_y \end{Bmatrix} = \begin{bmatrix} C & S \\ -S & C \end{bmatrix} \begin{Bmatrix} d_x \\ d_y \end{Bmatrix} \tag{3.3.16}$$

where $C = \cos\theta$ and $S = \sin\theta$.

Equation (3.3.16) relates the global displacement $\underline{d}$ to the local displacement $\hat{\underline{d}}$. The matrix

$$\begin{bmatrix} C & S \\ -S & C \end{bmatrix} \tag{3.3.17}$$

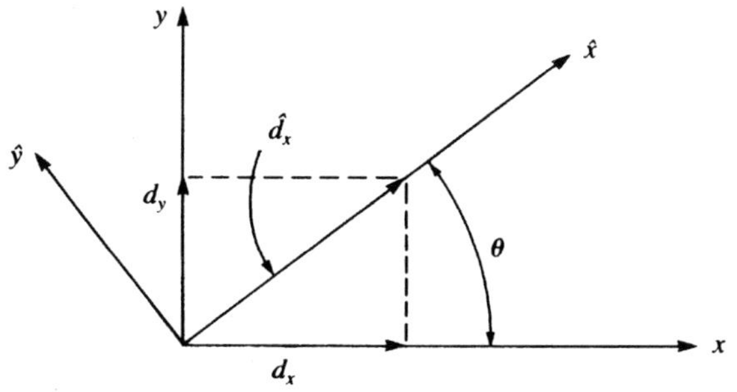

Figure 3–9 Relationship between local and global displacements

is called the *transformation* (or *rotation*) *matrix*. For an additional description of this matrix, see Appendix A. It will be used in Section 3.4 to develop the global stiffness matrix for an arbitrarily oriented bar element and to transform global nodal displacements and forces to local ones.

Now, for the case of $\hat{d}_y = 0$, we have, from Eq. (3.3.1),

$$d_x\mathbf{i} + d_y\mathbf{j} = \hat{d}_x\hat{\mathbf{i}} \tag{3.3.18}$$

Figure 3–9 shows $\hat{d}_x$ expressed in terms of global x and y components. Using trigonometry and Figure 3–9, we then obtain the magnitude of $\hat{d}_x$ as

$$\hat{d}_x = Cd_x + Sd_y \tag{3.3.19}$$

Equation (3.3.19) is equivalent to equation 1 of Eq. (3.3.16).

Example 3.2

The global nodal displacements at node 2 have been determined to be $d_{2x} = 0.1$ in. and $d_{2y} = 0.2$ in. for the bar element shown in Figure 3–10. Determine the local $\hat{x}$ displacement at node 2.

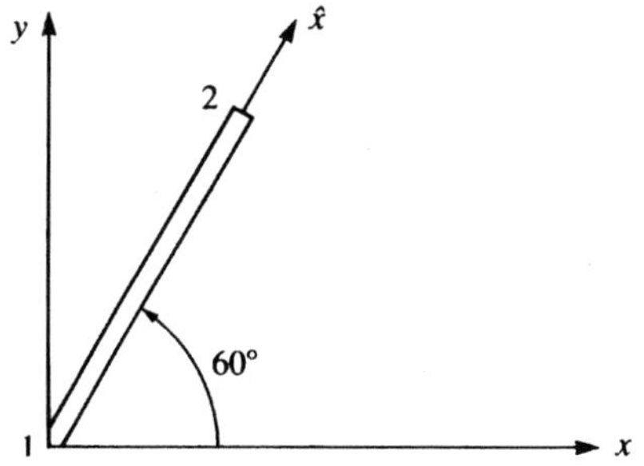

Figure 3–10 Bar element

Using Eq. (3.3.19), we obtain

$$\hat{d}_{2x} = (\cos 60^\circ)(0.1) + (\sin 60^\circ)(0.2) = 0.223 \text{ in.}$$ ■

3.4 Global Stiffness Matrix

We will now use the transformation relationship Eq. (3.3.16) to obtain the global stiffness matrix for a bar element. We need the global stiffness matrix of each element to assemble the total global stiffness matrix of the structure. We have shown in Eq. (3.1.13) that for a bar element in the local coordinate system,

$$\begin{Bmatrix} \hat{f}_{1x} \\ \hat{f}_{2x} \end{Bmatrix} = \frac{AE}{L}\begin{bmatrix} 1 & -1 \\ -1 & 1 \end{bmatrix}\begin{Bmatrix} \hat{d}_{1x} \\ \hat{d}_{2x} \end{Bmatrix} \tag{3.4.1}$$

or

$$\underline{\hat{f}} = \underline{\hat{k}}\,\underline{\hat{d}} \tag{3.4.2}$$

We now want to relate the global element nodal forces $\underline{f}$ to the global nodal displacements $\underline{d}$ for a bar element arbitrarily oriented with respect to the global axes as was shown in Figure 3–2. This relationship will yield the global stiffness matrix $\underline{k}$ of the element. That is, we want to find a matrix $\underline{k}$ such that

$$\begin{Bmatrix} f_{1x} \\ f_{1y} \\ f_{2x} \\ f_{2y} \end{Bmatrix} = \underline{k}\begin{Bmatrix} d_{1x} \\ d_{1y} \\ d_{2x} \\ d_{2y} \end{Bmatrix} \tag{3.4.3}$$

or, in simplified matrix form, Eq. (3.4.3) becomes

$$\underline{f} = \underline{k}\,\underline{d} \tag{3.4.4}$$

We observe from Eq. (3.4.3) that a total of four components of force and four of displacement arise when global coordinates are used. However, a total of two components of force and two of displacement appear for the local-coordinate representation of a spring or a bar, as shown by Eq. (3.4.1). By using relationships between local and global force components and between local and global displacement components, we will be able to obtain the global stiffness matrix. We know from transformation relationship Eq. (3.3.15) that

$$\begin{aligned} \hat{d}_{1x} &= d_{1x}\cos\theta + d_{1y}\sin\theta \\ \hat{d}_{2x} &= d_{2x}\cos\theta + d_{2y}\sin\theta \end{aligned} \tag{3.4.5}$$

In matrix form, Eqs. (3.4.5) can be written as

$$\begin{Bmatrix} \hat{d}_{1x} \\ \hat{d}_{2x} \end{Bmatrix} = \begin{bmatrix} C & S & 0 & 0 \\ 0 & 0 & C & S \end{bmatrix}\begin{Bmatrix} d_{1x} \\ d_{1y} \\ d_{2x} \\ d_{2y} \end{Bmatrix} \tag{3.4.6}$$

or as

$$\underline{\hat{d}} = \underline{T}^{*}\underline{d} \tag{3.4.7}$$

where

$$\underline{T}^{*} = \begin{bmatrix} C & S & 0 & 0 \\ 0 & 0 & C & S \end{bmatrix} \tag{3.4.8}$$

Similarly, because forces transform in the same manner as displacements, we have

$$\begin{Bmatrix} \hat{f}_{1x} \\ \hat{f}_{2x} \end{Bmatrix} = \begin{bmatrix} C & S & 0 & 0 \\ 0 & 0 & C & S \end{bmatrix} \begin{Bmatrix} f_{1x} \\ f_{1y} \\ f_{2x} \\ f_{2y} \end{Bmatrix} \tag{3.4.9}$$

Using Eq. (3.4.8), we can write Eq. (3.4.9) as

$$\underline{\hat{f}} = \underline{T}^{*}\underline{f} \tag{3.4.10}$$

Now, substituting Eq. (3.4.7) into Eq. (3.4.2), we obtain

$$\underline{\hat{f}} = \underline{\hat{k}}\underline{T}^{*}\underline{d} \tag{3.4.11}$$

and using Eq. (3.4.10) in Eq. (3.4.11) yields

$$\underline{T}^{*}\underline{f} = \underline{\hat{k}}\underline{T}^{*}\underline{d} \tag{3.4.12}$$

However, to write the final expression relating global nodal forces to global nodal displacements for an element, we must invert $\underline{T}^{*}$ in Eq. (3.4.12). This is not immediately possible because $\underline{T}^{*}$ is not a square matrix. Therefore, we must expand $\underline{\hat{d}}$, $\underline{\hat{f}}$, and $\underline{\hat{k}}$ to the order that is consistent with the use of global coordinates even though $\hat{f}_{1y}$ and $\hat{f}_{2y}$ are zero. Using Eq. (3.3.16) for each nodal displacement, we thus obtain

$$\begin{Bmatrix} \hat{d}_{1x} \\ \hat{d}_{1y} \\ \hat{d}_{2x} \\ \hat{d}_{2y} \end{Bmatrix} = \begin{bmatrix} C & S & 0 & 0 \\ -S & C & 0 & 0 \\ 0 & 0 & C & S \\ 0 & 0 & -S & C \end{bmatrix} \begin{Bmatrix} d_{1x} \\ d_{1y} \\ d_{2x} \\ d_{2y} \end{Bmatrix} \tag{3.4.13}$$

or

$$\underline{\hat{d}} = \underline{T}\underline{d} \tag{3.4.14}$$

where

$$\underline{T} = \begin{bmatrix} C & S & 0 & 0 \\ -S & C & 0 & 0 \\ 0 & 0 & C & S \\ 0 & 0 & -S & C \end{bmatrix} \tag{3.4.15}$$

Similarly, we can write

$$\underline{\hat{f}} = \underline{T}\underline{f} \tag{3.4.16}$$

because forces are like displacements—both are vectors. Also, $\underline{\hat{k}}$ must be expanded to a 4×4 matrix. Therefore, Eq. (3.4.1) in expanded form becomes

$$\begin{Bmatrix} \hat{f}_{1x} \\ \hat{f}_{1y} \\ \hat{f}_{2x} \\ \hat{f}_{2y} \end{Bmatrix} = \frac{AE}{L} \begin{bmatrix} 1 & 0 & -1 & 0 \\ 0 & 0 & 0 & 0 \\ -1 & 0 & 1 & 0 \\ 0 & 0 & 0 & 0 \end{bmatrix} \begin{Bmatrix} \hat{d}_{1x} \\ \hat{d}_{1y} \\ \hat{d}_{2x} \\ \hat{d}_{2y} \end{Bmatrix} \tag{3.4.17}$$

In Eq. (3.4.17), because $\hat{f}_{1y}$ and $\hat{f}_{2y}$ are zero, rows of zeros corresponding to the row numbers $\hat{f}_{1y}$ and $\hat{f}_{2y}$ appear in $\hat{\underline{k}}$. Now, using Eqs. (3.4.14) and (3.4.16) in Eq. (3.4.2), we obtain

$$\underline{T}\underline{f} = \hat{\underline{k}}\underline{T}\underline{d} \tag{3.4.18}$$

Equation (3.4.18) is Eq. (3.4.12) expanded. Premultiplying both sides of Eq. (3.4.18) by $\underline{T}^{-1}$, we have

$$\underline{f} = \underline{T}^{-1}\hat{\underline{k}}\underline{T}\underline{d} \tag{3.4.19}$$

where $\underline{T}^{-1}$ is the *inverse* of $\underline{T}$. However, it can be shown (see Problem 3.28) that

$$\underline{T}^{-1} = \underline{T}^{T} \tag{3.4.20}$$

where $\underline{T}^{T}$ is the *transpose* of $\underline{T}$. The property of square matrices such as $\underline{T}$ given by Eq. (3.4.20) defines $\underline{T}$ to be an orthogonal matrix. For more about orthogonal matrices, see Appendix A. The transformation matrix $\underline{T}$ between rectangular coordinate frames is orthogonal. This property of $\underline{T}$ is used throughout this text. Substituting Eq. (3.4.20) into Eq. (3.4.19), we obtain

$$\underline{f} = \underline{T}^{T}\hat{\underline{k}}\underline{T}\underline{d} \tag{3.4.21}$$

Equating Eqs. (3.4.4) and (3.4.21), we obtain the global stiffness matrix for an element as

$$\underline{k} = \underline{T}^{T}\hat{\underline{k}}\underline{T} \tag{3.4.22}$$

Substituting Eq. (3.4.15) for $\underline{T}$ and the expanded form of $\hat{\underline{k}}$ given in Eq. (3.4.17) into Eq. (3.4.22), we obtain $\underline{k}$ given in explicit form by

$$\underline{k} = \frac{AE}{L}\begin{bmatrix} C^2 & CS & -C^2 & -CS \\ & S^2 & -CS & -S^2 \\ & & C^2 & CS \\ \text{Symmetry} & & & S^2 \end{bmatrix} \tag{3.4.23}$$

Now, because the trial displacement function Eq. (3.1.1) was assumed piecewise-continuous element by element, the stiffness matrix for each element can be summed by using the direct stiffness method to obtain

$$\sum_{e=1}^{N} \underline{k}^{(e)} = \underline{K} \tag{3.4.24}$$

where $\underline{K}$ is the total stiffness matrix and N is the total number of elements. Similarly, each element global nodal force matrix can be summed such that

$$\sum_{e=1}^{N} \underline{f}^{(e)} = \underline{F} \tag{3.4.25}$$

$\underline{K}$ now relates the global nodal forces $\underline{F}$ to the global nodal displacements $\underline{d}$ for the whole structure by

$$\underline{F} = \underline{K}\underline{d} \tag{3.4.26}$$

Example 3.3

For the bar element shown in Figure 3–11, evaluate the global stiffness matrix with respect to the x-y coordinate system. Let the bar's cross-sectional area equal 2 in.^{2}, length equal 60 in., and modulus of elasticity equal 30×10^6 psi. The angle the bar makes with the x axis is 30°.

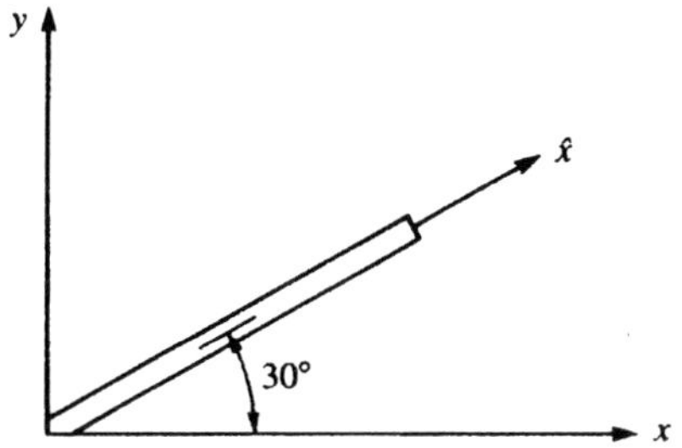

Figure 3–11 Bar element for stiffness matrix evaluation

To evaluate the global stiffness matrix $\underset{\sim}{k}$ for a bar, we use Eq. (3.4.23) with angle θ defined to be positive when measured counterclockwise from x to $\hat{x}$. Therefore,

$$\theta = 30^\circ \qquad C = \cos 30^\circ = \frac{\sqrt{3}}{2} \qquad S = \sin 30^\circ = \frac{1}{2}$$

$$\underset{\sim}{k} = \frac{(2)(30 \times 10^6)}{60} \begin{bmatrix} \frac{3}{4} & \frac{\sqrt{3}}{4} & \frac{-3}{4} & \frac{-\sqrt{3}}{4} \\ & \frac{1}{4} & \frac{-\sqrt{3}}{4} & \frac{-1}{4} \\ & & \frac{3}{4} & \frac{\sqrt{3}}{4} \\ \text{Symmetry} & & & \frac{1}{4} \end{bmatrix} \frac{\text{lb}}{\text{in.}} \tag{3.4.27}$$

Simplifying Eq. (3.4.27), we have

$$\underset{\sim}{k} = 10^6 \begin{bmatrix} 0.75 & 0.433 & -0.75 & -0.433 \\ & 0.25 & -0.433 & -0.25 \\ & & 0.75 & 0.433 \\ \text{Symmetry} & & & 0.25 \end{bmatrix} \frac{\text{lb}}{\text{in.}} \tag{3.4.28}$$

■

3.5 Computation of Stress for a Bar in the *x*-*y* Plane

We will now consider the determination of the stress in a bar element. For a bar, the local forces are related to the local displacements by Eq. (3.1.13) or Eq. (3.4.17). This equation is repeated here for convenience.

$$\begin{Bmatrix} \hat{f}_{1x} \\ \hat{f}_{2x} \end{Bmatrix} = \frac{AE}{L}\begin{bmatrix} 1 & -1 \\ -1 & 1 \end{bmatrix}\begin{Bmatrix} \hat{d}_{1x} \\ \hat{d}_{2x} \end{Bmatrix} \tag{3.5.1}$$

The usual definition of axial tensile stress is axial force divided by cross-sectional area. Therefore, axial stress is

$$\sigma = \frac{\hat{f}_{2x}}{A} \tag{3.5.2}$$

where $\hat{f}_{2x}$ is used because it is the axial force that pulls on the bar as shown in Figure 3–12. By Eq. (3.5.1),

$$\hat{f}_{2x} = \frac{AE}{L}[-1 \quad 1]\begin{Bmatrix} \hat{d}_{1x} \\ \hat{d}_{2x} \end{Bmatrix} \tag{3.5.3}$$

Therefore, combining Eqs. (3.5.2) and (3.5.3) yields

$$\underline{\sigma} = \frac{E}{L}[-1 \quad 1]\underline{\hat{d}} \tag{3.5.4}$$

Now, using Eq. (3.4.7), we obtain

$$\underline{\sigma} = \frac{E}{L}[-1 \quad 1]\underline{T}^*\underline{d} \tag{3.5.5}$$

Equation (3.5.5) can be expressed in simpler form as

$$\underline{\sigma} = \underline{C}'\underline{d} \tag{3.5.6}$$

where, when we use Eq. (3.4.8),

$$\underline{C}' = \frac{E}{L}[-1 \quad 1]\begin{bmatrix} C & S & 0 & 0 \\ 0 & 0 & C & S \end{bmatrix} \tag{3.5.7}$$

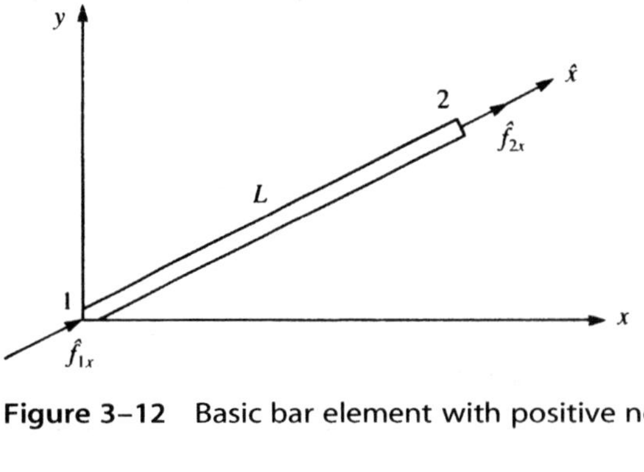

Figure 3–12 Basic bar element with positive nodal forces

After multiplying the matrices in Eq. (3.5.7), we have

$$\underline{C}' = \frac{E}{L}[-C \quad -S \quad C \quad S] \tag{3.5.8}$$

Example 3.4

For the bar shown in Figure 3–13, determine the axial stress. Let $A = 4 \times 10^{-4}$ m^2, $E = 210$ GPa, and $L = 2$ m, and let the angle between x and $\hat{x}$ be 60°. Assume the global displacements have been previously determined to be $d_{1x} = 0.25$ mm, $d_{1y} = 0.0$, $d_{2x} = 0.50$ mm, and $d_{2y} = 0.75$ mm.

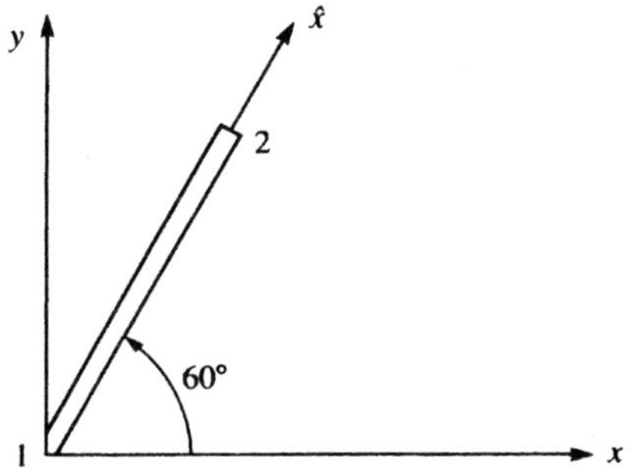

Figure 3–13 Bar element for stress evaluation

We can use Eq. (3.5.6) to evaluate the axial stress. Therefore, we first calculate $\underline{C}'$ from Eq. (3.5.8) as

$$\underline{C}' = \frac{210 \times 10^6 \text{ kN/m}^2}{2 \text{ m}} \begin{bmatrix} \frac{-1}{2} & \frac{-\sqrt{3}}{2} & \frac{1}{2} & \frac{\sqrt{3}}{2} \end{bmatrix} \tag{3.5.9}$$

where we have used $C = \cos 60° = \frac{1}{2}$ and $S = \sin 60° = \sqrt{3}/2$ in Eq. (3.5.9). Now $\underline{d}$ is given by

$$\underline{d} = \begin{Bmatrix} d_{1x} \\ d_{1y} \\ d_{2x} \\ d_{2y} \end{Bmatrix} = \begin{Bmatrix} 0.25 \times 10^{-3} \text{ m} \\ 0.0 \\ 0.50 \times 10^{-3} \text{ m} \\ 0.75 \times 10^{-3} \text{ m} \end{Bmatrix} \tag{3.5.10}$$

Using Eqs. (3.5.9) and (3.5.10) in Eq. (3.5.6), we obtain the bar axial stress as

$$\sigma_x = \frac{210 \times 10^6}{2} \begin{bmatrix} \frac{-1}{2} & \frac{-\sqrt{3}}{2} & \frac{1}{2} & \frac{\sqrt{3}}{2} \end{bmatrix} \begin{Bmatrix} 0.25 \\ 0.0 \\ 0.50 \\ 0.75 \end{Bmatrix} \times 10^{-3}$$

$$= 81.32 \times 10^3 \text{ kN/m}^2 = 81.32 \text{ MPa}$$

■

3.6 Solution of a Plane Truss

We will now illustrate the use of equations developed in Sections 3.4 and 3.5, along with the direct stiffness method of assembling the total stiffness matrix and equations, to solve the following plane truss example problems. *A* **plane truss** *is a structure composed of bar elements that all lie in a common plane and are connected by frictionless pins.* The plane truss also must have loads acting only in the common plane and all loads must be applied at the nodes or joints.

Example 3.5

For the plane truss composed of the three elements shown in Figure 3–14 subjected to a downward force of 10,000 lb applied at node 1, determine the x and y displacements at node 1 and the stresses in each element. Let $E = 30 \times 10^6$ psi and $A = 2$ in.^{2} for all elements. The lengths of the elements are shown in the figure.

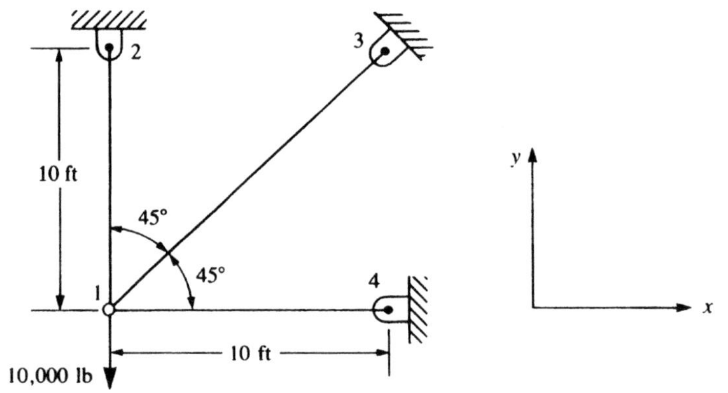

Figure 3–14 Plane truss

First, we determine the global stiffness matrices for each element by using Eq. (3.4.23). This requires determination of the angle θ between the global x axis and the local $\hat{x}$ axis for each element. In this example, the direction of the $\hat{x}$ axis for each element is taken in the direction *from* node 1 *to* the other node. The node numbering is arbitrary for each element. However, once the direction is chosen, the angle θ is then established as positive when measured counterclockwise from positive x to $\hat{x}$. For element 1, the local $\hat{x}$ axis is directed from node 1 to node 2; therefore, $\theta^{(1)} = 90°$. For element 2, the local $\hat{x}$ axis is directed from node 1 to node 3 and $\theta^{(2)} = 45°$. For element 3, the local $\hat{x}$ axis is directed from node 1 to node 4 and $\theta^{(3)} = 0°$. It is convenient to construct Table 3–1 to aid in determining each element stiffness matrix.

There are a total of eight nodal components of displacement, or degrees of freedom, for the truss before boundary constraints are imposed. Thus the order of the

Table 3–1 Data for the truss of Figure 3–14

Element	$\theta°$	C	S	C^2	S^2	CS
1	90°	0	1	0	1	0
2	45°	$\sqrt{2}/2$	$\sqrt{2}/2$	$\frac{1}{2}$	$\frac{1}{2}$	$\frac{1}{2}$
3	0°	1	0	1	0	0

total stiffness matrix must be 8 × 8. We could then expand the $\underline{k}$ matrix for each element to the order 8 × 8 by adding rows and columns of zeros as explained in the first part of Section 2.4. Alternatively, we could label the rows and columns of each element stiffness matrix according to the displacement components associated with it as explained in the latter part of Section 2.4. Using this latter approach, we construct the total stiffness matrix $\underline{K}$ simply by adding terms from the individual element stiffness matrices into their corresponding locations in $\underline{K}$. This approach will be used here and throughout this text.

For element 1, using Eq. (3.4.23), along with Table 3–1 for the direction cosines, we obtain

$$\underline{k}^{(1)} = \frac{(30 \times 10^6)(2)}{120} \begin{bmatrix} 0 & 0 & 0 & 0 \\ 0 & 1 & 0 & -1 \\ 0 & 0 & 0 & 0 \\ 0 & -1 & 0 & 1 \end{bmatrix} \quad \begin{matrix} d_{1x} & d_{1y} & d_{2x} & d_{2y} \end{matrix} \tag{3.6.1}$$

Similarly, for element 2, we have

$$\underline{k}^{(2)} = \frac{(30 \times 10^6)(2)}{120 \times \sqrt{2}} \begin{bmatrix} 0.5 & 0.5 & -0.5 & -0.5 \\ 0.5 & 0.5 & -0.5 & -0.5 \\ -0.5 & -0.5 & 0.5 & 0.5 \\ -0.5 & -0.5 & 0.5 & 0.5 \end{bmatrix} \quad \begin{matrix} d_{1x} & d_{1y} & d_{3x} & d_{3y} \end{matrix} \tag{3.6.2}$$

and for element 3, we have

$$\underline{k}^{(3)} = \frac{(30 \times 10^6)(2)}{120} \begin{bmatrix} 1 & 0 & -1 & 0 \\ 0 & 0 & 0 & 0 \\ -1 & 0 & 1 & 0 \\ 0 & 0 & 0 & 0 \end{bmatrix} \quad \begin{matrix} d_{1x} & d_{1y} & d_{4x} & d_{4y} \end{matrix} \tag{3.6.3}$$

The common factor of $30 \times 10^6 \times 2/120$ (= 500,000) can be taken from each of Eqs. (3.6.1)–(3.6.3). After adding terms from the individual element stiffness matrices into their corresponding locations in $\underline{K}$, we obtain the total stiffness matrix as

$$\underline{K} = (500{,}000) \begin{bmatrix} 1.354 & 0.354 & 0 & 0 & -0.354 & -0.354 & -1 & 0 \\ 0.354 & 1.354 & 0 & -1 & -0.354 & -0.354 & 0 & 0 \\ 0 & 0 & 0 & 0 & 0 & 0 & 0 & 0 \\ 0 & -1 & 0 & 1 & 0 & 0 & 0 & 0 \\ -0.354 & -0.354 & 0 & 0 & 0.354 & 0.354 & 0 & 0 \\ -0.354 & -0.354 & 0 & 0 & 0.354 & 0.354 & 0 & 0 \\ -1 & 0 & 0 & 0 & 0 & 0 & 1 & 0 \\ 0 & 0 & 0 & 0 & 0 & 0 & 0 & 0 \end{bmatrix} \tag{3.6.4}$$

(columns: d_{1x}, d_{1y}, d_{2x}, d_{2y}, d_{3x}, d_{3y}, d_{4x}, d_{4y})

The global $\underline{K}$ matrix, Eq. (3.6.4), relates the global forces to the global displacements. We thus write the total structure stiffness equations, accounting for the applied force at node 1 and the boundary constraints at nodes 2–4 as follows:

$$\begin{Bmatrix} 0 \\ -10{,}000 \\ F_{2x} \\ F_{2y} \\ F_{3x} \\ F_{3y} \\ F_{4x} \\ F_{4y} \end{Bmatrix} = (500{,}000) \begin{bmatrix} 1.354 & 0.354 & 0 & 0 & -0.354 & -0.354 & -1 & 0 \\ 0.354 & 1.354 & 0 & -1 & -0.354 & -0.354 & 0 & 0 \\ 0 & 0 & 0 & 0 & 0 & 0 & 0 & 0 \\ 0 & -1 & 0 & 1 & 0 & 0 & 0 & 0 \\ -0.354 & -0.354 & 0 & 0 & 0.354 & 0.354 & 0 & 0 \\ -0.354 & -0.354 & 0 & 0 & 0.354 & 0.354 & 0 & 0 \\ -1 & 0 & 0 & 0 & 0 & 0 & 1 & 0 \\ 0 & 0 & 0 & 0 & 0 & 0 & 0 & 0 \end{bmatrix}$$

$$\times \begin{Bmatrix} d_{1x} \\ d_{1y} \\ d_{2x} = 0 \\ d_{2y} = 0 \\ d_{3x} = 0 \\ d_{3y} = 0 \\ d_{4x} = 0 \\ d_{4y} = 0 \end{Bmatrix} \tag{3.6.5}$$

We could now use the partitioning scheme described in the first part of Section 2.5 to obtain the equations used to determine unknown displacements d_{1x} and d_{1y}—that is, partition the first two equations from the third through the eighth in Eq. (3.6.5). Alternatively, we could eliminate rows and columns in the total stiffness matrix corresponding to zero displacements as previously described in the latter part of Section 2.5. Here we will use the latter approach; that is, we eliminate rows and column 3–8 in Eq. (3.6.5) because those rows and columns correspond to zero displacements.

(Remember, this direct approach must be modified for nonhomogeneous boundary conditions as was indicated in Section 2.5.) We then obtain

$$\begin{Bmatrix} 0 \\ -10{,}000 \end{Bmatrix} = (500{,}000)\begin{bmatrix} 1.354 & 0.354 \\ 0.354 & 1.354 \end{bmatrix}\begin{Bmatrix} d_{1x} \\ d_{1y} \end{Bmatrix} \tag{3.6.6}$$

Equation (3.6.6) can now be solved for the displacements by multiplying both sides of the matrix equation by the inverse of the 2×2 stiffness matrix or by solving the two equations simultaneously. Using either procedure for solution yields the displacements

$$d_{1x} = 0.414 \times 10^{-2} \text{ in.} \qquad d_{1y} = -1.59 \times 10^{-2} \text{ in.}$$

The minus sign in the d_{1y} result indicates that the displacement component in the y direction at node 1 is in the direction opposite that of the positive y direction based on the assumed global coordinates, that is, a downward displacement occurs at node 1.

Using Eq. (3.5.6) and Table 3–1, we determine the stresses in each element as follows:

$$\sigma^{(1)} = \frac{30 \times 10^6}{120}[0 \quad -1 \quad 0 \quad 1]\begin{Bmatrix} d_{1x} = 0.414 \times 10^{-2} \\ d_{1y} = -1.59 \times 10^{-2} \\ d_{2x} = 0 \\ d_{2y} = 0 \end{Bmatrix} = 3965 \text{ psi}$$

$$\sigma^{(2)} = \frac{30 \times 10^6}{120\sqrt{2}}\begin{bmatrix} \frac{-\sqrt{2}}{2} & \frac{-\sqrt{2}}{2} & \frac{\sqrt{2}}{2} & \frac{\sqrt{2}}{2} \end{bmatrix}\begin{Bmatrix} d_{1x} = 0.414 \times 10^{-2} \\ d_{1y} = -1.59 \times 10^{-2} \\ d_{3x} = 0 \\ d_{3y} = 0 \end{Bmatrix}$$

$$= 1471 \text{ psi}$$

$$\sigma^{(3)} = \frac{30 \times 10^6}{120}[-1 \quad 0 \quad 1 \quad 0]\begin{Bmatrix} d_{1x} = 0.414 \times 10^{-2} \\ d_{1y} = -1.59 \times 10^{-2} \\ d_{4x} = 0 \\ d_{4y} = 0 \end{Bmatrix} = -1035 \text{ psi}$$

We now verify our results by examining force equilibrium at node 1; that is, summing forces in the global x and y directions, we obtain

$$\sum F_x = 0 \qquad (1471 \text{ psi})(2 \text{ in}^2)\frac{\sqrt{2}}{2} - (1035 \text{ psi})(2 \text{ in}^2) = 0$$

$$\sum F_y = 0 \qquad (3965 \text{ psi})(2 \text{ in}^2) + (1471 \text{ psi})(2 \text{ in}^2)\frac{\sqrt{2}}{2} - 10{,}000 = 0 \qquad \blacksquare$$

Example 3.6

For the two-bar truss shown in Figure 3–15, determine the displacement in the y direction of node 1 and the axial force in each element. A force of $P = 1000$ kN is applied at node 1 in the positive y direction while node 1 settles an amount $\delta = 50$ mm in the negative x direction. Let $E = 210$ GPa and $A = 6.00 \times 10^{-4}$ m^2 for each element. The lengths of the elements are shown in the figure.

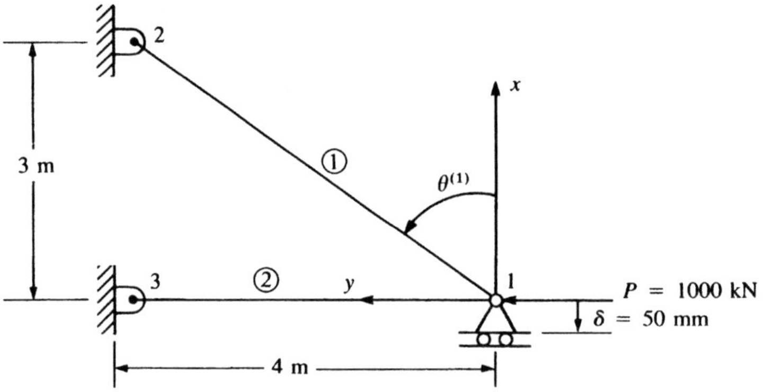

Figure 3–15 Two-bar truss

We begin by using Eq. (3.4.23) to determine each element stiffness matrix.

Element 1

$$\cos\theta^{(1)} = \frac{3}{5} = 0.60 \qquad \sin\theta^{(1)} = \frac{4}{5} = 0.80$$

$$\underline{k}^{(1)} = \frac{(6.0 \times 10^{-4}\ \text{m}^2)(210 \times 10^6\ \text{kN/m}^2)}{5\ \text{m}} \begin{bmatrix} 0.36 & 0.48 & -0.36 & -0.48 \\ & 0.64 & -0.48 & -0.64 \\ & & 0.36 & 0.48 \\ \text{Symmetry} & & & 0.64 \end{bmatrix} \tag{3.6.7}$$

Simplifying Eq. (3.6.7), we obtain

$$\underline{k}^{(1)} = (25{,}200) \begin{bmatrix} 0.36 & 0.48 & -0.36 & -0.48 \\ & 0.64 & -0.48 & -0.64 \\ & & 0.36 & 0.48 \\ \text{Symmetry} & & & 0.64 \end{bmatrix} \tag{3.6.8}$$

(columns labelled d_{1x}, d_{1y}, d_{2x}, d_{2y})

Element 2

$$\cos\theta^{(2)} = 0.0 \qquad \sin\theta^{(2)} = 1.0$$

$$\underline{k}^{(2)} = \frac{(6.0 \times 10^{-4})(210 \times 10^{6})}{4} \begin{bmatrix} 0 & 0 & 0 & 0 \\ & 1 & 0 & -1 \\ & & 0 & 0 \\ \text{Symmetry} & & & 1 \end{bmatrix} \tag{3.6.9}$$

$$\underline{k}^{(2)} = (25{,}200) \begin{bmatrix} \overset{d_{1x}}{0} & \overset{d_{1y}}{0} & \overset{d_{3x}}{0} & \overset{d_{3y}}{0} \\ & 1.25 & 0 & -1.25 \\ & & 0 & 0 \\ \text{Symmetry} & & & 1.25 \end{bmatrix} \tag{3.6.10}$$

where, for computational simplicity, Eq. (3.6.10) is written with the same factor (25,200) in front of the matrix as Eq. (3.6.8). Superimposing the element stiffness matrices, Eqs. (3.6.8) and (3.6.10), we obtain the global $\underline{K}$ matrix and relate the global forces to global displacements by

$$\begin{Bmatrix} F_{1x} \\ F_{1y} \\ F_{2x} \\ F_{2y} \\ F_{3x} \\ F_{3y} \end{Bmatrix} = (25{,}200) \begin{bmatrix} 0.36 & 0.48 & -0.36 & -0.48 & 0 & 0 \\ & 1.89 & -0.48 & -0.64 & 0 & -1.25 \\ & & 0.36 & 0.48 & 0 & 0 \\ & & & 0.64 & 0 & 0 \\ & & & & 0 & 0 \\ \text{Symmetry} & & & & & 1.25 \end{bmatrix} \begin{Bmatrix} d_{1x} \\ d_{1y} \\ d_{2x} \\ d_{2y} \\ d_{3x} \\ d_{3y} \end{Bmatrix} \tag{3.6.11}$$

We can again partition equations with known displacements and then simultaneously solve those associated with unknown displacements. To do this partitioning, we consider the boundary conditions given by

$$d_{1x} = \delta \qquad d_{2x} = 0 \qquad d_{2y} = 0 \qquad d_{3x} = 0 \qquad d_{3y} = 0 \tag{3.6.12}$$

Therefore, using Eqs. (3.6.12), we partition equation 2 from equations 1, 3, 4, 5, and 6 of Eq. (3.6.11) and are left with

$$P = 25{,}200(0.48\delta + 1.89 d_{1y}) \tag{3.6.13}$$

where $F_{1y} = P$ and $d_{1x} = \delta$ were substituted into Eq. (3.6.13). Expressing Eq. (3.6.13) in terms of P and δ allows these two influences on d_{1y} to be clearly separated. Solving Eq. (3.6.13) for d_{1y}, we have

$$d_{1y} = 0.000021P - 0.254\delta \tag{3.6.14}$$

Now, substituting the numerical values $P = 1000$ kN and $\delta = -0.05$ m into Eq. (3.6.14), we obtain

$$d_{1y} = 0.0337 \text{ m} \tag{3.6.15}$$

where the positive value indicates horizontal displacement to the left.

The local element forces are obtained by using Eq. (3.4.11). We then have the following.

Element 1

$$\begin{Bmatrix} \hat{f}_{1x} \\ \hat{f}_{2x} \end{Bmatrix} = (25{,}200)\begin{bmatrix} 1 & -1 \\ -1 & 1 \end{bmatrix}\begin{bmatrix} 0.60 & 0.80 & 0 & 0 \\ 0 & 0 & 0.60 & 0.80 \end{bmatrix}\begin{Bmatrix} d_{1x} = -0.05 \\ d_{1y} = 0.0337 \\ d_{2x} = 0 \\ d_{2y} = 0 \end{Bmatrix} \tag{3.6.16}$$

Performing the matrix triple product in Eq. (3.6.16) yields

$$\hat{f}_{1x} = -76.6 \text{ kN} \qquad \hat{f}_{2x} = 76.6 \text{ kN} \tag{3.6.17}$$

Element 2

$$\begin{Bmatrix} \hat{f}_{1x} \\ \hat{f}_{3x} \end{Bmatrix} = (31{,}500)\begin{bmatrix} 1 & -1 \\ -1 & 1 \end{bmatrix}\begin{bmatrix} 0 & 1 & 0 & 0 \\ 0 & 0 & 0 & 1 \end{bmatrix}\begin{Bmatrix} d_{1x} = -0.05 \\ d_{1y} = 0.0337 \\ d_{3x} = 0 \\ d_{3y} = 0 \end{Bmatrix} \tag{3.6.18}$$

Performing the matrix triple product in Eq. (3.6.18), we obtain

$$\hat{f}_{1x} = 1061 \text{ kN} \qquad \hat{f}_{3x} = -1061 \text{ kN} \tag{3.6.19}$$

Verification of the computations by checking that equilibrium is satisfied at node 1 is left to your discretion. ■

Example 3.7

To illustrate how we can combine spring and bar elements in one structure, we now solve the two-bar truss supported by a spring shown in Figure 3–16. Both bars have $E = 210$ GPa and $A = 5.0 \times 10^{-4}\ \text{m}^2$. Bar one has a length of 5 m and bar two a length of 10 m. The spring stiffness is $k = 2000$ kN/m.

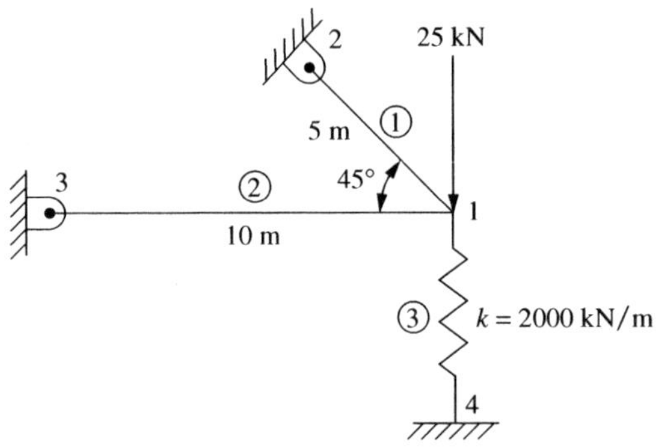

Figure 3–16 Two-bar truss with spring support

We begin by using Eq. (3.4.23) to determine each element stiffness matrix.

Element 1

$$\theta^{(1)} = 135^\circ, \qquad \cos\theta^{(1)} = -\sqrt{2}/2, \qquad \sin\theta^{(1)} = \sqrt{2}/2$$

$$\underline{k}^{(1)} = \frac{(5.0\times 10^{-4}\,\text{m}^2)(210\times 10^6\,\text{kN/m}^2)}{5\,\text{m}}\begin{bmatrix} 0.5 & -0.5 & -0.5 & 0.5 \\ -0.5 & 0.5 & 0.5 & -0.5 \\ -0.5 & 0.5 & 0.5 & -0.5 \\ 0.5 & -0.5 & -0.5 & 0.5 \end{bmatrix} \tag{3.6.20}$$

Simplifying Eq. (3.6.20), we obtain

$$\underline{k}^{(1)} = 105\times 10^5\begin{bmatrix} 1 & -1 & -1 & 1 \\ -1 & 1 & 1 & -1 \\ -1 & 1 & 1 & -1 \\ 1 & -1 & -1 & 1 \end{bmatrix} \tag{3.6.21}$$

Element 2

$$\theta^{(2)} = 180^\circ, \qquad \cos\theta^{(2)} = -1.0, \qquad \sin\theta^{(2)} = 0$$

$$\underline{k}^{(2)} = \frac{(5\times 10^{-4}\,\text{m}^2)(210\times 10^6\,\text{kN/m}^2)}{10\,\text{m}}\begin{bmatrix} 1 & 0 & -1 & 0 \\ 0 & 0 & 0 & 0 \\ -1 & 0 & 1 & 0 \\ 0 & 0 & 0 & 0 \end{bmatrix} \tag{3.6.22}$$

Simplifying Eq. (3.6.22), we obtain

$$\underline{k}^{(2)} = 105\times 10^5\begin{bmatrix} 1 & 0 & -1 & 0 \\ 0 & 0 & 0 & 0 \\ -1 & 0 & 1 & 0 \\ 0 & 0 & 0 & 0 \end{bmatrix} \tag{3.6.23}$$

Element 3

$$\theta^{(3)} = 270^\circ, \qquad \cos\theta^{(3)} = 0, \qquad \sin\theta^{(3)} = 1.0$$

Using Eq. (3.4.23) but replacing AE/L with the spring constant k, we obtain the stiffness matrix of the spring as

$$\underline{k}^{(3)} = 20\times 10^5\begin{bmatrix} 0 & 0 & 0 & 0 \\ 0 & 1 & 0 & -1 \\ 0 & 0 & 0 & 0 \\ 0 & -1 & 0 & 1 \end{bmatrix} \tag{3.6.24}$$

Applying the boundary conditions, we have

$$d_{2x} = d_{2y} = d_{3x} = d_{3y} = d_{4x} = d_{4y} = 0 \tag{3.6.25}$$

Using the boundary conditions in Eq. (3.6.25), the reduced assembled global equations are given by:

$$\begin{Bmatrix} F_{1x} = 0 \\ F_{1y} = -25\,\text{kN} \end{Bmatrix} = 10^5 \begin{bmatrix} 210 & -105 \\ -105 & 125 \end{bmatrix} \begin{Bmatrix} d_{1x} \\ d_{1y} \end{Bmatrix} \tag{3.6.26}$$

Solving Eq. (3.6.26) for the global displacements, we obtain

$$d_{1x} = -1.724 \times 10^{-3}\,\text{m} \qquad d_{1y} = -3.448 \times 10^{-3}\,\text{m} \tag{3.6.27}$$

We can obtain the stresses in the bar elements by using Eq. (3.5.6) as

$$\sigma^{(1)} = \frac{210 \times 10^3\,\text{MN/m}^2}{5\,\text{m}} [0.707 \quad -0.707 \quad -0.707 \quad 0.707] \begin{Bmatrix} -1.724 \times 10^{-3} \\ -3.448 \times 10^{-3} \\ 0 \\ 0 \end{Bmatrix}$$

Simplifying, we obtain

$$\sigma^{(1)} = 51.2\,\text{MPa}\,(T)$$

Similarly, we obtain the stress in element two as

$$\sigma^{(2)} = \frac{210 \times 10^3\,\text{MN/m}^2}{10\,\text{m}} [1.0 \quad 0 \quad -1.0 \quad 0] \begin{Bmatrix} -1.724 \times 10^{-3} \\ -3.448 \times 10^{-3} \\ 0 \\ 0 \end{Bmatrix}$$

Simplifying, we obtain

$$\sigma^{(2)} = -36.2\,\text{MPa}\,(\text{C})$$

■

3.7 Transformation Matrix and Stiffness Matrix for a Bar in Three-Dimensional Space

We will now derive the transformation matrix necessary to obtain the general stiffness matrix of a bar element arbitrarily oriented in three-dimensional space as shown in Figure 3–17. Let the coordinates of node 1 be taken as x_1, y_1, and z_1, and let those of node 2 be taken as x_2, y_2, and z_2. Also, let θ_x, θ_y, and θ_z be the angles measured from the global x, y, and z axes, respectively, to the local $\hat{x}$ axis. Here $\hat{x}$ is directed along the element from node 1 to node 2. We must now determine $\underline{T}^*$ such that $\underline{\hat{d}} = \underline{T}^* \underline{d}$. We begin the derivation of $\underline{T}^*$ by considering the vector $\hat{\mathbf{d}} = \mathbf{d}$ expressed in three dimensions as

$$\hat{d}_x \hat{\mathbf{i}} + \hat{d}_y \hat{\mathbf{j}} + \hat{d}_z \hat{\mathbf{k}} = d_x \mathbf{i} + d_y \mathbf{j} + d_z \mathbf{k} \tag{3.7.1}$$

where $\hat{\mathbf{i}}$, $\hat{\mathbf{j}}$, and $\hat{\mathbf{k}}$ are unit vectors associated with the local $\hat{x}$, $\hat{y}$, and $\hat{z}$ axes, respectively, and $\mathbf{i}$, $\mathbf{j}$, and $\mathbf{k}$ are unit vectors associated with the global x, y, and z axes. Taking the

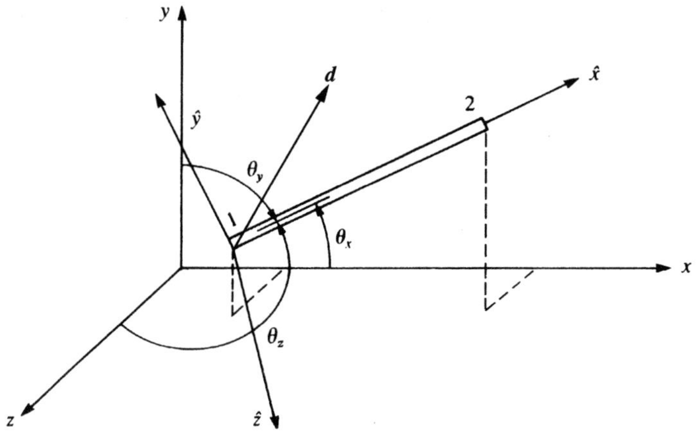

Figure 3–17 Bar in three-dimensional space

dot product of Eq. (3.7.1) with $\hat{\mathbf{i}}$, we have

$$\hat{d}_x + 0 + 0 = d_x(\hat{\mathbf{i}} \cdot \mathbf{i}) + d_y(\hat{\mathbf{i}} \cdot \mathbf{j}) + d_z(\hat{\mathbf{i}} \cdot \mathbf{k}) \tag{3.7.2}$$

and, by definition of the dot product,

$$\begin{aligned} \hat{\mathbf{i}} \cdot \mathbf{i} &= \frac{x_2 - x_1}{L} = C_x \\ \hat{\mathbf{i}} \cdot \mathbf{j} &= \frac{y_2 - y_1}{L} = C_y \\ \hat{\mathbf{i}} \cdot \mathbf{k} &= \frac{z_2 - z_1}{L} = C_z \end{aligned} \tag{3.7.3}$$

where $$L = [(x_2 - x_1)^2 + (y_2 - y_1)^2 + (z_2 - z_1)^2]^{1/2}$$

and $$C_x = \cos\theta_x \qquad C_y = \cos\theta_y \qquad C_z = \cos\theta_z \tag{3.7.4}$$

Here C_x, C_y, and C_z are the projections of $\hat{\mathbf{i}}$ on $\mathbf{i}, \mathbf{j}$, and $\mathbf{k}$, respectively. Therefore, using Eqs. (3.7.3) in Eq. (3.7.2), we have

$$\hat{d}_x = C_x d_x + C_y d_y + C_z d_z \tag{3.7.5}$$

For a vector in space directed along the $\hat{x}$ axis, Eq. (3.7.5) gives the components of that vector in the global x, y, and z directions. Now, using Eq. (3.7.5), we can write $\underline{\hat{d}} = \underline{T}^* \underline{d}$ in explicit form as

$$\begin{Bmatrix} \hat{d}_{1x} \\ \hat{d}_{2x} \end{Bmatrix} = \begin{bmatrix} C_x & C_y & C_z & 0 & 0 & 0 \\ 0 & 0 & 0 & C_x & C_y & C_z \end{bmatrix} \begin{Bmatrix} d_{1x} \\ d_{1y} \\ d_{1z} \\ d_{2x} \\ d_{2y} \\ d_{2z} \end{Bmatrix} \tag{3.7.6}$$

where
$$\underline{T}^* = \begin{bmatrix} C_x & C_y & C_z & 0 & 0 & 0 \\ 0 & 0 & 0 & C_x & C_y & C_z \end{bmatrix} \tag{3.7.7}$$

is the transformation matrix, which enables the local displacement matrix $\hat{\underline{d}}$ to be expressed in terms of displacement components in the global coordinate system.

We showed in Section 3.4 that the global stiffness matrix (the stiffness matrix for a bar element referred to global axes) is given in general by $\underline{k} = \underline{T}^T\hat{\underline{k}}\underline{T}$. This equation will now be used to express the general form of the stiffness matrix of a bar arbitrarily oriented in space. In general, we must expand the transformation matrix in a manner analogous to that done in expanding $\underline{T}^*$ to $\underline{T}$ in Section 3.4. However, the same result will be obtained here by simply using $\underline{T}^*$, defined by Eq. (3.7.7), in place of $\underline{T}$. Then $\underline{k}$ is obtained by using the equation $\underline{k} = (\underline{T}^*)^T\hat{\underline{k}}\underline{T}^*$ as follows:

$$\underline{k} = \begin{bmatrix} C_x & 0 \\ C_y & 0 \\ C_z & 0 \\ 0 & C_x \\ 0 & C_y \\ 0 & C_z \end{bmatrix} \frac{AE}{L}\begin{bmatrix} 1 & -1 \\ -1 & 1 \end{bmatrix}\begin{bmatrix} C_x & C_y & C_z & 0 & 0 & 0 \\ 0 & 0 & 0 & C_x & C_y & C_z \end{bmatrix} \tag{3.7.8}$$

Simplifying Eq. (3.7.8), we obtain the explicit form of $\underline{k}$ as

$$\underline{k} = \frac{AE}{L}\begin{bmatrix} C_x^2 & C_xC_y & C_xC_z & -C_x^2 & -C_xC_y & -C_xC_z \\ & C_y^2 & C_yC_z & -C_xC_y & -C_y^2 & -C_yC_z \\ & & C_z^2 & -C_xC_z & -C_yC_z & -C_z^2 \\ & & & C_x^2 & C_xC_y & C_xC_z \\ & & & & C_y^2 & C_yC_z \\ \text{Symmetry} & & & & & C_z^2 \end{bmatrix} \tag{3.7.9}$$

You should verify Eq. (3.7.9). First, expand $\underline{T}^*$ to a 6×6 square matrix in a manner similar to that done in Section 3.4 for the two-dimensional case. Then expand $\hat{\underline{k}}$ to a 6×6 matrix by adding appropriate rows and columns of zeros (for the $\hat{d}_z$ terms) to Eq. (3.4.17). Finally, perform the matrix triple product $\underline{k} = \underline{T}^T\hat{\underline{k}}\underline{T}$ (see Problem 3.44).

Equation (3.7.9) is the basic form of the stiffness matrix for a bar element arbitrarily oriented in three-dimensional space. We will now analyze a simple space truss to illustrate the concepts developed in this section. We will show that the direct stiffness method provides a simple procedure for solving space truss problems.

Example 3.8

Analyze the space truss shown in Figure 3–18. The truss is composed of four nodes, whose coordinates (in inches) are shown in the figure, and three elements, whose cross-sectional areas are given in the figure. The modulus of elasticity $E = 1.2 \times 10^6$ psi for all elements. A load of 1000 lb is applied at node 1 in the negative z direction. Nodes 2–4 are supported by ball-and-socket joints and thus constrained from movement in

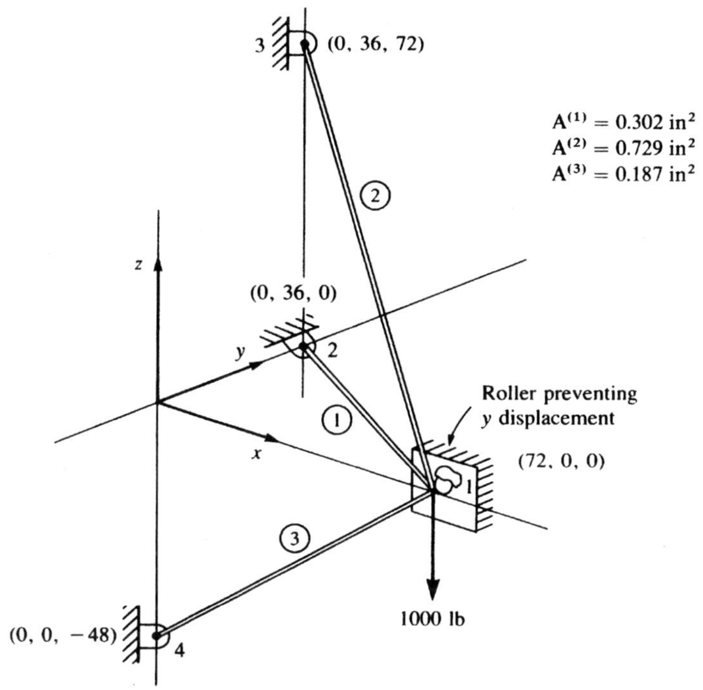

Figure 3–18 Space truss

the x, y, and z directions. Node 1 is constrained from movement in the y direction by the roller shown in Figure 3–18.

Using Eq. (3.7.9), we will now determine the stiffness matrices of the three elements in Figure 3–18. To simplify the numerical calculations, we first express $\underline{k}$ for each element, given by Eq. (3.7.9), in the form

$$\underline{k} = \frac{AE}{L}\left[\begin{array}{c|c} \underline{\lambda} & -\underline{\lambda} \\ \hline -\underline{\lambda} & \underline{\lambda} \end{array}\right] \tag{3.7.10}$$

where $\underline{\lambda}$ is a 3×3 submatrix defined by

$$\underline{\lambda} = \begin{bmatrix} C_x^2 & C_x C_y & C_x C_z \\ C_y C_x & C_y^2 & C_y C_z \\ C_z C_x & C_z C_y & C_z^2 \end{bmatrix} \tag{3.7.11}$$

Therefore, determining $\underline{\lambda}$ will sufficiently describe $\underline{k}$.

Element 3

The direction cosines of element 3 are given, in general, by

$$C_x = \frac{x_4 - x_1}{L^{(3)}} \qquad C_y = \frac{y_4 - y_1}{L^{(3)}} \qquad C_z = \frac{z_4 - z_1}{L^{(3)}} \tag{3.7.12}$$

where the notation x_i, y_i, and z_i is used to denote the coordinates of each node, and $L^{(e)}$ denotes the element length. From the coordinate information given in Figure 3–18, we obtain the length and the direction cosines as

$$L^{(3)} = [(-72.0)^2 + (-48.0)^2]^{1/2} = 86.5 \text{ in.}$$

$$C_x = \frac{-72.0}{86.5} = -0.833 \qquad C_y = 0 \qquad C_z = \frac{-48.0}{86.5} = -0.550 \tag{3.7.13}$$

Using the results of Eqs. (3.7.13) in Eq. (3.7.11) yields

$$\underline{\lambda} = \begin{bmatrix} 0.69 & 0 & 0.46 \\ 0 & 0 & 0 \\ 0.46 & 0 & 0.30 \end{bmatrix} \tag{3.7.14}$$

and, from Eq. (3.7.10),

$$\underline{k}^{(3)} = \frac{(0.187)(1.2 \times 10^6)}{86.5} \overset{\begin{matrix} d_{1x}d_{1y}d_{1z} & d_{4x}d_{4y}d_{4z} \end{matrix}}{\left[\begin{array}{c|c} \underline{\lambda} & -\underline{\lambda} \\ \hline -\underline{\lambda} & \underline{\lambda} \end{array}\right]} \tag{3.7.15}$$

Element 1

Similarly, for element 1, we obtain

$$L^{(1)} = 80.5 \text{ in.}$$

$$C_x = -0.89 \qquad C_y = 0.45 \qquad C_z = 0$$

$$\underline{\lambda} = \begin{bmatrix} 0.79 & -0.40 & 0 \\ -0.40 & 0.20 & 0 \\ 0 & 0 & 0 \end{bmatrix}$$

and

$$\underline{k}^{(1)} = \frac{(0.302)(1.2 \times 10^6)}{80.5} \overset{\begin{matrix} d_{1x}d_{1y}d_{1z} & d_{2x}d_{2y}d_{2z} \end{matrix}}{\left[\begin{array}{c|c} \underline{\lambda} & -\underline{\lambda} \\ \hline -\underline{\lambda} & \underline{\lambda} \end{array}\right]} \tag{3.7.16}$$

Element 2

Finally, for element 2, we obtain

$$L^{(2)} = 108 \text{ in.}$$

$$C_x = -0.667 \qquad C_y = 0.33 \qquad C_z = 0.667$$

$$\underline{\lambda} = \begin{bmatrix} 0.45 & -0.22 & -0.45 \\ -0.22 & 0.11 & 0.22 \\ -0.45 & 0.22 & 0.45 \end{bmatrix}$$

and

$$\underset{\sim}{k}^{(2)} = \frac{(0.729)(1.2 \times 10^6)}{108} \begin{array}{c} d_{1x} d_{1y} d_{1z} \quad d_{3x} d_{3y} d_{3z} \\ \left[\begin{array}{c|c} \underset{\sim}{\lambda} & -\underset{\sim}{\lambda} \\ \hline -\underset{\sim}{\lambda} & \underset{\sim}{\lambda} \end{array}\right] \end{array} \tag{3.7.17}$$

Using the zero-displacement boundary conditions $d_{1y} = 0, d_{2x} = d_{2y} = d_{2z} = 0, d_{3x} = d_{3y} = d_{3z} = 0$, and $d_{4x} = d_{4y} = d_{4z} = 0$, we can cancel the corresponding rows and columns of each element stiffness matrix. After canceling appropriate rows and columns in Eqs. (3.7.15)–(3.7.17) and then superimposing the resulting element stiffness matrices, we have the total stiffness matrix for the truss as

$$\underset{\sim}{K} = \begin{array}{c} \quad d_{1x} \qquad d_{1z} \\ \begin{bmatrix} 9000 & -2450 \\ -2450 & 4450 \end{bmatrix} \end{array} \tag{3.7.18}$$

The global stiffness equations are then expressed by

$$\begin{Bmatrix} 0 \\ -1000 \end{Bmatrix} = \begin{bmatrix} 9000 & -2450 \\ -2450 & 4450 \end{bmatrix} \begin{Bmatrix} d_{1x} \\ d_{1z} \end{Bmatrix} \tag{3.7.19}$$

Solving Eq. (3.7.19) for the displacements, we obtain

$$\begin{aligned} d_{1x} &= -0.072 \text{ in.} \\ d_{1z} &= -0.264 \text{ in.} \end{aligned} \tag{3.7.20}$$

where the minus signs in the displacements indicate these displacements to be in the negative x and z directions.

We will now determine the stress in each element. The stresses are determined by using Eq. (3.5.6) expanded to three dimensions. Thus, for an element with nodes i and j, Eq. (3.5.6) expanded to three dimensions becomes

$$\underset{\sim}{\sigma} = \frac{E}{L}[-C_x \quad -C_y \quad -C_z \quad C_x \quad C_y \quad C_z] \begin{Bmatrix} d_{ix} \\ d_{iy} \\ d_{iz} \\ d_{jx} \\ d_{jy} \\ d_{jz} \end{Bmatrix} \tag{3.7.21}$$

Derive Eq. (3.7.21) in a manner similar to that used to derive Eq. (3.5.6) (see Problem 3.45, for instance). For element 3, using Eqs. (3.7.13) for the direction cosines, along with the proper length and modulus of elasticity, we obtain the stress as

$$\underset{\sim}{\sigma}^{(3)} = \frac{1.2 \times 10^6}{86.5}[0.83 \quad 0 \quad 0.55 \quad -0.83 \quad 0 \quad -0.55] \begin{Bmatrix} -0.072 \\ 0 \\ -0.264 \\ 0 \\ 0 \\ 0 \end{Bmatrix} \tag{3.7.22}$$

Simplifying Eq. (3.7.22), we find that the result is

$$\sigma^{(3)} = -2850 \text{ psi}$$

where the negative sign in the answer indicates a compressive stress. The stresses in the other elements can be determined in a manner similar to that used for element 3. For brevity's sake, we will not show the calculations but will merely list these stresses:

$$\sigma^{(1)} = -945 \text{ psi} \qquad \sigma^{(2)} = 1440 \text{ psi}$$ ■

Example 3.9

Analyze the space truss shown in Figure 3–19. The truss is composed of four nodes, whose coordinates (in meters) are shown in the figure, and three elements, whose cross-sectional areas are all $10 \times 10^{-4}\,\text{m}^2$. The modulus of elasticity $E = 210$ GPa for all the elements. A load of 20 kN is applied at node 1 in the global x-direction. Nodes 2–4 are pin supported and thus constrained from movement in the x, y, and z directions.

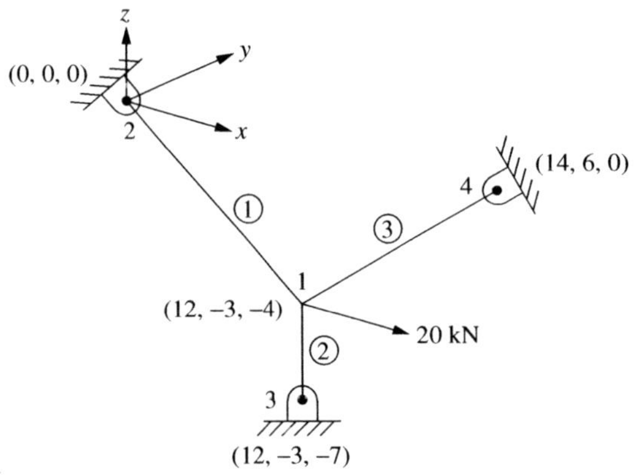

Figure 3–19 Space truss

First calculate the element lengths using the distance formula and coordinates given in Figure 3–19 as

$$L^{(1)} = [(0 - 12)^2 + (0 - (-3))^2 + (0 - (-4))^2]^{1/2} = 13\,\text{m}$$

$$L^{(2)} = [(12 - 12)^2 + (-3 + 3)^2 + (-7 + 4)^2]^{1/2} = 3\,\text{m}$$

$$L^{(3)} = [(14 - 12)^2 + (6 + 3)^2 + (0 + 4)^2]^{1/2} = 10.05\,\text{m}$$

For convenience, set up a table of direction cosines, where the local $\hat{x}$ axis is taken from node 1 to 2, from 1 to 3 and from 1 to 4 for elements 1, 2, and 3, respectively.

Element Number	$C_x = \frac{x_j - x_i}{L^{(1)}}$	$C_y = \frac{y_j - y_i}{L^{(2)}}$	$C_z = \frac{z_j - z_i}{L^{(3)}}$
1	$-12/13$	3/13	4/13
2	0	0	-1
3	2/10.05	9/10.05	4/10.05

Now set up a table of products of direction cosines as indicated by the definition of $\underline{\lambda}$ defined by Eq. (3.7.11) as

Element Number	C_x^2	C_xC_y	C_xC_z	C_y^2	C_yC_z	C_z^2
1	0.852	-0.213	-0.284	0.053	-0.071	0.095
2	0	0	0	0	0	1
3	0.040	0.178	0.079	0.802	0.356	0.158

Using Eq. (3.7.11), we express $\underline{\lambda}$ for each element as

$$\underline{\lambda}^{(1)} = \begin{bmatrix} 0.852 & -0.213 & -0.284 \\ -0.213 & 0.053 & 0.071 \\ -0.284 & 0.071 & 0.095 \end{bmatrix} \quad \underline{\lambda}^{(2)} = \begin{bmatrix} 0 & 0 & 0 \\ 0 & 0 & 0 \\ 0 & 0 & 1 \end{bmatrix} \quad \underline{\lambda}^{(3)} = \begin{bmatrix} 0.040 & 0.178 & 0.079 \\ 0.128 & 0.802 & 0.356 \\ 0.079 & 0.356 & 0.158 \end{bmatrix} \tag{3.7.23}$$

The boundary conditions are given by

$$d_{2x} = d_{2y} = d_{2z} = 0, \quad d_{3x} = d_{3y} = d_{3z} = 0, \quad d_{4x} = d_{4y} = d_{4z} = 0 \tag{3.7.24}$$

Using the stiffness matrix expressed in terms of $\underline{\lambda}$ in the form of Eq. (3.7.10), we obtain each stiffness matrix as

$$\underline{k}^{(1)} = \frac{AE}{13}\left[\begin{array}{c|c} \underline{\lambda}^{(1)} & -\underline{\lambda}^{(1)} \\ \hline -\underline{\lambda}^{(1)} & \underline{\lambda}^{(1)} \end{array}\right] \quad \underline{k}^{(2)} = \frac{AE}{3}\left[\begin{array}{c|c} \underline{\lambda}^{(2)} & -\underline{\lambda}^{(2)} \\ \hline -\underline{\lambda}^{(2)} & \underline{\lambda}^{(2)} \end{array}\right] \quad \underline{k}^{(3)} = \frac{AE}{10.05}\left[\begin{array}{c|c} \underline{\lambda}^{(3)} & -\underline{\lambda}^{(3)} \\ \hline -\underline{\lambda}^{(3)} & \underline{\lambda}^{(3)} \end{array}\right] \tag{3.7.25}$$

Applying the boundary conditions and canceling appropriate rows and columns associated with each zero displacement boundary condition in Eqs. (3.7.25) and then superimposing the resulting element stiffness matrices, we have the total stiffness matrix for the truss as

$$\underline{K} = 210 \times 10^3 \begin{bmatrix} 69.519 & 1.327 & -13.985 \\ 1.327 & 83.879 & 40.885 \\ -13.985 & 40.885 & 356.363 \end{bmatrix} \text{kN/m} \tag{3.7.26}$$

The global stiffness equations are then expressed by

$$\begin{Bmatrix} 20\,\text{kN} \\ 0 \\ 0 \end{Bmatrix} = 210 \times 10^3 \begin{bmatrix} 69.519 & 1.327 & -13.985 \\ 1.327 & 83.879 & 40.885 \\ -13.985 & 40.885 & 356.363 \end{bmatrix} \begin{Bmatrix} d_{1x} \\ d_{1y} \\ d_{1z} \end{Bmatrix} \tag{3.7.27}$$

Solving for the displacements, we obtain

$$\begin{aligned} d_{1x} &= 1.383 \times 10^{-3}\ \text{m} \\ d_{1y} &= -5.119 \times 10^{-5}\ \text{m} \\ d_{1z} &= 6.015 \times 10^{-5}\ \text{m} \end{aligned} \tag{3.7.28}$$

We now determine the element stresses using Eq. (3.7.21) as

$$\sigma^{(1)} = \frac{210 \times 10^6}{13}[\,12/13 \quad -3/13 \quad -4/13 \quad -12/13 \quad 3/13 \quad 4/13\,]\begin{Bmatrix} 1.383 \times 10^{-3} \\ -5.119 \times 10^{-5} \\ 6.015 \times 10^{-5} \\ 0 \\ 0 \\ 0 \end{Bmatrix} \tag{3.7.29}$$

Simplifying Eq. (3.7.29), we obtain upon converting to MPa units

$$\sigma^{(1)} = 20.51\ \text{MPa} \tag{3.7.30}$$

The stress in the other elements can be found in a similar manner as

$$\sigma^{(2)} = 4.21\ \text{MPa} \qquad \sigma^{(3)} = -5.29\ \text{MPa} \tag{3.7.31}$$

The negative sign in Eq. (3.7.31) indicates a compressive stress in element 3. ■

3.8 Use of Symmetry in Structure

Different types of symmetry may exist in a structure. These include reflective or mirror, skew, axial, and cyclic. Here we introduce the most common type of symmetry, reflective symmetry. Axial symmetry occurs when a solid of revolution is generated by rotating a plane shape about an axis in the plane. These axisymmetric bodies are common, and hence their analysis is considered in Chapter 9.

In many instances, we can use reflective symmetry to facilitate the solution of a problem. **Reflective symmetry** *means correspondence in size, shape, and position of loads; material properties; and boundary conditions that are on opposite sides of a dividing line or plane.* The use of symmetry allows us to consider a reduced problem instead of the actual problem. Thus, the order of the total stiffness matrix and total set of stiffness equations can be reduced. Longhand solution time is then reduced, and computer solution time for large-scale problems is substantially decreased. Example 3.10 will be used to illustrate reflective symmetry. Additional examples of the use of symmetry are presented in Chapter 4 for beams and in Chapter 7 for plane problems.

Example 3.10

Solve the plane truss problem shown in Figure 3–20. The truss is composed of eight elements and five nodes as shown. A vertical load of $2P$ is applied at node 4. Nodes 1 and 5 are pin supports. Bar elements 1, 2, 7, and 8 have axial stiffnesses of $\sqrt{2}AE$, and bars 3–6 have axial stiffness of AE. Here again, A and E represent the cross-sectional area and modulus of elasticity of a bar.

In this problem, we will use a plane of symmetry. The vertical plane perpendicular to the plane truss passing through nodes 2, 4, and 3 is the plane of reflective symmetry because identical geometry, material, loading, and boundary conditions occur at the corresponding locations on opposite sides of this plane. For loads such as $2P$, occurring in the plane of symmetry, half of the total load must be applied to the reduced structure. For elements occurring in the plane of symmetry, half of the cross-sectional area must be used in the reduced structure. Furthermore, for nodes in the plane of symmetry, the displacement components normal to the plane of symmetry must be set to zero in the reduced structure; that is, we set $d_{2x} = 0, d_{3x} = 0$, and $d_{4x} = 0$. Figure 3–21 shows the reduced structure to be used to analyze the plane truss of Figure 3–20.

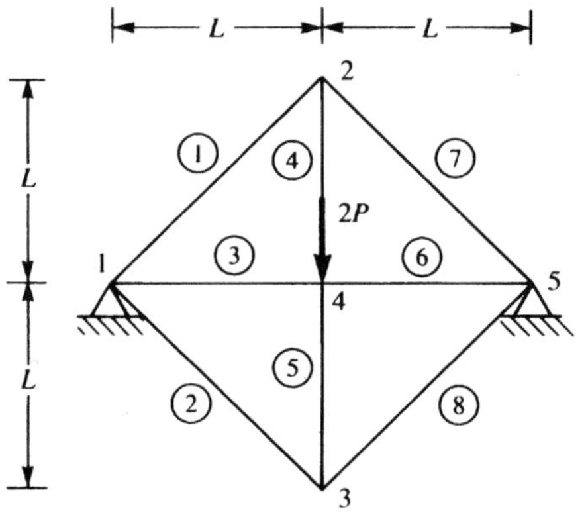

Figure 3–20 Plane truss

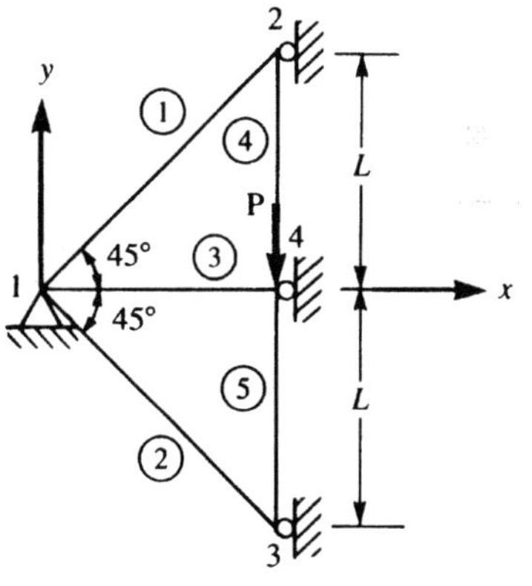

Figure 3–21 Truss of Figure 3–20 reduced by symmetry

We begin the solution of the problem by determining the angles θ for each bar element. For instance, for element 1, assuming $\hat{x}$ to be directed from node 1 to node 2, we obtain $\theta^{(1)} = 45°$. Table 3–2 is used in determining each element stiffness matrix.

There are a total of eight nodal components of displacement for the truss before boundary constraints are imposed. Therefore, $\underline{K}$ must be of order 8×8. For element 1,

Table 3–2 Data for the truss of Figure 3–21

Element	$\theta°$	C	S	C^2	S^2	CS
1	45°	$\sqrt{2}/2$	$\sqrt{2}/2$	1/2	1/2	1/2
2	315°	$\sqrt{2}/2$	$-\sqrt{2}/2$	1/2	1/2	−1/2
3	0°	1	0	1	0	0
4	90°	0	1	0	1	0
5	90°	0	1	0	1	0

using Eq. (3.4.23) along with Table 3–2 for the direction cosines, we obtain

$$\underline{k}^{(1)} = \frac{\sqrt{2}AE}{\sqrt{2}L} \begin{array}{c} \begin{matrix} d_{1x} & d_{1y} & d_{2x} & d_{2y} \end{matrix} \\ \begin{bmatrix} \frac{1}{2} & \frac{1}{2} & -\frac{1}{2} & -\frac{1}{2} \\ \frac{1}{2} & \frac{1}{2} & -\frac{1}{2} & -\frac{1}{2} \\ -\frac{1}{2} & -\frac{1}{2} & \frac{1}{2} & \frac{1}{2} \\ -\frac{1}{2} & -\frac{1}{2} & \frac{1}{2} & \frac{1}{2} \end{bmatrix} \end{array} \tag{3.8.1}$$

Similarly, for elements 2–5, we obtain

$$\underline{k}^{(2)} = \frac{\sqrt{2}AE}{\sqrt{2}L} \begin{array}{c} \begin{matrix} d_{1x} & d_{1y} & d_{3x} & d_{3y} \end{matrix} \\ \begin{bmatrix} \frac{1}{2} & -\frac{1}{2} & -\frac{1}{2} & \frac{1}{2} \\ -\frac{1}{2} & \frac{1}{2} & \frac{1}{2} & -\frac{1}{2} \\ -\frac{1}{2} & \frac{1}{2} & \frac{1}{2} & -\frac{1}{2} \\ \frac{1}{2} & -\frac{1}{2} & -\frac{1}{2} & \frac{1}{2} \end{bmatrix} \end{array} \tag{3.8.2}$$

$$\underline{k}^{(3)} = \frac{AE}{L} \begin{array}{c} \begin{matrix} d_{1x} & d_{1y} & d_{4x} & d_{4y} \end{matrix} \\ \begin{bmatrix} 1 & 0 & -1 & 0 \\ 0 & 0 & 0 & 0 \\ -1 & 0 & 1 & 0 \\ 0 & 0 & 0 & 0 \end{bmatrix} \end{array} \tag{3.8.3}$$

$$\underline{k}^{(4)} = \frac{AE}{L} \begin{array}{c} \begin{matrix} d_{4x} & d_{4y} & d_{2x} & d_{2y} \end{matrix} \\ \begin{bmatrix} 0 & 0 & 0 & 0 \\ 0 & \frac{1}{2} & 0 & -\frac{1}{2} \\ 0 & 0 & 0 & 0 \\ 0 & -\frac{1}{2} & 0 & \frac{1}{2} \end{bmatrix} \end{array} \tag{3.8.4}$$

$$\underline{k}^{(5)} = \frac{AE}{L} \begin{array}{c} \begin{matrix} d_{3x} & d_{3y} & d_{4x} & d_{4y} \end{matrix} \\ \begin{bmatrix} 0 & 0 & 0 & 0 \\ 0 & \frac{1}{2} & 0 & -\frac{1}{2} \\ 0 & 0 & 0 & 0 \\ 0 & -\frac{1}{2} & 0 & \frac{1}{2} \end{bmatrix} \end{array} \tag{3.8.5}$$

where, in Eqs. (3.8.1)–(3.8.5), the column labels indicate the degrees of freedom associated with each element. Also, because elements 4 and 5 lie in the plane of symmetry, half of their original areas have been used in Eqs. (3.8.4) and (3.8.5).

We will limit the solution to determining the displacement components. Therefore, considering the boundary constraints that result in zero-displacement components, we can immediately obtain the reduced set of equations by eliminating rows and columns in each element stiffness matrix corresponding to a zero-displacement component. That is, because $d_{1x} = 0$ and $d_{1y} = 0$ (owing to the pin support at node 1 in Figure 3–21) and $d_{2x} = 0, d_{3x} = 0$, and $d_{4x} = 0$ (owing to the symmetry condition), we can cancel rows and columns corresponding to these displacement components in each element stiffness matrix before assembling the total stiffness matrix. The resulting set of stiffness equations is

$$\frac{AE}{L}\begin{bmatrix} 1 & 0 & -\frac{1}{2} \\ 0 & 1 & -\frac{1}{2} \\ -\frac{1}{2} & -\frac{1}{2} & 1 \end{bmatrix}\begin{Bmatrix} d_{2y} \\ d_{3y} \\ d_{4y} \end{Bmatrix} = \begin{Bmatrix} 0 \\ 0 \\ -P \end{Bmatrix} \tag{3.8.6}$$

On solving Eq. (3.8.6) for the displacements, we obtain

$$d_{2y} = \frac{-PL}{AE} \qquad d_{3y} = \frac{-PL}{AE} \qquad d_{4y} = \frac{-2PL}{AE} \tag{3.8.7}$$

■

The ideas presented regarding the use of symmetry should be used sparingly and cautiously in problems of vibration and buckling. For instance, a structure such as a simply supported beam has symmetry about its center but has antisymmetric vibration modes as well as symmetric vibration modes. This will be shown in Chapter 16. If only half the beam were modeled using reflective symmetry conditions, the support conditions would permit only the symmetric vibration modes.

3.9 Inclined, or Skewed, Supports

In the preceding sections, the supports were oriented such that the resulting boundary conditions on the displacements were in the global directions.

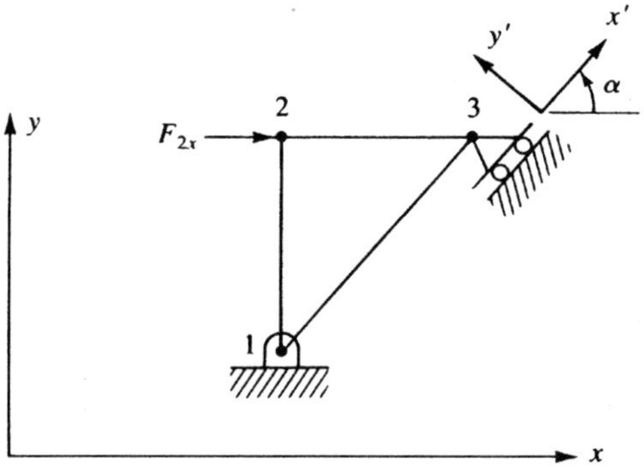

Figure 3–22 Plane truss with inclined boundary conditions at node 3

However, if a support is inclined, or skewed, at an angle α from the global x axis, as shown at node 3 in the plane truss of Figure 3–22, the resulting boundary conditions on the displacements are not in the global x-y directions but are in the local x'-y' directions. We will now describe two methods used to handle inclined supports.

In the first method, to account for inclined boundary conditions, we must perform a transformation of the global displacements at node 3 only into the local nodal coordinate system x'-y', while keeping all other displacements in the x-y global system. We can then enforce the zero-displacement boundary condition d'_{3y} in the force/displacement equations and, finally, solve the equations in the usual manner.

The transformation used is analogous to that for transforming a vector from local to global coordinates. For the plane truss, we use Eq. (3.3.16) applied to node 3 as follows:

$$\begin{Bmatrix} d'_{3x} \\ d'_{3y} \end{Bmatrix} = \begin{bmatrix} \cos\alpha & \sin\alpha \\ -\sin\alpha & \cos\alpha \end{bmatrix} \begin{Bmatrix} d_{3x} \\ d_{3y} \end{Bmatrix} \tag{3.9.1}$$

Rewriting Eq. (3.9.1), we have

$$\{d'_3\} = [t_3]\{d_3\} \tag{3.9.2}$$

where

$$[t_3] = \begin{bmatrix} \cos\alpha & \sin\alpha \\ -\sin\alpha & \cos\alpha \end{bmatrix} \tag{3.9.3}$$

We now write the transformation for the entire nodal displacement vector as

$$\{d'\} = [T_1]\{d\} \tag{3.9.4}$$

or

$$\{d\} = [T_1]^T\{d'\} \tag{3.9.5}$$

where the transformation matrix for the entire truss is the 6×6 matrix

$$[T_1] = \begin{bmatrix} [I] & [0] & [0] \\ [0] & [I] & [0] \\ [0] & [0] & [t_3] \end{bmatrix} \tag{3.9.6}$$

Each submatrix in Eq. (3.9.6) (the identity matrix $[I]$, the null matrix $[0]$, and matrix $[t_3]$ has the same 2×2 order, that order in general being equal to the number of degrees of freedom at each node.

To obtain the desired displacement vector with global displacement components at nodes 1 and 2 and local displacement components at node 3, we use Eq. (3.9.5) to obtain

$$\begin{Bmatrix} d_{1x} \\ d_{1y} \\ d_{2x} \\ d_{2y} \\ d_{3x} \\ d_{3y} \end{Bmatrix} = \begin{bmatrix} [I] & [0] & [0] \\ [0] & [I] & [0] \\ [0] & [0] & [t_3]^T \end{bmatrix} \begin{Bmatrix} d'_{1x} \\ d'_{1y} \\ d'_{2x} \\ d'_{2y} \\ d'_{3x} \\ d'_{3y} \end{Bmatrix} \tag{3.9.7}$$

In Eq. (3.9.7), we observe that only the node 3 global components are transformed, as indicated by the placement of the $[t_3]^T$ matrix. We denote the square matrix in Eq. (3.9.7) by $[T_1]^T$. In general, we place a 2×2 $[t]$ matrix in $[T_1]$ wherever the transformation from global to local displacements is needed (where skewed supports exist).

Upon considering Eqs. (3.9.5) and (3.9.6), we observe that only node 3 components of $\{d\}$ are really transformed to local (skewed) axes components. This transformation is indeed necessary whenever the local axes x'-y' fixity directions are known.

Furthermore, the global force vector can also be transformed by using the same transformation as for $\{d'\}$:

$$\{f'\} = [T_1]\{f\} \tag{3.9.8}$$

In global coordinates, we then have

$$\{f\} = [K]\{d\} \tag{3.9.9}$$

Premultiplying Eq. (3.9.9) by $[T_1]$, we have

$$[T_1]\{f\} = [T_1][K]\{d\} \tag{3.9.10}$$

For the truss in Figure 3–22, the left side of Eq. (3.9.10) is

$$\begin{bmatrix} [I] & [0] & [0] \\ [0] & [I] & [0] \\ [0] & [0] & [t_3] \end{bmatrix} \begin{Bmatrix} f_{1x} \\ f_{1y} \\ f_{2x} \\ f_{2y} \\ f_{3x} \\ f_{3y} \end{Bmatrix} = \begin{Bmatrix} f_{1x} \\ f_{1y} \\ f_{2x} \\ f_{2y} \\ f'_{3x} \\ f'_{3y} \end{Bmatrix} \tag{3.9.11}$$

where the fact that local forces transform similarly to Eq. (3.9.2) as

$$\{f'_3\} = [t_3]\{f_3\} \tag{3.9.12}$$

has been used in Eq. (3.9.11). From Eq. (3.9.11), we see that only the node 3 components of $\{f\}$ have been transformed to the local axes components, as desired.

Using Eq. (3.9.5) in Eq. (3.9.10), we have

$$[T_1]\{f\} = [T_1][K][T_1]^T\{d'\} \tag{3.9.13}$$

Using Eq. (3.9.11), we find that the form of Eq. (3.9.13) becomes

$$\begin{Bmatrix} F_{1x} \\ F_{1y} \\ F_{2x} \\ F_{2y} \\ F'_{3x} \\ F'_{3y} \end{Bmatrix} = [T_1][K][T_1]^T \begin{Bmatrix} d_{1x} \\ d_{1y} \\ d_{2x} \\ d_{2y} \\ d'_{3x} \\ d'_{3y} \end{Bmatrix} \tag{3.9.14}$$

as $d_{1x} = d'_{1x}, d_{1y} = d'_{1y}, d_{2x} = d'_{2x}$, and $d_{2y} = d'_{2y}$ from Eq. (3.9.7). Equation (3.9.14) is the desired form that allows all known global and inclined boundary conditions to be enforced. The global forces now result in the left side of Eq. (3.9.14). To solve Eq. (3.9.14), first perform the matrix triple product $[T_1][K][T_1]^T$. Then invoke the following boundary conditions (for the truss in Figure 3–22):

$$d_{1x} = 0 \qquad d_{1y} = 0 \qquad d'_{3y} = 0 \tag{3.9.15}$$

Then substitute the known value of the applied force F_{2x} along with $F_{2y} = 0$ and $F'_{3x} = 0$ into Eq. (3.9.14). Finally, partition the equations with known displacements—here equations 1, 2, and 6 of Eq. (3.9.14)—and then simultaneously solve those associated with the unknown displacements d_{2x}, d_{2y}, and d'_{3x}.

After solving for the displacements, return to Eq. (3.9.14) to obtain the global reactions F_{1x} and F_{1y} and the inclined roller reaction F'_{3y}.

Example 3.11

For the plane truss shown in Figure 3–23, determine the displacements and reactions. Let $E = 210$ GPa, $A = 6.00 \times 10^{-4}$ m^2 for elements 1 and 2, and $A = 6\sqrt{2} \times 10^{-4}$ m^2 for element 3.

We begin by using Eq. (3.4.23) to determine each element stiffness matrix.

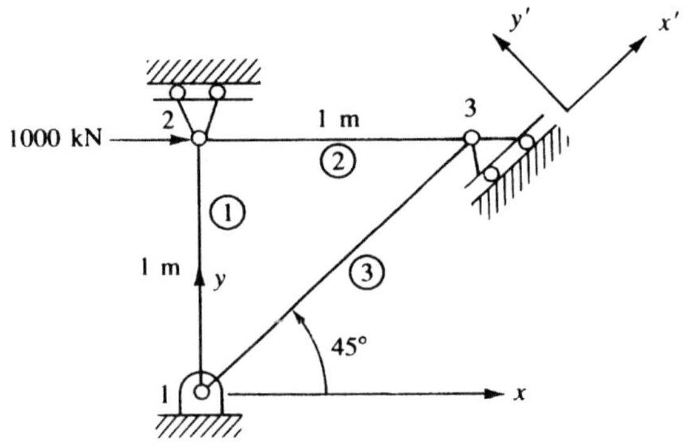

Figure 3–23 Plane truss with inclined support

Element 1

$$\cos\theta = 0 \qquad \sin\theta = 1$$

$$\underline{k}^{(1)} = \frac{(6.0 \times 10^{-4}\ \text{m}^2)(210 \times 10^9\ \text{N/m}^2)}{1\ \text{m}} \begin{array}{c} \begin{matrix} d_{1x} & d_{1y} & d_{2x} & d_{2y} \end{matrix} \\ \begin{bmatrix} 0 & 0 & 0 & 0 \\ & 1 & 0 & -1 \\ & & 0 & 0 \\ \text{Symmetry} & & & 1 \end{bmatrix} \end{array} \tag{3.9.16}$$

Element 2

$$\cos\theta = 1 \qquad \sin\theta = 0$$

$$\underline{k}^{(2)} = \frac{(6.0 \times 10^{-4}\ \text{m}^2)(210 \times 10^9\ \text{N/m}^2)}{1\ \text{m}} \begin{array}{c} \begin{matrix} d_{2x} & d_{2y} & d_{3x} & d_{3y} \end{matrix} \\ \begin{bmatrix} 1 & 0 & -1 & 0 \\ & 0 & 0 & 0 \\ & & 1 & 0 \\ \text{Symmetry} & & & 0 \end{bmatrix} \end{array} \tag{3.9.17}$$

Element 3

$$\cos\theta = \frac{\sqrt{2}}{2} \qquad \sin\theta = \frac{\sqrt{2}}{2}$$

$$\underline{k}^{(3)} = \frac{(6\sqrt{2} \times 10^{-4}\ \text{m}^2)(210 \times 10^9\ \text{N/m}^2)}{\sqrt{2}\ \text{m}} \begin{array}{c} \begin{matrix} d_{1x} & d_{1y} & d_{3x} & d_{3y} \end{matrix} \\ \begin{bmatrix} 0.5 & 0.5 & -0.5 & -0.5 \\ & 0.5 & -0.5 & -0.5 \\ & & 0.5 & 0.5 \\ \text{Symmetry} & & & 0.5 \end{bmatrix} \end{array} \tag{3.9.18}$$

Using the direct stiffness method on Eqs. (3.9.16)–(3.9.18), we obtain the global $\underline{K}$ matrix as

$$\underline{K} = 1260 \times 10^5\ \text{N/m} \begin{bmatrix} 0.5 & 0.5 & 0 & 0 & -0.5 & -0.5 \\ & 1.5 & 0 & -1 & -0.5 & -0.5 \\ & & 1 & 0 & -1 & 0 \\ & & & 1 & 0 & 0 \\ & & & & 1.5 & 0.5 \\ \text{Symmetry} & & & & & 0.5 \end{bmatrix} \tag{3.9.19}$$

Next we obtain the transformation matrix $\underline{T}_1$ using Eq. (3.9.6) to transform the global displacements at node 3 into local nodal coordinates x'-y'. In using Eq. (3.9.6), the angle α is 45°.

$$[T_1] = \begin{bmatrix} 1 & 0 & 0 & 0 & 0 & 0 \\ 0 & 1 & 0 & 0 & 0 & 0 \\ 0 & 0 & 1 & 0 & 0 & 0 \\ 0 & 0 & 0 & 1 & 0 & 0 \\ 0 & 0 & 0 & 0 & \sqrt{2}/2 & \sqrt{2}/2 \\ 0 & 0 & 0 & 0 & -\sqrt{2}/2 & \sqrt{2}/2 \end{bmatrix} \tag{3.9.20}$$

Next we use Eq. (3.9.14) (in general, we would use Eq. (3.9.13)) to express the assembled equations. First define $\underline{K}^* = \underline{T}_1\underline{K}\underline{T}_1^T$ and evaluate in steps as follows:

$$\underline{T}_1\underline{K} = 1260 \times 10^5 \begin{bmatrix} 0.5 & 0.5 & 0 & 0 & -0.5 & -0.5 \\ 0.5 & 1.5 & 0 & -1 & -0.5 & -0.5 \\ 0 & 0 & 1 & 0 & -1 & 0 \\ 0 & -1 & 0 & 1 & 0 & 0 \\ -0.707 & -0.707 & -0.707 & 0 & 1.414 & 0.707 \\ 0 & 0 & 0.707 & 0 & -0.707 & 0 \end{bmatrix} \tag{3.9.21}$$

and

$$\underline{T}_1\underline{K}\underline{T}_1^T = 1260 \times 10^5 \text{ N/m} \begin{bmatrix} 0.5 & 0.5 & 0 & 0 & -0.707 & 0 \\ 0.5 & 1.5 & 0 & -1 & -0.707 & 0 \\ 0 & 0 & 1 & 0 & -0.707 & 0.707 \\ 0 & -1 & 0 & 1 & 0 & 0 \\ -0.707 & -0.707 & -0.707 & 0 & 1.500 & -0.500 \\ 0 & 0 & 0.707 & 0 & -0.500 & 0.500 \end{bmatrix}$$

(columns: d_{1x}, d_{1y}, d_{2x}, d_{2y}, d'_{3x}, d'_{3y})

(3.9.22)

Applying the boundary conditions, $d_{1x} = d_{1y} = d_{2y} = d'_{3y} = 0$, to Eq. (3.9.22), we obtain

$$\begin{Bmatrix} F_{2x} = 1000 \text{ kN} \\ F'_{3x} = 0 \end{Bmatrix} = (126 \times 10^3 \text{ kN/m}) \begin{bmatrix} 1 & -0.707 \\ -0.707 & 1.50 \end{bmatrix} \begin{Bmatrix} d_{2x} \\ d'_{3x} \end{Bmatrix} \tag{3.9.23}$$

Solving Eq. (3.9.23) for the displacements yields

$$d_{2x} = 11.91 \times 10^{-3} \text{ m} \tag{3.9.24}$$

$$d'_{3x} = 5.613 \times 10^{-3} \text{ m}$$

Postmultiplying the known displacement vector times Eq. (3.9.22) (see Eq. (3.9.14), we obtain the reactions as

$$\begin{aligned} F_{1x} &= -500 \text{ kN} \\ F_{1y} &= -500 \text{ kN} \\ F_{2y} &= 0 \\ F'_{3y} &= 707 \text{ kN} \end{aligned} \tag{3.9.25}$$

The free-body diagram of the truss with the reactions is shown in Figure 3–24. You can easily verify that the truss is in equilibrium. ■

In the second method used to handle skewed boundary conditions, we use a boundary element of large stiffness to constrain the desired displacement. This is the method used in some computer programs [9].

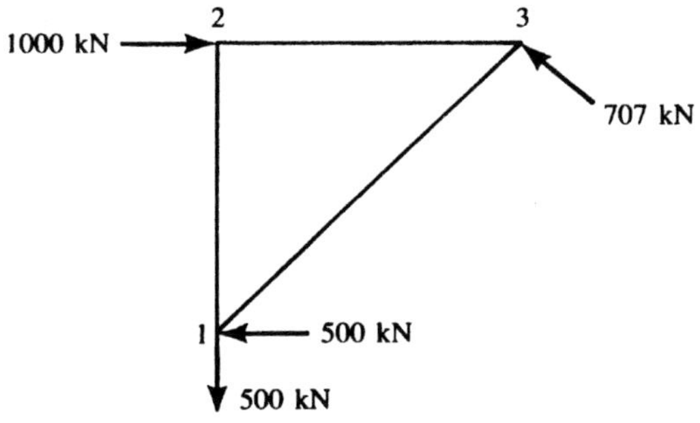

Figure 3–24 Free-body diagram of the truss of Figure 3–23

Boundary elements are used to specify nonzero displacements and rotations to nodes. They are also used to evaluate reactions at rigid and flexible supports. Boundary elements are two-node elements. The line defined by the two nodes specifies the direction along which the force reaction is evaluated or the displacement is specified. In the case of moment reaction, the line specifies the axis about which the moment is evaluated and the rotation is specified.

We consider boundary elements that are used to obtain reaction forces (rigid boundary elements) or specify translational displacements (displacement boundary elements) as truss elements with only one nonzero translational stiffness. Boundary elements used to either evaluate reaction moments or specify rotations behave like beam elements with only one nonzero stiffness corresponding to the rotational stiffness about the specified axis.

The elastic boundary elements are used to model flexible supports and to calculate reactions at skewed or inclined boundaries. Consult Reference [9] for more details about using boundary elements.

3.10 Potential Energy Approach to Derive Bar Element Equations

We now present the principle of minimum potential energy to derive the bar element equations. Recall from Section 2.6 that the total potential energy π_p was defined as the sum of the internal strain energy U and the potential energy of the external forces Ω:

$$\pi_p = U + \Omega \tag{3.10.1}$$

To evaluate the strain energy for a bar, we consider only the work done by the internal forces during deformation. Because we are dealing with a one-dimensional bar, the internal force doing work is given in Figure 3–25 as $\sigma_x(\Delta y)(\Delta z)$, due only to normal stress σ_x. The displacement of the x face of the element is $\Delta x(\varepsilon_x)$; the displacement of the $x + \Delta x$ face is $\Delta x(\varepsilon_x + d\varepsilon_x)$. The change in displacement is then

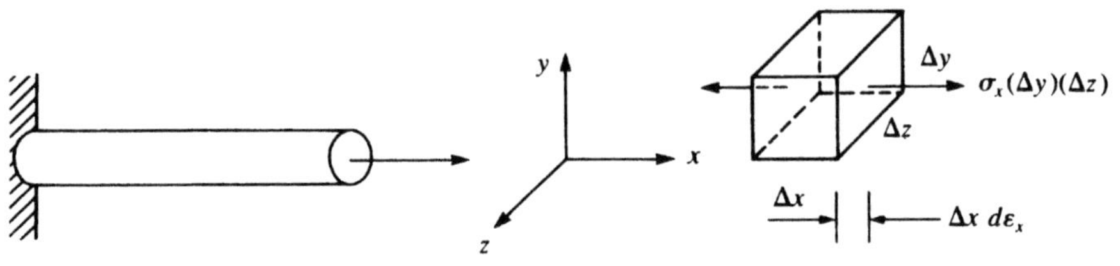

Figure 3–25 Internal force in a one-dimensional bar

$\Delta x\, d\varepsilon_x$, where $d\varepsilon_x$ is the differential change in strain occurring over length Δx. The differential internal work (or strain energy) dU is the internal force multiplied by the displacement through which the force moves, given by

$$dU = \sigma_x(\Delta y)(\Delta z)(\Delta x)\, d\varepsilon_x \tag{3.10.2}$$

Rearranging and letting the volume of the element approach zero, we obtain, from Eq. (3.10.2),

$$dU = \sigma_x\, d\varepsilon_x\, dV \tag{3.10.3}$$

For the whole bar, we then have

$$U = \iiint_V \left\{ \int_0^{\varepsilon_x} \sigma_x\, d\varepsilon_x \right\} dV \tag{3.10.4}$$

Now, for a linear-elastic (Hooke's law) material as shown in Figure 3–26, we see that $\sigma_x = E\varepsilon_x$. Hence substituting this relationship into Eq. (3.10.4), integrating with respect to ε_x, and then resubstituting σ_x for $E\varepsilon_x$, we have

$$U = \frac{1}{2}\iiint_V \sigma_x \varepsilon_x\, dV \tag{3.10.5}$$

as the expression for the strain energy for one-dimensional stress.

The potential energy of the external forces, being opposite in sign from the external work expression because the potential energy of external forces is lost when the

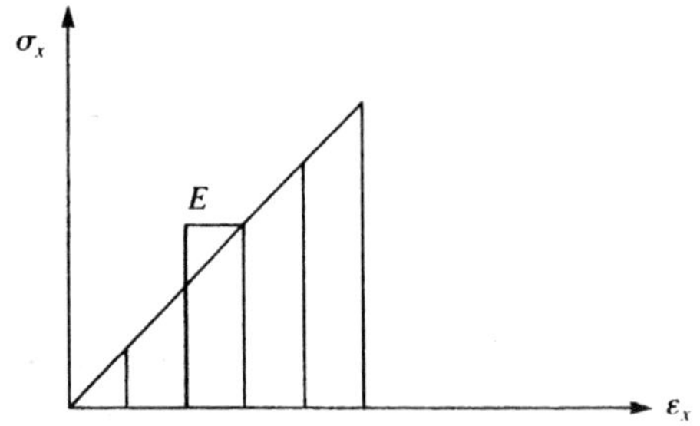

Figure 3–26 Linear-elastic (Hooke's law) material

work is done by the external forces, is given by

$$\Omega = -\iiint_V \hat{X}_b \hat{u}\, dV - \iint_{S_1} \hat{T}_x \hat{u}_s\, dS - \sum_{i=1}^{M} \hat{f}_{ix} \hat{d}_{ix} \tag{3.10.6}$$

where the first, second, and third terms on the right side of Eq. (3.10.6) represent the potential energy of (1) body forces $\hat{X}_b$, typically from the self-weight of the bar (in units of force per unit volume) moving through displacement function $\hat{u}$, (2) surface loading or traction $\hat{T}_x$, typically from distributed loading acting along the surface of the element (in units of force per unit surface area) moving through displacements $\hat{u}_s$, where $\hat{u}_s$ are the displacements occurring over surface S_1, and (3) nodal concentrated forces $\hat{f}_{ix}$ moving through nodal displacements $\hat{d}_{ix}$. The forces $\hat{X}_b$, $\hat{T}_x$, and $\hat{f}_{ix}$ are considered to act in the local $\hat{x}$ direction of the bar as shown in Figure 3–27. In Eqs. (3.10.5) and (3.10.6), V is the volume of the body and S_1 is the part of the surface S on which surface loading acts. For a bar element with two nodes and one degree of freedom per node, $M = 2$.

We are now ready to describe the finite element formulation of the bar element equations by using the principle of minimum potential energy.

The finite element process seeks a minimum in the potential energy within the constraint of an assumed displacement pattern within each element. The greater the number of degrees of freedom associated with the element (usually meaning increasing the number of nodes), the more closely will the solution approximate the true one and ensure complete equilibrium (provided the true displacement can, in the limit, be approximated). An approximate finite element solution found by using the stiffness method will always provide an approximate value of potential energy greater than or equal to the correct one. This method also results in a structure behavior that is predicted to be physically stiffer than, or at best to have the same stiffness as, the actual one. This is explained by the fact that the structure model is allowed to displace only into shapes defined by the terms of the assumed displacement field within each element of the structure. The correct shape is usually only approximated by the assumed field, although the correct shape can be the same as the assumed field. The assumed field effectively constrains the structure from deforming in its natural manner. This constraint effect stiffens the predicted behavior of the structure.

Apply the following steps when using the principle of minimum potential energy to derive the finite element equations.

1. Formulate an expression for the total potential energy.
2. Assume the displacement pattern to vary with a finite set of undetermined parameters (here these are the nodal displacements d_{ix}), which are substituted into the expression for total potential energy.
3. Obtain a set of simultaneous equations minimizing the total potential energy with respect to these nodal parameters. These resulting equations represent the element equations.

The resulting equations are the approximate (or possibly exact) equilibrium equations whose solution for the nodal parameters seeks to minimize the potential energy when back-substituted into the potential energy expression. The preceding

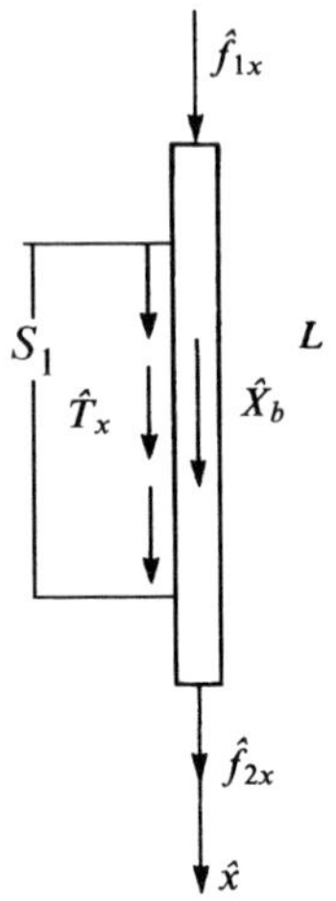

Figure 3–27 General forces acting on a one-dimensional bar

three steps will now be followed to derive the bar element equations and stiffness matrix.

Consider the bar element of length L, with constant cross-sectional area A, shown in Figure 3–27. Using Eqs. (3.10.5) and (3.10.6), we find that the total potential energy, Eq. (3.10.1), becomes

$$\pi_p = \frac{A}{2}\int_0^L \sigma_x \varepsilon_x \, d\hat{x} - \hat{f}_{1x}\hat{d}_{1x} - \hat{f}_{2x}\hat{d}_{2x} - \iint_{S_1} \hat{u}_s \hat{T}_x \, dS - \iiint_V \hat{u}\hat{X}_b \, dV \tag{3.10.7}$$

because A is a constant and variables σ_x and ε_x at most vary with $\hat{x}$.

From Eqs. (3.1.3) and (3.1.4), we have the axial displacement function expressed in terms of the shape functions and nodal displacements by

$$\hat{u} = [N]\{\hat{d}\} \qquad \hat{u}_s = [N_S]\{\hat{d}\} \tag{3.10.8}$$

where

$$[N] = \left[1 - \frac{\hat{x}}{L} \quad \frac{\hat{x}}{L}\right] \tag{3.10.9}$$

$[N_S]$ is the shape function matrix evaluated over the surface that the distributed surface traction acts and

$$\{\hat{d}\} = \begin{Bmatrix} \hat{d}_{1x} \\ \hat{d}_{2x} \end{Bmatrix} \tag{3.10.10}$$

Then, using the strain/displacement relationship $\varepsilon_x = d\hat{u}/d\hat{x}$, we can write the axial strain as

$$\{\varepsilon_x\} = \left[-\frac{1}{L} \quad \frac{1}{L}\right]\{\hat{d}\} \tag{3.10.11}$$

or

$$\{\varepsilon_x\} = [B]\{\hat{d}\} \tag{3.10.12}$$

where we define

$$[B] = \begin{bmatrix} -\dfrac{1}{L} & \dfrac{1}{L} \end{bmatrix} \tag{3.10.13}$$

The axial stress/strain relationship is given by

$$\{\sigma_x\} = [D]\{\varepsilon_x\} \tag{3.10.14}$$

where

$$[D] = [E] \tag{3.10.15}$$

for the one-dimensional stress/strain relationship and E is the modulus of elasticity. Now, by Eq. (3.10.12), we can express Eq. (3.10.14) as

$$\{\sigma_x\} = [D][B]\{\hat{d}\} \tag{3.10.16}$$

Using Eq. (3.10.7) expressed in matrix notation form, we have the total potential energy given by

$$\pi_p = \frac{A}{2}\int_0^L \{\sigma_x\}^T\{\varepsilon_x\}\,d\hat{x} - \{\hat{d}\}^T\{P\} - \iint_{S_1}\{\hat{u}_s\}^T\{\hat{T}_x\}\,dS - \iiint_V \{\hat{u}\}^T\{\hat{X}_b\}\,dV \tag{3.10.17}$$

where $\{P\}$ now represents the concentrated nodal loads and where in general both $\underline{\sigma}_x$ and $\underline{\varepsilon}_x$ are column matrices. For proper matrix multiplication, we must place the transpose on $\{\sigma_x\}$. Similarly, $\{\hat{u}\}$ and $\{\hat{T}_x\}$ in general are column matrices, so for proper matrix multiplication, $\{\hat{u}\}$ is transposed in Eq. (3.10.17).

Using Eqs. (3.10.8), (3.10.12), and (3.10.16) in Eq. (3.10.17), we obtain

$$\begin{aligned}\pi_p = {} & \frac{A}{2}\int_0^L \{\hat{d}\}^T[B]^T[D]^T[B]\{\hat{d}\}\,d\hat{x} - \{\hat{d}\}^T\{P\} \\ & - \iint_{S_1}\{\hat{d}\}^T[N_S]^T\{\hat{T}_x\}\,dS - \iiint_V \{\hat{d}\}^T[N]^T\{\hat{X}_b\}\,dV\end{aligned} \tag{3.10.18}$$

In Eq. (3.10.18), π_p is seen to be a function of $\{\hat{d}\}$; that is, $\pi_p = \pi_p(\hat{d}_{1x}, \hat{d}_{2x})$. However, $[B]$ and $[D]$, Eqs. (3.10.13) and (3.10.15), and the nodal degrees of freedom $\hat{d}_{1x}$ and $\hat{d}_{2x}$ are not functions of $\hat{x}$. Therefore, integrating Eq. (3.10.18) with respect to $\hat{x}$ yields

$$\pi_p = \frac{AL}{2}\{\hat{d}\}^T[B]^T[D]^T[B]\{\hat{d}\} - \{\hat{d}\}^T\{\hat{f}\} \tag{3.10.19}$$

where

$$\{\hat{f}\} = \{P\} + \iint_{S_1}[N_S]^T\{\hat{T}_x\}\,dS + \iiint_V [N]^T\{\hat{X}_b\}\,dV \tag{3.10.20}$$

From Eq. (3.10.20), we observe three separate types of load contributions from concentrated nodal forces, surface tractions, and body forces, respectively. We define these surface tractions and body-force matrices as

$$\{\hat{f}_s\} = \iint_{S_1} [N_S]^T \{\hat{T}_x\}\, dS \tag{3.10.20a}$$

$$\{\hat{f}_b\} = \iiint_V [N]^T \{\hat{X}_b\}\, dV \tag{3.10.20b}$$

The expression for $\{\hat{f}\}$ given by Eq. (3.10.20) then describes how certain loads can be considered to best advantage.

Loads calculated by Eqs. (3.10.20a) and (3.10.20b) are called consistent because they are based on the same shape functions $[N]$ used to calculate the element stiffness matrix. The loads calculated by Eq. (3.10.20a) and (3.10.20b) are also statically equivalent to the original loading; that is, both $\{\hat{f}_s\}$ and $\{\hat{f}_b\}$ and the original loads yield the same resultant force and same moment about an arbitrarily chosen point.

The minimization of π_p with respect to each nodal displacement requires that

$$\frac{\partial \pi_p}{\partial \hat{d}_{1x}} = 0 \qquad \text{and} \qquad \frac{\partial \pi_p}{\partial \hat{d}_{2x}} = 0 \tag{3.10.21}$$

Now we explicitly evaluate π_p given by Eq. (3.10.19) to apply Eq. (3.10.21). We define the following for convenience:

$$\{U^*\} = \{\hat{d}\}^T [B]^T [D]^T [B] \{\hat{d}\} \tag{3.10.22}$$

Using Eqs. (3.10.10), (3.10.13), and (3.10.15) in Eq. (3.10.22) yields

$$\{U^*\} = [\hat{d}_{1x} \quad \hat{d}_{2x}] \begin{Bmatrix} -\frac{1}{L} \\ \frac{1}{L} \end{Bmatrix} [E] \begin{bmatrix} -\frac{1}{L} & \frac{1}{L} \end{bmatrix} \begin{Bmatrix} \hat{d}_{1x} \\ \hat{d}_{2x} \end{Bmatrix} \tag{3.10.23}$$

Simplifying Eq. (3.10.23), we obtain

$$U^* = \frac{E}{L^2} (\hat{d}_{1x}^2 - 2\hat{d}_{1x}\hat{d}_{2x} + \hat{d}_{2x}^2) \tag{3.10.24}$$

Also, the explicit expression for $\{\hat{d}\}^T \{\hat{f}\}$ is

$$\{\hat{d}\}^T \{\hat{f}\} = \hat{d}_{1x} \hat{f}_{1x} + \hat{d}_{2x} \hat{f}_{2x} \tag{3.10.25}$$

Therefore, using Eqs. (3.10.24) and (3.10.25) in Eq. (3.10.19) and then applying Eqs. (3.10.21), we obtain

$$\frac{\partial \pi_p}{\partial \hat{d}_{1x}} = \frac{AL}{2} \left[\frac{E}{L^2} (2\hat{d}_{1x} - 2\hat{d}_{2x}) \right] - \hat{f}_{1x} = 0 \tag{3.10.26}$$

and
$$\frac{\partial \pi_p}{\partial \hat{d}_{2x}} = \frac{AL}{2}\left[\frac{E}{L^2}(-2\hat{d}_{1x} + 2\hat{d}_{2x})\right] - \hat{f}_{2x} = 0$$

In matrix form, we express Eqs. (3.10.26) as

$$\frac{\partial \pi_p}{\partial \{\hat{d}\}} = \frac{AE}{L}\begin{bmatrix} 1 & -1 \\ -1 & 1 \end{bmatrix}\begin{Bmatrix} \hat{d}_{1x} \\ \hat{d}_{2x} \end{Bmatrix} - \begin{Bmatrix} \hat{f}_{1x} \\ \hat{f}_{2x} \end{Bmatrix} = \begin{Bmatrix} 0 \\ 0 \end{Bmatrix} \tag{3.10.27}$$

or, because $\{\hat{f}\} = [\hat{k}]\{\hat{d}\}$, we have the stiffness matrix for the bar element obtained from Eq. (3.10.27) as

$$[\hat{k}] = \frac{AE}{L}\begin{bmatrix} 1 & -1 \\ -1 & 1 \end{bmatrix} \tag{3.10.28}$$

As expected, Eq. (3.10.28) is identical to the stiffness matrix obtained in Section 3.1.

Finally, instead of the cumbersome process of explicitly evaluating π_p, we can use the matrix differentiation as given by Eq. (2.6.12) and apply it directly to Eq. (3.10.19) to obtain

$$\frac{\partial \pi_p}{\partial \{\hat{d}\}} = AL[B]^T[D][B]\{\hat{d}\} - \{\hat{f}\} = 0 \tag{3.10.29}$$

where $[D]^T = [D]$ has been used in writing Eq. (3.10.29). The result of the evaluation of $AL[B]^T[D][B]$ is then equal to $[\hat{k}]$ given by Eq. (3.10.28). Throughout this text, we will use this matrix differentiation concept (also see Appendix A), which greatly simplifies the task of evaluating $[\hat{k}]$.

To illustrate the use of Eq. (3.10.20a) to evaluate the equivalent nodal loads for a bar subjected to axial loading traction $\hat{T}_x$, we now solve Example 3.12.

Example 3.12

A bar of length L is subjected to a linearly distributed axial loading that varies from zero at node 1 to a maximum at node 2 (Figure 3–28). Determine the energy equivalent nodal loads.

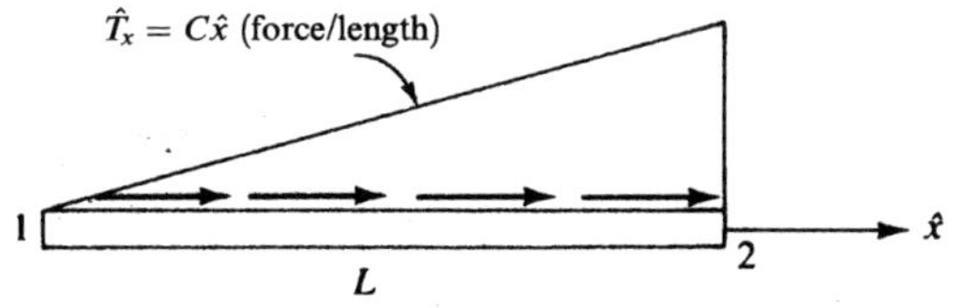

Figure 3–28 Element subjected to linearly varying axial load

Using Eq. (3.10.20a) and shape functions from Eq. (3.10.9), we solve for the energy equivalent nodal forces of the distributed loading as follows:

$$\{\hat{f}_0\} = \begin{Bmatrix} \hat{f}_{1x} \\ \hat{f}_{2x} \end{Bmatrix} = \iint_{S_1} [N]^T \{\hat{T}_x\}\, dS = \int_0^L \begin{Bmatrix} 1 - \dfrac{\hat{x}}{L} \\ \dfrac{\hat{x}}{L} \end{Bmatrix} \{C\hat{x}\}\, d\hat{x} \tag{3.10.30}$$

$$= \begin{Bmatrix} \dfrac{C\hat{x}^2}{2} - \dfrac{C\hat{x}^3}{3L} \\ \dfrac{C\hat{x}^3}{3L} \end{Bmatrix}_0^L$$

$$= \begin{Bmatrix} \dfrac{CL^2}{6} \\ \dfrac{CL^2}{3} \end{Bmatrix} \tag{3.10.31}$$

where the integration was carried out over the length of the bar, because $\hat{T}_x$ is in units of force/length.

Note that the total load is the area under the load distribution given by

$$F = \frac{1}{2}(L)(CL) = \frac{CL^2}{2} \tag{3.10.32}$$

Therefore, comparing Eq. (3.10.31) with (3.10.32), we find that the equivalent nodal loads for a linearly varying load are

$$\begin{aligned} \hat{f}_{1x} &= \frac{1}{3}F = \text{one-third of the total load} \\ \hat{f}_{2x} &= \frac{2}{3}F = \text{two-thirds of the total load} \end{aligned} \tag{3.10.33}$$

In summary, for the simple two-noded bar element subjected to a linearly varying load (triangular loading), place one-third of the total load at the node where the distributed loading begins (zero end of the load) and two-thirds of the total load at the node where the peak value of the distributed load ends. ■

We now illustrate (Example 3.13) a complete solution for a bar subjected to a surface traction loading.

Example 3.13

For the rod loaded axially as shown in Figure 3–29, determine the axial displacement and axial stress. Let $E = 30 \times 10^6$ psi, $A = 2$ in.^{2}, and $L = 60$ in. Use (a) one and (b) two elements in the finite element solutions. (In Section 3.11 one-, two-, four-, and eight-element solutions will be presented from the computer program Algor [9].

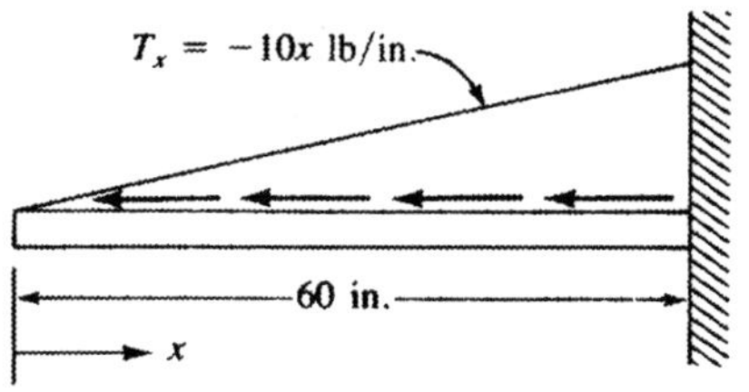

Figure 3–29 Rod subjected to triangular load distribution

(a) One-element solution (Figure 3–30).

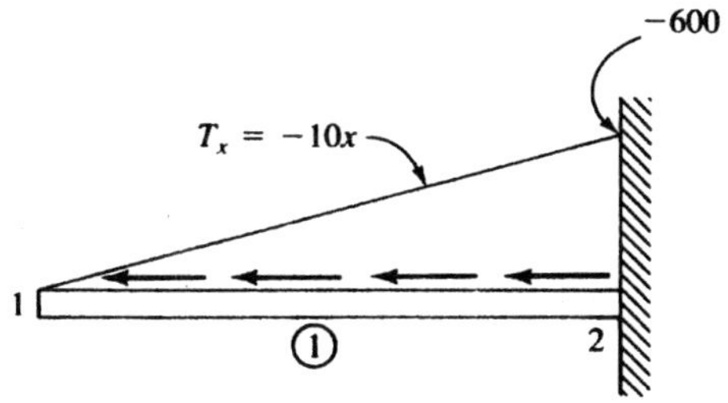

Figure 3–30 One-element model

From Eq. (3.10.20a), the distributed load matrix is evaluated as follows:

$$\{F_0\} = \int_0^L [N]^T \{T_x\}\, dx \tag{3.10.34}$$

where T_x is a line load in units of pounds per inch and $\hat{\underline{f}}_0 = \underline{F}_0$ as $\underline{x} = \hat{\underline{x}}$. Therefore, using Eq. (3.1.4) for $[N]$ in Eq. (3.10.34), we obtain

$$\{F_0\} = \int_0^L \begin{Bmatrix} 1 - \dfrac{x}{L} \\ \dfrac{x}{L} \end{Bmatrix} \{-10x\}\, dx \tag{3.10.35}$$

or

$$\begin{Bmatrix} F_{1x} \\ F_{2x} \end{Bmatrix} = \begin{Bmatrix} \dfrac{-10L^2}{2} + \dfrac{10L^2}{3} \\ \dfrac{-10L^2}{3} \end{Bmatrix} = \begin{Bmatrix} \dfrac{-10L^2}{6} \\ \dfrac{-10L^2}{3} \end{Bmatrix} = \begin{Bmatrix} \dfrac{-10(60)^2}{6} \\ \dfrac{-10(60)^2}{3} \end{Bmatrix}$$

or

$$F_{1x} = -6000 \text{ lb} \qquad F_{2x} = -12{,}000 \text{ lb} \tag{3.10.36}$$

Using Eq. (3.10.33), we could have determined the same forces at nodes 1 and 2—that is, one-third of the total load is at node 1 and two-thirds of the total load is at node 2.

Using Eq. (3.10.28), we find that the stiffness matrix is given by

$$k^{(1)} = 10^6 \begin{bmatrix} 1 & -1 \\ -1 & 1 \end{bmatrix}$$

The element equations are then

$$10^6 \begin{bmatrix} 1 & -1 \\ -1 & 1 \end{bmatrix} \begin{Bmatrix} d_{1x} \\ 0 \end{Bmatrix} = \begin{Bmatrix} -6000 \\ R_{2x} - 12{,}000 \end{Bmatrix} \tag{3.10.37}$$

Solving Eq. 1 of Eq. (3.10.37), we obtain

$$d_{1x} = -0.006 \text{ in.} \tag{3.10.38}$$

The stress is obtained from Eq. (3.10.14) as

$$\begin{aligned} \{\sigma_x\} &= [D]\{\varepsilon_x\} \\ &= E[B]\{d\} \\ &= E\begin{bmatrix} -\dfrac{1}{L} & \dfrac{1}{L} \end{bmatrix} \begin{Bmatrix} d_{1x} \\ d_{2x} \end{Bmatrix} \\ &= E\left(\frac{d_{2x} - d_{1x}}{L}\right) \\ &= 30 \times 10^6 \left(\frac{0 + 0.006}{60}\right) \\ &= 3000 \text{ psi } (T) \end{aligned} \tag{3.10.39}$$

(b) Two-element solution (Figure 3–31).

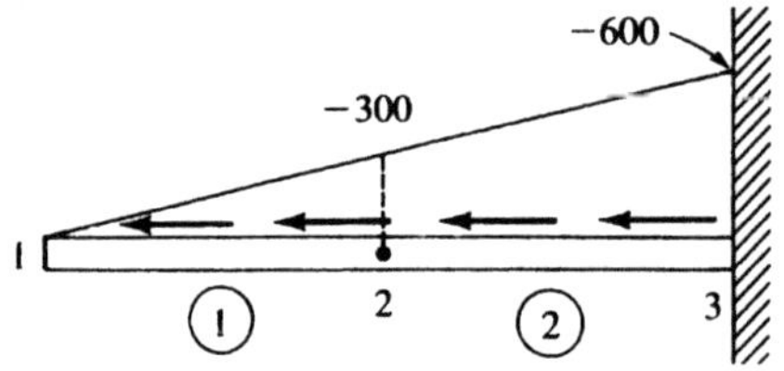

Figure 3–31 Two-element model

We first obtain the element forces. For element 2, we divide the load into a uniform part and a triangular part. For the uniform part, half the total uniform load is placed at each node associated with the element. Therefore, the total uniform part is

$$(30 \text{ in.})(-300 \text{ lb/in.}) = -9000 \text{ lb}$$

and using Eq. (3.10.33) for the triangular part of the load, we have, for element 2,

$$\begin{Bmatrix} f_{2x}^{(2)} \\ f_{3x}^{(2)} \end{Bmatrix} = \begin{Bmatrix} -[\frac{1}{2}(9000) + \frac{1}{3}(4500)] \\ -[\frac{1}{2}(9000) + \frac{2}{3}(4500)] \end{Bmatrix} = \begin{Bmatrix} -6000 \text{ lb} \\ -7500 \text{ lb} \end{Bmatrix} \tag{3.10.40}$$

For element 1, the total force is from the triangle-shaped distributed load only and is given by

$$\tfrac{1}{2}(30 \text{ in.})(-300 \text{ lb/in.}) = -4500 \text{ lb}$$

On the basis of Eq. (3.10.33), this load is separated into nodal forces as shown:

$$\begin{Bmatrix} f_{1x}^{(1)} \\ f_{2x}^{(1)} \end{Bmatrix} = \begin{Bmatrix} \frac{1}{3}(-4500) \\ \frac{2}{3}(-4500) \end{Bmatrix} = \begin{Bmatrix} -1500 \text{ lb} \\ -3000 \text{ lb} \end{Bmatrix} \tag{3.10.41}$$

The final nodal force matrix is then

$$\begin{Bmatrix} F_{1x} \\ F_{2x} \\ F_{3x} \end{Bmatrix} = \begin{Bmatrix} -1500 \\ -6000 - 3000 \\ R_{3x} - 7500 \end{Bmatrix} \tag{3.10.42}$$

The element stiffness matrices are now

$$\underline{k}^{(1)} = \underline{k}^{(2)} = \frac{AE}{L/2}\overset{\begin{matrix} 1 & 2 \\ 2 & 3 \end{matrix}}{\begin{bmatrix} 1 & -1 \\ -1 & 1 \end{bmatrix}} = (2 \times 10^6)\overset{\begin{matrix} 1 & 2 \\ 2 & 3 \end{matrix}}{\begin{bmatrix} 1 & -1 \\ -1 & 1 \end{bmatrix}} \tag{3.10.43}$$

The assembled global stiffness matrix is

$$\underline{K} = (2 \times 10^6)\begin{bmatrix} 1 & -1 & 0 \\ -1 & 2 & -1 \\ 0 & -1 & 1 \end{bmatrix} \frac{\text{lb}}{\text{in.}} \tag{3.10.44}$$

The assembled global equations are then

$$(2 \times 10^6)\begin{bmatrix} 1 & -1 & 0 \\ -1 & 2 & -1 \\ 0 & -1 & 1 \end{bmatrix}\begin{Bmatrix} d_{1x} \\ d_{2x} \\ d_{3x} = 0 \end{Bmatrix} = \begin{Bmatrix} -1500 \\ -9000 \\ R_{3x} - 7500 \end{Bmatrix} \tag{3.10.45}$$

where the boundary condition $d_{3x} = 0$ has been substituted into Eq. (3.10.45). Now, solving equations 1 and 2 of Eq. (3.10.45), we obtain

$$\begin{aligned} d_{1x} &= -0.006 \text{ in.} \\ d_{2x} &= -0.00525 \text{ in.} \end{aligned} \tag{3.10.46}$$

The element stresses are as follows:

Element 1

$$\begin{aligned} \sigma_x &= E\begin{bmatrix} -\frac{1}{30} & \frac{1}{30} \end{bmatrix}\begin{Bmatrix} d_{1x} = -0.006 \\ d_{2x} = -0.00525 \end{Bmatrix} \\ &= 750 \text{ psi } (T) \end{aligned} \tag{3.10.47}$$

Element 2

$$\sigma_x = E\left[-\frac{1}{30} \quad \frac{1}{30}\right]\begin{Bmatrix} d_{2x} = -0.00525 \\ d_{3x} = 0 \end{Bmatrix}$$

$$= 5250 \text{ psi } (T) \qquad (3.10.48)$$

■

3.11 Comparison of Finite Element Solution to Exact Solution for Bar

We will now compare the finite element solutions for Example 3.13 using one, two, four, and eight elements to model the bar element and the exact solution. The exact solution for displacement is obtained by solving the equation

$$u = \frac{1}{AE}\int_0^x P(x)\,dx \qquad (3.11.1)$$

where, using the following free-body diagram,

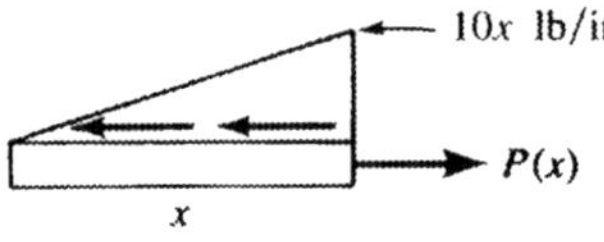

we have $$P(x) = \tfrac{1}{2}x(10x) = 5x^2 \text{ lb} \qquad (3.11.2)$$

Therefore, substituting Eq. (3.11.2) into Eq. (3.11.1), we have

$$u = \frac{1}{AE}\int_0^x 5x^2\,dx$$

$$= \frac{5x^3}{3AE} + C_1 \qquad (3.11.3)$$

Now, applying the boundary condition at $x = L$, we obtain

$$u(L) = 0 = \frac{5L^3}{3AE} + C_1$$

or $$C_1 = -\frac{5L^3}{3AE} \qquad (3.11.4)$$

Substituting Eq. (3.11.4) into Eq. (3.11.3) makes the final expression for displacement

$$u = \frac{5}{3AE}(x^3 - L^3) \qquad (3.11.5)$$

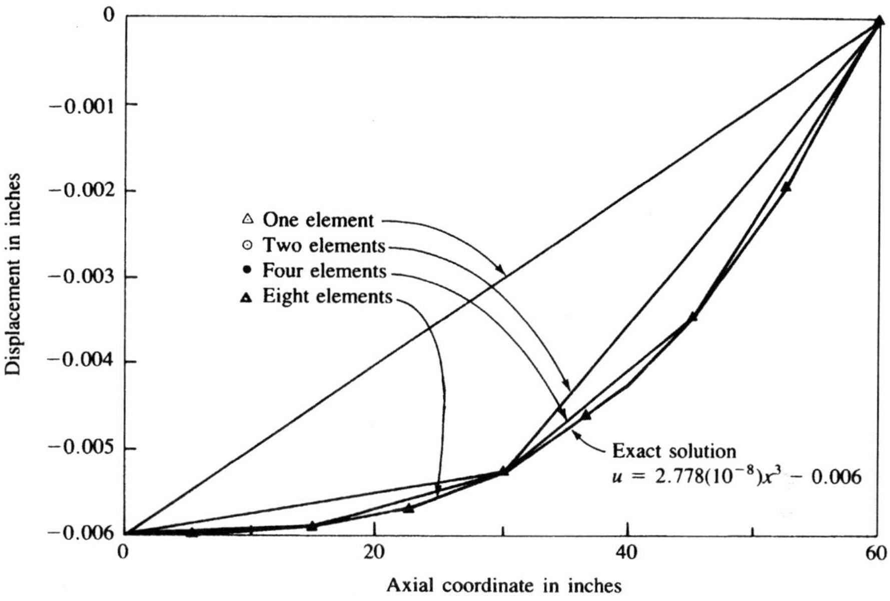

Figure 3–32 Comparison of exact and finite element solutions for axial displacement (along length of bar)

Substituting $A = 2$ in.^{2}, $E = 30 \times 10^6$ psi, and $L = 60$ in. into Eq. (3.11.5), we obtain

$$u = 2.778 \times 10^{-8} x^3 - 0.006 \tag{3.11.6}$$

The exact solution for axial stress is obtained by solving the equation

$$\sigma(x) = \frac{P(x)}{A} = \frac{5x^2}{2 \text{ in}^2} = 2.5x^2 \text{ psi} \tag{3.11.7}$$

Figure 3–32 shows a plot of Eq. (3.11.6) along with the finite element solutions (part of which were obtained in Example 3.13). Some conclusions from these results follow.

1. The finite element solutions match the exact solution at the node points. The reason why these nodal values are correct is that the element nodal forces were calculated on the basis of being energy-equivalent to the distributed load based on the assumed linear displacement field within each element. (For uniform cross-sectional bars and beams, the nodal degrees of freedom are exact. In general, computed nodal degrees of freedom are not exact.)
2. Although the node values for displacement match the exact solution, the values at locations between the nodes are poor using few elements (see one- and two-element solutions) because we used a linear displacement function within each element, whereas the exact solution, Eq. (3.11.6), is a cubic function. However, because we use increasing

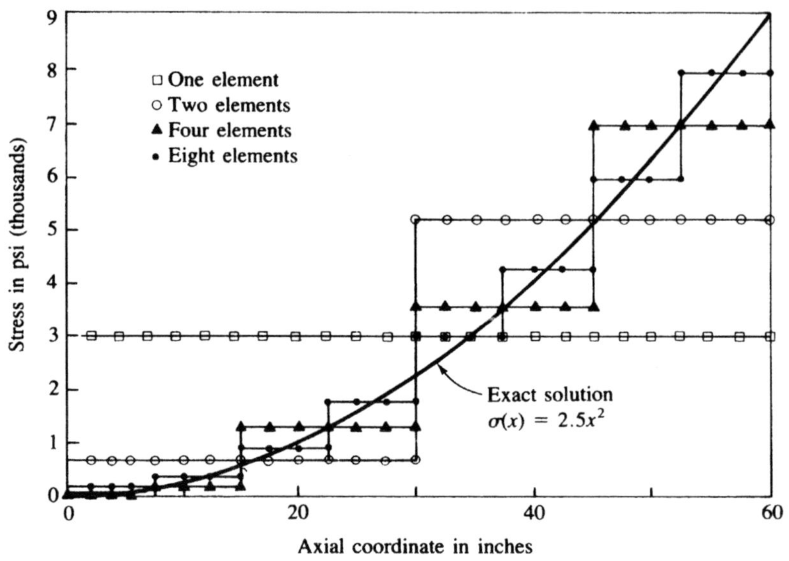

Figure 3–33 Comparison of exact and finite element solutions for axial stress (along length of bar)

numbers of elements, the finite element solution converges to the exact solution (see the four- and eight-element solutions in Figure 3–32).

3. The stress is derived from the slope of the displacement curve as $\sigma = E\varepsilon = E(du/dx)$. Therefore, by the finite element solution, because u is a linear function in each element, axial stress is constant in each element. It then takes even more elements to model the first derivative of the displacement function or, equivalently, the axial stress. This is shown in Figure 3–33, where the best results occur for the eight-element solution.
4. The best approximation of the stress occurs at the midpoint of the element, not at the nodes (Figure 3–33). This is because the derivative of displacement is better predicted between the nodes than at the nodes.
5. The stress is not continuous across element boundaries. Therefore, equilibrium is not satisfied across element boundaries. Also, equilibrium within each element is, in general, not satisfied. This is shown in Figure 3–34 for element 1 in the two-element solution and element 1 in the eight-element solution [in the eight-element solution the forces are obtained from the Algor computer code [9]]. As the number of elements used increases, the discontinuity in the stress decreases across element boundaries, and the approximation of equilibrium improves.

Finally, in Figure 3–35, we show the convergence of axial stress at the fixed end ($x = L$) as the number of elements increases.

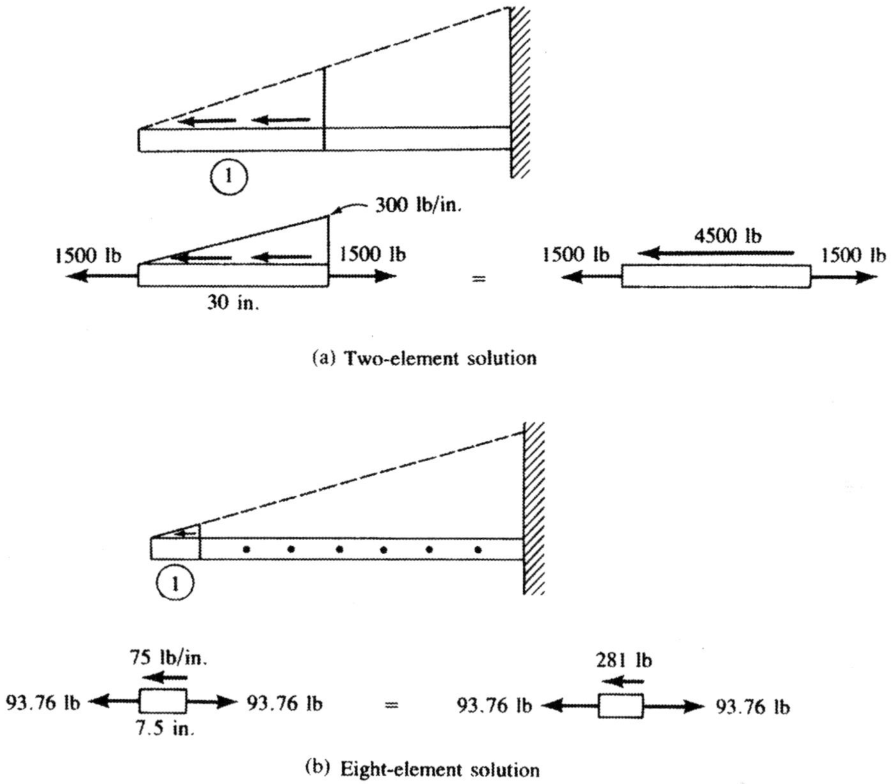

Figure 3–34 Free-body diagram of element 1 in both two- and eight-element models, showing that equilibrium is not satisfied

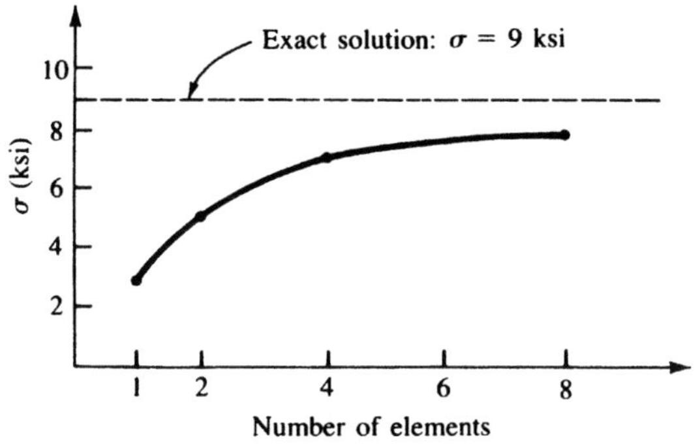

Figure 3–35 Axial stress at fixed end as number of elements increases

However, if we formulate the problem in a customary general way, as described in detail in Chapter 4 for beams subjected to distributed loading, we can obtain the exact stress distribution with any of the models used. That is, letting $\underline{\hat{f}} = \underline{\hat{k}}\underline{\hat{d}} - \underline{\hat{f}}_0$, where $\underline{\hat{f}}_0$ is the initial nodal replacement force system of the distributed load on each element, we subtract the initial replacement force system from the $\underline{\hat{k}}\underline{\hat{d}}$ result. This yields the nodal forces in each element. For example, considering element 1 of the two-element model, we have [see also Eqs. (3.10.33) and (3.10.41)]

$$\underline{\hat{f}}_0 = \begin{Bmatrix} -1500 \text{ lb} \\ -3000 \text{ lb} \end{Bmatrix}$$

Using $\underline{\hat{f}} = \underline{\hat{k}}\underline{\hat{d}} - \underline{\hat{f}}_0$, we obtain

$$\underline{\hat{f}} = \frac{2(30 \times 10^6)}{(30 \text{ in.})} \begin{bmatrix} 1 & -1 \\ -1 & 1 \end{bmatrix} \begin{Bmatrix} -0.006 \text{ in.} \\ -0.00525 \text{ in.} \end{Bmatrix} - \begin{Bmatrix} -1500 \text{ lb} \\ -3000 \text{ lb} \end{Bmatrix}$$

$$= \begin{Bmatrix} -1500 + 1500 \\ 1500 + 3000 \end{Bmatrix} = \begin{Bmatrix} 0 \\ 4500 \end{Bmatrix}$$

as the actual nodal forces. Drawing a free-body diagram of element 1, we have

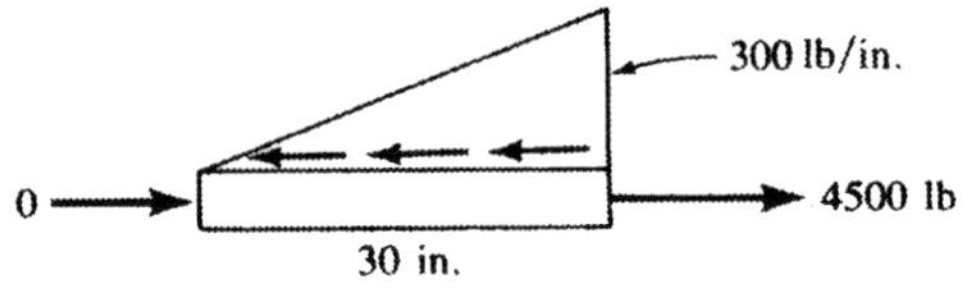

$$\sum F_x = 0: \quad -\tfrac{1}{2}(300 \text{ lb/in.})(30 \text{ in.}) + 4500 \text{ lb} = 0$$

For other kinds of elements (other than beams), this adjustment is ignored in practice. The adjustment is less important for plane and solid elements than for beams. Also, these adjustments are more difficult to formulate for an element of general shape.

3.12 Galerkin's Residual Method and Its Use to Derive the One-Dimensional Bar Element Equations

General Formulation

We developed the bar finite element equations by the direct method in Section 3.1 and by the potential energy method (one of a number of variational methods) in Section 3.10. In fields other than structural/solid mechanics, it is quite probable that a variational principle, analogous to the principle of minimum potential energy, for instance, may not be known or even exist. In some flow problems in fluid mechanics and in mass transport problems (Chapter 13), we often have only the differential equation and boundary conditions available. However, the finite element method can still be applied.

The methods of weighted residuals applied directly to the differential equation can be used to develop the finite element equations. In this section, we describe Galerkin's residual method in general and then apply it to the bar element. This development provides the basis for later applications of Galerkin's method to the beam element in Chapter 4 and to the nonstructural heat-transfer element (specifically, the one-dimensional combined conduction, convection, and mass transport element described in Chapter 13). Because of the mass transport phenomena, the variational formulation is not known (or certainly is difficult to obtain), so Galerkin's method is necessarily applied to develop the finite element equations.

There are a number of other residual methods. Among them are collocation, least squares, and subdomain as described in Section 3.13. (For more on these methods, see Reference [5].)

In weighted residual methods, a trial or approximate function is chosen to approximate the independent variable, such as a displacement or a temperature, in a problem defined by a differential equation. This trial function will not, in general, satisfy the governing differential equation. Thus substituting the trial function into the differential equation results in a residual over the whole region of the problem as follows:

$$\iiint_V R\, dV = \text{minimum} \tag{3.12.1}$$

In the residual method, we require that a weighted value of the residual be a minimum over the whole region. The weighting functions allow the weighted integral of residuals to go to zero. If we denote the weighting function by W, the general form of the weighted residual integral is

$$\iiint_V RW\, dV = 0 \tag{3.12.2}$$

Using Galerkin's method, we choose the interpolation function, such as Eq. (3.1.3), in terms of N_i shape functions for the independent variable in the differential equation. In general, this substitution yields the residual $R \neq 0$. By the Galerkin criterion, the shape functions N_i are chosen to play the role of the weighting functions W. Thus for each i, we have

$$\iiint_V RN_i\, dV = 0 \qquad (i = 1, 2, \ldots, n) \tag{3.12.3}$$

Equation (3.12.3) results in a total of n equations. Equation (3.12.3) applies to points within the region of a body without reference to boundary conditions such as specified applied loads or displacements. To obtain boundary conditions, we apply integration by parts to Eq. (3.12.3), which yields integrals applicable for the region and its boundary.

Bar Element Formulation

We now illustrate Galerkin's method to formulate the bar element stiffness equations. We begin with the basic differential equation, without distributed load, derived in

Section 3.1 as

$$\frac{d}{d\hat{x}}\left(AE\frac{d\hat{u}}{d\hat{x}}\right) = 0 \tag{3.12.4}$$

where constants A and E are now assumed. The residual R is now defined to be Eq. (3.12.4). Applying Galerkin's criterion [Eq. (3.12.3)] to Eq. (3.12.4), we have

$$\int_0^L \frac{d}{d\hat{x}}\left(AE\frac{d\hat{u}}{d\hat{x}}\right)N_i\,d\hat{x} = 0 \qquad (i = 1, 2) \tag{3.12.5}$$

We now apply integration by parts to Eq. (3.12.5). Integration by parts is given in general by

$$\int u\,dv = uv - \int v\,du \tag{3.12.6}$$

where u and v are simply variables in the general equation. Letting

$$\begin{aligned} u &= N_i \qquad & du &= \frac{dN_i}{d\hat{x}}\,d\hat{x} \\ dv &= \frac{d}{d\hat{x}}\left(AE\frac{d\hat{u}}{d\hat{x}}\right)d\hat{x} \qquad & v &= AE\frac{d\hat{u}}{d\hat{x}} \end{aligned} \tag{3.12.7}$$

in Eq. (3.12.5) and integrating by parts according to Eq. (3.12.6), we find that Eq. (3.12.5) becomes

$$\left(N_i AE\frac{d\hat{u}}{d\hat{x}}\right)\Bigg|_0^L - \int_0^L AE\frac{d\hat{u}}{d\hat{x}}\frac{dN_i}{d\hat{x}}\,d\hat{x} = 0 \tag{3.12.8}$$

where the integration by parts introduces the boundary conditions.

Recall that, because $\hat{u} = [N]\{\hat{d}\}$, we have

$$\frac{d\hat{u}}{d\hat{x}} = \frac{dN_1}{d\hat{x}}\hat{d}_{1x} + \frac{dN_2}{d\hat{x}}\hat{d}_{2x} \tag{3.12.9}$$

or, when Eqs. (3.1.4) are used for $N_1 = 1 - \hat{x}/L$ and $N_2 = \hat{x}/L$,

$$\frac{d\hat{u}}{d\hat{x}} = \begin{bmatrix} -\frac{1}{L} & \frac{1}{L} \end{bmatrix}\begin{Bmatrix} \hat{d}_{1x} \\ \hat{d}_{2x} \end{Bmatrix} \tag{3.12.10}$$

Using Eq. (3.12.10) in Eq. (3.12.8), we then express Eq. (3.12.8) as

$$AE\int_0^L \frac{dN_i}{d\hat{x}}\begin{bmatrix} -\frac{1}{L} & \frac{1}{L} \end{bmatrix}d\hat{x}\begin{Bmatrix} \hat{d}_{1x} \\ \hat{d}_{2x} \end{Bmatrix} = \left(N_i AE\frac{d\hat{u}}{d\hat{x}}\right)\Bigg|_0^L \qquad (i = 1, 2) \tag{3.12.11}$$

Equation (3.12.11) is really two equations (one for $N_i = N_1$ and one for $N_i = N_2$). First, using the weighting function $N_i = N_1$, we have

$$AE\int_0^L \frac{dN_1}{d\hat{x}}\begin{bmatrix} -\frac{1}{L} & \frac{1}{L} \end{bmatrix}d\hat{x}\begin{Bmatrix} \hat{d}_{1x} \\ \hat{d}_{2x} \end{Bmatrix} = \left(N_1 AE\frac{d\hat{u}}{d\hat{x}}\right)\Bigg|_0^L \tag{3.12.12}$$

Substituting for $dN_1/d\hat{x}$, we obtain

$$AE\int_0^L \begin{bmatrix} -\dfrac{1}{L} \end{bmatrix} \begin{bmatrix} -\dfrac{1}{L} & \dfrac{1}{L} \end{bmatrix} d\hat{x} \begin{Bmatrix} \hat{d}_{1x} \\ \hat{d}_{2x} \end{Bmatrix} = \hat{f}_{1x} \tag{3.12.13}$$

where $\hat{f}_{1x} = AE(d\hat{u}/d\hat{x})$ because $N_1 = 1$ at $x = 0$ and $N_1 = 0$ at $x = L$. Evaluating Eq. (3.12.13) yields

$$\frac{AE}{L}(\hat{d}_{1x} - \hat{d}_{2x}) = \hat{f}_{1x} \tag{3.12.14}$$

Similarly, using $N_i = N_2$, we obtain

$$AE\int_0^L \begin{bmatrix} \dfrac{1}{L} \end{bmatrix} \begin{bmatrix} -\dfrac{1}{L} & \dfrac{1}{L} \end{bmatrix} d\hat{x} \begin{Bmatrix} \hat{d}_{1x} \\ \hat{d}_{2x} \end{Bmatrix} = \left(N_2 AE \frac{d\hat{u}}{d\hat{x}} \right) \Bigg|_0^L \tag{3.12.15}$$

Simplifying Eq. (3.12.15) yields

$$\frac{AE}{L}(\hat{d}_{2x} - \hat{d}_{1x}) = \hat{f}_{2x} \tag{3.12.16}$$

where $\hat{f}_{2x} = AE(d\hat{u}/d\hat{x})$ because $N_2 = 1$ at $x = L$ and $N_2 = 0$ at $x = 0$. Equations (3.12.14) and (3.12.16) are then seen to be the same as Eqs. (3.1.13) and (3.10.27) derived, respectively, by the direct and the variational method.

3.13 Other Residual Methods and Their Application to a One-Dimensional Bar Problem

As indicated in Section 3.12 when describing Galerkin's residual method, weighted residual methods are based on assuming an approximate solution to the governing differential equation for the given problem. The assumed or trial solution is typically a displacement or a temperature function that must be made to satisfy the initial and boundary conditions of the problem. This trial solution will not, in general, satisfy the governing differential equation. Thus, substituting the trial function into the differential equation will result in some residuals or errors. Each residual method requires the error to vanish over some chosen intervals or at some chosen points. To demonstrate this concept, we will solve the problem of a rod subjected to a triangular load distribution as shown in Figure 3–29 (see Section 3.10) for which we also have an exact solution for the axial displacement given by Eq. (3.11.5) in Section 3.11. We will illustrate four common weighted residual methods: *collocation, subdomain, least squares,* and *Galerkin's method.*

It is important to note that the primary intent in this section is to introduce you to the general concepts of these other weighted residual methods through a simple

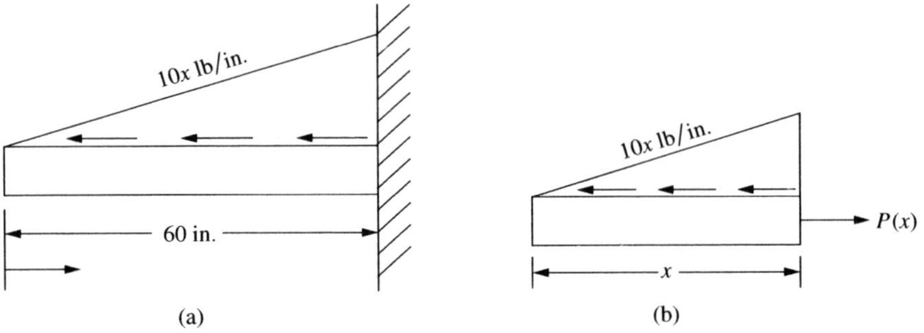

Figure 3–36 (a) Rod subjected to triangular load distribution and (b) free-body diagram of section of rod

example. You should note that we will assume a displacement solution that will in general yield an approximate solution (in our example the assumed displacement function yields an exact solution) over the whole domain of the problem (the rod previously solved in Section 13.10). As you have seen already for the spring and bar elements, we have assumed a linear function over each spring or bar element, and then combined the element solutions as was illustrated in Section 3.10 for the same rod solved in this section. It is common practice to use the simple linear function in each element of a finite element model, with an increasing number of elements used to model the rod yielding a closer and closer approximation to the actual displacement as seen in Figure 3–32.

For clarity's sake, Figure 3–36(a) shows the problem we are solving, along with a free-body diagram of a section of the rod with the internal axial force $P(x)$ shown in Figure 3–36(b).

The governing differential equation for the axial displacement, u, is given by

$$\left(AE\frac{du}{dx}\right) - P(x) = 0 \tag{3.13.1}$$

where the internal axial force is $P(x) = 5x^2$. The boundary condition is $u(x = L) = 0$.

The method of weighted residuals requires us to assume an approximation function for the displacement. This approximate solution must satisfy the boundary condition of the problem. Here we assume the following function:

$$u(x) = c_1(x - L) + c_2(x - L)^2 + c_3(x - L)^3 \tag{3.13.2}$$

where c_1, c_2 and c_3 are unknown coefficients. Equation (3.13.2) also satisfies the boundary condition given by $u(x = L) = 0$.

Substituting Eq. (3.13.2) for u into the governing differential equation, Eq. (3.13.1), results in the following error function, R:

$$AE[c_1 + 2c_2(x - L) + 3c_3(x - L)^2] - 5x^2 = R \tag{3.13.3}$$

We now illustrate how to solve the governing differential equation by the four weighted residual methods.

Collocation Method

The *collocation method* requires that the error or residual function, R, be forced to zero at as many points as there are unknown coefficients. Equation (3.13.2) has three unknown coefficients. Therefore, we will make the error function equal zero at three points along the rod. We choose the error function to go to zero at $x = 0$, $x = L/3$, and $x = 2L/3$ as follows:

$$
\begin{aligned}
&R(c, x = 0) = 0 = AE[c_1 + 2c_2(-L) + 3c_3(-L)^2] = 0 \\
&R(c, x = L/3) = 0 = AE[c_1 + 2c_2(-2L/3) + 3c_3(-2L/3)^2] - 5(L/3)^2 = 0 \qquad (3.13.4) \\
&R(c, x = 2L/3) = 0 = AE[c_1 + 2c_2(-L/3) + 3c_3(-L/3)^2] - 5(2L/3)^2 = 0
\end{aligned}
$$

The three linear equations, Eq. (3.13.4), can now be solved for the unknown coefficients, c_1, c_2 and c_3. The result is

$$c_1 = 5L^2/(AE) \quad c_2 = 5L/(AE) \quad c_3 = 5/(3AE) \qquad (3.13.5)$$

Substituting the numerical values, $A = 2$, $E = 30 \times 10^6$, and $L = 60$ into Eq. (3.13.5), we obtain the c's as:

$$c_1 = 3 \times 10^{-4}, \quad c_2 = 5 \times 10^{-6}, \quad c_3 = 2.778 \times 10^{-8} \qquad (3.13.6)$$

Substituting the numerical values for the coefficients given in Eq. (3.13.6) into Eq. (3.13.2), we obtain the final expression for the axial displacement as

$$u(x) = 3 \times 10^{-4}(x - L) + 5 \times 10^{-6}(x - L)^2 + 2.778 \times 10^{-8}(x - L)^3 \qquad (3.13.7)$$

Because we have chosen a cubic displacement function, Eq. (3.13.2), and the exact solution, Eq. (3.11.6), is also cubic, the collocation method yields the identical solution as the exact solution. The plot of the solution is shown in Figure 3–32 on page 121.

Subdomain Method

The subdomain method requires that the integral of the error or residual function over some selected subintervals be set to zero. The number of subintervals selected must equal the number of unknown coefficients. Because we have three unknown coefficients in the rod example, we must make the number of subintervals equal to three. We choose the subintervals from 0 to $L/3$, from $L/3$ to $2L/3$, and from $2L/3$ to L as follows:

$$
\begin{aligned}
&\int_0^{L/3} R\,dx = 0 = \int_0^{L/3} \{AE[c_1 + 2c_2(x - L) + 3c_3(x - L)^2] - 5x^2\}dx \\
&\int_{L/3}^{2L/3} R\,dx = 0 = \int_{L/3}^{2L/3} \{AE[c_1 + 2c_2(x - L) + 3c_3(x - L)^2] - 5x^2\}dx \qquad (3.13.8) \\
&\int_{2L/3}^{L} R\,dx = 0 = \int_{2L/3}^{L} \{AE[c_1 + 2c_2(x - L) + 3c_3(x - L)^2] - 5x^2\}dx
\end{aligned}
$$

where we have used Eq. (3.13.3) for R in Eqs. (3.13.8).

Integration of Eqs. (3.13.8) results in three simultaneous linear equations that can be solved for the coefficients c_1, c_2 and c_3. Using the numerical values for A, E, and L as previously done, the three coefficients are numerically identical to those given by Eq. (3.13.6). The resulting axial displacement is then identical to Eq. (3.13.7).

Least Squares Method

The least squares method requires the integral over the length of the rod of the error function squared to be minimized with respect to each of the unknown coefficients in the assumed solution, based on the following:

$$\frac{\partial}{\partial c_i}\left(\int_0^L R^2\,dx\right) = 0 \quad i = 1, 2, \ldots N \text{ (for } N \text{ unknown coefficients)} \tag{3.13.9}$$

or equivalently to

$$\int_0^L R\frac{\partial R}{\partial c_i}\,dx = 0 \tag{3.13.10}$$

Because we have three unknown coefficients in the approximate solution, we will perform the integration three times according to Eq. (3.13.10) with three resulting equations as follows:

$$\begin{aligned}
&\int_0^L \{AE[c_1 + 2c_2(x - L) + 3c_3(x - L)^2] - 5x^2\}AE\,dx = 0\\
&\int_0^L \{AE[c_1 + 2c_2(x - L) + 3c_3(x - L)^2] - 5x^2\}AE2(x - L)\,dx = 0\\
&\int_0^L \{AE[c_1 + 2c_2(x - L) + 3c_3(x - L)^2] - 5x^2\}AE3(x - L)^2\,dx = 0
\end{aligned} \tag{3.13.11}$$

In the first, second, and third of Eqs. (3.13.11), respectively, we have used the following partial derivatives:

$$\frac{\partial R}{\partial c_1} = AE, \qquad \frac{\partial R}{\partial c_2} = AE2(x - L), \qquad \frac{\partial R}{\partial c_3} = AE3(x - L)^2 \tag{3.13.12}$$

Integration of Eqs. (3.13.11) yields three linear equations that are solved for the three coefficients. The numerical values of the coefficients again are identical to those of Eq. (3.13.6). Hence, the solution is identical to the exact solution.

Galerkin's Method

Galerkin's method requires the error to be orthogonal[1] to some weighting functions W_i as given previously by Eq. (3.12.2). For the rod example, this integral becomes

$$\int_0^L R W_i \, dx = 0 \qquad I = 1, 2, \ldots, N \tag{3.13.13}$$

The weighting functions are chosen to be a part of the approximate solution. Because we have three unknown constants in the approximate solution, we need to generate three equations. Recall that the assumed solution is the cubic given by Eq. (3.13.2); therefore, we select the weighting functions to be

$$W_1 = x - L \qquad W_2 = (x - L)^2 \qquad W_3 = (x - L)^3 \tag{3.13.14}$$

Using the weighting functions from Eq. (3.13.14) successively in Eq. (3.13.13), we generate the following three equations:

$$\begin{aligned}
&\int_0^L \{AE[c_1 + 2c_2(x - L) + 3c_3(x - L)^2] - 5x^2\}(x - L)\, dx = 0 \\
&\int_0^L \{AE[c_1 + 2c_2(x - L) + 3c_3(x - L)^2] - 5x^2\}(x - L)^2\, dx = 0 \\
&\int_0^L \{AE[c_1 + 2c_2(x - L) + 3c_3(x - L)^2] - 5x^2\}(x - L)^3\, dx = 0
\end{aligned} \tag{3.13.15}$$

Integration of Eqs. (3.13.15) results in three linear equations that can be solved for the unknown coefficients. The numerical values are the same as those given by Eq. (3.13.6). Hence, the solution is identical to the exact solution.

In conclusion, because we assumed the approximate solution in the form of a cubic in x and the exact solution is also a cubic in x, all residual methods have yielded the exact solution. The purpose of this section has still been met to illustrate the four common residual methods to obtain an approximate (or exact in this example) solution to a known differential equation. The exact solution is shown by Eq. (3.11.6) and in Figure 3–32 in Section 3.11.

[1] The use of the word orthogonal in this context is a generalization of its use with respect to vectors. Here the ordinary scalar product is replaced by an integral in Eq. (3.13.13). In Eq. (3.13.13), the functions $u(x) = R$ and $v(x) = W_i$ are said to be orthogonal on the interval $0 \le x \le L$ if $\int_0^L u(x)v(x)\,dx$ equals 0.

References

[1] Turner, M. J., Clough, R. W., Martin, H. C., and Topp, L. J., "Stiffness and Deflection Analysis of Complex Structures," *Journal of the Aeronautical Sciences*, Vol. 23, No. 9, Sept. 1956, pp. 805–824.
[2] Martin, H. C., "Plane Elasticity Problems and the Direct Stiffness Method," *The Trend in Engineering*, Vol. 13, Jan. 1961, pp. 5–19.
[3] Melosh, R. J., "Basis for Derivation of Matrices for the Direct Stiffness Method," *Journal of the American Institute of Aeronautics and Astronautics*, Vol. 1, No. 7, July 1963, pp. 1631–1637.
[4] Oden, J. T., and Ripperger, E. A., *Mechanics of Elastic Structures*, 2nd ed., McGraw-Hill, New York, 1981.
[5] Finlayson, B. A., *The Method of Weighted Residuals and Variational Principles*, Academic Press, New York, 1972.
[6] Zienkiewicz, O. C., *The Finite Element Method*, 3rd ed., McGraw-Hill, London, 1977.
[7] Cook, R. D., Malkus, D. S., Plesha, M. E., and Witt, R. J., Concepts and Applications of Finite Element Analysis, 4th ed., Wiley, New York, 2002.
[8] Forray, M. J., *Variational Calculus in Science and Engineering*, McGraw-Hill, New York, 1968.
[9] Linear Stress and Dynamics Reference Division, Docutech On-Line Documentation, Algor Interactive Systems, Pittsburgh, PA.

Problems

3.1 **a.** Compute the total stiffness matrix $\underline{K}$ of the assemblage shown in Figure P3–1 by superimposing the stiffness matrices of the individual bars. Note that $\underline{K}$ should be in terms of $A_1, A_2, A_3, E_1, E_2, E_3, L_1, L_2$, and L_3. Here A, E, and L are generic symbols used for cross-sectional area, modulus of elasticity, and length, respectively.

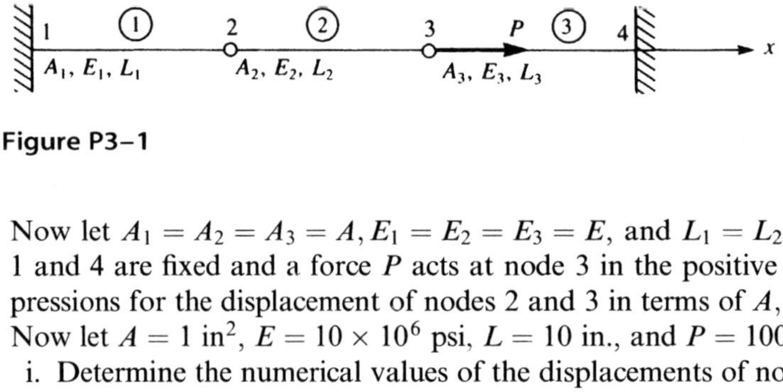

Figure P3–1

b. Now let $A_1 = A_2 = A_3 = A, E_1 = E_2 = E_3 = E$, and $L_1 = L_2 = L_3 = L$. If nodes 1 and 4 are fixed and a force P acts at node 3 in the positive x direction, find expressions for the displacement of nodes 2 and 3 in terms of A, E, L, and P.

c. Now let $A = 1$ in^2, $E = 10 \times 10^6$ psi, $L = 10$ in., and $P = 1000$ lb.

i. Determine the numerical values of the displacements of nodes 2 and 3.
ii. Determine the numerical values of the reactions at nodes 1 and 4.
iii. Determine the stresses in elements 1–3.

3.2–3.11 For the bar assemblages shown in Figures P3–2–P3–11, determine the nodal displacements, the forces in each element, and the reactions. Use the direct stiffness method for these problems.

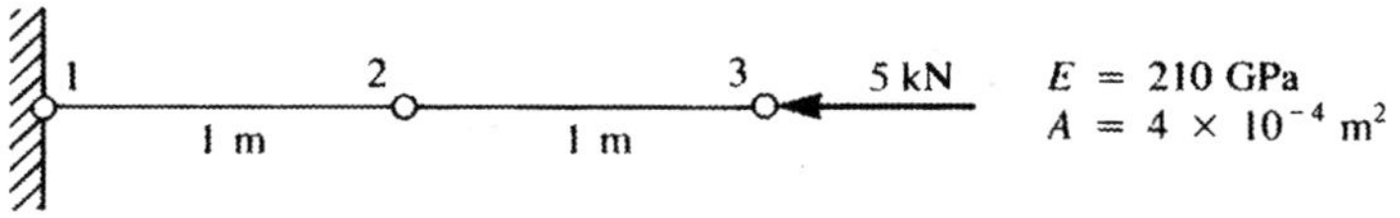

Figure P3–2

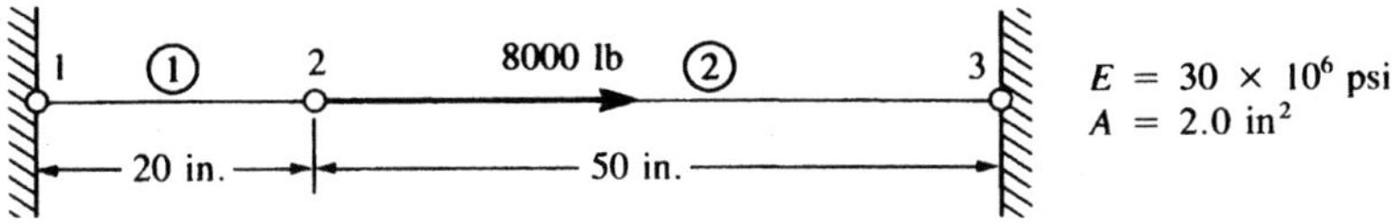

Figure P3–3

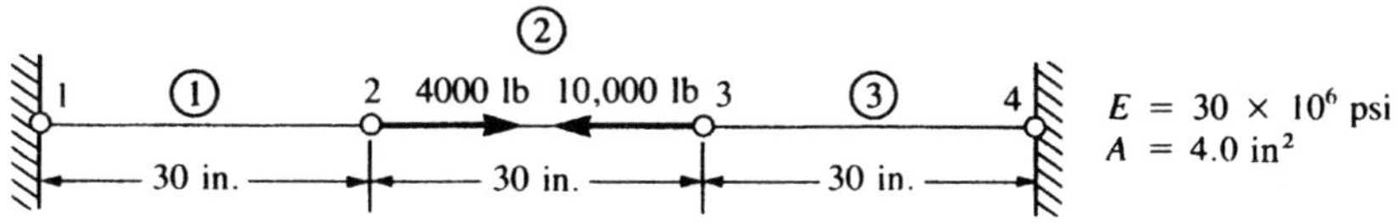

Figure P3–4

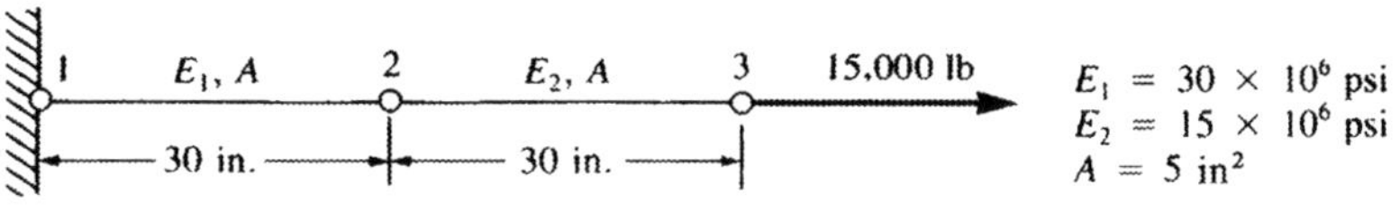

Figure P3–5

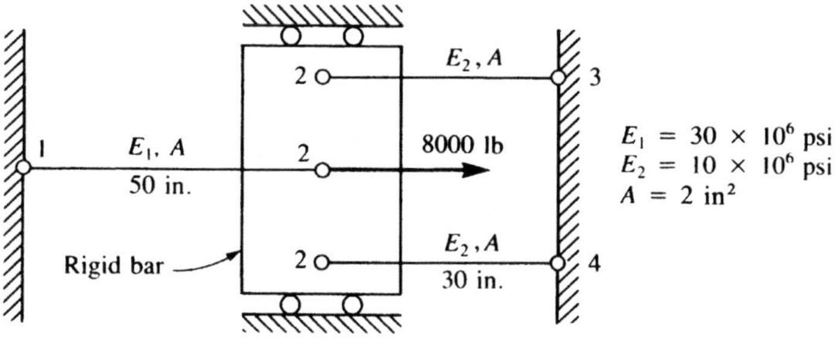

Figure P3–6

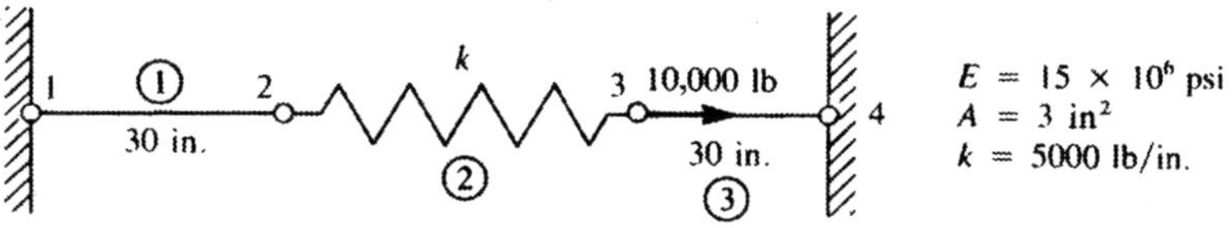

Figure P3–7

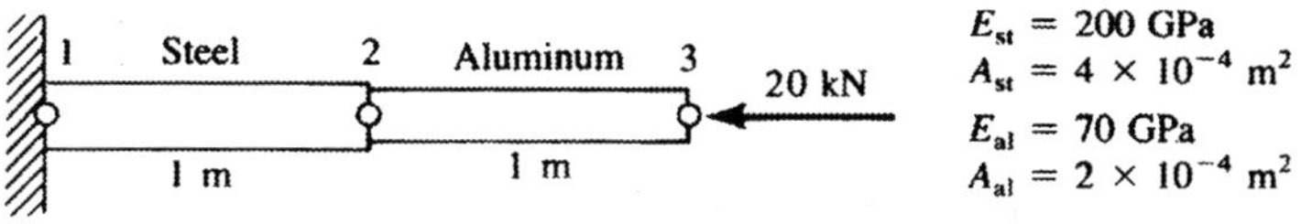

Figure P3–8

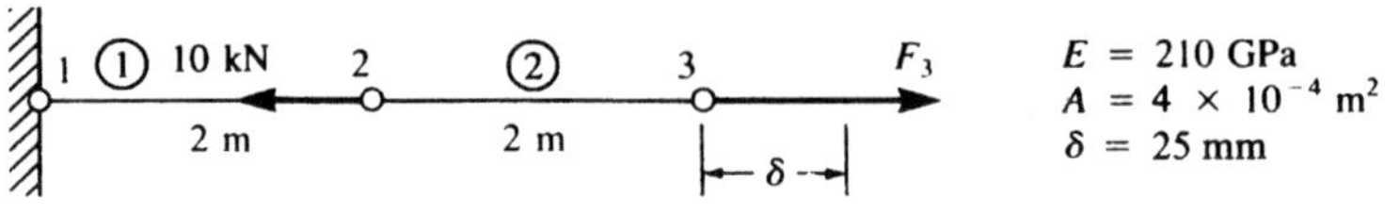

Figure P3–9

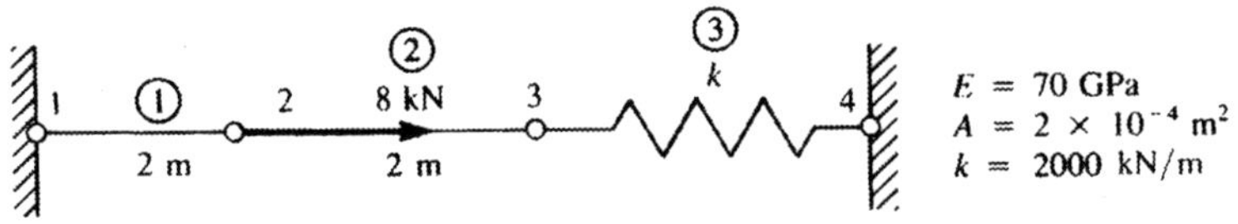

Figure P3–10

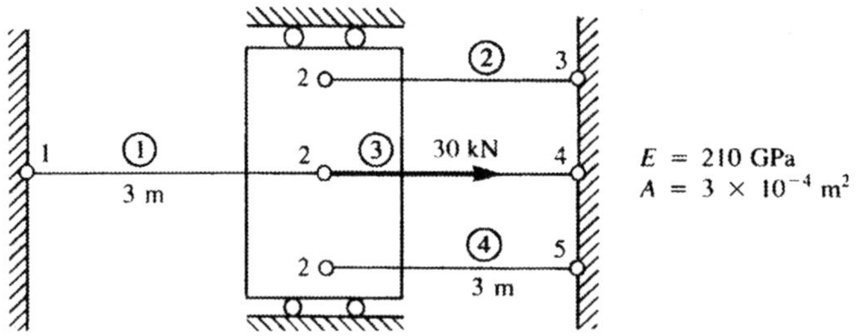

Figure P3–11

3.12 Solve for the axial displacement and stress in the tapered bar shown in Figure P3–12 using one and then two constant-area elements. Evaluate the area at the center of each element length. Use that area for each element. Let $A_0 = 2$ in^2, $L = 20$ in., $E = 10 \times 10^6$ psi, and $P = 1000$ lb. Compare your finite element solutions with the exact solution.

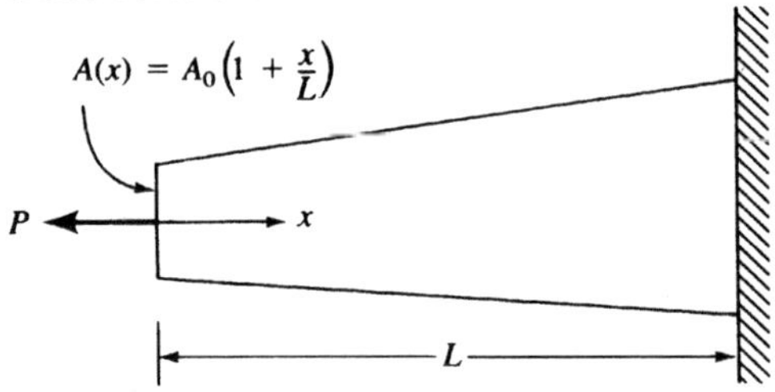

Figure P3–12

3.13 Determine the stiffness matrix for the bar element with end nodes and midlength node shown in Figure P3–13. Let axial displacement $u = a_1 + a_2 x + a_3 x^2$. (This is a higher-order element in that strain now varies linearly through the element.)

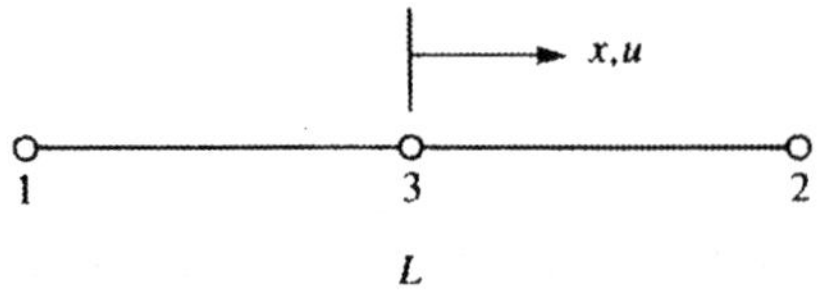

Figure P3–13

3.14 Consider the following displacement function for the two-noded bar element:

$$u = a + bx^2$$

Is this a valid displacement function? Discuss why or why not.

3.15 For each of the bar elements shown in Figure P3–15, evaluate the global x-y stiffness matrix.

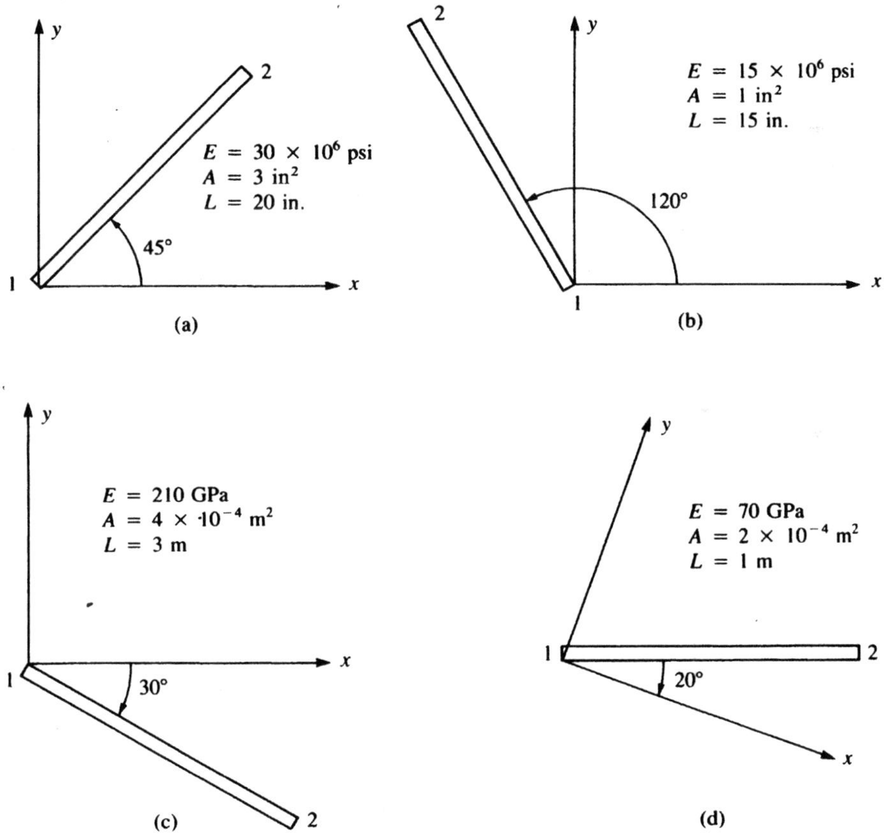

Figure P3–15

3.16 For the bar elements shown in Figure P3–16, the global displacements have been determined to be $d_{1x} = 0.5$ in., $d_{1y} = 0.0$, $d_{2x} = 0.25$ in., and $d_{2y} = 0.75$ in. Determine the local $\hat{x}$ displacements at each end of the bars. Let $E = 12 \times 10^6$ psi, $A = 0.5$ in^2, and $L = 60$ in. for each element.

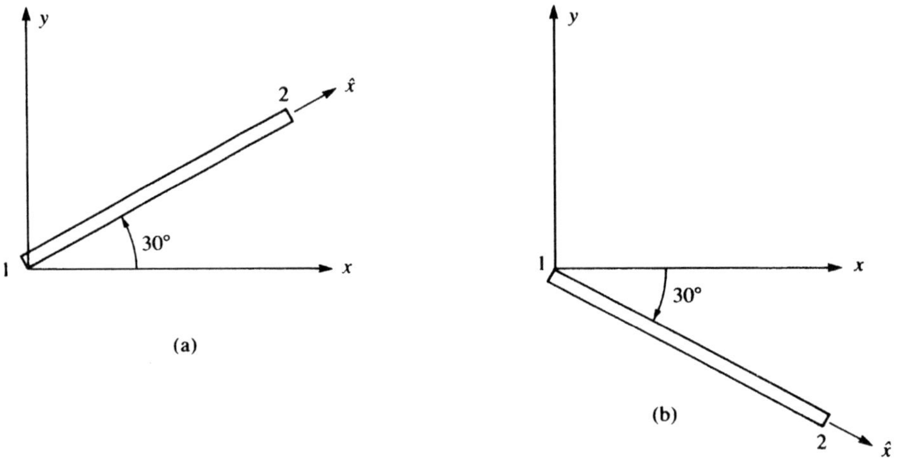

Figure P3–16

3.17 For the bar elements shown in Figure P3–17, the global displacements have been determined to be $d_{1x} = 0.0, d_{1y} = 2.5$ mm, $d_{2x} = 5.0$ mm, and $d_{2y} = 3.0$ mm. Determine the local $\hat{x}$ displacements at the ends of each bar. Let $E = 210$ GPa, $A = 10 \times 10^{-4}$ m^2, and $L = 3$ m for each element.

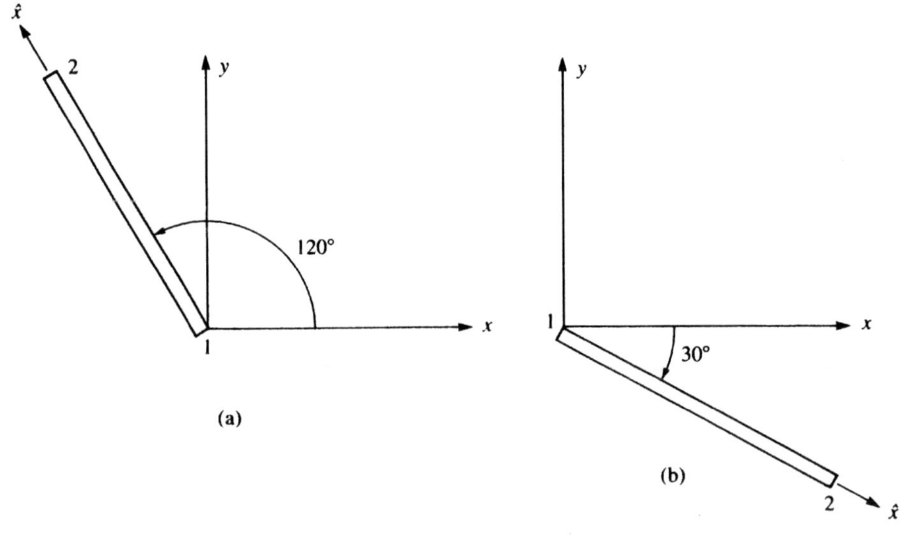

Figure P3–17

3.18 Using the method of Section 3.5, determine the axial stress in each of the bar elements shown in Figure P3–18.

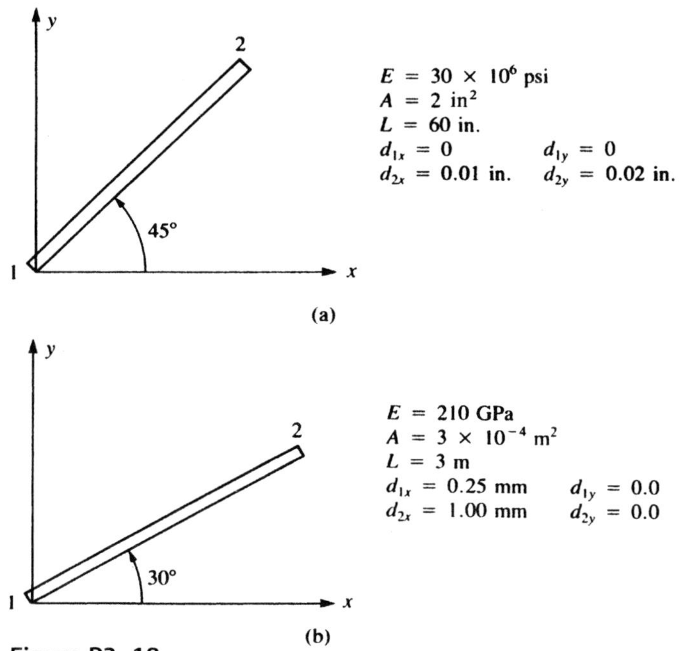

Figure P3–18

3.19 **a.** Assemble the stiffness matrix for the assemblage shown in Figure P3–19 by superimposing the stiffness matrices of the springs. Here k is the stiffness of each spring.

b. Find the x and y components of deflection of node 1.

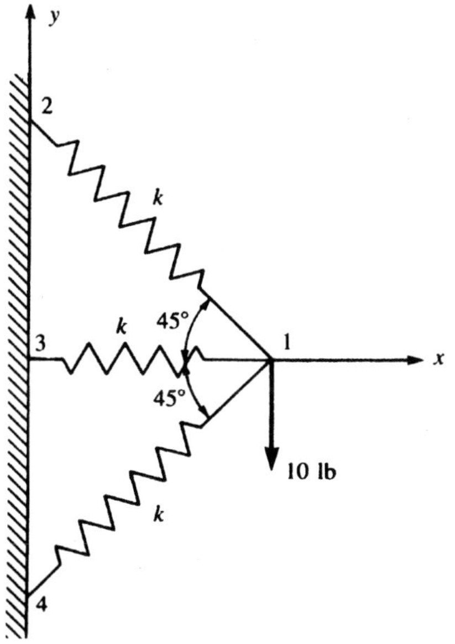

Figure P3–19

3.20 For the plane truss structure shown in Figure P3–20, determine the displacement of node 2 using the stiffness method. Also determine the stress in element 1. Let $A = 5$ in^2, $E = 1 \times 10^6$ psi, and $L = 100$ in.

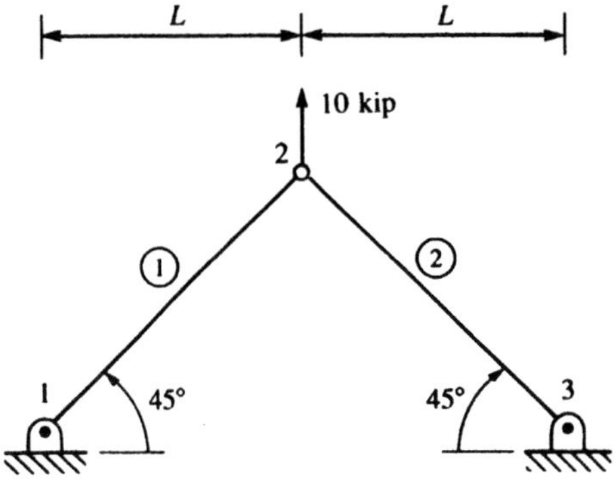

Figure P3–20

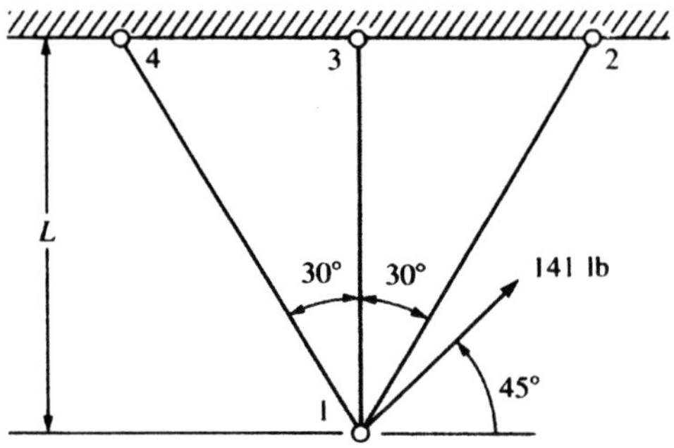

Figure P3–21

3.21 Find the horizontal and vertical displacements of node 1 for the truss shown in Figure P3–21. Assume AE is the same for each element.

3.22 For the truss shown in Figure P3–22 solve for the horizontal and vertical components of displacement at node 1 and determine the stress in each element. Also verify force equilibrium at node 1. All elements have $A_1 = 1$ in.^{2} and $E = 10 \times 10^6$ psi. Let $L = 100$ in.

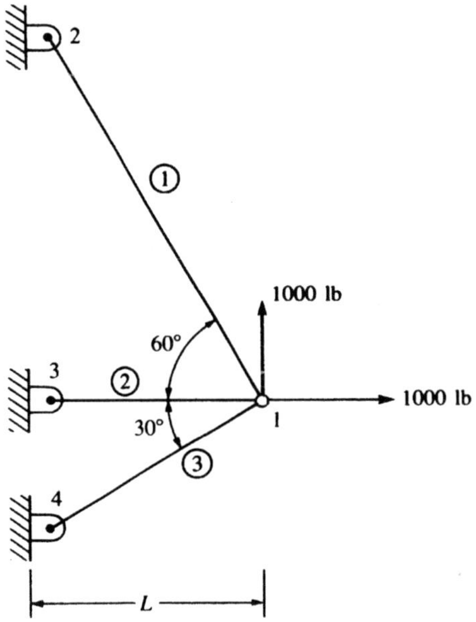

Figure P3–22

3.23 For the truss shown in Figure P3–23, solve for the horizontal and vertical components of displacement at node 1. Also determine the stress in element 1. Let $A = 1$ in^2, $E = 10.0 \times 10^6$ psi, and $L = 100$ in.

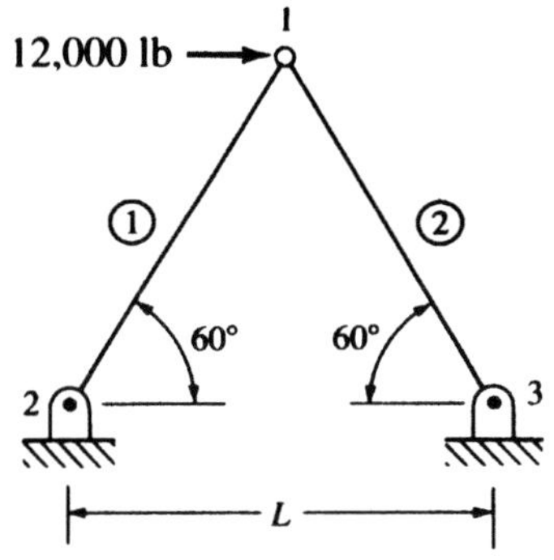

Figure P3–23

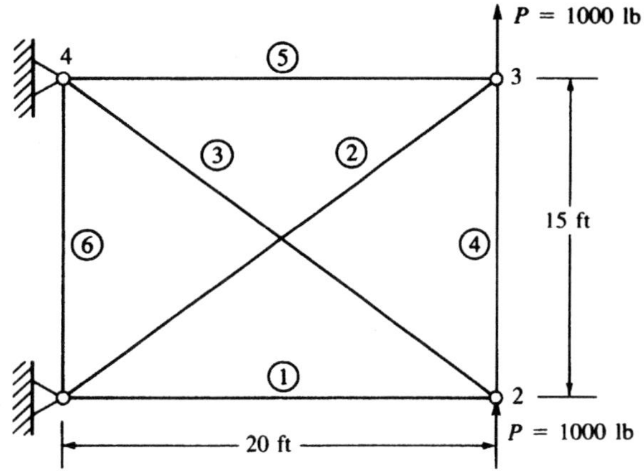

Figure P3–24

3.24 Determine the nodal displacements and the element forces for the truss shown in Figure P3–24. Assume all elements have the same AE.

3.25 Now remove the element connecting nodes 2 and 4 in Figure P3–24. Then determine the nodal displacements and element forces.

3.26 Now remove *both* cross elements in Figure P3–24. Can you determine the nodal displacements? If not, why?

3.27 Determine the displacement components at node 3 and the element forces for the plane truss shown in Figure P3–27. Let $A = 3$ in^2 and $E = 30 \times 10^6$ psi for all elements. Verify force equilibrium at node 3.

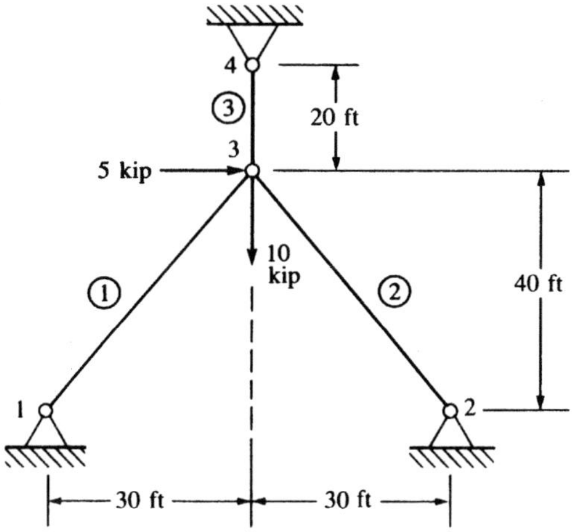

Figure P3–27

3.28 Show that for the transformation matrix $\underline{T}$ of Eq. (3.4.15), $\underline{T}^T = \underline{T}^{-1}$ and hence Eq. (3.4.21) is indeed correct, thus also illustrating that $\underline{k} = \underline{T}^T \underline{\hat{k}} \underline{T}$ is the expression for the global stiffness matrix for an element.

3.29–3.30 For the plane trusses shown in Figures P3–29 and P3–30, determine the horizontal and vertical displacements of node 1 and the stresses in each element. All elements have $E = 210$ GPa and $A = 4.0 \times 10^{-4}$ m^2.

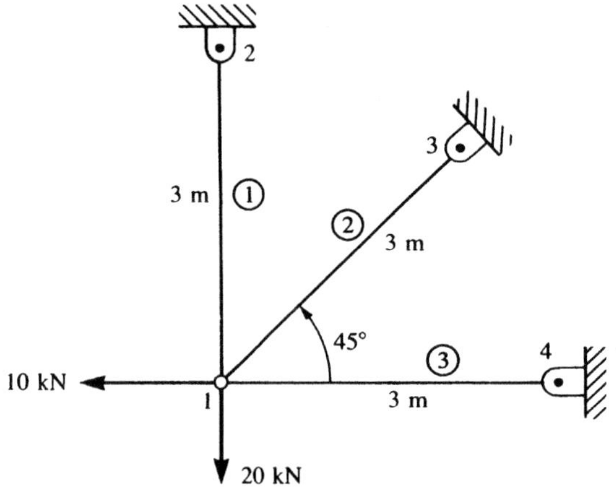

Figure P3–29

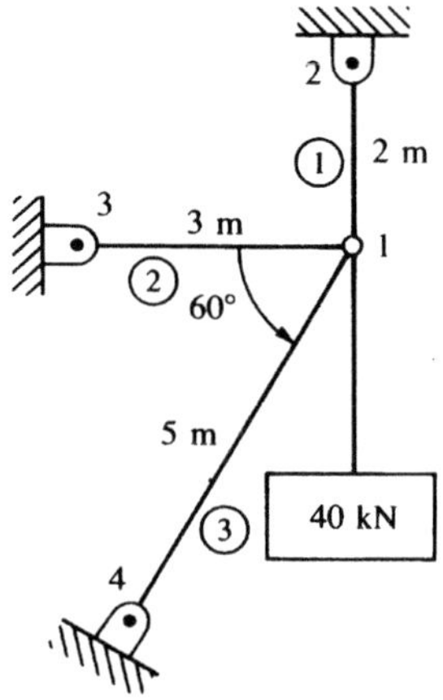

Figure P3–30

3.31 Remove element 1 from Figure P3–30 and solve the problem. Compare the displacements and stresses to the results for Problem 3.30.

3.32 For the plane truss shown in Figure P3–32, determine the nodal displacements, the element forces and stresses, and the support reactions. All elements have $E = 70$ GPa and $A = 3.0 \times 10^{-4}$ m^2. Verify force equilibrium at nodes 2 and 4. Use symmetry in your model.

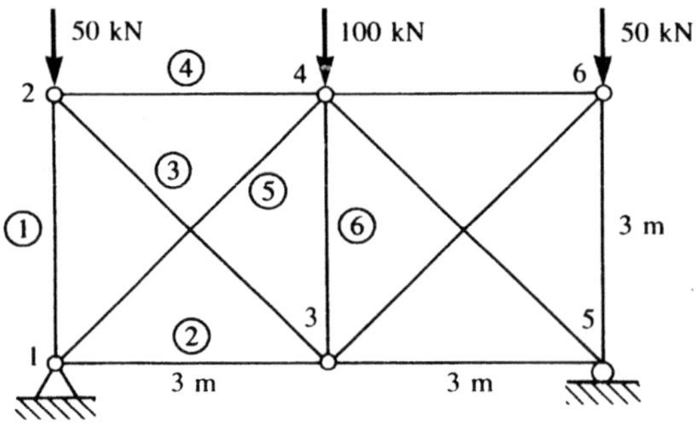

Figure P3–32

3.33 For the plane trusses supported by the spring at node 1 in Figure P3–33 (a) and (b), determine the nodal displacements and the stresses in each element. Let $E = 210$ GPa and $A = 5.0 \times 10^{-4}$ m^2 for both truss elements.

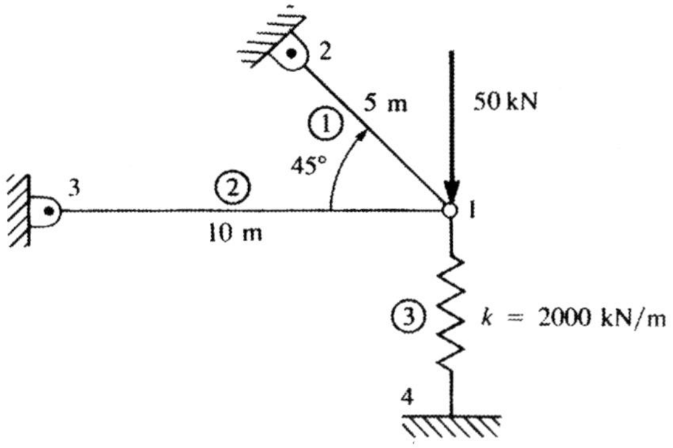

Figure P3–33(a)

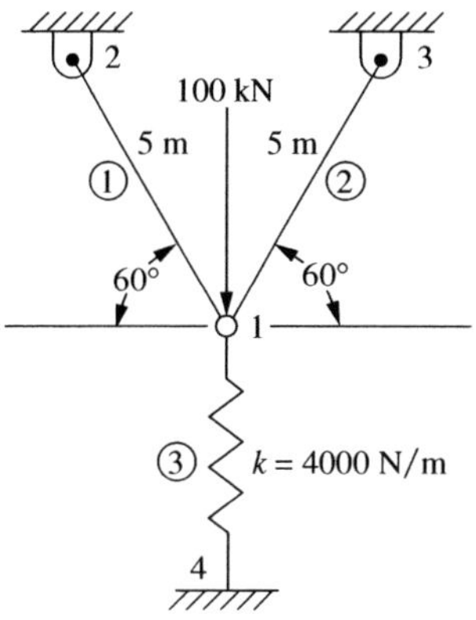

Figure P3–33(b)

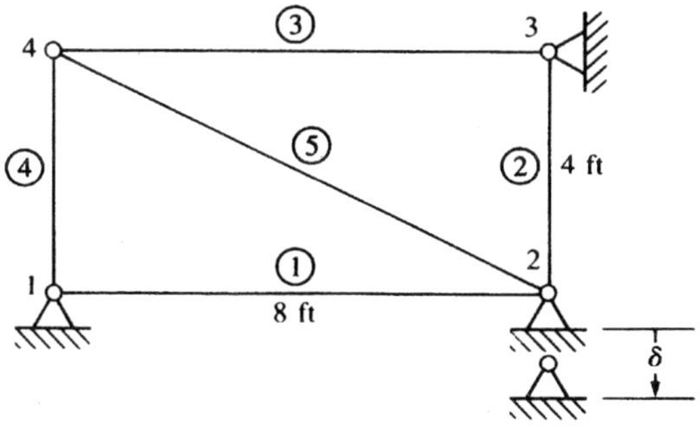

Figure P3–34

3.34 For the plane truss shown in Figure P3–34, node 2 settles an amount $\delta = 0.05$ in. Determine the forces and stresses in each element due to this settlement. Let $E = 30 \times 10^6$ psi and $A = 2$ in^2 for each element.

3.35 For the symmetric plane truss shown in Figure P3–35, determine (a) the deflection of node 1 and (b) the stress in element 1. AE/L for element 3 is twice AE/L for the other

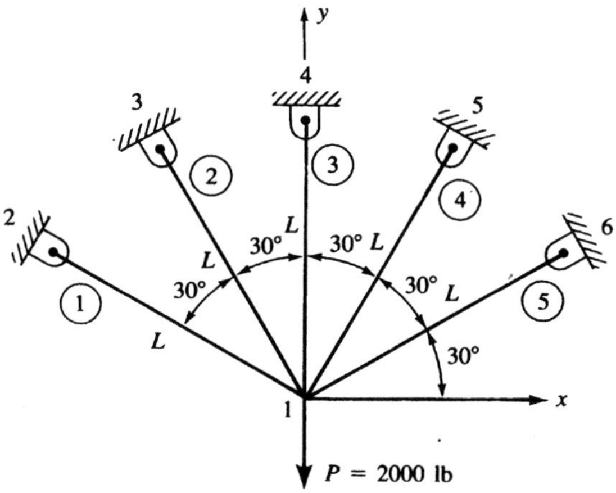

Figure P3–35

elements. Let $AE/L = 10^6$ lb/in. Then let $A = 1$ in^2, $L = 10$ in., and $E = 10 \times 10^6$ psi to obtain numerical results.

3.36–3.37 For the space truss elements shown in Figures P3–36 and P3–37, the global displacements at node 1 have been determined to be $d_{1x} = 0.1$ in., $d_{1y} = 0.2$ in., and $d_{1z} = 0.15$ in. Determine the displacement along the local $\hat{x}$ axis at node 1 of the elements. The coordinates, in inches, are shown in the figures.

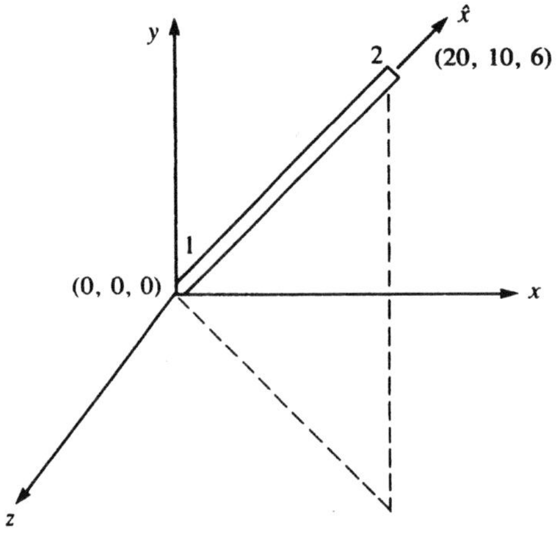

Figure P3–36

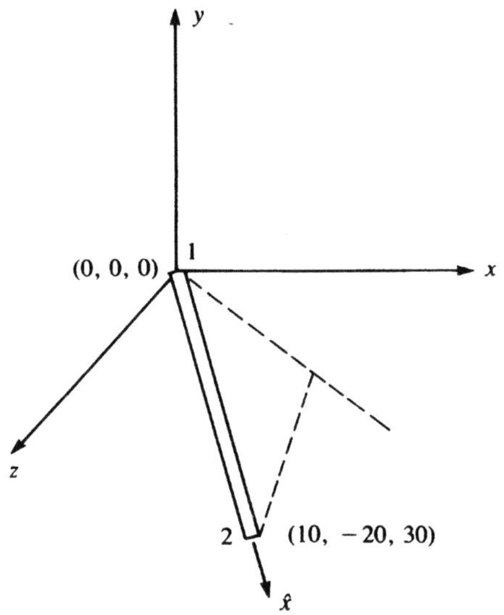

Figure P3–37

3.38–3.39 For the space truss elements shown in Figures P3–38 and P3–39, the global displacements at node 2 have been determined to be $d_{2x} = 5$ mm, $d_{2y} = 10$ mm, and

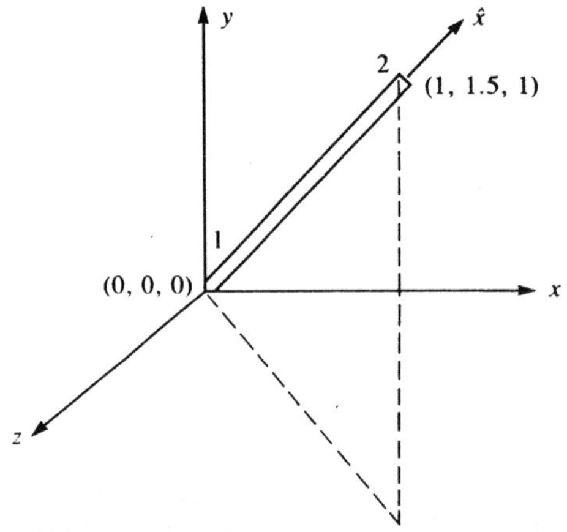

Figure P3–38

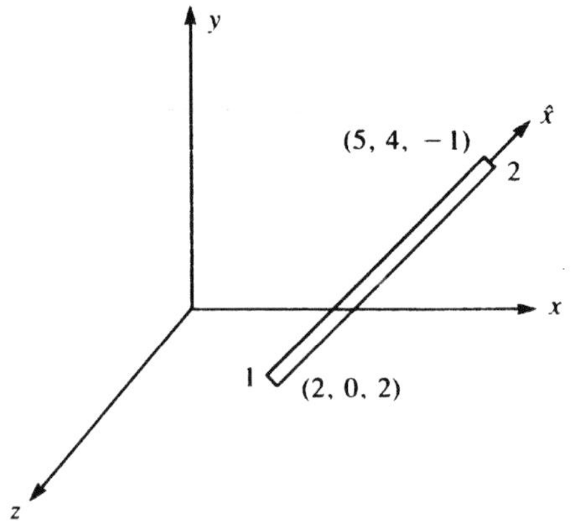

Figure P3–39

$d_{2z} = 15$ mm. Determine the displacement along the local $\hat{x}$ axis at node 2 of the elements. The coordinates, in meters, are shown in the figures.

3.40–3.41 For the space trusses shown in Figures P3–40 and P3–41, determine the nodal displacements and the stresses in each element. Let $E = 210$ GPa and $A = 10 \times 10^{-4}$ m^2 for all elements. Verify force equilibrium at node 1. The coordinates of each node, in meters, are shown in the figure. All supports are ball-and-socket joints.

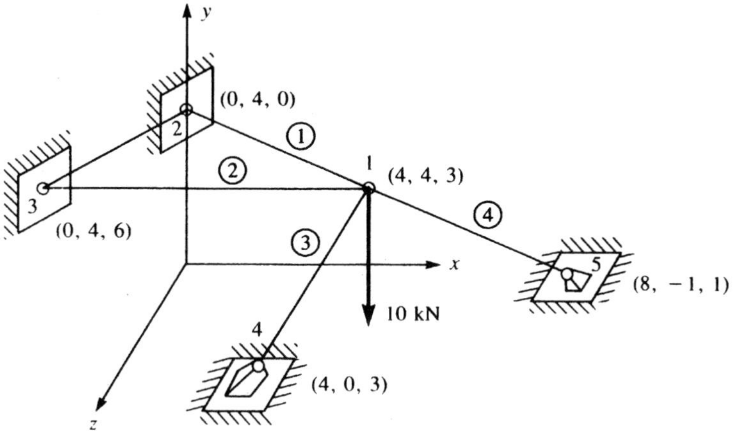

Figure P3–40

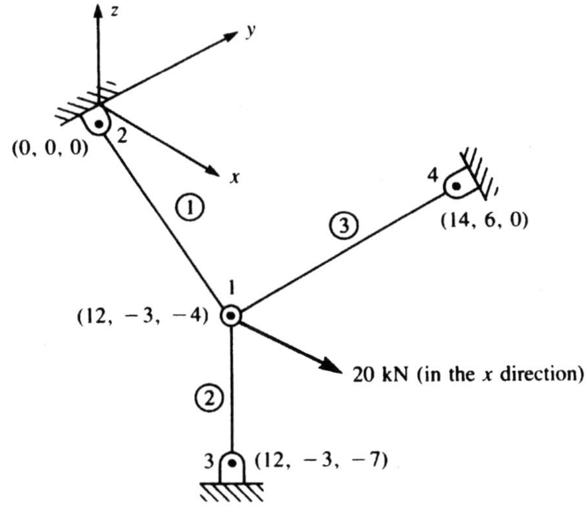

Figure P3–41

3.42 For the space truss subjected to a 1000-lb load in the x direction, as shown in Figure P3–42, determine the displacement of node 5. Also determine the stresses in each element. Let $A = 4$ in^2 and $E = 30 \times 10^6$ psi for all elements. The coordinates of each node, in inches, are shown in the figure. Nodes 1–4 are supported by ball-and-socket joints (fixed supports).

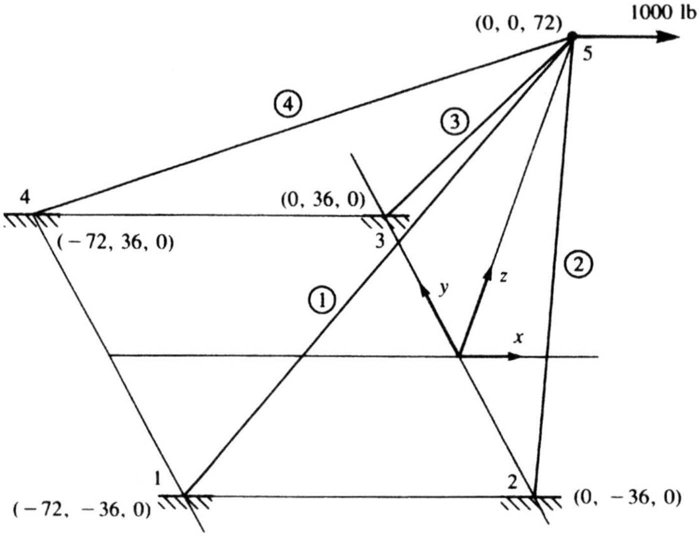

Figure P3–42

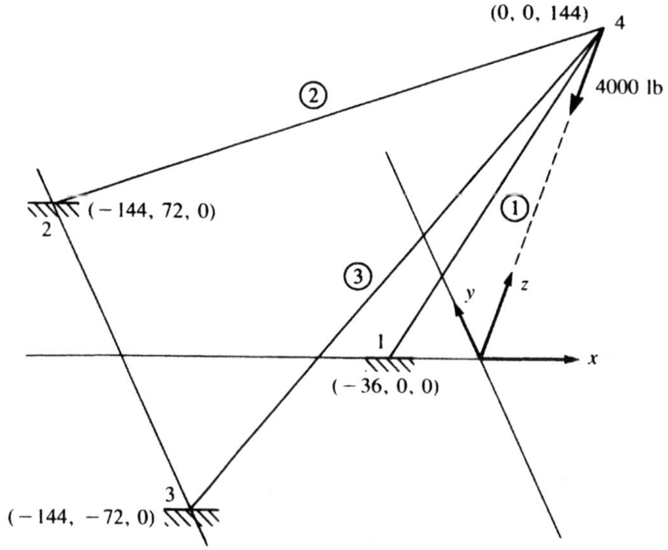

Figure P3–43

3.43 For the space truss subjected to the 4000-lb load acting as shown in Figure P3–43, determine the displacement of node 4. Also determine the stresses in each element. Let

$A = 6\ \text{in}^2$ and $E = 30 \times 10^6$ psi for all elements. The coordinates of each node, in inches, are shown in the figure. Nodes 1–3 are supported by ball-and-socket joints (fixed supports).

3.44 Verify Eq. (3.7.9) for $\underline{k}$ by first expanding $\underline{T}^*$, given by Eq. (3.7.7), to a 6×6 square matrix in a manner similar to that done in Section 3.4 for the two-dimensional case. Then expand $\underline{\hat{k}}$ to a 6×6 matrix by adding appropriate rows and columns of zeros (for the $\hat{d}_z$ terms) to Eq. (3.4.17). Finally, perform the matrix triple product $\underline{k} = \underline{T}^T \underline{\hat{k}} \underline{T}$.

3.45 Derive Eq. (3.7.21) for stress in space truss elements by a process similar to that used to derive Eq. (3.5.6) for stress in a plane truss element.

3.46 For the truss shown in Figure P3–46, use symmetry to determine the displacements of the nodes and the stresses in each element. All elements have $E = 30 \times 10^6$ psi. Elements 1, 2, 4, and 5 have $A = 10\ \text{in}^2$ and element 3 has $A = 20\ \text{in}^2$.

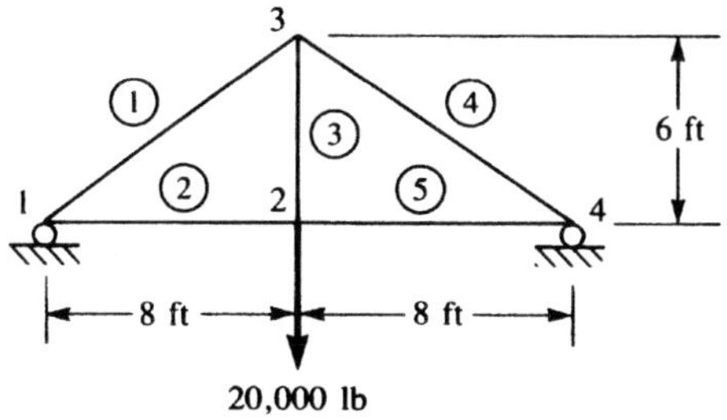

Figure P3–46

3.47 All elements of the structure in Figure P3–47 have the same AE except element 1, which has an axial stiffness of $2AE$. Find the displacements of the nodes and the stresses in elements 2, 3, and 4 by using symmetry. Check equilibrium at node 4. You might want to use the results obtained from the stiffness matrix of Problem 3.24.

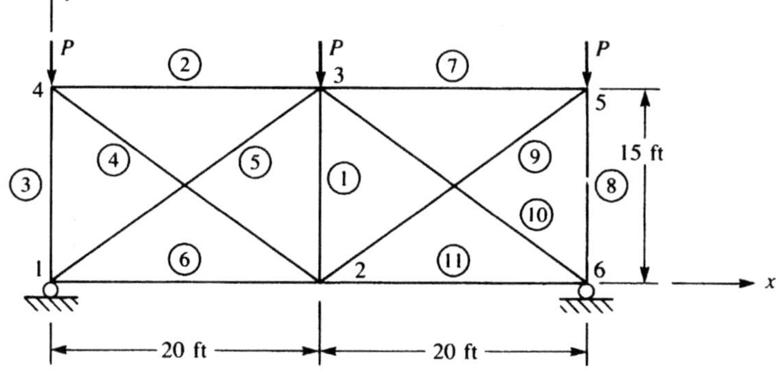

Figure P3–47

3.48 For the roof truss shown in Figure P3–48, use symmetry to determine the displacements of the nodes and the stresses in each element. All elements have $E = 210$ GPa and $A = 10 \times 10^{-4}$ m^2.

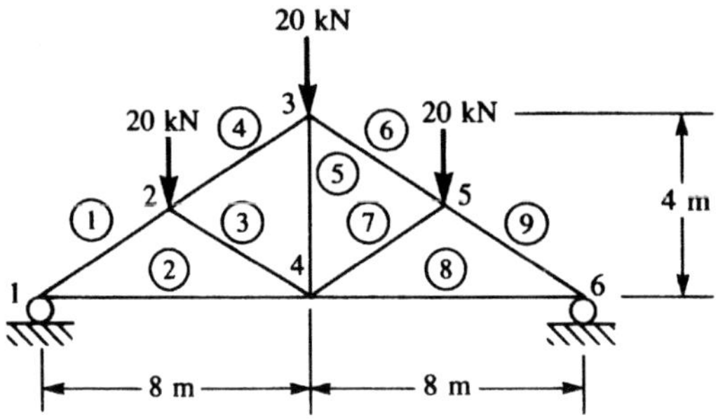

Figure P3–48

3.49–3.51 For the plane trusses with inclined supports shown in Figures P3–49—P3–51, solve for the nodal displacements and element stresses in the bars. Let $A = 2$ in^2, $E = 30 \times 10^6$ psi, and $L = 30$ in. for each truss.

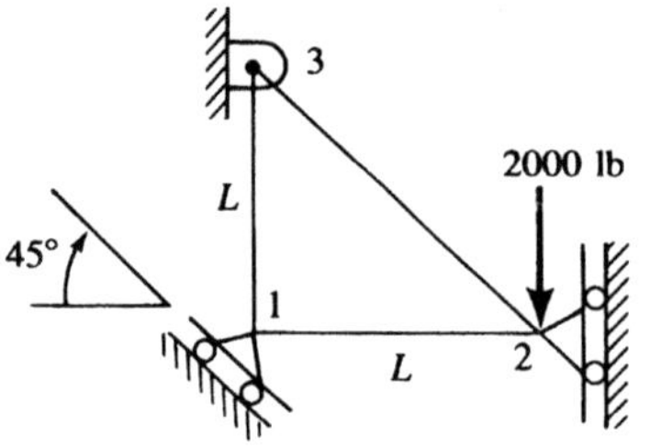

Figure P3–49

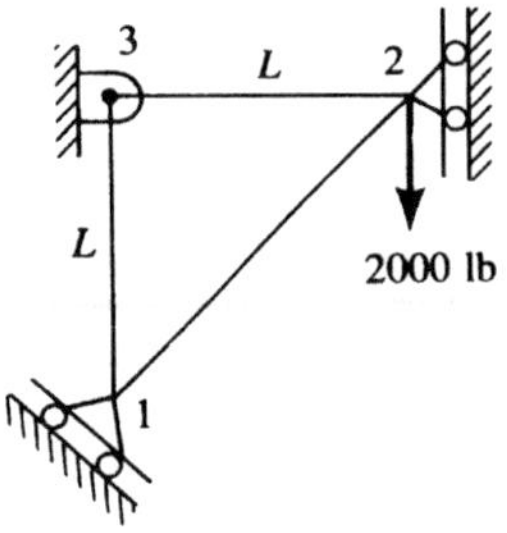

Figure P3–50

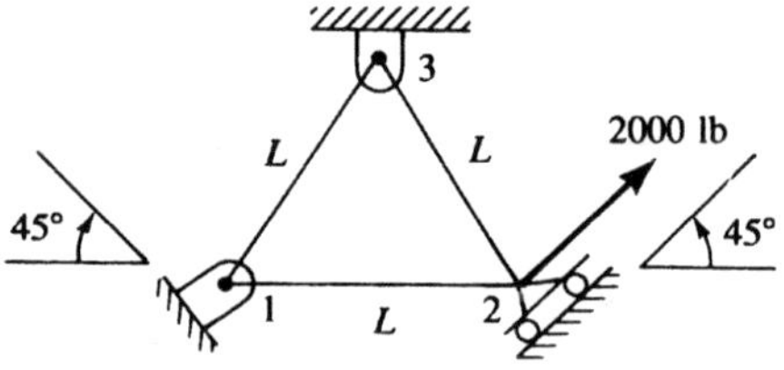

Figure P3–51

3.52 Use the principle of minimum potential energy developed in Section 3.10 to solve the bar problems shown in Figure P3–52. That is, plot the total potential energy for variations in the displacement of the free end of the bar to determine the minimum potential energy. Observe that the displacement that yields the minimum potential energy also yields the stable equilibrium position. Use displacement increments of

0.002 in., beginning with $x = -0.004$. Let $E = 30 \times 10^6$ psi and $A = 2$ in^2 for the bars.

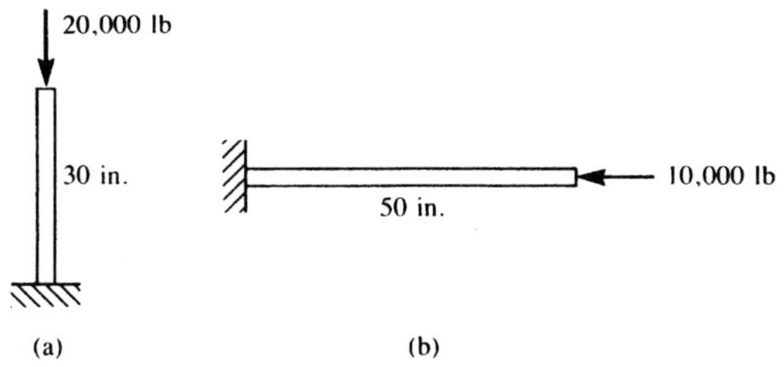

Figure P3–52

3.53 Derive the stiffness matrix for the nonprismatic bar shown in Figure P3–53 using the principle of minimum potential energy. Let E be constant.

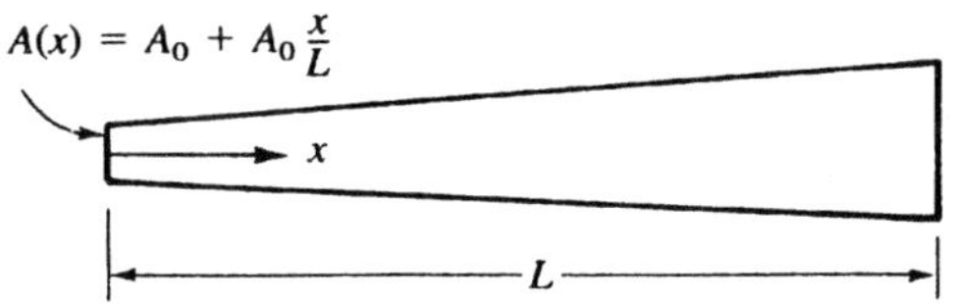

Figure P3–53

3.54 For the bar subjected to the linear varying axial load shown in Figure P3–54, determine the nodal displacements and axial stress distribution using (a) two equal-length elements and (b) four equal-length elements. Let $A = 2$ in.^{2} and $E = 30 \times 10^6$ psi. Compare the finite element solution with an exact solution.

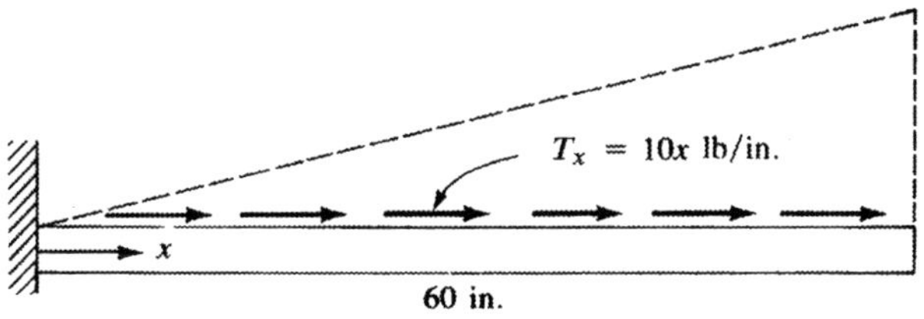

Figure P3–54

3.55 For the bar subjected to the uniform line load in the axial direction shown in Figure P3–55, determine the nodal displacements and axial stress distribution using (a) two equal-length elements and (b) four equal-length elements. Compare the finite element results with an exact solution. Let $A = 2$ in^2 and $E = 30 \times 10^6$ psi.

3.56 For the bar fixed at both ends and subjected to the uniformly distributed loading shown in Figure P3–56, determine the displacement at the middle of the bar and the stress in the bar. Let $A = 2$ in^2 and $E = 30 \times 10^6$ psi.

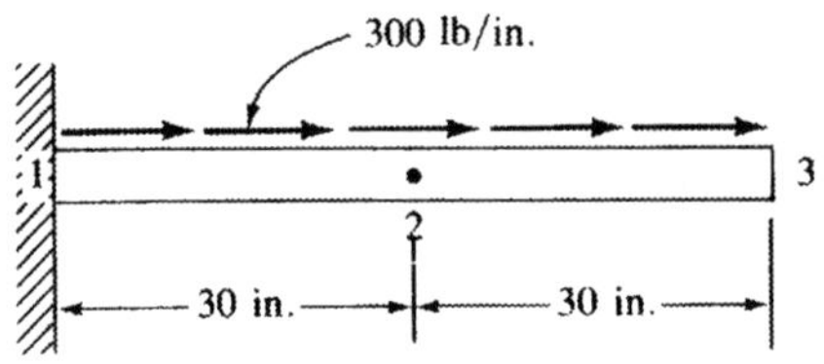

Figure P3–55

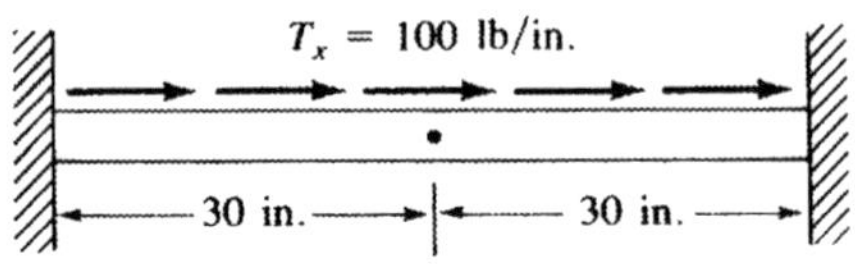

Figure P3–56

3.57 For the bar hanging under its own weight shown in Figure P3–57, determine the nodal displacements using (a) two equal-length elements and (b) four equal-length elements. Let $A = 2$ in^2, $E = 30 \times 10^6$ psi, and weight density $\rho_w = 0.283$ lb/in^3. (*Hint:* The internal force is a function of x. Use the potential energy approach.)

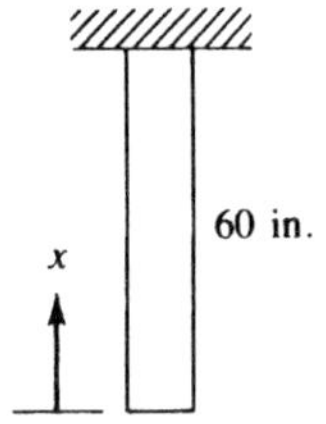

Figure P3–57

3.58 Determine the energy equivalent nodal forces for the axial distributed loading shown acting on the bar elements in Figure P3–58.

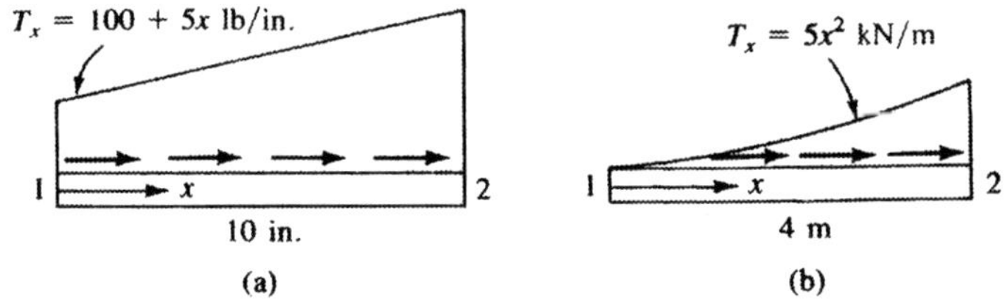

Figure P3–58

3.59 Solve problem 3.55 for the axial displacement in the bar using collocation, subdomain, least squares, and Galerkin's methods. Choose a quadratic polynomial $u(x) = c_1x + c_2x^2$ in each method. Compare these weighted residual method solutions to the exact solution.

3.60 For the tapered bar shown with cross sectional areas $A_1 = 2$ in.^{2} and $A_2 = 1$ in.^{2} at each end, use the collocation, subdomain, least squares, and Galerkin's methods to obtain the displacement in the bar. Compare these weighted residual solutions to the exact solution. Choose a cubic polynomial $u(x) = c_1x + c_2x^2 + c_3x^3$.

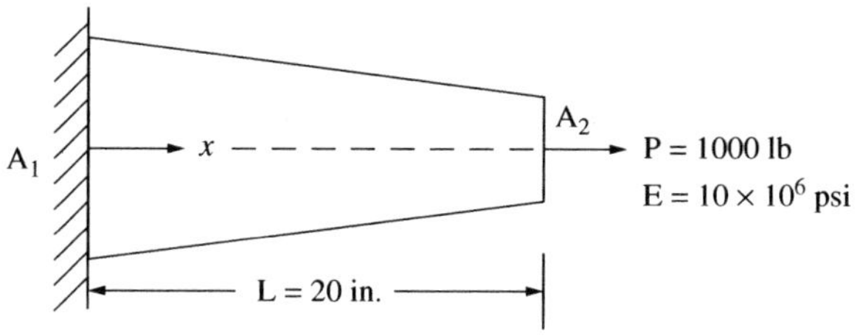

Figure P3–60

3.61 For the bar shown in Figure P 3–61 subjected to the linear varying axial load, determine the displacements and stresses using (a) one and then two finite element models and (b) the collocation, subdomain, least squares, and Galerkin's methods assuming a cubic polynomial of the form $u(x) = c_1 x + c_2 x^2 + c_3 x^3$.

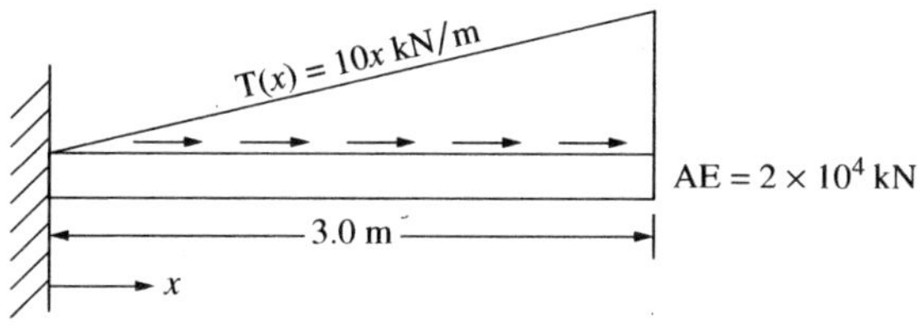

Figure P3–61

3.62–3.67  Use a computer program to solve the truss design problems shown in Figures P3. 62–3.67. Determine the single most critical cross-sectional area based on maximum allowable yield strength or buckling strength (based on either Euler's or Johnson's formula as relevant) using a factor of safety (FS) listed next to each truss. Recommend a common structural shape and size for each truss. List the largest three nodal displacements and their locations. Also include a plot of the deflected shape of the truss and a principal stress plot.

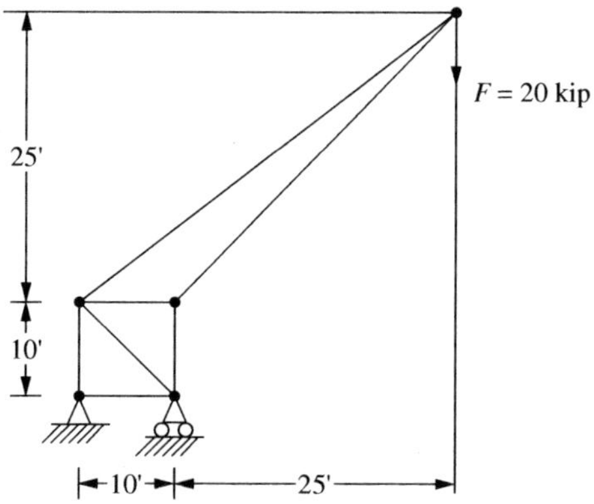

Figure P3–62 Derrick truss (FS = 4.0)

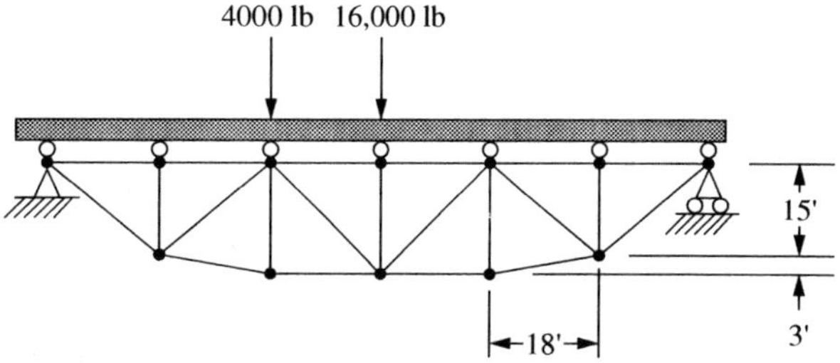

Figure P3–63 Truss bridge (FS = 3.0)

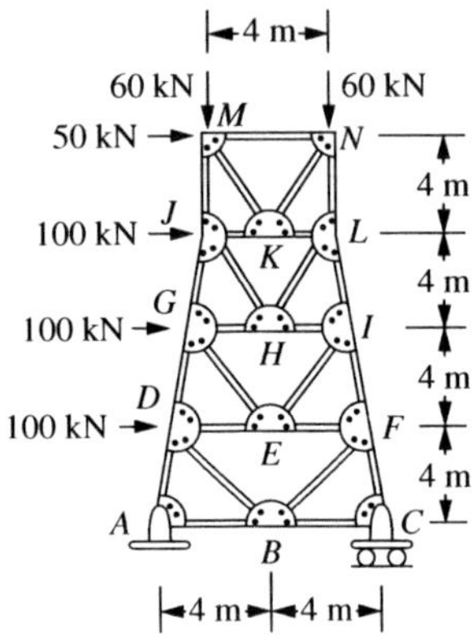

Figure P3–64 Tower (FS = 2.5)

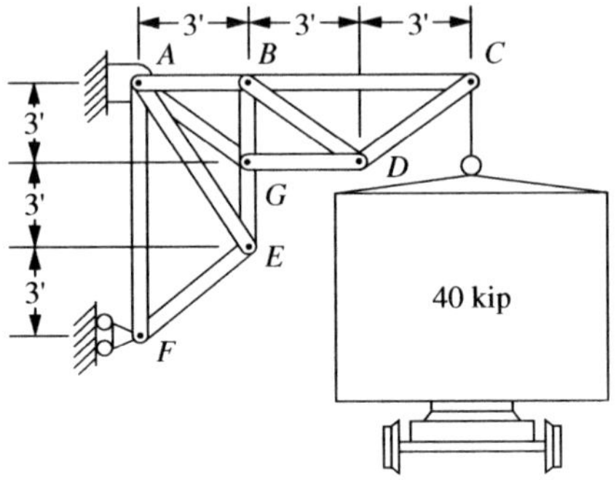

Figure P3–65 Boxcar lift (FS = 3.0)

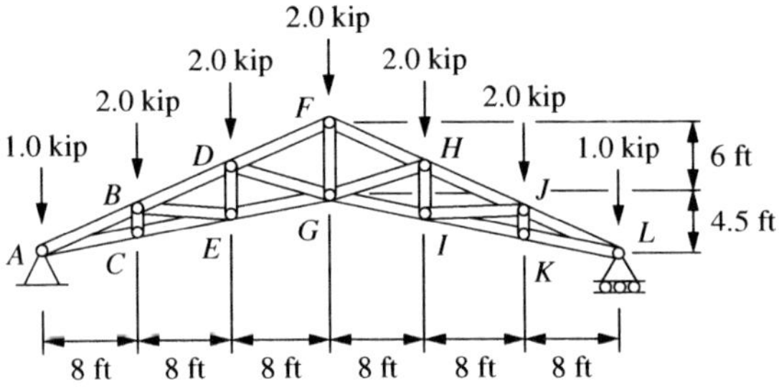

Figure P3–66 Howe scissors roof truss (FS = 2.0)

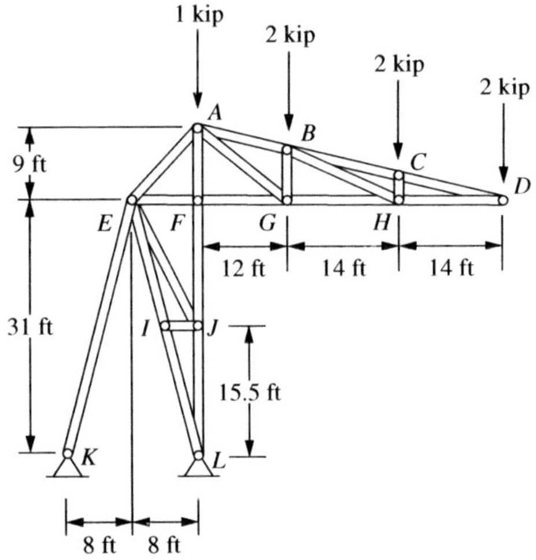

Figure P3–67 Stadium roof truss (FS = 3.0)

CHAPTER 4

Development of Beam Equations

Introduction

We begin this chapter by developing the stiffness matrix for the bending of a beam element, the most common of all structural elements as evidenced by its prominence in buildings, bridges, towers, and many other structures. The beam element is considered to be straight and to have constant cross-sectional area. We will first derive the beam element stiffness matrix by using the principles developed for simple beam theory.

We will then present simple examples to illustrate the assemblage of beam element stiffness matrices and the solution of beam problems by the direct stiffness method presented in Chapter 2. The solution of a beam problem illustrates that the degrees of freedom associated with a node are a transverse displacement and a rotation. We will include the nodal shear forces and bending moments and the resulting shear force and bending moment diagrams as part of the total solution.

Next, we will discuss procedures for handling distributed loading, because beams and frames are often subjected to distributed loading as well as concentrated nodal loading. We will follow the discussion with solutions of beams subjected to distributed loading and compare a finite element solution to an exact solution for a beam subjected to a distributed loading.

We will then develop the beam element stiffness matrix for a beam element with a nodal hinge and illustrate the solution of a beam with an internal hinge.

To further acquaint you with the potential energy approach for developing stiffness matrices and equations, we will again develop the beam bending element equations using this approach. We hope to increase your confidence in this approach. It will be used throughout much of this text to develop stiffness matrices and equations for more complex elements, such as two-dimensional (plane) stress, axisymmetric, and three-dimensional stress.

Finally, the Galerkin residual method is applied to derive the beam element equations.

The concepts presented in this chapter are prerequisite to understanding the concepts for frame analysis presented in Chapter 5.

4.1 Beam Stiffness

In this section, we will derive the stiffness matrix for a simple beam element. A **beam** *is a long, slender structural member generally subjected to transverse loading that produces significant bending effects as opposed to twisting or axial effects.* This bending deformation is measured as a transverse displacement and a rotation. Hence, the degrees of freedom considered per node are a transverse displacement and a rotation (as opposed to only an axial displacement for the bar element of Chapter 3).

Consider the beam element shown in Figure 4–1. The beam is of length L with axial local coordinate $\hat{x}$ and transverse local coordinate $\hat{y}$. The local transverse nodal displacements are given by $\hat{d}_{iy}$'s and the rotations by $\hat{\phi}_i$'s. The local nodal forces are given by $\hat{f}_{iy}$'s and the bending moments by $\hat{m}_i$'s as shown. We initially neglect all axial effects.

At all nodes, the following sign conventions are used:

1. Moments are positive in the counterclockwise direction.
2. Rotations are positive in the counterclockwise direction.
3. Forces are positive in the positive $\hat{y}$ direction.
4. Displacements are positive in the positive $\hat{y}$ direction.

Figure 4–2 indicates the sign conventions used in simple beam theory for positive shear forces $\hat{V}$ and bending moments $\hat{m}$.

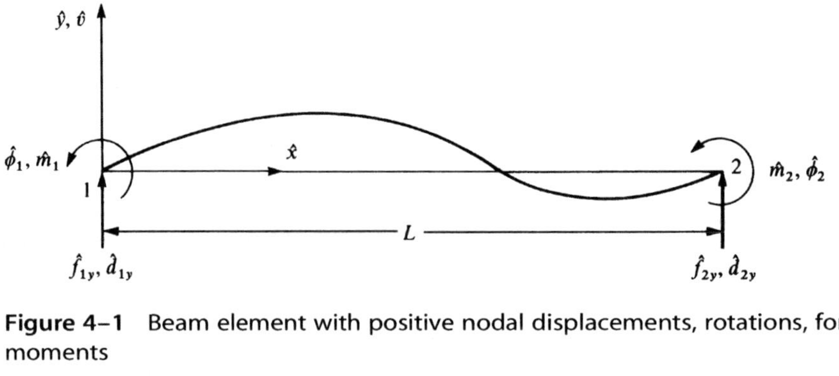

Figure 4–1 Beam element with positive nodal displacements, rotations, forces, and moments

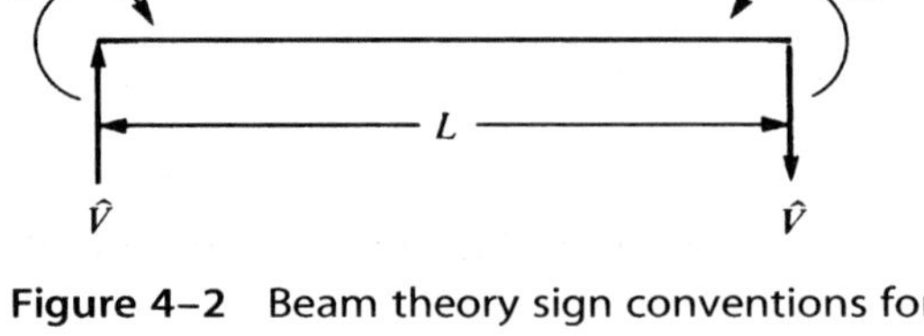

Figure 4–2 Beam theory sign conventions for shear forces and bending moments

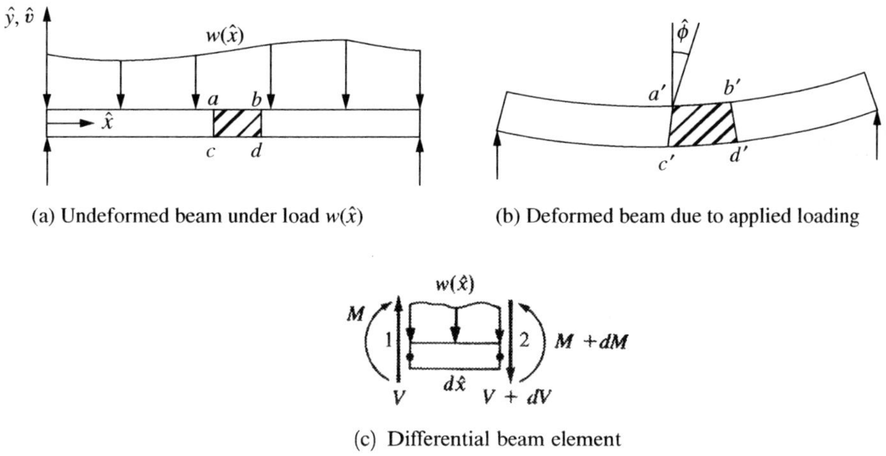

Figure 4–3 Beam under distributed load

Beam Stiffness Matrix Based on Euler-Bernouli Beam Theory (Considering Bending Deformations Only)

The differential equation governing elementary linear-elastic beam behavior [1] (called the Euler-Bernoulli beam as derived by Euler and Bernoulli) is based on plane cross sections perpendicular to the longitudinal centroidal axis of the beam before bending occurs remaining plane and perpendicular to the longitudinal axis after bending occurs. This is illustrated in Figure 4–3, where a plane through vertical line $a-c$ (Figure 4–3(a)) is perpendicular to the longitudinal $\hat{x}$ axis before bending, and this same plane through $a'-c'$ (rotating through angle $\hat{\phi}$ in Figure 4–3(b)) remains perpendicular to the bent $\hat{x}$ axis after bending. This occurs in practice only when a pure couple or constant moment exists in the beam. However it is a reasonable assumption that yields equations that quite accurately predict beam behavior for most practical beams.

The differential equation is derived as follows. Consider the beam shown in Figure 4–3 subjected to a distributed loading $w(\hat{x})$ (force/length). From force and moment equilibrium of a differential element of the beam, shown in Figure 4–3(c), we have

$$\Sigma F_y = 0\text{:} \; V - (V + dV) - w(\hat{x})\,dx = 0 \tag{4.1.1a}$$

Or, simplifying Eq. (4.1.1a), we obtain

$$-w\,d\hat{x} - dV = 0 \qquad \text{or} \qquad w = -\frac{dV}{d\hat{x}} \tag{4.1.1b}$$

$$\Sigma M_2 = 0\text{:} \; -V\,dx + dM + w(\hat{x})\,d\hat{x}\left(\frac{d\hat{x}}{2}\right) = 0 \qquad \text{or} \qquad V = \frac{dM}{d\hat{x}} \tag{4.1.1c}$$

The final form of Eq. (4.1.1c), relating the shear force to the bending moment, is obtained by dividing the left equation by $d\hat{x}$ and then taking the limit of the equation as $d\hat{x}$ approaches 0. The $w(\hat{x})$ term then disappears.

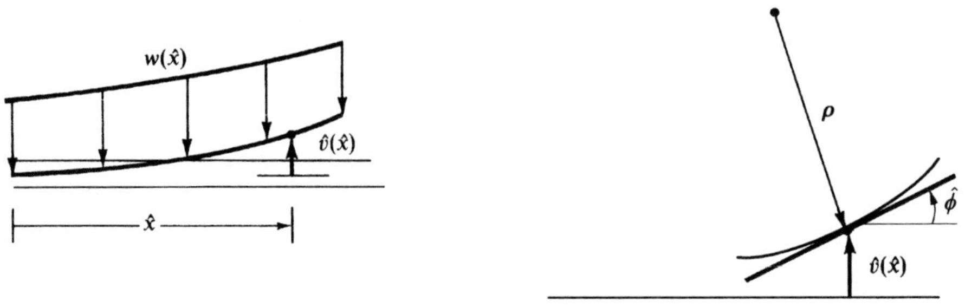

(a) Portion of deflected curve of beam (b) Radius of deflected curve at $\hat{v}(\hat{x})$

Figure 4–4 Deflected curve of beam

Also, the curvature κ of the beam is related to the moment by

$$\kappa = \frac{1}{\rho} = \frac{M}{EI} \tag{4.1.1d}$$

where ρ is the radius of the deflected curve shown in Figure 4–4b, $\hat{v}$ is the transverse displacement function in the $\hat{y}$ direction (see Figure 4–4a), E is the modulus of elasticity, and I is the principal moment of inertia about the $\hat{z}$ axis (where the $\hat{z}$ axis is perpendicular to the $\hat{x}$ and $\hat{y}$ axes).

The curvature for small slopes $\hat{\phi} = d\hat{v}/d\hat{x}$ is given by

$$\kappa = \frac{d^2\hat{v}}{d\hat{x}^2} \tag{4.1.1e}$$

Using Eq. (4.1.1e) in (4.1.1d), we obtain

$$\frac{d^2\hat{v}}{d\hat{x}^2} = \frac{M}{EI} \tag{4.1.1f}$$

Solving Eq. (4.1.1f) for M and substituting this result into (4.1.1c) and (4.1.1b), we obtain

$$\frac{d^2}{d\hat{x}^2}\left(EI\frac{d^2\hat{v}}{d\hat{x}^2}\right) = -w(\hat{x}) \tag{4.1.1g}$$

For constant EI and only nodal forces and moments, Eq. (4.1.1g) becomes

$$EI\frac{d^4\hat{v}}{d\hat{x}^4} = 0 \tag{4.1.1h}$$

We will now follow the steps outlined in Chapter 1 to develop the stiffness matrix and equations for a beam element and then to illustrate complete solutions for beams.

Step 1 Select the Element Type

Represent the beam by labeling nodes at each end and in general by labeling the element number (Figure 4–1).

Step 2 Select a Displacement Function

Assume the transverse displacement variation through the element length to be

$$\hat{v}(\hat{x}) = a_1\hat{x}^3 + a_2\hat{x}^2 + a_3\hat{x} + a_4 \tag{4.1.2}$$

The complete cubic displacement function Eq. (4.1.2) is appropriate because there are four total degrees of freedom (a transverse displacement $\hat{d}_{iy}$ and a small rotation $\hat{\phi}_i$ at each node). The cubic function also satisfies the basic beam differential equation—further justifying its selection. In addition, the cubic function also satisfies the conditions of displacement and slope continuity at nodes shared by two elements.

Using the same procedure as described in Section 2.2, we express $\hat{v}$ as a function of the nodal degrees of freedom $\hat{d}_{1y}$, $\hat{d}_{2y}$, $\hat{\phi}_1$, and $\hat{\phi}_2$ as follows:

$$\begin{aligned}
\hat{v}(0) &= \hat{d}_{1y} = a_4 \\
\frac{d\hat{v}(0)}{d\hat{x}} &= \hat{\phi}_1 = a_3 \\
\hat{v}(L) &= \hat{d}_{2y} = a_1L^3 + a_2L^2 + a_3L + a_4 \\
\frac{d\hat{v}(L)}{d\hat{x}} &= \hat{\phi}_2 = 3a_1L^2 + 2a_2L + a_3
\end{aligned} \tag{4.1.3}$$

where $\hat{\phi} = d\hat{v}/d\hat{x}$ for the assumed small rotation $\hat{\phi}$. Solving Eqs. (4.1.3) for a_1 through a_4 in terms of the nodal degrees of freedom and substituting into Eq. (4.1.2), we have

$$\begin{aligned}
\hat{v} = &\left[\frac{2}{L^3}(\hat{d}_{1y} - \hat{d}_{2y}) + \frac{1}{L^2}(\hat{\phi}_1 + \hat{\phi}_2)\right]\hat{x}^3 \\
&+ \left[-\frac{3}{L^2}(\hat{d}_{1y} - \hat{d}_{2y}) - \frac{1}{L}(2\hat{\phi}_1 + \hat{\phi}_2)\right]\hat{x}^2 + \hat{\phi}_1\hat{x} + \hat{d}_{1y}
\end{aligned} \tag{4.1.4}$$

In matrix form, we express Eq. (4.1.4) as

$$\hat{v} = [N]\{\hat{d}\} \tag{4.1.5}$$

where

$$\{\hat{d}\} = \begin{Bmatrix} \hat{d}_{1y} \\ \hat{\phi}_1 \\ \hat{d}_{2y} \\ \hat{\phi}_2 \end{Bmatrix} \tag{4.1.6a}$$

and where

$$[N] = [N_1 \quad N_2 \quad N_3 \quad N_4] \tag{4.1.6b}$$

and

$$\begin{aligned}
N_1 &= \frac{1}{L^3}(2\hat{x}^3 - 3\hat{x}^2L + L^3) \qquad & N_2 &= \frac{1}{L^3}(\hat{x}^3L - 2\hat{x}^2L^2 + \hat{x}L^3) \\
N_3 &= \frac{1}{L^3}(-2\hat{x}^3 + 3\hat{x}^2L) \qquad & N_4 &= \frac{1}{L^3}(\hat{x}^3L - \hat{x}^2L^2)
\end{aligned} \tag{4.1.7}$$

N_1, N_2, N_3, and N_4 are called the **shape functions** for a beam element. These cubic shape (or interpolation) functions are known as *Hermite cubic interpolation (or cubic* *spline) functions.* For the beam element, $N_1 = 1$ when evaluated at node 1 and $N_1 = 0$ when evaluated at node 2. Because N_2 is associated with $\hat{\phi}_1$, we have, from the second of Eqs. (4.1.7), $(dN_2/d\hat{x}) = 1$ when evaluated at node 1. Shape functions N_3 and N_4 have analogous results for node 2.

Step 3 Define the Strain/Displacement and Stress/Strain Relationships

Assume the following axial strain/displacement relationship to be valid:

$$\varepsilon_x(\hat{x}, \hat{y}) = \frac{d\hat{u}}{d\hat{x}} \tag{4.1.8}$$

where $\hat{u}$ is the axial displacement function. From the deformed configuration of the beam shown in Figure 4–5, we relate the axial displacement to the transverse displacement by

$$\hat{u} = -\hat{y}\frac{d\hat{v}}{d\hat{x}} \tag{4.1.9}$$

where we should recall from elementary beam theory [1] the basic assumption that cross sections of the beam (such as cross section $ABCD$) that are planar before bending deformation remain planar after deformation and, in general, rotate through a small angle $(d\hat{v}/d\hat{x})$. Using Eq. (4.1.9) in Eq. (4.1.8), we obtain

$$\varepsilon_x(\hat{x}, \hat{y}) = -\hat{y}\frac{d^2\hat{v}}{d\hat{x}^2} \tag{4.1.10}$$

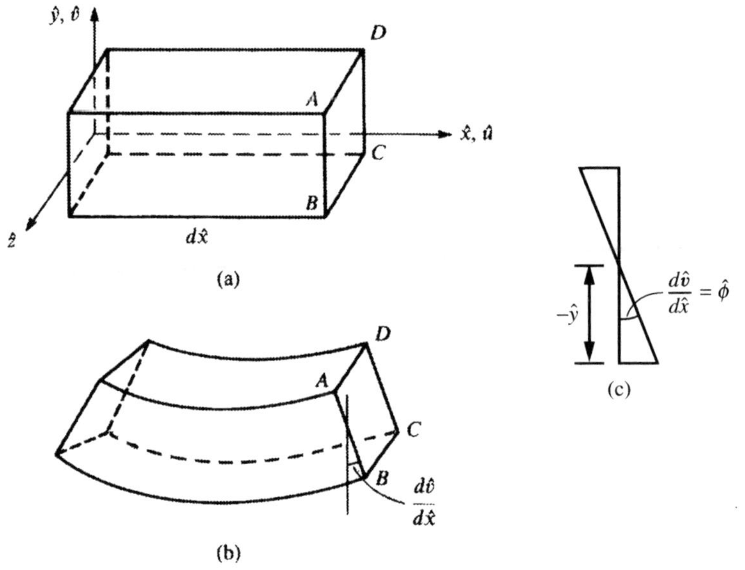

Figure 4–5 Beam segment (a) before deformation and (b) after deformation; (c) angle of rotation of cross section *ABCD*

From elementary beam theory, the bending moment and shear force are related to the transverse displacement function. Because we will use these relationships in the derivation of the beam element stiffness matrix, we now present them as

$$\hat{m}(\hat{x}) = EI\frac{d^2\hat{v}}{d\hat{x}^2} \qquad \hat{V} = EI\frac{d^3\hat{v}}{d\hat{x}^3} \tag{4.1.11}$$

Step 4 Derive the Element Stiffness Matrix and Equations

First, derive the element stiffness matrix and equations using a direct equilibrium approach. We now relate the nodal and beam theory sign conventions for shear forces and bending moments (Figures 4–1 and 4–2), along with Eqs. (4.1.4) and (4.1.11), to obtain

$$\begin{aligned}
\hat{f}_{1y} &= \hat{V} = EI\frac{d^3\hat{v}(0)}{d\hat{x}^3} = \frac{EI}{L^3}(12\hat{d}_{1y} + 6L\hat{\phi}_1 - 12\hat{d}_{2y} + 6L\hat{\phi}_2) \\
\hat{m}_1 &= -\hat{m} = -EI\frac{d^2\hat{v}(0)}{d\hat{x}^2} = \frac{EI}{L^3}(6L\hat{d}_{1y} + 4L^2\hat{\phi}_1 - 6L\hat{d}_{2y} + 2L^2\hat{\phi}_2) \\
\hat{f}_{2y} &= -\hat{V} = -EI\frac{d^3\hat{v}(L)}{d\hat{x}^3} = \frac{EI}{L^3}(-12\hat{d}_{1y} - 6L\hat{\phi}_1 + 12\hat{d}_{2y} - 6L\hat{\phi}_2) \\
\hat{m}_2 &= \hat{m} = EI\frac{d^2\hat{v}(L)}{d\hat{x}^2} = \frac{EI}{L^3}(6L\hat{d}_{1y} + 2L^2\hat{\phi}_1 - 6L\hat{d}_{2y} + 4L^2\hat{\phi}_2)
\end{aligned} \tag{4.1.12}$$

where the minus signs in the second and third of Eqs. (4.1.12) are the result of opposite nodal and beam theory positive bending moment conventions at node 1 and opposite nodal and beam theory positive shear force conventions at node 2 as seen by comparing Figures 4–1 and 4–2. Equations (4.1.12) relate the nodal forces to the nodal displacements. In matrix form, Eqs. (4.1.12) become

$$\begin{Bmatrix} \hat{f}_{1y} \\ \hat{m}_1 \\ \hat{f}_{2y} \\ \hat{m}_2 \end{Bmatrix} = \frac{EI}{L^3}\begin{bmatrix} 12 & 6L & -12 & 6L \\ 6L & 4L^2 & -6L & 2L^2 \\ -12 & -6L & 12 & -6L \\ 6L & 2L^2 & -6L & 4L^2 \end{bmatrix}\begin{Bmatrix} \hat{d}_{1y} \\ \hat{\phi}_1 \\ \hat{d}_{2y} \\ \hat{\phi}_2 \end{Bmatrix} \tag{4.1.13}$$

where the stiffness matrix is then

$$\underline{\hat{k}} = \frac{EI}{L^3}\begin{bmatrix} 12 & 6L & -12 & 6L \\ 6L & 4L^2 & -6L & 2L^2 \\ -12 & -6L & 12 & -6L \\ 6L & 2L^2 & -6L & 4L^2 \end{bmatrix} \tag{4.1.14}$$

Equation (4.1.13) indicates that $\underline{\hat{k}}$ relates transverse forces and bending moments to transverse displacements and rotations, whereas axial effects have been neglected.

In the beam element stiffness matrix (Eq. (4.1.14) derived in this section, it is assumed that the beam is long and slender; that is, the length, L, to depth, h, dimension ratio of the beam is large. In this case, the deflection due to bending that is predicted by using the stiffness matrix from Eq. (4.1.14) is quite adequate. However, for short, deep beams the transverse shear deformation can be significant and can have the same order of magnitude contribution to the total deformation of the beam. This is seen by the expressions for the bending and shear contributions to the deflection of a beam, where the bending contribution is of order $(L/h)^3$, whereas the shear contribution is only of order (L/h). A general rule for rectangular cross-section beams, is that for a length at least eight times the depth, the transverse shear deflection is less than five percent of the bending deflection [4]. Castigliano's method for finding beam and frame deflections is a convenient way to include the effects of the transverse shear term as shown in [4]. The derivation of the stiffness matrix for a beam including the transverse shear deformation contribution is given in a number of references [5–8]. The inclusion of the shear deformation in beam theory with application to vibration problems was developed by Timoshenko and is known as the Timoshenko beam [9–10].

Beam Stiffness Matrix Based on Timoshenko Beam Theory (Including Transverse Shear Deformation)

The shear deformation beam theory is derived as follows. Instead of plane sections remaining plane after bending occurs as shown previously in Figure 4–5, the shear deformation (deformation due to the shear force V) is now included. Referring to Figure 4–6, we observe a section of a beam of differential length $d\hat{x}$ with the cross section assumed to remain plane but no longer perpendicular to the neutral axis

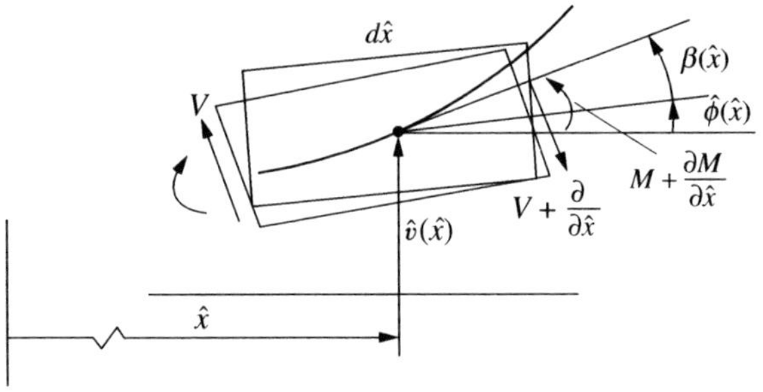

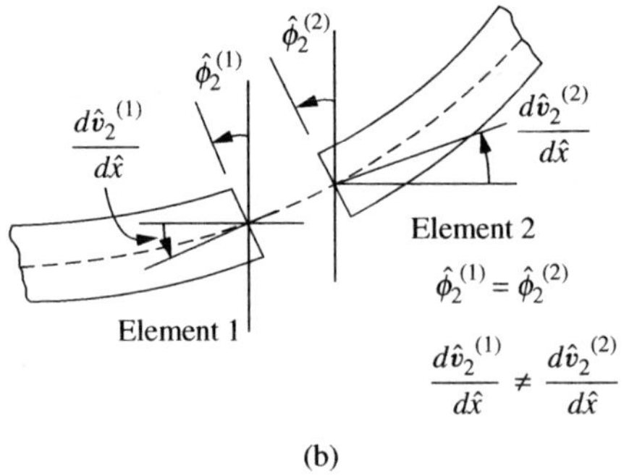

Figure 4–6 (a) Element of Timoshenko beam showing shear deformation. Cross sections are no longer perpendicular to the neutral axis line. (b) Two beam elements meeting at node 2

($\hat{x}$ axis) due to the inclusion of the shear force resulting in a rotation term indicated by β. The total deflection of the beam at a point $\hat{x}$ now consists of two parts, one caused by bending and one by shear force, so that the slope of the deflected curve at point $\hat{x}$ is now given by

$$\frac{d\hat{v}}{d\hat{x}} = \hat{\phi}(\hat{x}) + \beta(\hat{x}) \tag{4.1.15a}$$

where rotation due to bending moment and due to transverse shear force are given, respectively, by $\hat{\phi}(\hat{x})$ and $\beta(\hat{x})$.

We assume as usual that the linear deflection and angular deflection (slope) are small.

The relation between bending moment and bending deformation (curvature) is now

$$M(\hat{x}) = EI\,\frac{d\hat{\phi}(\hat{x})}{d\hat{x}} \tag{4.1.15b}$$

and the relation between the shear force and shear deformation (rotation due to shear) (shear strain) is given by

$$V(\hat{x}) = k_s AG\beta(\hat{x}) \tag{4.1.15c}$$

The difference in $d\hat{v}/d\hat{x}$ and $\hat{\phi}$ represents the shear strain $\gamma_{yz}(=\beta)$ of the beam as

$$\gamma_{yz} = \frac{d\hat{v}}{d\hat{x}} - \hat{\phi} \tag{4.1.15d}$$

Now consider the differential element in Figure 4–3c and Eqs. (4.1.1b) and (4.1.1c) obtained from summing transverse forces and then summing bending moments. We now substitute Eq. (4.1.15c) for V and Eq. (4.1.15b) for M into Eqs. (4.1.1b) and (4.1.1c) along with β from Eq. (4.1.15a) to obtain the two governing differential equations as

$$\frac{d}{d\hat{x}}\left[k_s AG\left(\frac{d\hat{v}}{d\hat{x}} - \hat{\phi}\right)\right] = -w \tag{4.1.15e}$$

$$\frac{d}{d\hat{x}}\left(EI\frac{d\hat{\phi}}{d\hat{x}}\right) + k_s AG\left(\frac{d\hat{v}}{d\hat{x}} - \hat{\phi}\right) = 0 \tag{4.1.15f}$$

To derive the stiffness matrix for the beam element including transverse shear deformation, we assume the transverse displacement to be given by the cubic function in Eq. (4.1.2). In a manner similar to [8], we choose transverse shear strain γ consistent with the cubic polynomial for $\hat{v}(\hat{x})$, such that γ is a constant given by

$$\gamma = c \tag{4.1.15g}$$

Using the cubic displacement function for $\hat{v}$, the slope relation given by Eq. (4.1.15a), and the shear strain Eq. (4.1.15g), along with the bending moment-curvature relation, Eq. (4.1.15b) and the shear force-shear strain relation Eq. (4.1.15c), in the bending moment–shear force relation Eq. (4.1.1c), we obtain

$$c = 6a_1 g \tag{4.1.15h}$$

where $g = EI/k_s AG$ and $k_s A$ is the shear area. Shear areas, A_s vary with cross-section shapes. For instance, for a rectangular shape A_s is taken as 5/6 times the cross section A, for a solid circular cross section it is taken as 0.9 times the cross section, for a wide-flange cross section it is taken as the thickness of the web times the full depth of the wide-flange, and for thin-walled cross sections it is taken as two times the product of the thickness of the wall times the depth of the cross section.

Using Eqs. (4.1.2), (4.1.15a), (4.1.15g), and (4.1.15h) allow ϕ to be expressed as a polynomial in $\hat{x}$ as follows:

$$\hat{\phi} = a_3 + 2a_2\hat{x} + (3\hat{x}^2 + 6g)a_1 \tag{4.1.15i}$$

Using Eqs. (4.1.2) and (4.1.15i), we can now express the coefficients a_1 through a_4 in terms of the nodal displacements $\hat{d}_{1y}$ and $\hat{d}_{2y}$ and rotations $\hat{\phi}_1$ and $\hat{\phi}_2$ of the beam at the ends $\hat{x} = 0$ and $\hat{x} = L$ as previously done to obtain Eq. (4.1.4) when shear deformation was neglected. The expressions for a_1 through a_4 are then given as follows:

$$\begin{aligned}
a_1 &= \frac{2\hat{d}_{1y} + L\hat{\phi}_1 - 2\hat{d}_{2y} + L\hat{\phi}_2}{L(L^2 + 12g)} \\
a_2 &= \frac{-3L\hat{d}_{1y} - (2L^2 + 6g)\hat{\phi}_1 + 3L\hat{d}_{2y} + (-L^2 + 6g)\hat{\phi}_2}{L(L^2 + 12g)} \\
a_3 &= \frac{-12g\hat{d}_{1y} + (L^3 + 6gL)\hat{\phi}_1 + 12g\hat{d}_{2y} - 6gL\hat{\phi}_2}{L(L^2 + 12g)} \\
a_4 &= \hat{d}_{1y}
\end{aligned} \tag{4.1.15j}$$

Substituting these a's into Eq. (4.1.2), we obtain

$$\begin{aligned}
\hat{v} = {} & \frac{2\hat{d}_{1y} + L\hat{\phi}_1 - 2\hat{d}_{2y} + L\hat{\phi}_2}{L(L^2 + 12g)}\hat{x}^3 \\
& \frac{-3L\hat{d}_{1y} - (2L^2 + 6g)\hat{\phi}_1 + 3L\hat{d}_{2y} + (-L^2 + 6g)\hat{\phi}^2}{L(L^2 + 12g)}\hat{x}^2 \\
& \frac{-12g\hat{d}_{1y} + (L^3 + 6gL)\hat{\phi}_1 + 12g\hat{d}_{2y} - 6gL\hat{\phi}_2}{L(L^2 + 12g)}\hat{x} + \hat{d}_{1y}
\end{aligned} \tag{4.1.15k}$$

In a manner similar to step 4 used to derive the stiffness matrix for the beam element without shear deformation included, we have

$$\begin{aligned}
\hat{f}_{1y} &= \hat{V}(0) = 6EIa_1 = \frac{EI(12\hat{d}_{1y} + 6L\hat{\phi}_1 - 12\hat{d}_{2y} + 6L\hat{\phi}_2)}{L(L^2 + 12g)} \\
\hat{m}_1 &= -\hat{m}(0) = -2EIa_2 = \frac{EI[6L\hat{d}_{1y} + (4L^2 + 12g)\hat{\phi}_1 - 6L\hat{d}_{2y} + (2L^2 - 12g)\hat{\phi}_2]}{L(L^2 + 12g)} \\
\hat{f}_{2y} &= -\hat{V}(L) = \frac{EI(-12\hat{d}_{1y} - 6L\hat{\phi}_1 + 12\hat{d}_{2y} - 6L\hat{\phi}_2)}{L(L^2 + 12g)} \\
\hat{m}_2 &= \hat{m}(L) = \frac{EI[6L\hat{d}_{1y} + (2L^2 - 12g)\hat{\phi}_1 - 6L\hat{d}_{2y} + (4L^2 + 12g)\hat{\phi}_2]}{L(L^2 + 12g)}
\end{aligned} \tag{4.1.15l}$$

where again the minus signs in the second and third of Eqs. (4.1.15l) are the result of opposite nodal and beam theory positive moment conventions at node 1 and opposite nodal and beam theory positive shear force conventions at node 2, as seen by comparing Figures 4–2 and 4–7. In matrix form Eqs (4.1.15l) become

$$\begin{Bmatrix} \hat{f}_{1y} \\ \hat{m}_1 \\ \hat{f}_{2y} \\ \hat{m}_2 \end{Bmatrix} = \frac{EI}{L(L^2+12g)} \begin{bmatrix} 12 & 6L & -12 & 6L \\ 6L & (4L^2+12g) & -6L & (2L^2-12g) \\ -12 & -6L & 12 & -6L \\ 6L & (2L^2-12g) & -6L & (4L^2+12g) \end{bmatrix} \begin{Bmatrix} \hat{d}_{1y} \\ \hat{\phi}_1 \\ \hat{d}_{2y} \\ \hat{\phi}_2 \end{Bmatrix} \tag{4.1.15m}$$

where the stiffness matrix including both bending and shear deformation is then given by

$$\underline{\hat{k}} = \frac{EI}{L(L^2+12g)} \begin{bmatrix} 12 & 6L & -12 & 6L \\ 6L & (4L^2+12g) & -6L & (2L^2-12g) \\ 12 & -6L & 12 & -6L \\ 6L & (2L^2-12g) & -6L & (4L^2+12g) \end{bmatrix} \tag{4.1.15n}$$

In Eq. (4.1.15n) remember that g represents the transverse shear term, and if we set $g = 0$, we obtain Eq. (4.1.14) for the beam stiffness matrix, neglecting transverse shear deformation. To more easily see the effect of the shear correction factor, we define the nondimensional shear correction term as $\varphi = 12EI/(k_s AGL^2) = 12g/L^2$ and rewrite the stiffness matrix as

$$\underline{\hat{k}} = \frac{EI}{L^3(1+\varphi)} \begin{bmatrix} 12 & 6L & -12 & 6L \\ 6L & (4+\varphi)L^2 & -6L & (2-\varphi)L^2 \\ -12 & -6L & 12 & -6L \\ 6L & (2-\varphi)L^2 & -6L & (4+\varphi)L^2 \end{bmatrix} \tag{4.1.15o}$$

Most commercial computer programs, such as [11], will include the shear deformation by having you input the shear area, $A_s = k_s A$.

4.2 Example of Assemblage of Beam Stiffness Matrices

Step 5 Assemble the Element Equations to Obtain the Global Equations and Introduce Boundary Conditions

Consider the beam in Figure 4–7 as an example to illustrate the procedure for assemblage of beam element stiffness matrices. Assume EI to be constant throughout the beam. A force of 1000 lb and a moment of 1000 lb-ft are applied to the beam at midlength. The left end is a fixed support and the right end is a pin support.

First, we discretize the beam into two elements with nodes 1–3 as shown. We include a node at midlength because applied force and moment exist at midlength and, at this time, loads are assumed to be applied only at nodes. (Another procedure for handling loads applied on elements will be discussed in Section 4.4.)

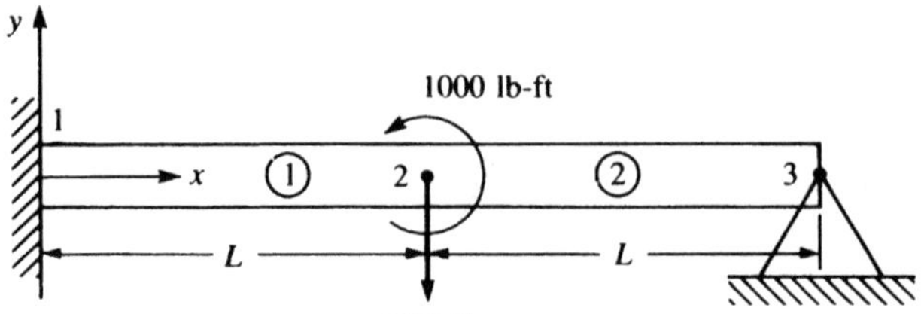

Figure 4–7 Fixed hinged beam subjected to a force and a moment

Using Eq. (4.1.14), we find that the global stiffness matrices for the two elements are now given by

$$\underline{k}^{(1)} = \frac{EI}{L^3}\begin{array}{c} \begin{matrix} d_{1y} & \phi_1 & d_{2y} & \phi_2 \end{matrix} \\ \begin{bmatrix} 12 & 6L & -12 & 6L \\ 6L & 4L^2 & -6L & 2L^2 \\ -12 & -6L & 12 & -6L \\ 6L & 2L^2 & -6L & 4L^2 \end{bmatrix} \end{array} \tag{4.2.1}$$

and

$$\underline{k}^{(2)} = \frac{EI}{L^3}\begin{array}{c} \begin{matrix} d_{2y} & \phi_2 & d_{3y} & \phi_3 \end{matrix} \\ \begin{bmatrix} 12 & 6L & -12 & 6L \\ 6L & 4L^2 & -6L & 2L^2 \\ -12 & -6L & 12 & -6L \\ 6L & 2L^2 & -6L & 4L^2 \end{bmatrix} \end{array} \tag{4.2.2}$$

where the degrees of freedom associated with each beam element are indicated by the usual labels above the columns in each element stiffness matrix. Here the local coordinate axes for each element coincide with the global x and y axes of the whole beam. Consequently, the local and global stiffness matrices are identical, so hats (ˆ) are not needed in Eqs. (4.2.1) and (4.2.2).

The total stiffness matrix can now be assembled for the beam by using the direct stiffness method. When the total (global) stiffness matrix has been assembled, the external global nodal forces are related to the global nodal displacements. Through direct superposition and Eqs. (4.2.1) and (4.2.2), the governing equations for the beam are thus given by

$$\begin{Bmatrix} F_{1y} \\ M_1 \\ F_{2y} \\ M_2 \\ F_{3y} \\ M_3 \end{Bmatrix} = \frac{EI}{L^3}\begin{bmatrix} 12 & 6L & -12 & 6L & 0 & 0 \\ 6L & 4L^2 & -6L & 2L^2 & 0 & 0 \\ -12 & -6L & 12+12 & -6L+6L & -12 & 6L \\ 6L & 2L^2 & -6L+6L & 4L^2+4L^2 & -6L & 2L^2 \\ 0 & 0 & -12 & -6L & 12 & -6L \\ 0 & 0 & 6L & 2L^2 & -6L & 4L^2 \end{bmatrix}\begin{Bmatrix} d_{1y} \\ \phi_1 \\ d_{2y} \\ \phi_2 \\ d_{3y} \\ \phi_3 \end{Bmatrix} \tag{4.2.3}$$

Now considering the boundary conditions, or constraints, of the fixed support at node 1 and the hinge (pinned) support at node 3, we have

$$\phi_1 = 0 \qquad d_{1y} = 0 \qquad d_{3y} = 0 \tag{4.2.4}$$

On considering the third, fourth, and sixth equations of Eqs. (4.2.3) corresponding to the rows with unknown degrees of freedom and using Eqs. (4.2.4), we obtain

$$\begin{Bmatrix} -1000 \\ 1000 \\ 0 \end{Bmatrix} = \frac{EI}{L^3} \begin{bmatrix} 24 & 0 & 6L \\ 0 & 8L^2 & 2L^2 \\ 6L & 2L^2 & 4L^2 \end{bmatrix} \begin{Bmatrix} d_{2y} \\ \phi_2 \\ \phi_3 \end{Bmatrix} \tag{4.2.5}$$

where $F_{2y} = -1000$ lb, $M_2 = 1000$ lb-ft, and $M_3 = 0$ have been substituted into the reduced set of equations. We could now solve Eq. (4.2.5) simultaneously for the unknown nodal displacement d_{2y} and the unknown nodal rotations ϕ_2 and ϕ_3. We leave the final solution for you to obtain. Section 4.3 provides complete solutions to beam problems.

4.3 Examples of Beam Analysis Using the Direct Stiffness Method

We will now perform complete solutions for beams with various boundary supports and loads to illustrate further the use of the equations developed in Section 4.1.

Example 4.1

Using the direct stiffness method, solve the problem of the propped cantilever beam subjected to end load P in Figure 4–8. The beam is assumed to have constant EI and length $2L$. It is supported by a roller at midlength and is built in at the right end.

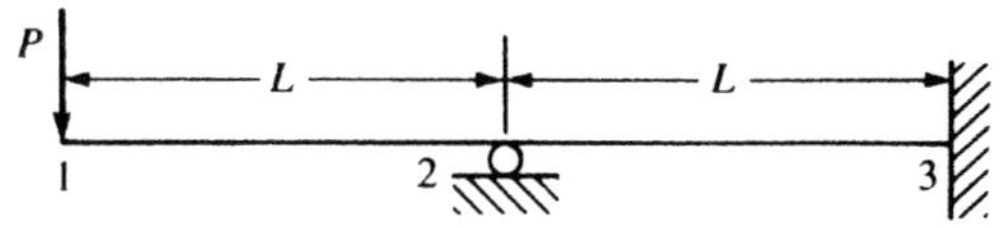

Figure 4–8 Propped cantilever beam

We have discretized the beam and established global coordinate axes as shown in Figure 4–8. We will determine the nodal displacements and rotations, the reactions, and the complete shear force and bending moment diagrams.

Using Eq. (4.1.14) for each element, along with superposition, we obtain the structure total stiffness matrix by the same method as described in Section 4.2 for obtaining the stiffness matrix in Eq. (4.2.3). The $\underline{K}$ is

$$\underline{K} = \frac{EI}{L^3} \begin{bmatrix} \overset{d_{1y}}{12} & \overset{\phi_1}{6L} & \overset{d_{2y}}{-12} & \overset{\phi_2}{6L} & \overset{d_{3y}}{0} & \overset{\phi_3}{0} \\ & 4L^2 & -6L & 2L^2 & 0 & 0 \\ & & 12+12 & -6L+6L & -12 & 6L \\ & & & 4L^2+4L^2 & -6L & 2L^2 \\ & & & & 12 & -6L \\ \text{Symmetry} & & & & & 4L^2 \end{bmatrix} \tag{4.3.1}$$

The governing equations for the beam are then given by

$$\begin{Bmatrix} F_{1y} \\ M_1 \\ F_{2y} \\ M_2 \\ F_{3y} \\ M_3 \end{Bmatrix} = \frac{EI}{L^3} \begin{bmatrix} 12 & 6L & -12 & 6L & 0 & 0 \\ 6L & 4L^2 & -6L & 2L^2 & 0 & 0 \\ -12 & -6L & 24 & 0 & -12 & 6L \\ 6L & 2L^2 & 0 & 8L^2 & -6L & 2L^2 \\ 0 & 0 & -12 & -6L & 12 & -6L \\ 0 & 0 & 6L & 2L^2 & -6L & 4L^2 \end{bmatrix} \begin{Bmatrix} d_{1y} \\ \phi_1 \\ d_{2y} \\ \phi_2 \\ d_{3y} \\ \phi_3 \end{Bmatrix} \tag{4.3.2}$$

On applying the boundary conditions

$$d_{2y} = 0 \qquad d_{3y} = 0 \qquad \phi_3 = 0 \tag{4.3.3}$$

and partitioning the equations associated with unknown displacements [the first, second, and fourth equations of Eqs. (4.3.2)] from those equations associated with known displacements in the usual manner, we obtain the final set of equations for a longhand solution as

$$\begin{Bmatrix} -P \\ 0 \\ 0 \end{Bmatrix} = \frac{EI}{L^3} \begin{bmatrix} 12 & 6L & 6L \\ 6L & 4L^2 & 2L^2 \\ 6L & 2L^2 & 8L^2 \end{bmatrix} \begin{Bmatrix} d_{1y} \\ \phi_1 \\ \phi_2 \end{Bmatrix} \tag{4.3.4}$$

where $F_{1y} = -P$, $M_1 = 0$, and $M_2 = 0$ have been used in Eq. (4.3.4). We will now solve Eq. (4.3.4) for the nodal displacement and nodal slopes. We obtain the transverse displacement at node 1 as

$$d_{1y} = -\frac{7PL^3}{12EI} \tag{4.3.5}$$

where the minus sign indicates that the displacement of node 1 is downward.

The slopes are

$$\phi_1 = \frac{3PL^2}{4EI} \qquad \phi_2 = \frac{PL^2}{4EI} \tag{4.3.6}$$

where the positive signs indicate counterclockwise rotations at nodes 1 and 2.

We will now determine the global nodal forces. To do this, we substitute the known global nodal displacements and rotations, Eqs. (4.3.5) and (4.3.6), into Eq. (4.3.2). The resulting equations are

$$\begin{Bmatrix} F_{1y} \\ M_1 \\ F_{2y} \\ M_2 \\ F_{3y} \\ M_3 \end{Bmatrix} = \frac{EI}{L^3} \begin{bmatrix} 12 & 6L & -12 & 6L & 0 & 0 \\ 6L & 4L^2 & -6L & 2L^2 & 0 & 0 \\ -12 & -6L & 24 & 0 & -12 & 6L \\ 6L & 2L^2 & 0 & 8L^2 & -6L & 2L^2 \\ 0 & 0 & -12 & -6L & 12 & -6L \\ 0 & 0 & 6L & 2L^2 & -6L & 4L^2 \end{bmatrix} \begin{Bmatrix} -\dfrac{7PL^3}{12EI} \\ \dfrac{3PL^2}{4EI} \\ 0 \\ \dfrac{PL^2}{4EI} \\ 0 \\ 0 \end{Bmatrix} \tag{4.3.7}$$

Multiplying the matrices on the right-hand side of Eq. (4.3.7), we obtain the global nodal forces and moments as

$$\begin{array}{lll} F_{1y} = -P & M_1 = 0 & F_{2y} = \frac{5}{2}P \\ M_2 = 0 & F_{3y} = -\frac{3}{2}P & M_3 = \frac{1}{2}PL \end{array} \tag{4.3.8}$$

The results of Eqs. (4.3.8) can be interpreted as follows: The value of $F_{1y} = -P$ is the applied force at node 1, as it must be. The values of F_{2y}, F_{3y}, and M_3 are the reactions from the supports as felt by the beam. The moments M_1 and M_2 are zero because no applied or reactive moments are present on the beam at node 1 or node 2.

It is generally necessary to determine the local nodal forces associated with each element of a large structure to perform a stress analysis of the entire structure. We will thus consider the forces in element 1 of this example to illustrate this concept (element 2 can be treated similarly). Using Eqs. (4.3.5) and (4.3.6) in the $\underline{\hat{f}} = \underline{\hat{k}}\underline{\hat{d}}$ equation for element 1 [also see Eq. (4.1.13)], we have

$$\begin{Bmatrix} \hat{f}_{1y} \\ \hat{m}_1 \\ \hat{f}_{2y} \\ \hat{m}_2 \end{Bmatrix} = \frac{EI}{L^3} \begin{bmatrix} 12 & 6L & -12 & 6L \\ 6L & 4L^2 & -6L & 2L^2 \\ -12 & -6L & 12 & -6L \\ 6L & 2L^2 & -6L & 4L^2 \end{bmatrix} \begin{Bmatrix} -\dfrac{7PL^3}{12EI} \\ \dfrac{3PL^2}{4EI} \\ 0 \\ \dfrac{PL^2}{4EI} \end{Bmatrix} \tag{4.3.9}$$

where, again, because the local coordinate axes of the element coincide with the global axes of the whole beam, we have used the relationships $\underline{d} = \underline{\hat{d}}$ and $\underline{k} = \underline{\hat{k}}$ (that is, the local nodal displacements are also the global nodal displacements, and so forth). Equation (4.3.9) yields

$$\hat{f}_{1y} = -P \qquad \hat{m}_1 = 0 \qquad \hat{f}_{2y} = P \qquad \hat{m}_2 = -PL \tag{4.3.10}$$

A free-body diagram of element 1, shown in Figure 4–9(a), should help you to understand the results of Eqs. (4.3.10). The figure shows a nodal transverse force of negative P at node 1 and of positive P and negative moment PL at node 2. These values are consistent with the results given by Eqs. (4.3.10). For completeness, the free-body diagram of element 2 is shown in Figure 4–9(b). We can easily verify the element nodal forces by writing an equation similar to Eq. (4.3.9).

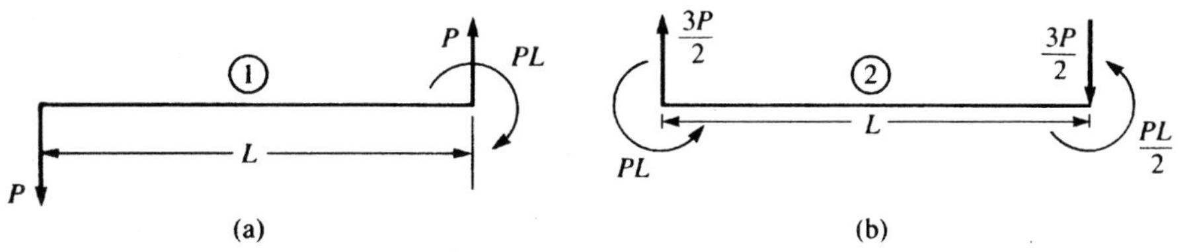

Figure 4–9 Free-body diagrams showing forces and moments on (a) element 1 and (b) element 2

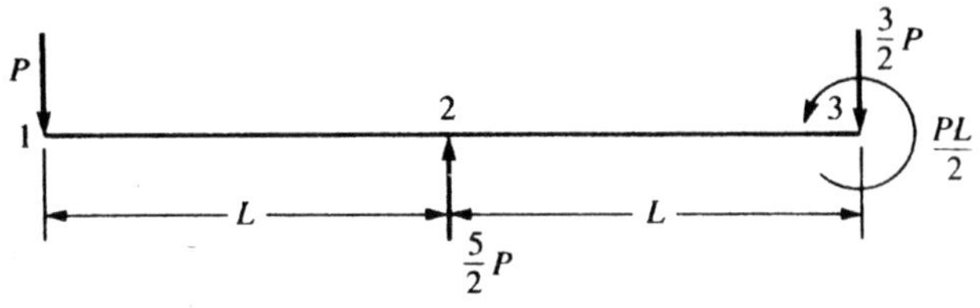

Figure 4–10 Nodal forces and moment on the beam

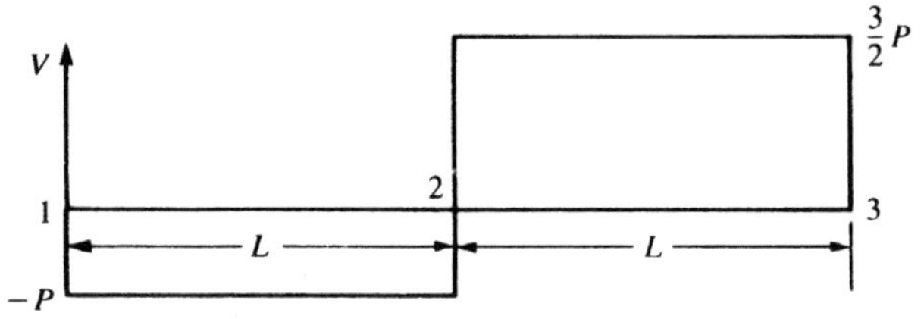

Figure 4–11 Shear force diagram for the beam of Figure 4–10

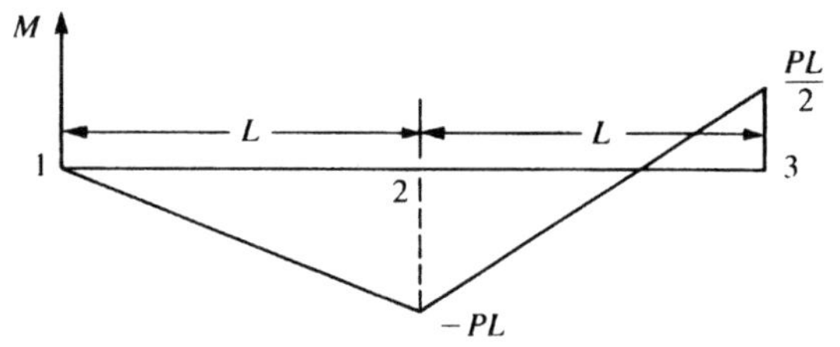

Figure 4–12 Bending moment diagram for the beam of Figure 4–10

From the results of Eqs. (4.3.8), the nodal forces and moments for the whole beam are shown on the beam in Figure 4–10. Using the beam sign conventions established in Section 4.1, we obtain the shear force V and bending moment M diagrams shown in Figures 4–11 and 4–12. ■

In general, for complex beam structures, we will use the element local forces to determine the shear force and bending moment diagrams for each element. We can then use these values for design purposes. Chapter 5 will further discuss this concept as used in computer codes.

Example 4.2

Determine the nodal displacements and rotations, global nodal forces, and element forces for the beam shown in Figure 4–13. We have discretized the beam as indicated by the node numbering. The beam is fixed at nodes 1 and 5 and has a roller support at node 3. Vertical loads of 10,000 lb each are applied at nodes 2 and 4. Let $E = 30 \times 10^6$ psi and $I = 500$ in^4 throughout the beam.

We must have consistent units; therefore, the 10-ft lengths in Figure 4–13 will be converted to 120 in. during the solution. Using Eq. (4.1.10), along with superposition of the four beam element stiffness matrices, we obtain the global stiffness matrix

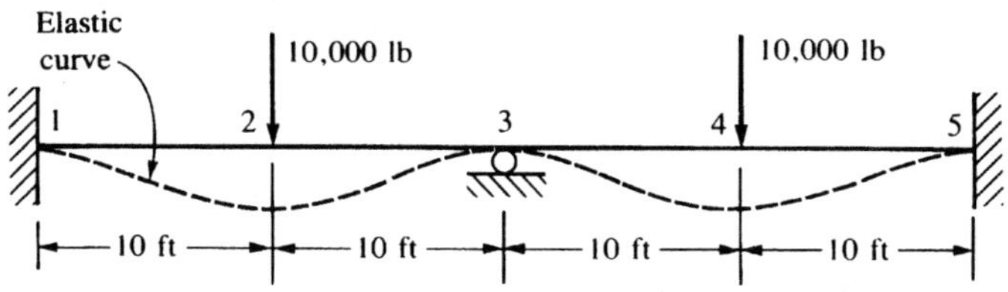

Figure 4–13 Beam example

and the global equations as given in Eq. (4.3.11). Here the lengths of each element are the same. Thus, we can factor an L out of the superimposed stiffness matrix.

$$\begin{Bmatrix} F_{1y} \\ M_1 \\ F_{2y} \\ M_2 \\ F_{3y} \\ M_3 \\ F_{4y} \\ M_4 \\ F_{5y} \\ M_5 \end{Bmatrix} = \frac{EI}{L^3} \begin{array}{c} \begin{matrix} d_{1y} & \phi_1 & d_{2y} & \phi_2 & d_{3y} & \phi_3 & d_{4y} & \phi_4 & d_{5y} & \phi_5 \end{matrix} \\ \begin{bmatrix} 12 & 6L & -12 & 6L & 0 & 0 & 0 & 0 & 0 & 0 \\ 6L & 4L^2 & -6L & 2L^2 & 0 & 0 & 0 & 0 & 0 & 0 \\ -12 & -6L & 12+12 & -6L+6L & -12 & 6L & 0 & 0 & 0 & 0 \\ 6L & 2L^2 & -6L+6L & 4L^2+4L^2 & -6L & 2L^2 & 0 & 0 & 0 & 0 \\ 0 & 0 & -12 & -6L & 12+12 & -6L+6L & -12 & 6L & 0 & 0 \\ 0 & 0 & 6L & 2L^2 & -6L+6L & 4L^2+4L^2 & -6L & 2L^2 & 0 & 0 \\ 0 & 0 & 0 & 0 & -12 & -6L & 12+12 & -6L+6L & -12 & 6L \\ 0 & 0 & 0 & 0 & 6L & 2L^2 & -6L+6L & 4L^2+4L^2 & -6L & 2L^2 \\ 0 & 0 & 0 & 0 & 0 & 0 & -12 & -6L & 12 & -6L \\ 0 & 0 & 0 & 0 & 0 & 0 & 6L & 2L^2 & -6L & 4L^2 \end{bmatrix} \end{array} \begin{Bmatrix} d_{1y} \\ \phi_1 \\ d_{2y} \\ \phi_2 \\ d_{3y} \\ \phi_3 \\ d_{4y} \\ \phi_4 \\ d_{5y} \\ \phi_5 \end{Bmatrix} \tag{4.3.11}$$

For a longhand solution, we reduce Eq. (4.3.11) in the usual manner by application of the boundary conditions

$$d_{1y} = \phi_1 = d_{3y} = d_{5y} = \phi_5 = 0$$

The resulting equation is

$$\begin{Bmatrix} -10{,}000 \\ 0 \\ 0 \\ -10{,}000 \\ 0 \end{Bmatrix} = \frac{EI}{L^3} \begin{bmatrix} 24 & 0 & 6L & 0 & 0 \\ 0 & 8L^2 & 2L^2 & 0 & 0 \\ 6L & 2L^2 & 8L^2 & -6L & 2L^2 \\ 0 & 0 & -6L^2 & 24 & 0 \\ 0 & 0 & 2L^2 & 0 & 8L^2 \end{bmatrix} \begin{Bmatrix} d_{2y} \\ \phi_2 \\ \phi_3 \\ d_{4y} \\ \phi_4 \end{Bmatrix} \tag{4.3.12}$$

The rotations (slopes) at nodes 2–4 are equal to zero because of symmetry in loading, geometry, and material properties about a plane perpendicular to the beam length and passing through node 3. Therefore, $\phi_2 = \phi_3 = \phi_4 = 0$, and we can further reduce Eq. (4.3.12) to

$$\begin{Bmatrix} -10{,}000 \\ -10{,}000 \end{Bmatrix} = \frac{EI}{L^3} \begin{bmatrix} 24 & 0 \\ 0 & 24 \end{bmatrix} \begin{Bmatrix} d_{2y} \\ d_{4y} \end{Bmatrix} \tag{4.3.13}$$

Solving for the displacements using $L = 120$ in., $E = 30 \times 10^6$ psi, and $I = 500$ in.^{4} in Eq. (4.3.13), we obtain

$$d_{2y} = d_{4y} = -0.048 \text{ in.} \tag{4.3.14}$$

as expected because of symmetry.

As observed from the solution of this problem, the greater the static redundancy (degrees of static indeterminacy or number of unknown forces and moments that cannot be determined by equations of statics), the smaller the kinematic redundancy (unknown nodal degrees of freedom, such as displacements or slopes)—hence, the fewer the number of unknown degrees of freedom to be solved for. Moreover, the use of symmetry, when applicable, reduces the number of unknown degrees of freedom even further. We can now back-substitute the results from Eq. (4.3.14), along with the numerical values for E, I, and L, into Eq. (4.3.12) to determine the global nodal forces as

$$\begin{aligned} F_{1y} &= 5000 \text{ lb} & M_1 &= 25{,}000 \text{ lb-ft} \\ F_{2y} &= 10{,}000 \text{ lb} & M_2 &= 0 \\ F_{3y} &= 10{,}000 \text{ lb} & M_3 &= 0 \\ F_{4y} &= 10{,}000 \text{ lb} & M_4 &= 0 \\ F_{5y} &= 5000 \text{ lb} & M_5 &= -25{,}000 \text{ lb-ft} \end{aligned} \tag{4.3.15}$$

Once again, the global nodal forces (and moments) at the support nodes (nodes 1, 3, and 5) can be interpreted as the reaction forces, and the global nodal forces at nodes 2 and 4 are the applied nodal forces.

However, for large structures we must obtain the local element shear force and bending moment at each node end of the element because these values are used in the design/analysis process. We will again illustrate this concept for the element connecting nodes 1 and 2 in Figure 4–13. Using the local equations for this element, for which all nodal displacements have now been determined, we obtain

$$\begin{Bmatrix} \hat{f}_{1y} \\ \hat{m}_1 \\ \hat{f}_{2y} \\ \hat{m}_2 \end{Bmatrix} = \frac{EI}{L^3} \begin{bmatrix} 12 & 6L & -12 & 6L \\ 6L & 4L^2 & -6L & 2L^2 \\ -12 & -6L & 12 & -6L \\ 6L & 2L^2 & -6L & 4L^2 \end{bmatrix} \begin{Bmatrix} \hat{d}_{1y} = 0 \\ \hat{\phi}_1 = 0 \\ \hat{d}_{2y} = -0.048 \\ \hat{\phi}_2 = 0 \end{Bmatrix} \tag{4.3.16}$$

Simplifying Eq. (4.3.16), we have

$$\begin{Bmatrix} \hat{f}_{1y} \\ \hat{m}_1 \\ \hat{f}_{2y} \\ \hat{m}_2 \end{Bmatrix} = \begin{Bmatrix} 5000 \text{ lb} \\ 25{,}000 \text{ lb-ft} \\ -5000 \text{ lb} \\ 25{,}000 \text{ lb-ft} \end{Bmatrix} \tag{4.3.17}$$

If you wish, you can draw a free-body diagram to confirm the equilibrium of the element. ■

Finally, you should note that because of reflective symmetry about a vertical plane passing through node 3, we could have initially considered one-half of this beam and used the following model. The fixed support at node 3 is due to the slope being zero at node 3 because of the symmetry in the loading and support conditions.

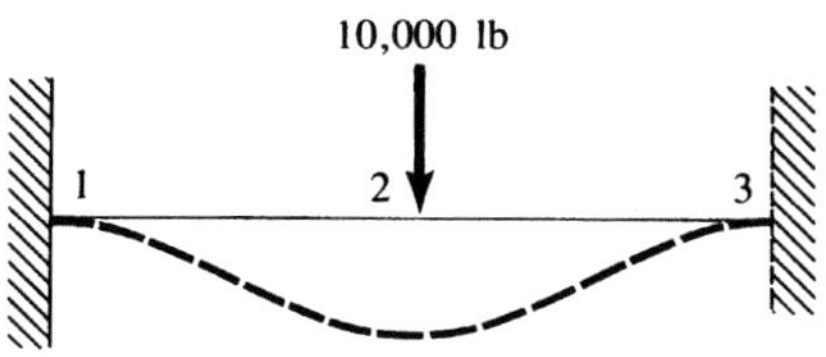

Example 4.3

Determine the nodal displacements and rotations and the global and element forces for the beam shown in Figure 4–14. We have discretized the beam as shown by the node numbering. The beam is fixed at node 1, has a roller support at node 2, and has an elastic spring support at node 3. A downward vertical force of $P = 50$ kN is applied at node 3. Let $E = 210$ GPa and $I = 2 \times 10^{-4}$ m^4 throughout the beam, and let $k = 200$ kN/m.

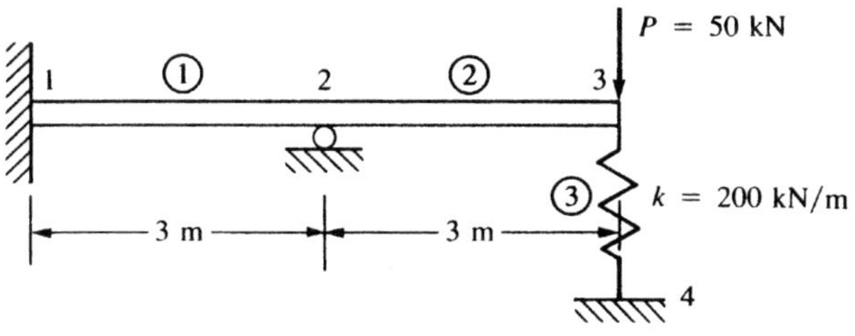

Figure 4–14 Beam example

Using Eq. (4.1.14) for each beam element and Eq. (2.2.18) for the spring element as well as the direct stiffness method, we obtain the structure stiffness matrix as

$$\underline{K} = \frac{EI}{L^3} \begin{array}{c} \begin{matrix} d_{1y} & \phi_1 & d_{2y} & \phi_2 & d_{3y} & \phi_3 & d_{4y} \end{matrix} \\ \left[\begin{array}{ccccccc} 12 & 6L & -12 & 6L & 0 & 0 & 0 \\ & 4L^2 & -6L & 2L^2 & 0 & 0 & 0 \\ & & 24 & 0 & -12 & 6L & 0 \\ & & & 8L^2 & -6L & 2L^2 & 0 \\ & & & & 12 + \dfrac{kL^3}{EI} & -6L & -\dfrac{kL^3}{EI} \\ & & & & & 4L^2 & 0 \\ \text{Symmetry} & & & & & & \dfrac{kL^3}{EI} \end{array} \right] \end{array} \qquad (4.3.18a)$$

where the spring stiffness matrix $\underline{k}_s$ given below by Eq. (4.3.18b) has been directly added into the global stiffness matrix corresponding to its degrees of freedom at nodes 3 and 4.

$$\underline{k}_s = \begin{matrix} d_{3y} & d_{4y} \end{matrix} \atop \begin{bmatrix} k & -k \\ -k & k \end{bmatrix} \tag{4.3.18b}$$

It is easier to solve the problem using the general variables, later making numerical substitutions into the final displacement expressions. The governing equations for the beam are then given by

$$\begin{Bmatrix} F_{1y} \\ M_1 \\ F_{2y} \\ M_2 \\ F_{3y} \\ M_3 \\ F_{4y} \end{Bmatrix} = \frac{EI}{L^3} \begin{bmatrix} 12 & 6L & -12 & 6L & 0 & 0 & 0 \\ & 4L^2 & -6L & 2L^2 & 0 & 0 & 0 \\ & & 24 & 0 & -12 & 6L & 0 \\ & & & 8L^2 & -6L & 2L^2 & 0 \\ & & & & 12+k' & -6L & -k' \\ & & & & & 4L^2 & 0 \\ \text{Symmetry} & & & & & & k' \end{bmatrix} \begin{Bmatrix} d_{1y} \\ \phi_1 \\ d_{2y} \\ \phi_2 \\ d_{3y} \\ \phi_3 \\ d_{4y} \end{Bmatrix} \tag{4.3.19}$$

where $k' = kL^3/(EI)$ is used to simplify the notation. We now apply the boundary conditions

$$d_{1y} = 0 \qquad \phi_1 = 0 \qquad d_{2y} = 0 \qquad d_{4y} = 0 \tag{4.3.20}$$

We delete the first three equations and the seventh equation (corresponding to the boundary conditions given by Eq. (4.3.20)) of Eqs. (4.3.19). The remaining three equations are

$$\begin{Bmatrix} 0 \\ -P \\ 0 \end{Bmatrix} = \frac{EI}{L^3} \begin{bmatrix} 8L^2 & -6L & 2L^2 \\ -6L & 12+k' & -6L \\ 2L^2 & -6L & 4L^2 \end{bmatrix} \begin{Bmatrix} \phi_2 \\ d_{3y} \\ \phi_3 \end{Bmatrix} \tag{4.3.21}$$

Solving Eqs. (4.3.21) simultaneously for the displacement at node 3 and the rotations at nodes 2 and 3, we obtain

$$d_{3y} = -\frac{7PL^3}{EI}\left(\frac{1}{12+7k'}\right) \qquad \phi_2 = -\frac{3PL^2}{EI}\left(\frac{1}{12+7k'}\right)$$
$$\phi_3 = -\frac{9PL^2}{EI}\left(\frac{1}{12+7k'}\right) \tag{4.3.22}$$

The influence of the spring stiffness on the displacements is easily seen in Eq. (4.3.22). Solving for the numerical displacements using $P = 50$ kN, $L = 3$ m, $E = 210$ GPa ($= 210 \times 10^6$ kN/m^2), $I = 2 \times 10^{-4}$ m^4, and $k' = 0.129$ in Eq. (4.3.22), we obtain

$$d_{3y} = \frac{-7(50\text{ kN})(3\text{ m})^3}{(210 \times 10^6\text{ kN/m}^2)(2 \times 10^{-4}\text{ m}^4)}\left(\frac{1}{12+7(0.129)}\right) = -0.0174\text{ m} \tag{4.3.23}$$

Similar substitutions into Eq. (4.3.26) yield

$$\phi_2 = -0.00249\text{ rad} \qquad \phi_3 = -0.00747\text{ rad} \tag{4.3.24}$$

We now back-substitute the results from Eqs. (4.3.23) and (4.3.24), along with numerical values for P, E, I, L, and k', into Eq. (4.3.19) to obtain the global nodal forces as

$$F_{1y} = -69.9 \text{ kN} \quad M_1 = -69.7 \text{ kN}\cdot\text{m}$$
$$F_{2y} = 116.4 \text{ kN} \quad M_2 = 0.0 \text{ kN}\cdot\text{m} \tag{4.3.25}$$
$$F_{3y} = -50.0 \text{ kN} \quad M_3 = 0.0 \text{ kN}\cdot\text{m}$$

For the beam-spring structure, an additional global force F_{4y} is determined at the base of the spring as follows:

$$F_{4y} = -d_{3y}k = (0.0174)200 = 3.5 \text{ kN} \tag{4.3.26}$$

This force provides the additional global y force for equilibrium of the structure.

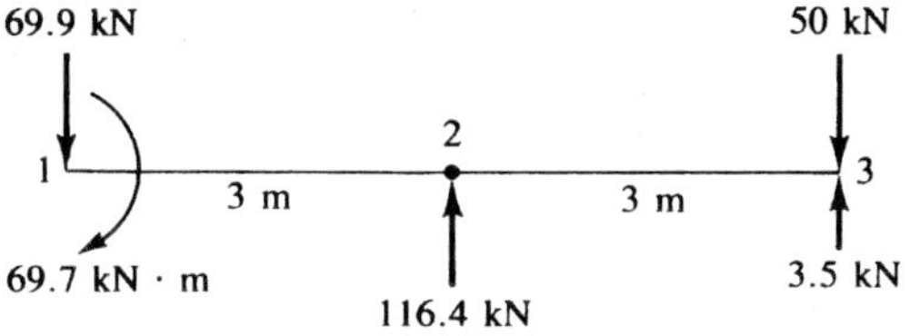

Figure 4–15 Free-body diagram of beam of Figure 4–14

A free-body diagram, including the forces and moments from Eqs. (4.3.25) and (4.3.26) acting on the beam, is shown in Figure 4–15. ■

Example 4.4

Determine the displacement and rotation under the force and moment located at the center of the beam shown in Figure 4–16. The beam has been discretized into the two elements shown in Figure 4–16. The beam is fixed at each end. A downward force of 10 kN and an applied moment of 20 kN-m act at the center of the beam. Let $E = 210$ GPa and $I = 4 \times 10^{-4}\ m^4$ throughout the beam length.

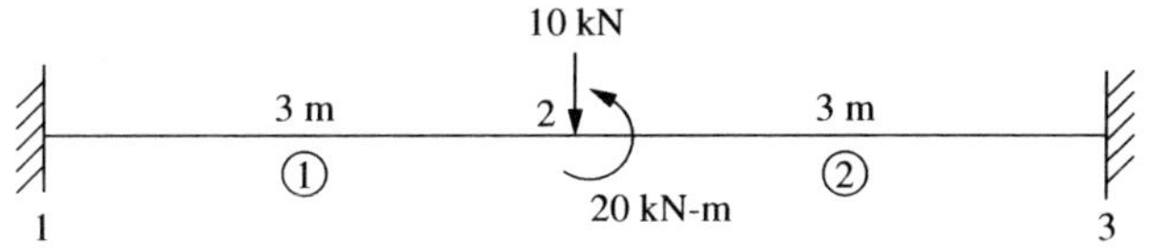

Figure 4–16 Fixed-fixed beam subjected to applied force and moment

Using Eq. (4.1.14) for each beam element with $L = 3$ m, we obtain the element stiffness matrices as follows:

$$\underline{k}^{(1)} = \frac{EI}{L^3}\begin{bmatrix} 12 & 6L & -12 & 6L \\ & 4L^2 & -6L & 2L^2 \\ & & 12 & -6L \\ \text{Symmetry} & & & 4L^2 \end{bmatrix} \quad (\text{columns: } d_{1y}, \phi_1, d_{2y}, \phi_2)$$

$$\underline{k}^{(2)} = \frac{EI}{L^3}\begin{bmatrix} 12 & 6L & -12 & 6L \\ & 4L^2 & -6L & 2L^2 \\ & & 12 & -6L \\ \text{Symmetry} & & & 4L^2 \end{bmatrix} \quad (\text{columns: } d_{2y}, \phi_2, d_{3y}, \phi_3) \tag{4.3.27}$$

The boundary conditions are given by

$$d_{1y} = \phi_1 = d_{3y} = \phi_3 = 0 \tag{4.3.28}$$

The global forces are $F_{2y} = -10{,}000$ N and $M_2 = 20{,}000$ N-m.

Applying the global forces and boundary conditions, Eq. (4.3.28), and assembling the global stiffness matrix using the direct stiffness method and Eqs. (4.3.27), we obtain the global equations as:

$$\begin{Bmatrix} -10{,}000 \\ 20{,}000 \end{Bmatrix} = \frac{(210 \times 10^9)(4 \times 10^{-4})}{3^3} \begin{bmatrix} 24 & 0 \\ 0 & 8(3^2) \end{bmatrix} \begin{Bmatrix} d_{2y} \\ \phi_2 \end{Bmatrix} \tag{4.3.29}$$

Solving Eq. (4.3.29) for the displacement and rotation, we obtain

$$d_{2y} = -1.339 \times 10^{-4} \text{ m and } \phi_2 = 8.928 \times 10^{-5} \text{ rad} \tag{4.3.30}$$

Using the local equations for each element, we obtain the local nodal forces and moments for element one as follows:

$$\begin{Bmatrix} f_{1y}^{(1)} \\ m_1^{(1)} \\ f_{2y}^{(1)} \\ m_2^{(1)} \end{Bmatrix} = \frac{(210 \times 10^9)(4 \times 10^{-4})}{3^3} \begin{bmatrix} 12 & 6(3) & -12 & 6(3) \\ 6(3) & 4(3^2) & -6(3) & 2(3^2) \\ -12 & -6(3) & 12 & -6(3) \\ 6(3) & 2(3^2) & -6(3) & 4(3^2) \end{bmatrix} \begin{Bmatrix} 0 \\ 0 \\ -1.3339 \times 10^{-4} \\ 8.928 \times 10^{-5} \end{Bmatrix} \tag{4.3.31}$$

Simplifying Eq. (4.3.31), we have

$$f_{1y}^{(1)} = 10{,}000 \text{ N}, \quad m_1^{(1)} = 12{,}500 \text{ N-m}, \quad f_{2y}^{(1)} = -10{,}000 \text{ N}, \quad m_2^{(1)} = 17{,}500 \text{ N-m} \tag{4.3.32}$$

Similarly, for element two the local nodal forces and moments are

$$f_{2y}^{(2)} = 0, \quad m_2^{(2)} = 2500 \text{ N-m}, \quad f_{3y}^{(2)} = 0, \quad m_3^{(2)} = -2500 \text{ N-m} \tag{4.3.33}$$

Using the results from Eqs. (4.3.32) and (4.3.33), we show the local forces and moments acting on each element in Figure 4–16 as follows:

Using the results from Eqs. (4.3.32) and (4.3.33), or Figure 4–17, we obtain the shear force and bending moment diagrams for each element as shown in Figure 4–18.

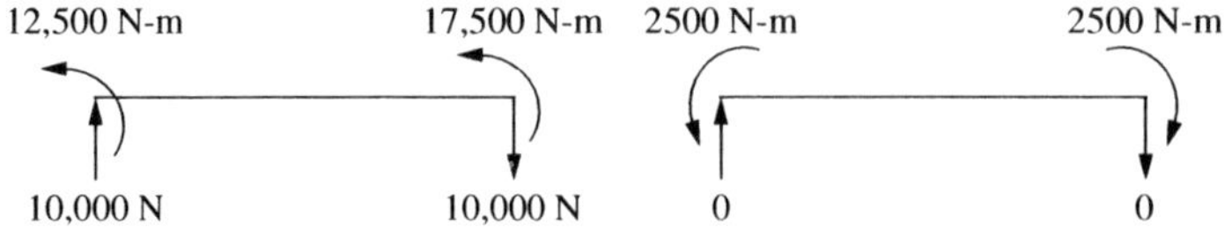

Figure 4–17 Nodal forces and moments acting on each element of Figure 4–15

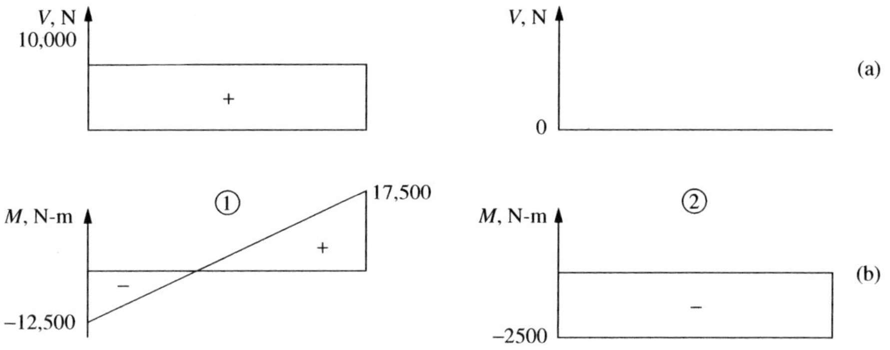

Figure 4–18 Shear force (a) and bending moment (b) diagrams for each element

■

Example 4.5

To illustrate the effects of shear deformation along with the usual bending deformation, we now solve the simple beam shown in Figure 4–19. We will use the beam stiffness matrix given by Eq. (4.1.15o) that includes both the bending and shear deformation contributions for deformation in the x–y plane. The beam is simply supported with a concentrated load of 10,000 N applied at mid-span. We let material properties be $E = 207$ GPa and $G = 80$ GPa. The beam width and height are $b = 25$ mm and $h = 50$ mm, respectively.

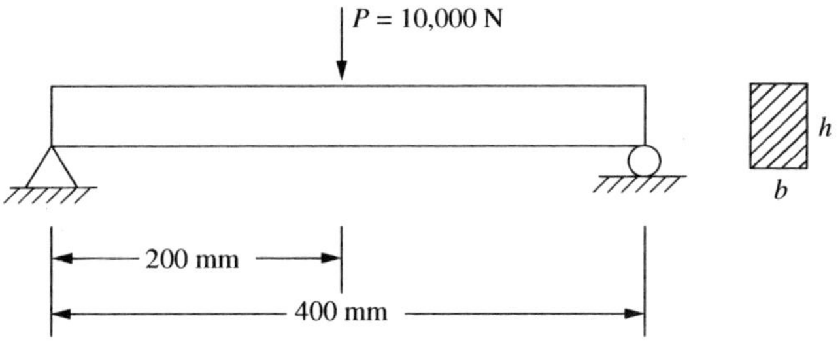

Figure 4–19 Simple beam subjected to concentrated load at center of span

We will use symmetry to simplify the solution. Therefore, only one half of the beam will be considered with the slope at the center forced to be zero. Also, one half of the concentrated load is then used. The model with symmetry enforced is shown in Figure 4–20.

The finite element model will consist of only one beam element. Using Eq. (4.1.15o) for the Timoshenko beam element stiffness matrix, we obtain the global

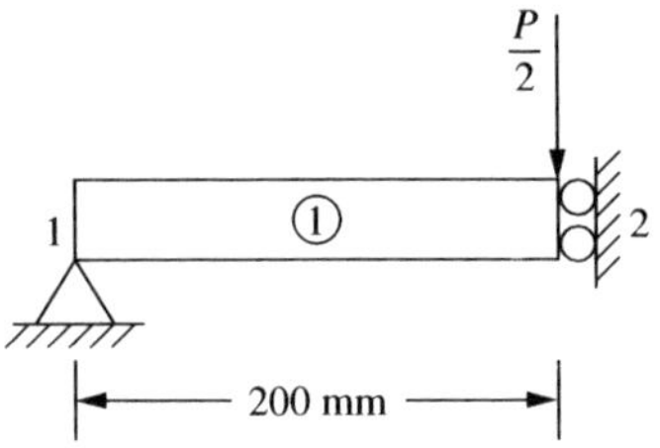

Figure 4–20 Beam with symmetry enforced

equations as

$$\frac{EI}{L^3(1+\varphi)}\begin{bmatrix} 12 & 6L & -12 & 6L \\ 6L & (4+\varphi)L^2 & -6L & (2-\varphi)L^2 \\ -12 & -6L & 12 & -6L \\ 6L & (2-\varphi)L^2 & -6L & (4+\varphi)L^2 \end{bmatrix}\begin{Bmatrix} d_{1y}=0 \\ \phi_1 \\ d_{2y} \\ \phi_2 = 0 \end{Bmatrix} = \begin{Bmatrix} F_{1y} \\ 0 \\ -P/2 \\ 0 \end{Bmatrix} \tag{4.3.34}$$

Note that the boundary conditions given by $d_{1y} = 0$ and $\phi_2 = 0$ have been included in Eq. (4.3.34).

Using the second and third equations of Eq. (4.3.34) whose rows are associated with the two unknowns, ϕ_1 and d_{2y}, we obtain

$$d_{2y} = \frac{-PL^3(4+\varphi)}{24EI} \quad \text{and} \quad \phi_1 = \frac{-PL^2}{4EI} \tag{4.3.35}$$

As the beam is rectangular in cross section, the moment of inertia is

$$I = bh^3/12$$

Substituting the numerical values for b and h, we obtain I as

$$I = 0.26 \times 10^{-6} \text{ m}^4$$

The shear correction factor is given by

$$\varphi = \frac{12EI}{k_s AGL^2}$$

and k_s for a rectangular cross section is given by $k_s = 5/6$.

Substituting numerical values for E, I, G, L, and k_s, we obtain

$$\varphi = \frac{12 \times 207 \times 10^9 \times 0.26 \times 10^{-6}}{5/6 \times 0.025 \times 0.05 \times 80 \times 10^9 \times 0.2^2} = 0.1938$$

Substituting for $P = 10{,}000$ N, $L = 0.2$ m, and $\varphi = 0.1938$ into Eq. (4.3.35), we obtain the displacement at the mid-span as

$$d_{2y} = -2.597 \times 10^{-4} \text{ m} \tag{4.3.36}$$

If we let $l =$ the whole length of the beam, then $l = 2L$ and we can substitute $L = l/2$ into Eq. (4.3.35) to obtain the displacement in terms of the whole length of the beam as

$$d_{2y} = \frac{-Pl^3(4+\varphi)}{192EI} \tag{4.3.37}$$

For long slender beams with l about 10 or more times the beam depth, h, the transverse shear correction term φ is small and can be neglected. Therefore, Eq. (4.3.37) becomes

$$d_{2y} = \frac{-Pl^3}{48EI} \tag{4.3.38}$$

Equation (4.3.38) is the classical beam deflection formula for a simply supported beam subjected to a concentrated load at mid-span.

Using Eq. (4.3.38), the deflection is obtained as

$$d_{2y} = 2.474 \times 10^{-4} \text{ m} \tag{4.3.39}$$

Comparing the deflections obtained using the shear-correction factor with the deflection predicted using the beam-bending contribution only, we obtain

$$\% \text{ change} = \frac{2.597 - 2.474}{2.474} \times 100 = 4.97\% \text{ difference}$$

■

4.4 Distributed Loading

Beam members can support distributed loading as well as concentrated nodal loading. Therefore, we must be able to account for distributed loading. Consider the fixed-fixed beam subjected to a uniformly distributed loading w shown in Figure 4–21. The reactions, determined from structural analysis theory [2], are shown in Figure 4–22. These reactions are called *fixed-end reactions*. In general, **fixed-end reactions** are those reactions at the ends of an element if the ends of the element are assumed to be fixed—that is, if displacements and rotations are prevented. (Those of you who are unfamiliar with the analysis of indeterminate structures should assume these reactions as given and proceed with the rest of the discussion; we will develop these results in a subsequent presentation of the work-equivalence method.) Therefore, guided by the results from structural analysis for the case of a uniformly distributed load, we replace the load by concentrated nodal forces and moments tending to have the same

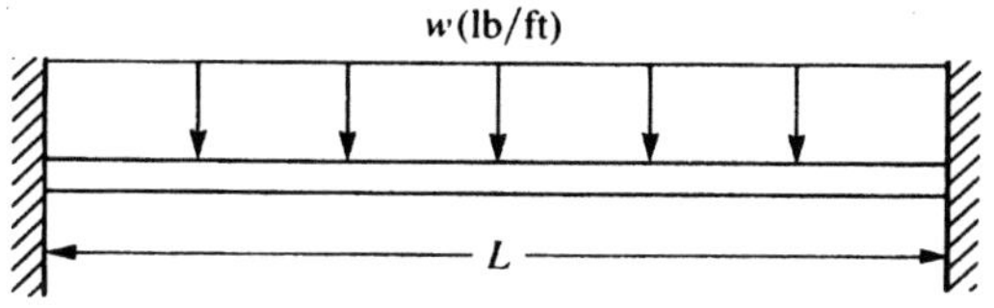

Figure 4–21 Fixed-fixed beam subjected to a uniformly distributed load

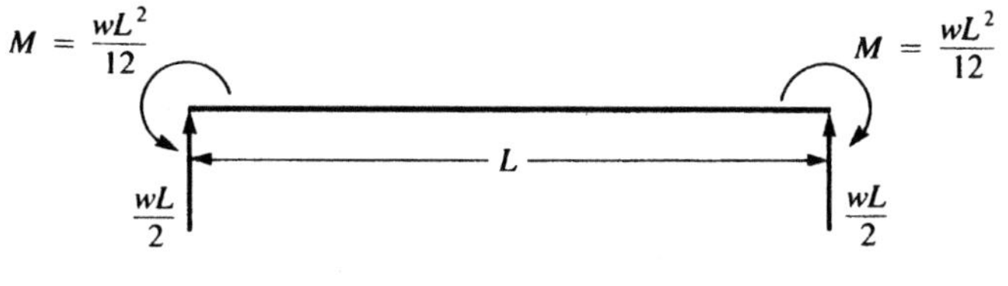

Figure 4–22 Fixed-end reactions for the beam of Figure 4–20

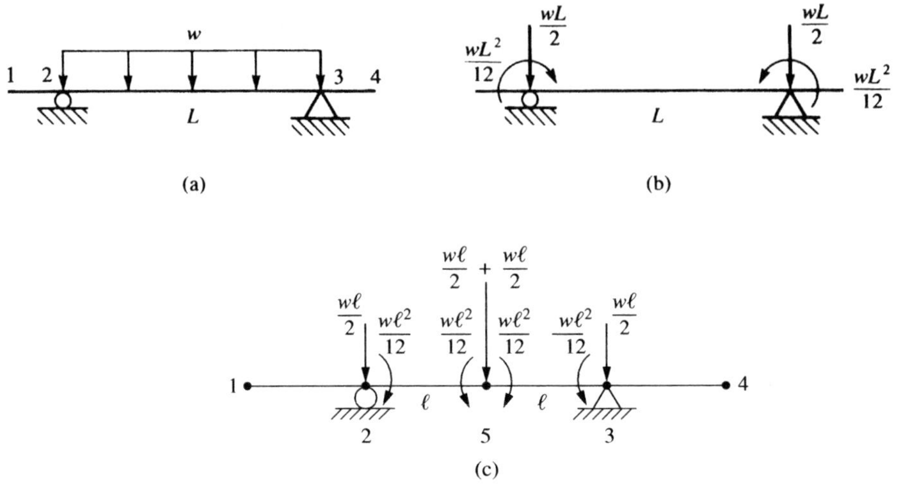

Figure 4–23 (a) Beam with a distributed load, (b) the equivalent nodal force system, and (c) the enlarged beam (for clarity's sake) with equivalent nodal force system when node 5 is added to the midspan

effect on the beam as the actual distributed load. Figure 4–23 illustrates this idea for a beam. We have replaced the uniformly distributed load by a statically equivalent force system consisting of a concentrated nodal force and moment at each end of the member carrying the distributed load. That is, both the statically equivalent concentrated nodal forces and moments and the original distributed load have the same resultant force and same moment about an arbitrarily chosen point. These statically equivalent forces are always of opposite sign from the fixed-end reactions. If we want to analyze the behavior of loaded member 2–3 in better detail, we can place a node at midspan and use the same procedure just described for each of the two elements representing the horizontal member. That is, to determine the maximum deflection and maximum moment in the beam span, a node 5 is needed at midspan of beam segment 2–3, and work-equivalent forces and moments are applied to each element (from node 2 to node 5 and from node 5 to node 3) shown in Figure 4–23 (c).

Work-Equivalence Method

We can use the work-equivalence method to replace a distributed load by a set of discrete loads. This method is based on the concept that the work of the distributed load $w(\hat{x})$ in going through the displacement field $\hat{v}(\hat{x})$ is equal to the work done by nodal loads $\hat{f}_{iy}$ and $\hat{m}_i$ in going through nodal displacements $\hat{d}_{iy}$ and $\hat{\phi}_i$ for arbitrary nodal displacements. To illustrate the method, we consider the example shown in Figure 4–24. The work due to the distributed load is given by

$$W_{\text{distributed}} = \int_0^L w(\hat{x})\hat{v}(\hat{x})\,d\hat{x} \tag{4.4.1}$$

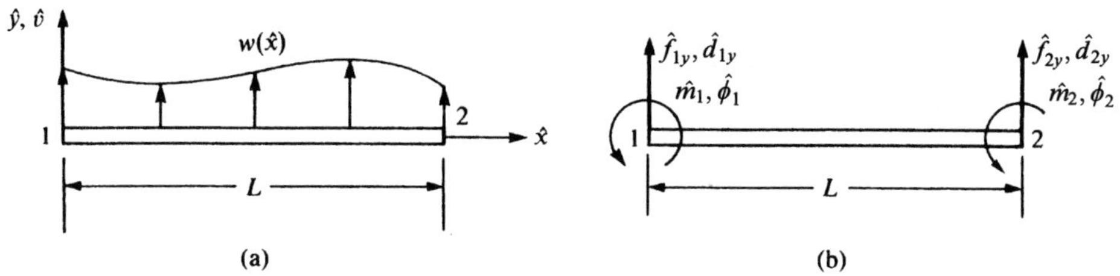

Figure 4–24 (a) Beam element subjected to a general load and (b) the statically equivalent nodal force system

where $\hat{v}(\hat{x})$ is the transverse displacement given by Eq. (4.1.4). The work due to the discrete nodal forces is given by

$$W_{\text{discrete}} = \hat{m}_1\hat{\phi}_1 + \hat{m}_2\hat{\phi}_2 + \hat{f}_{1y}\hat{d}_{1y} + \hat{f}_{2y}\hat{d}_{2y} \tag{4.4.2}$$

We can then determine the nodal moments and forces $\hat{m}_1, \hat{m}_2, \hat{f}_{1y}$, and $\hat{f}_{2y}$ used to replace the distributed load by using the concept of work equivalence—that is, by setting $W_{\text{distributed}} = W_{\text{discrete}}$ for arbitrary displacements $\hat{\phi}_1, \hat{\phi}_2, \hat{d}_{1y}$, and $\hat{d}_{2y}$.

Example of Load Replacement

To illustrate more clearly the concept of work equivalence, we will now consider a beam subjected to a specified distributed load. Consider the uniformly loaded beam shown in Figure 4–25(a). The support conditions are not shown because they are not relevant to the replacement scheme. By letting $W_{\text{discrete}} = W_{\text{distributed}}$ and by assuming arbitrary $\hat{\phi}_1, \hat{\phi}_2, \hat{d}_{1y}$, and $\hat{d}_{2y}$, we will find equivalent nodal forces $\hat{m}_1, \hat{m}_2, \hat{f}_{1y}$, and $\hat{f}_{2y}$. Figure 4–25(b) shows the nodal forces and moments directions as positive based on Figure 4–1.

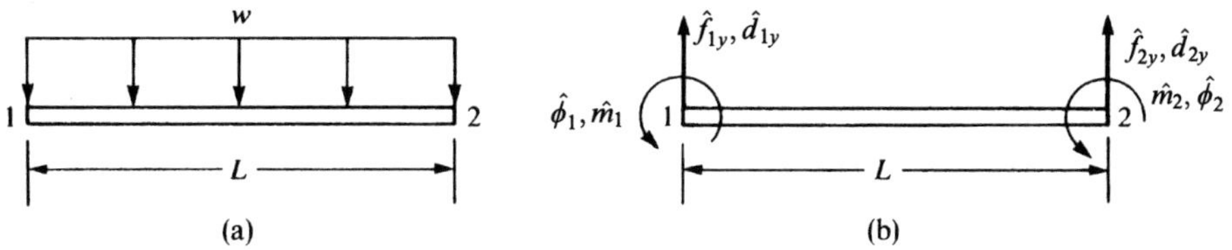

Figure 4–25 (a) Beam subjected to a uniformly distributed loading and (b) the equivalent nodal forces to be determined

Using Eqs. (4.4.1) and (4.4.2) for $W_{\text{distributed}} = W_{\text{discrete}}$, we have

$$\int_0^L w(\hat{x})\hat{v}(\hat{x})\,d\hat{x} = \hat{m}_1\hat{\phi}_1 + \hat{m}_2\hat{\phi}_2 + \hat{f}_{1y}\hat{d}_{1y} + \hat{f}_{2y}\hat{d}_{2y} \tag{4.4.3}$$

where $\hat{m}_1\hat{\phi}_1$ and $\hat{m}_2\hat{\phi}_2$ are the work due to concentrated nodal moments moving through their respective nodal rotations and $\hat{f}_{1y}\hat{d}_{1y}$ and $\hat{f}_{2y}\hat{d}_{2y}$ are the work due to the nodal forces moving through nodal displacements. Evaluating the left-hand side of

Eq. (4.4.3) by substituting $w(\hat{x}) = -w$ and $\hat{v}(\hat{x})$ from Eq. (4.1.4), we obtain the work due to the distributed load as

$$\int_0^L w(\hat{x})\hat{v}(\hat{x})\,d\hat{x} = -\frac{Lw}{2}(\hat{d}_{1y} - \hat{d}_{2y}) - \frac{L^2w}{4}(\hat{\phi}_1 + \hat{\phi}_2) - Lw(\hat{d}_{2y} - \hat{d}_{1y})$$
$$+\frac{L^2w}{3}(2\hat{\phi}_1 + \hat{\phi}_2) - \hat{\phi}_1\left(\frac{L^2w}{2}\right) - \hat{d}_{1y}(wL) \tag{4.4.4}$$

Now using Eqs. (4.4.3) and (4.4.4) for arbitrary nodal displacements, we let $\hat{\phi}_1 = 1$, $\hat{\phi}_2 = 0, \hat{d}_{1y} = 0$, and $\hat{d}_{2y} = 0$ and then obtain

$$\hat{m}_1(1) = -\left(\frac{L^2w}{4} - \frac{2}{3}L^2w + \frac{L^2}{2}w\right) = -\frac{wL^2}{12} \tag{4.4.5}$$

Similarly, letting $\hat{\phi}_1 = 0, \hat{\phi}_2 = 1, \hat{d}_{1y} = 0$, and $\hat{d}_{2y} = 0$ yields

$$\hat{m}_2(1) = -\left(\frac{L^2w}{4} - \frac{L^2w}{3}\right) = \frac{wL^2}{12} \tag{4.4.6}$$

Finally, letting all nodal displacements equal zero except first $\hat{d}_{1y}$ and then $\hat{d}_{2y}$, we obtain

$$\begin{aligned} \hat{f}_{1y}(1) &= -\frac{Lw}{2} + Lw - Lw = -\frac{Lw}{2} \\ \hat{f}_{2y}(1) &= \frac{Lw}{2} - Lw = -\frac{Lw}{2} \end{aligned} \tag{4.4.7}$$

We can conclude that, in general, for any given load function $w(\hat{x})$, we can multiply by $\hat{v}(\hat{x})$ and then integrate according to Eq. (4.4.3) to obtain the concentrated nodal forces (and/or moments) used to replace the distributed load. Moreover, we can obtain the load replacement by using the concept of fixed-end reactions from structural analysis theory. Tables of fixed-end reactions have been generated for numerous load cases and can be found in texts on structural analysis such as Reference [2]. A table of equivalent nodal forces has been generated in Appendix D of this text, guided by the fact that fixed-end reaction forces are of opposite sign from those obtained by the work equivalence method.

Hence, if a concentrated load is applied other than at the natural intersection of two elements, we can use the concept of equivalent nodal forces to replace the concentrated load by nodal concentrated values acting at the beam ends, instead of creating a node on the beam at the location where the load is applied. We provide examples of this procedure for handling concentrated loads on elements in beam Example 4.7 and in plane frame Example 5.3.

General Formulation

In general, we can account for distributed loads or concentrated loads acting on beam elements by starting with the following formulation application for a general structure:

$$\underline{F} = \underline{K}\underline{d} - \underline{F}_o \tag{4.4.8}$$

where $\underline{F}$ are the concentrated nodal forces and $\underline{F}_o$ are called the *equivalent nodal forces*, now expressed in terms of global-coordinate components, which are of such magnitude that they yield the same displacements at the nodes as would the distributed load. Using the table in Appendix D of equivalent nodal forces $\hat{\underline{f}}_o$ expressed in terms of local-coordinate components, we can express $\underline{F}_o$ in terms of global-coordinate components.

Recall from Section 3.10 the derivation of the element equations by the principle of minimum potential energy. Starting with Eqs. (3.10.19) and (3.10.20), the minimization of the total potential energy resulted in the same form of equation as Eq. (4.4.8) where $\underline{F}_o$ now represents the same work-equivalent force replacement system as given by Eq. (3.10.20a) for surface traction replacement. Also, $\underline{F} = \underline{P}$ [$\underline{P}$ from Eq. (3.10.20)] represents the global nodal concentrated forces. Because we now assume that concentrated nodal forces are not present ($\underline{F} = 0$), as we are solving beam problems with distributed loading only in this section, we can write Eq. (4.4.8) as

$$\underline{F}_o = \underline{K}\underline{d} \tag{4.4.9}$$

On solving for $\underline{d}$ in Eq. (4.4.9) and then substituting the global displacements $\underline{d}$ and equivalent nodal forces $\underline{F}_o$ into Eq. (4.4.8), we obtain the actual global nodal forces $\underline{F}$. For example, using the definition of $\hat{\underline{f}}_o$ and Eqs. (4.4.5)–(4.4.7) (or using load case 4 in Appendix D) for a uniformly distributed load w acting over a one-element beam, we have

$$\underline{F}_o = \begin{Bmatrix} \dfrac{-wL}{2} \\ \dfrac{-wL^2}{12} \\ \dfrac{-wL}{2} \\ \dfrac{wL^2}{12} \end{Bmatrix} \tag{4.4.10}$$

This concept can be applied on a local basis to obtain the local nodal forces $\hat{\underline{f}}$ in individual elements of structures by applying Eq. (4.4.8) locally as

$$\hat{\underline{f}} = \hat{\underline{k}}\hat{\underline{d}} - \hat{\underline{f}}_o \tag{4.4.11}$$

where $\hat{\underline{f}}_o$ are the equivalent local nodal forces.

Examples 4.6–4.8 illustrate the method of equivalent nodal forces for solving beams subjected to distributed and concentrated loadings. We will use global-coordinate notation in Examples 4.6–4.8—treating the beam as a general structure rather than as an element.

Example 4.6

For the cantilever beam subjected to the uniform load w in Figure 4–26, solve for the right-end vertical displacement and rotation and then for the nodal forces. Assume the beam to have constant EI throughout its length.

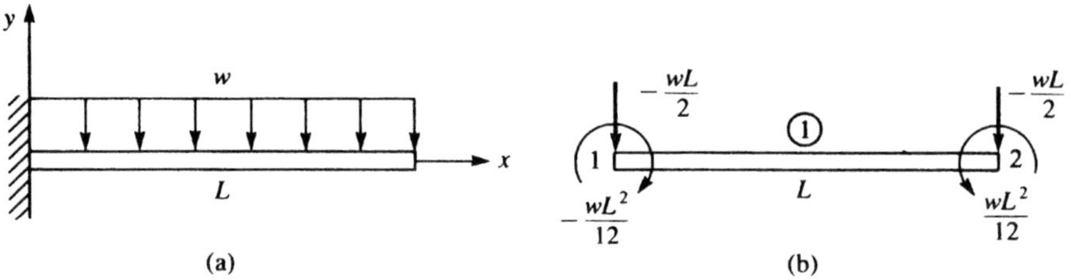

Figure 4–26 (a) Cantilever beam subjected to a uniformly distributed load and (b) the work equivalent nodal force system

We begin by discretizing the beam. Here only one element will be used to represent the whole beam. Next, the distributed load is replaced by its work-equivalent nodal forces as shown in Figure 4–26(b). The work-equivalent nodal forces are those that result from the uniformly distributed load acting over the whole beam given by Eq. (4.4.10). (Or see appropriate load case 4 in Appendix D.) Using Eq. (4.4.9) and the beam element stiffness matrix, and realizing $\underline{\hat{k}} = \underline{k}$ as the local $\hat{x}$ axis is coincident with the global x axis, we obtain

$$\frac{EI}{L^3}\begin{bmatrix} 12 & 6L & -12 & 6L \\ & 4L^2 & -6L^2 & 2L^2 \\ & & 12 & -6L \\ & & & 4L^2 \end{bmatrix}\begin{Bmatrix} d_{1y} \\ \phi_1 \\ d_{2y} \\ \phi_2 \end{Bmatrix} = \begin{Bmatrix} F_{1y} - \dfrac{wL}{2} \\ M_1 - \dfrac{wL^2}{12} \\ \dfrac{-wL}{2} \\ \dfrac{wL^2}{12} \end{Bmatrix} \tag{4.4.12}$$

where we have applied the work equivalent nodal forces and moments from Figure 4–26(b).

Applying the boundary conditions $d_{1y} = 0$ and $\phi_1 = 0$ to Eqs (4.4.12) and then partitioning off the third and fourth equations of Eq. (4.4.12), we obtain

$$\frac{EI}{L^3}\begin{bmatrix} 12 & -6L \\ -6L^2 & 4L^2 \end{bmatrix}\begin{Bmatrix} d_{2y} \\ \phi_2 \end{Bmatrix} = \begin{Bmatrix} -\dfrac{wL}{2} \\ \dfrac{wL^2}{12} \end{Bmatrix} \tag{4.4.13}$$

Solving Eq. (4.4.13) for the displacements, we obtain

$$\begin{Bmatrix} d_{2y} \\ \phi_2 \end{Bmatrix} = \frac{L}{6EI}\begin{bmatrix} 2L^2 & 3L \\ 3L & 6 \end{bmatrix}\begin{Bmatrix} \dfrac{-wL}{2} \\ \dfrac{wL^2}{12} \end{Bmatrix} \tag{4.4.14a}$$

Simplifying Eq. (4.4.14a), we obtain the displacement and rotation as

$$\begin{Bmatrix} d_{2y} \\ \phi_2 \end{Bmatrix} = \begin{Bmatrix} \dfrac{-wL^4}{8EI} \\ \dfrac{-wL^3}{6EI} \end{Bmatrix} \tag{4.4.14b}$$

The negative signs in the answers indicate that d_{2y} is downward and ϕ_2 is clockwise. In this case, the method of replacing the distributed load by discrete concentrated loads gives exact solutions for the displacement and rotation as could be obtained by classical methods, such as double integration [1]. This is expected, as the work-equivalence method ensures that the nodal displacement and rotation from the finite element method match those from an exact solution.

We will now illustrate the procedure for obtaining the global nodal forces. For convenience, we first define the product $\underline{K}\underline{d}$ to be $\underline{F}^{(e)}$, where $\underline{F}^{(e)}$ are called the *effective global nodal forces*. On using Eq. (4.4.14) for $\underline{d}$, we then have

$$\begin{Bmatrix} F_{1y}^{(e)} \\ M_1^{(e)} \\ F_{2y}^{(e)} \\ M_2^{(e)} \end{Bmatrix} = \frac{EI}{L^3} \begin{bmatrix} 12 & 6L & -12 & 6L \\ 6L & 4L^2 & -6L & 2L^2 \\ -12 & -6L & 12 & -6L \\ 6L & 2L^2 & -6L & 4L^2 \end{bmatrix} \begin{Bmatrix} 0 \\ 0 \\ \dfrac{-wL^4}{8EI} \\ \dfrac{-wL^3}{6EI} \end{Bmatrix} \qquad (4.4.15)$$

Simplifying Eq. (4.4.15), we obtain

$$\begin{Bmatrix} F_{1y}^{(e)} \\ M_1^{(e)} \\ F_{2y}^{(e)} \\ M_2^{(e)} \end{Bmatrix} = \begin{Bmatrix} \dfrac{wL}{2} \\ \dfrac{5wL^2}{12} \\ \dfrac{-wL}{2} \\ \dfrac{wL^2}{12} \end{Bmatrix} \qquad (4.4.16)$$

We then use Eqs. (4.4.10) and (4.4.16) in Eq. (4.4.8) ($\underline{F} = \underline{k}\,\underline{d} - \underline{F}_o$) to obtain the correct global nodal forces as

$$\begin{Bmatrix} F_{1y} \\ M_1 \\ F_{2y} \\ M_2 \end{Bmatrix} = \begin{Bmatrix} \dfrac{wL}{2} \\ \dfrac{5wL^2}{12} \\ \dfrac{-wL}{2} \\ \dfrac{wL^2}{12} \end{Bmatrix} - \begin{Bmatrix} \dfrac{-wL}{2} \\ \dfrac{-wL^2}{12} \\ \dfrac{-wL}{2} \\ \dfrac{wL^2}{12} \end{Bmatrix} = \begin{Bmatrix} wL \\ \dfrac{wL^2}{2} \\ 0 \\ 0 \end{Bmatrix} \qquad (4.4.17)$$

In Eq. (4.4.17), F_{1y} is the vertical force reaction and M_1 is the moment reaction as applied by the clamped support at node 1. The results for displacement given by Eq. (4.4.14b) and the global nodal forces given by Eq. (4.4.17) are sufficient to complete the solution of the cantilever beam problem.

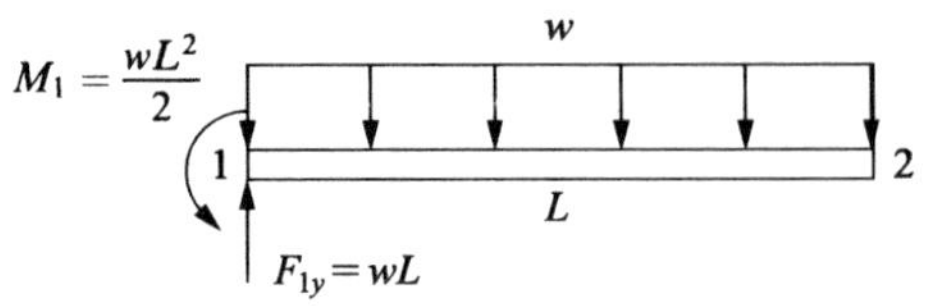

$$\sum F_y: F_{1y} - wL = 0$$

$$\sum M_1: \frac{wL^2}{2} - (wL)\frac{L}{2} = 0$$

Figure 4–26 (c) Free-body diagram and equations of equilibrium for beam of Figure 4–(26)a.

A free-body diagram of the beam using the reactions from Eq. (4.4.17) verifies both force and moment equilibrium as shown in Figure 4–26(c). ■

The nodal force and moment reactions obtained by Eq. (4.4.17) illustrate the importance of using Eq. (4.4.8) to obtain the correct global nodal forces and moments. By subtracting the work-equivalent force matrix, $\underline{F}_o$ from the product of $\underline{K}$ times $\underline{d}$, we obtain the correct reactions at node 1 as can be verified by simple static equilibrium equations. This verification validates the general method as follows:

1. Replace the distributed load by its work-equivalent as shown in Figure 4–26(b) to identify the nodal force and moment used in the solution.
2. Assemble the global force and stiffness matrices and global equations illustrated by Eq. (4.4.12).
3. Apply the boundary conditions to reduce the set of equations as done in previous problems and illustrated by Eq. (4.4.13) where the original four equations have been reduced to two equations to be solved for the unknown displacement and rotation.
4. Solve for the unknown displacement and rotation given by Eq. (4.4.14a) and Eq. (4.4.14b).
5. Use Eq. (4.4.8) as illustrated by Eq. (4.4.17) to obtain the final correct global nodal forces and moments. Those forces and moments at supports, such as the left end of the cantilever in Figure 4–26(a), will be the reactions.

We will solve the following example to illustrate the procedure for handling concentrated loads acting on beam elements at locations other than nodes.

Example 4.7

For the cantilever beam subjected to the concentrated load P in Figure 4–27, solve for the right-end vertical displacement and rotation and the nodal forces, including reactions, by replacing the concentrated load with equivalent nodal forces acting at each end of the beam. Assume EI constant throughout the beam.

We begin by discretizing the beam. Here only one element is used with nodes at each end of the beam. We then replace the concentrated load as shown in

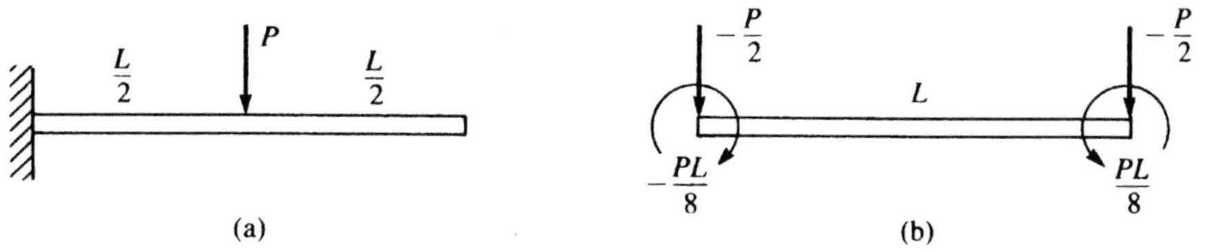

Figure 4–27 (a) Cantilever beam subjected to a concentrated load and (b) the equivalent nodal force replacement system

Figure 4–27(b) by using appropriate loading case 1 in Appendix D. Using Eq. (4.4.9) and the beam element stiffness matrix Eq. (4.1.14), we obtain

$$\frac{EI}{L^3}\begin{bmatrix} 12 & -6L \\ -6L & 4L^2 \end{bmatrix}\begin{Bmatrix} d_{2y} \\ \phi_2 \end{Bmatrix} = \begin{Bmatrix} \dfrac{-P}{2} \\ \dfrac{PL}{8} \end{Bmatrix} \tag{4.4.18}$$

where we have applied the nodal forces from Figure 4–27(b) and the boundary conditions $d_{1y} = 0$ and $\phi_1 = 0$ to reduce the number of matrix equations for the usual longhand solution. Solving Eq. (4.4.18) for the displacements, we obtain

$$\begin{Bmatrix} d_{2y} \\ \phi_2 \end{Bmatrix} = \frac{L}{6EI}\begin{bmatrix} 2L^2 & 3L \\ 3L & 6 \end{bmatrix}\begin{Bmatrix} \dfrac{-P}{2} \\ \dfrac{PL}{8} \end{Bmatrix} \tag{4.4.19}$$

Simplifying Eq. (4.4.19), we obtain the displacement and rotation as

$$\begin{Bmatrix} d_{2y} \\ \phi_2 \end{Bmatrix} = \begin{Bmatrix} \dfrac{-5PL^3}{48EI} \\ \dfrac{-PL^2}{8EI} \end{Bmatrix} \begin{matrix} \downarrow \\ \curvearrowright \end{matrix} \tag{4.4.20}$$

To obtain the unknown nodal forces, we begin by evaluating the effective nodal forces $\underline{F}^{(e)} = \underline{K}\underline{d}$ as

$$\begin{Bmatrix} F_{1y}^{(e)} \\ M_1^{(e)} \\ F_{2y}^{(e)} \\ M_2^{(e)} \end{Bmatrix} = \frac{EI}{L^3}\begin{bmatrix} 12 & 6L & -12 & 6L \\ 6L & 4L^2 & -6L & 2L^2 \\ -12 & -6L & 12 & -6L \\ 6L & 2L^2 & -6L & 4L^2 \end{bmatrix}\begin{Bmatrix} 0 \\ 0 \\ \dfrac{-5PL^3}{48EI} \\ \dfrac{-PL^2}{8EI} \end{Bmatrix} \tag{4.4.21}$$

Simplifying Eq. (4.4.21), we obtain

$$\begin{Bmatrix} F_{1y}^{(e)} \\ M_1^{(e)} \\ F_{2y}^{(e)} \\ M_2^{(e)} \end{Bmatrix} = \begin{Bmatrix} \dfrac{P}{2} \\ \dfrac{3PL}{8} \\ \dfrac{-P}{2} \\ \dfrac{PL}{8} \end{Bmatrix} \tag{4.4.22}$$

Then using Eq. (4.4.22) and the equivalent nodal forces from Figure 4–27(b) in Eq. (4.4.8), we obtain the correct nodal forces as

$$\begin{Bmatrix} F_{1y} \\ M_1 \\ F_{2y} \\ M_2 \end{Bmatrix} = \begin{Bmatrix} \dfrac{P}{2} \\ \dfrac{3PL}{8} \\ \dfrac{-P}{2} \\ \dfrac{PL}{8} \end{Bmatrix} - \begin{Bmatrix} \dfrac{-P}{2} \\ \dfrac{-PL}{8} \\ \dfrac{-P}{2} \\ \dfrac{PL}{8} \end{Bmatrix} = \begin{Bmatrix} P \\ \dfrac{PL}{2} \\ 0 \\ 0 \end{Bmatrix} \tag{4.4.23}$$

We can see from Eq. (4.4.23) that F_{1y} is equivalent to the vertical reaction force and M_1 is the reaction moment as applied by the clamped support at node 1.

Again, the reactions obtained by Eq. (4.4.23) can be verified to be correct by using static equilibrium equations to validate once more the correctness of the general formulation and procedures summarized in the steps given after Example 4.6. ■

To illustrate the procedure for handling concentrated nodal forces and distributed loads acting simultaneously on beam elements, we will solve the following example.

Example 4.8

For the cantilever beam subjected to the concentrated free-end load P and the uniformly distributed load w acting over the whole beam as shown in Figure 4–28, determine the free-end displacements and the nodal forces.

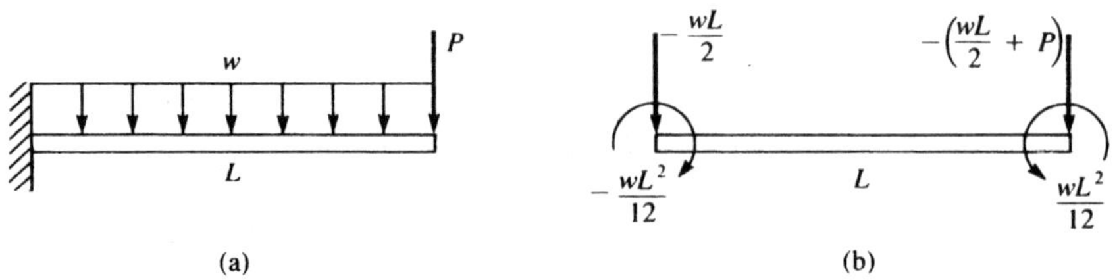

Figure 4–28 (a) Cantilever beam subjected to a concentrated load and a distributed load and (b) the equivalent nodal force replacement system

Once again, the beam is modeled using one element with nodes 1 and 2, and the distributed load is replaced as shown in Figure 4–28(b) using appropriate loading case 4 in Appendix D. Using the beam element stiffness Eq. (4.1.14), we obtain

$$\frac{EI}{L^3}\begin{bmatrix} 12 & -6L \\ -6L & 4L^2 \end{bmatrix}\begin{Bmatrix} d_{2y} \\ \phi_2 \end{Bmatrix} = \begin{Bmatrix} \dfrac{-wL}{2} - P \\ \dfrac{wL^2}{12} \end{Bmatrix} \tag{4.4.24}$$

where we have applied the nodal forces from Figure 4–28(b) and the boundary conditions $d_{1y} = 0$ and $\phi_1 = 0$ to reduce the number of matrix equations for the usual longhand solution. Solving Eq. (4.4.24) for the displacements, we obtain

$$\begin{Bmatrix} d_{2y} \\ \phi_2 \end{Bmatrix} = \begin{Bmatrix} \dfrac{-wL^4}{8EI} - \dfrac{PL^3}{3EI} \\ \dfrac{-wL^3}{6EI} - \dfrac{PL^2}{2EI} \end{Bmatrix} \begin{matrix} \downarrow \\ \curvearrowright \end{matrix} \tag{4.4.25}$$

Next, we obtain the effective nodal forces using $\underline{F}^{(e)} = \underline{K}\underline{d}$ as

$$\begin{Bmatrix} F_{1y}^{(e)} \\ M_1^{(e)} \\ F_{2y}^{(e)} \\ M_2^{(e)} \end{Bmatrix} = \frac{EI}{L^3}\begin{bmatrix} 12 & 6L & -12 & 6L \\ 6L & 4L^2 & -6L & 2L^2 \\ -12 & -6L & 12 & -6L \\ 6L & 2L^2 & -6L & 4L^2 \end{bmatrix}\begin{Bmatrix} 0 \\ 0 \\ \dfrac{-wL^4}{8EI} - \dfrac{PL^3}{3EI} \\ \dfrac{-wL^3}{6EI} - \dfrac{PL^2}{2EI} \end{Bmatrix} \tag{4.4.26}$$

Simplifying Eq. (4.4.26), we obtain

$$\begin{Bmatrix} F_{1y}^{(e)} \\ M_1^{(e)} \\ F_{2y}^{(e)} \\ M_2^{(e)} \end{Bmatrix} = \begin{Bmatrix} P + \dfrac{wL}{2} \\ PL + \dfrac{5wL^2}{12} \\ -P - \dfrac{wL}{2} \\ \dfrac{wL^2}{12} \end{Bmatrix} \tag{4.4.27}$$

Finally, subtracting the equivalent nodal force matrix [see Figure 4–27(b)] from the effective force matrix of Eq. (4.4.27), we obtain the correct nodal forces as

$$\begin{Bmatrix} F_{1y} \\ M_1 \\ F_{2y} \\ M_2 \end{Bmatrix} = \begin{Bmatrix} P + \dfrac{wL}{2} \\ PL + \dfrac{5wL^2}{12} \\ -P - \dfrac{wL}{2} \\ \dfrac{wL^2}{12} \end{Bmatrix} - \begin{Bmatrix} \dfrac{-wL}{2} \\ \dfrac{-wL^2}{12} \\ \dfrac{-wL}{2} \\ \dfrac{wL^2}{12} \end{Bmatrix} = \begin{Bmatrix} P + wL \\ PL + \dfrac{wL^2}{2} \\ -P \\ 0 \end{Bmatrix} \tag{4.4.28}$$

From Eq. (4.4.28), we see that F_{1y} is equivalent to the vertical reaction force, M_1 is the reaction moment at node 1, and F_{2y} is equal to the applied downward force P at node 2. [Remember that only the equivalent nodal force matrix is subtracted, not the original concentrated load matrix. This is based on the general formulation, Eq. (4.4.8).] ■

To generalize the work-equivalent method, we apply it to a beam with more than one element as shown in the following Example 4.9.

Example 4.9

For the fixed–fixed beam subjected to the linear varying distributed loading acting over the whole beam shown in Figure 4–29(a) determine the displacement and rotation at the center and the reactions.

The beam is now modeled using two elements with nodes 1, 2, and 3 and the distributed load is replaced as shown in Figure 4–29 (b) using the appropriate load cases 4 and 5 in Appendix D. Note that load case 5 is used for element one as it has only the linear varying distributed load acting on it with a high end value of $w/2$ as shown in Figure 4–29 (a), while both load cases 4 and 5 are used for element two as the distributed load is divided into a uniform part with magnitude $w/2$ and a linear varying part with magnitude at the high end of the load equal to $w/2$ also.

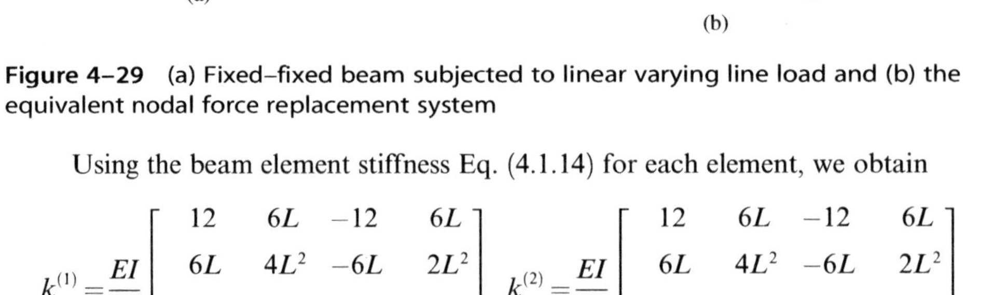

Figure 4–29 (a) Fixed–fixed beam subjected to linear varying line load and (b) the equivalent nodal force replacement system

Using the beam element stiffness Eq. (4.1.14) for each element, we obtain

$$\underline{k}^{(1)} = \frac{EI}{L^3}\begin{bmatrix} 12 & 6L & -12 & 6L \\ 6L & 4L^2 & -6L & 2L^2 \\ -12 & -6L & 12 & -6L \\ 6L & 2L^2 & -6L & 4L^2 \end{bmatrix} \quad \underline{k}^{(2)} = \frac{EI}{L^3}\begin{bmatrix} 12 & 6L & -12 & 6L \\ 6L & 4L^2 & -6L & 2L^2 \\ -12 & -6L & 12 & -6L \\ 6L & 2L^2 & -6L & 4L^2 \end{bmatrix} \tag{4.4.28}$$

The boundary conditions are $d_{1y} = 0$, $\phi_1 = 0$, $d_{3y} = 0$, and $\phi_3 = 0$. Using the direct stiffness method and Eqs. (4.4.28) to assemble the global stiffness matrix, and applying the boundary conditions, we obtain

$$\begin{Bmatrix} F_{2y} \\ M_2 \end{Bmatrix} = \begin{Bmatrix} \dfrac{-wL}{2} \\ \dfrac{-wL^2}{20} \end{Bmatrix} = \frac{EI}{L^3}\begin{bmatrix} 24 & 0 \\ 0 & 8L^2 \end{bmatrix} = \begin{Bmatrix} d_{2y} \\ \phi_2 \end{Bmatrix} \tag{4.4.29}$$

Solving Eq. (4.4.29) for the displacement and slope, we obtain

$$d_{2y} = \frac{-wL^4}{48EI} \qquad \phi_2 = \frac{-wL^3}{240EI} \tag{4.4.30}$$

Next, we obtain the effective nodal forces using $\underline{F}^{(e)} = \underline{K}\,\underline{d}$ as

$$\begin{Bmatrix} F_{1y}^{(e)} \\ M_1^{(e)} \\ F_{2y}^{(e)} \\ M_2^{(e)} \\ F_{3y}^{(e)} \\ M_3^{(e)} \end{Bmatrix} = \frac{EI}{L^3}\begin{bmatrix} 12 & 6L & -12 & 6L & 0 & 0 \\ 6L & 4L^2 & -6L & 2L^2 & 0 & 0 \\ -12 & -6L & 24 & 0 & -12 & 6L \\ 6L & 2L^2 & 0 & 8L^2 & -6L & 2L^2 \\ 0 & 0 & -12 & -6L & 12 & -6L \\ 0 & 0 & 6L & 2L^2 & -6L & 4L^2 \end{bmatrix}\begin{Bmatrix} 0 \\ 0 \\ \dfrac{-wL^4}{48EI} \\ \dfrac{-wL^3}{240EI} \\ 0 \\ 0 \end{Bmatrix} \tag{4.4.31}$$

Solving for the effective forces in Eq. (4.4.31), we obtain

$$\begin{aligned} F_{1y}^{(e)} &= \frac{9wL}{40} \qquad & M_1^{(e)} &= \frac{7wL^2}{60} \\ F_{2y}^{(e)} &= \frac{-wL}{2} \qquad & M_2^{(e)} &= \frac{-wL^2}{30} \\ F_{3y}^{(e)} &= \frac{11wL}{40} \qquad & M_3^{(e)} &= \frac{-2wL^2}{15} \end{aligned} \tag{4.4.32}$$

Finally, using Eq. (4.4.8) we subtract the equivalent nodal force matrix based on the equivalent load replacement shown in Figure 4–29(b) from the effective force matrix given by the results in Eq. (4.4.32), to obtain the correct nodal forces and moments as

$$\begin{Bmatrix} F_{1y} \\ M_1 \\ F_{2y} \\ M_2 \\ F_{3y} \\ M_3 \end{Bmatrix} = \begin{Bmatrix} \dfrac{9wL}{40} \\ \dfrac{7wL^2}{60} \\ \dfrac{-wL}{2} \\ \dfrac{-wL^2}{30} \\ \dfrac{11wL}{40} \\ \dfrac{-2wL^2}{15} \end{Bmatrix} - \begin{Bmatrix} \dfrac{-3wL}{40} \\ \dfrac{-wL^2}{60} \\ \dfrac{-wL}{2} \\ \dfrac{-wL^2}{30} \\ \dfrac{-17wL}{40} \\ \dfrac{wL^2}{15} \end{Bmatrix} = \begin{Bmatrix} \dfrac{12wL}{40} \\ \dfrac{8wL^2}{60} \\ 0 \\ 0 \\ \dfrac{28wL}{40} \\ \dfrac{-3wL^2}{15} \end{Bmatrix} \tag{4.4.33}$$

We used symbol L to represent one-half the length of the beam. If we replace L with the actual length $l = 2L$, we obtain the reactions for case 5 in Appendix D, thus verifying the correctness of our result.

In summary, for any structure in which an equivalent nodal force replacement is made, the actual nodal forces acting on the structure are determined by first evaluating the effective nodal forces $\underline{F}^{(e)}$ for the structure and then subtracting the equivalent nodal forces $\underline{F}_o$ for the structure, as indicated in Eq. (4.4.8). Similarly, for any element of a structure in which equivalent nodal force replacement is made, the actual local nodal forces acting on the element are determined by first evaluating the effective local nodal forces $\underline{\hat{f}}^{(e)}$ for the element and then subtracting the equivalent local nodal forces $\underline{\hat{f}}_o$ associated only with the element, as indicated in Eq. (4.4.11). We provide other examples of this procedure in plane frame Examples 5.2 and 5.3. ■

4.5 Comparison of the Finite Element Solution to the Exact Solution for a Beam

We will now compare the finite element solution to the exact classical beam theory solution for the cantilever beam shown in Figure 4–30 subjected to a uniformly distributed load. Both one- and two-element finite element solutions will be presented and compared to the exact solution obtained by the direct double-integration method. Let $E = 30 \times 10^6$ psi, $I = 100$ in^4, $L = 100$ in., and uniform load $w = 20$ lb/in.

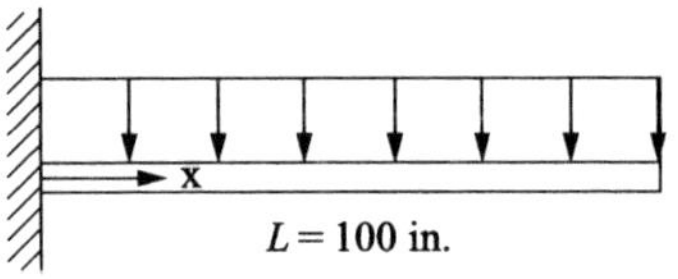

Figure 4–30 Cantilever beam subjected to uniformly distributed load

To obtain the solution from classical beam theory, we use the double-integration method [1]. Therefore, we begin with the moment-curvature equation

$$y'' = \frac{M(x)}{EI} \tag{4.5.1}$$

where the double prime superscript indicates differentiation with respect to x and M is expressed as a function of x by using a section of the beam as shown:

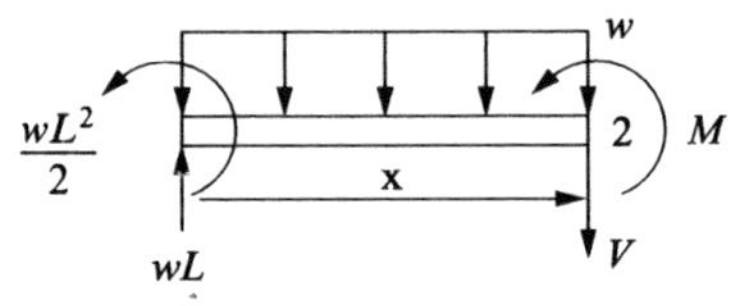

$$\Sigma F_y = 0\text{: } V(x) = wL - wx$$

$$\Sigma M_2 = 0\text{: } M(x) = \frac{-wL^2}{2} + wLx - (wx)\left(\frac{x}{2}\right) \tag{4.5.2}$$

Using Eq. (4.5.2) in Eq. (4.5.1), we have

$$y'' = \frac{1}{EI}\left(\frac{-wL^2}{2} + wLx - \frac{wx^2}{2}\right) \tag{4.5.3}$$

On integrating Eq. (4.5.3) with respect to x, we obtain an expression for the slope of the beam as

$$y' = \frac{1}{EI}\left(\frac{-wL^2x}{2} + \frac{wLx^2}{2} - \frac{wx^3}{6}\right) + C_1 \tag{4.5.4}$$

Integrating Eq. (4.5.4) with respect to x, we obtain the deflection expression for the beam as

$$y = \frac{1}{EI}\left(\frac{-wL^2x^2}{4} + \frac{wLx^3}{6} - \frac{wx^4}{24}\right) + C_1x + C_2 \tag{4.5.5}$$

Applying the boundary conditions $y = 0$ and $y' = 0$ at $x = 0$, we obtain

$$y'(0) = 0 = C_1 \qquad y(0) = 0 = C_2 \tag{4.5.6}$$

Using Eq. (4.5.6) in Eqs. (4.5.4) and (4.5.5), the final beam theory solution expressions for y' and y are then

$$y' = \frac{1}{EI}\left(\frac{-wx^3}{6} + \frac{wLx^2}{2} - \frac{wL^2x}{2}\right) \tag{4.5.7}$$

and

$$y = \frac{1}{EI}\left(\frac{-wx^4}{24} + \frac{wLx^3}{6} - \frac{wL^2x^2}{4}\right) \tag{4.5.8}$$

The one-element finite element solution for slope and displacement is given in variable form by Eqs. (4.4.14b). Using the numerical values of this problem in Eqs. (4.4.14b), we obtain the slope and displacement at the free end (node 2) as

$$\begin{aligned} \hat{\phi}_2 &= \frac{-wL^3}{6EI} = \frac{-(20\ \text{lb/in.})(100\ \text{in.})^3}{6(30 \times 10^6\ \text{psi})(100\ \text{in.}^4)} = -0.00111\ \text{rad} \\ \hat{d}_{2y} &= \frac{-wL^4}{8EI} = \frac{-(20\ \text{lb/in.})(100\ \text{in.})^4}{8(30 \times 10^6\ \text{psi})(100\ \text{in.}^4)} = -0.0833\ \text{in.} \end{aligned} \tag{4.5.9}$$

The slope and displacement given by Eq. (4.5.9) identically match the beam theory values, as Eqs. (4.5.7) and (4.5.8) evaluated at $x = L$ are identical to the variable form of the finite element solution given by Eqs. (4.4.14b). The reason why these nodal values from the finite element solution are correct is that the element nodal forces were calculated on the basis of being energy or work equivalent to the distributed load based on the assumed cubic displacement field within each beam element.

Values of displacement and slope at other locations along the beam for the finite element solution are obtained by using the assumed cubic displacement function [Eq. (4.1.4)] as

$$\hat{v}(x) = \frac{1}{L^3}(-2x^3 + 3x^2L)\hat{d}_{2y} + \frac{1}{L^3}(x^3L - x^2L^2)\hat{\phi}_2 \tag{4.5.10}$$

where the boundary conditions $\hat{d}_{1y} = \hat{\phi}_1 = 0$ have been used in Eq. (4.5.10). Using the numerical values in Eq. (4.5.10), we obtain the displacement at the midlength of the beam as

$$\begin{aligned}\hat{v}(x = 50 \text{ in.}) &= \frac{1}{(100 \text{ in.})^3}[-2(50 \text{ in.})^3 + 3(50 \text{ in.})^2(100 \text{ in.})](-0.0833 \text{ in.}) \\ &\quad + \frac{1}{(100 \text{ in.})^3}[(50 \text{ in.})^3(100 \text{ in.}) - (50 \text{ in.})^2(100 \text{ in.})^2] \\ &\quad \times (-0.00111 \text{ rad}) = -0.0278 \text{ in.}\end{aligned} \tag{4.5.11}$$

Using the beam theory [Eq. (4.5.8)], the deflection is

$$\begin{aligned}y(x = 50 \text{ in.}) &= \frac{20 \text{ lb/in.}}{30 \times 10^6 \text{ psi}(100 \text{ in.}^4)} \\ &\quad \times \left[\frac{-(50 \text{ in.})^4}{24} + \frac{(100 \text{ in.})(50 \text{ in.})^3}{6} - \frac{(100 \text{ in.})^2(50 \text{ in.})^2}{4}\right] \\ &= -0.0295 \text{ in.}\end{aligned} \tag{4.5.12}$$

We conclude that the beam theory solution for midlength displacement, $y = -0.0295$ in., is greater than the finite element solution for displacement, $\hat{v} = -0.0278$ in. In general, the displacements evaluated using the cubic function for $\hat{v}$ are lower as predicted by the finite element method than by the beam theory except at the nodes. This is always true for beams subjected to some form of distributed load that are modeled using the cubic displacement function. The exception to this result is at the nodes, where the beam theory and finite element results are identical because of the work-equivalence concept used to replace the distributed load by work-equivalent discrete loads at the nodes.

The beam theory solution predicts a quartic (fourth-order) polynomial expression for y [Eq. (4.5.5)] for a beam subjected to uniformly distributed loading, while the finite element solution $\hat{v}(x)$ assumes a cubic displacement behavior in each beam element under all load conditions. The finite element solution predicts a stiffer structure than the actual one. This is expected, as the finite element model forces the beam into specific modes of displacement and effectively yields a stiffer model than the actual structure. However, as more and more elements are used in the model, the finite element solution converges to the beam theory solution.

For the special case of a beam subjected to only nodal concentrated loads, the beam theory predicts a cubic displacement behavior, as the moment is a linear function and is integrated twice to obtain the resulting cubic displacement function. A simple verification of this cubic displacement behavior would be to solve the cantilevered beam subjected to an end load. In this special case, the finite element solution for displacement matches the beam theory solution for all locations along the beam length, as both functions $y(x)$ and $\hat{v}(x)$ are then cubic functions.

Monotonic convergence of the solution of a particular problem is discussed in Reference [3], and proof that compatible and complete displacement functions (as described in Section 3.2) used in the displacement formulation of the finite element method yield an upper bound on the true stiffness, hence a lower bound on the displacement of the problem, is discussed in Reference [3].

Under uniformly distributed loading, the beam theory solution predicts a quadratic moment and a linear shear force in the beam. However, the finite element solution using the cubic displacement function predicts a linear bending moment and a constant shear force within each beam element used in the model.

We will now determine the bending moment and shear force in the present problem based on the finite element method. The bending moment is given by

$$M = EIv'' = EI\frac{d^2(\underline{N}\underline{d})}{dx^2} = EI\frac{(d^2\underline{N})}{dx^2}\underline{d} \tag{4.5.13}$$

as $\underline{d}$ is not a function of x. Or in terms of the gradient matrix $\underline{B}$ we have

$$M = EI\underline{B}\underline{d} \tag{4.5.14}$$

where

$$\underline{B} = \frac{d^2\underline{N}}{dx^2} = \left[\left(-\frac{6}{L^2}+\frac{12x}{L^3}\right)\left(-\frac{4}{L}+\frac{6x}{L^2}\right)\left(\frac{6}{L^2}-\frac{12x}{L^3}\right)\left(-\frac{2}{L}+\frac{6x}{L^2}\right)\right] \tag{4.5.15}$$

The shape functions given by Eq. (4.1.7) are used to obtain Eq. (4.5.15) for the $\underline{B}$ matrix. For the single-element solution, the bending moment is then evaluated by substituting Eq. (4.5.15) for $\underline{B}$ into Eq. (4.5.14) and multiplying $\underline{B}$ by $\underline{d}$ to obtain

$$M = EI\left[\left(-\frac{6}{L^2}+\frac{12x}{L^3}\right)\hat{d}_{1x} + \left(-\frac{4}{L}+\frac{6x}{L^2}\right)\hat{\phi}_1 + \left(\frac{6}{L^2}-\frac{12x}{L^3}\right)\hat{d}_{2x} + \left(-\frac{2}{L}+\frac{6x}{L^2}\right)\hat{\phi}_2\right] \tag{4.5.16}$$

Evaluating the moment at the wall, $x = 0$, with $\hat{d}_{1x} = \hat{\phi}_1 = 0$, and $\hat{d}_{2x}$ and $\hat{\phi}_2$ given by Eq. (4.4.14) in Eq. (4.5.16), we have

$$M(x=0) = -\frac{10wL^2}{24} = -83{,}333 \text{ lb-in.} \tag{4.5.17}$$

Using Eq. (4.5.16) to evaluate the moment at $x = 50$ in., we have

$$M(x = 50 \text{ in.}) = -33{,}333 \text{ lb-in.} \tag{4.5.18}$$

Evaluating the moment at $x = 100$ in. by using Eq. (4.5.16) again, we obtain

$$M(x = 100 \text{ in.}) = 16{,}667 \text{ lb-in.} \tag{4.5.19}$$

The beam theory solution using Eq. (4.5.2) predicts

$$M(x=0) = \frac{-wL^2}{2} = -100{,}000 \text{ lb-in.} \tag{4.5.20}$$

$$M(x = 50 \text{ in.}) = -25{,}000 \text{ lb-in.}$$

and

$$M(x = 100 \text{ in.}) = 0$$

Figure 4–31(a)–(c) show the plots of the displacement variation, bending moment variation, and shear force variation through the beam length for the beam theory and the one-element finite element solutions. Again, the finite element solution for displacement matches the beam theory solution at the nodes but predicts smaller displacements (less deflection) at other locations along the beam length.

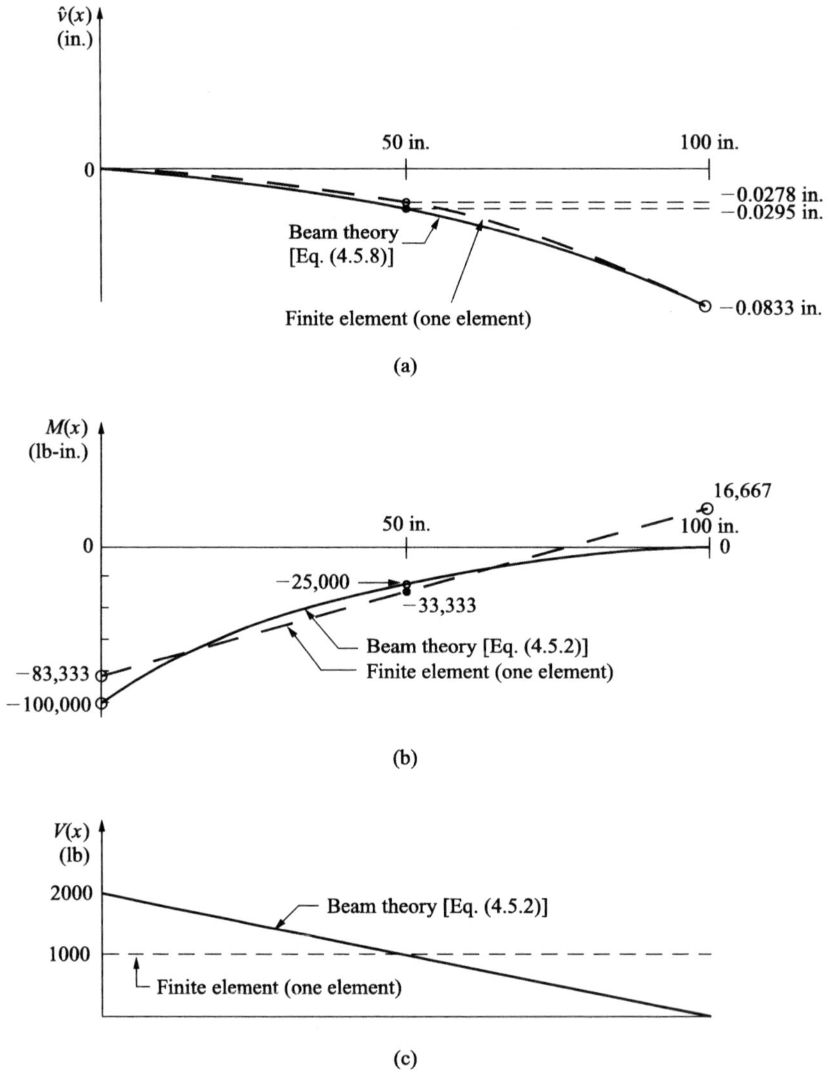

Figure 4–31 Comparison of beam theory and finite element results for a cantilever beam subjected to a uniformly distributed load: (a) displacement diagrams, (b) bending moment diagrams, and (c) shear force diagrams

The bending moment is derived by taking two derivatives on the displacement function. It then takes more elements to model the second derivative of the displacement function. Therefore, the finite element solution does not predict the bending moment as well as it does the displacement. For the uniformly loaded beam, the finite element model predicts a linear bending moment variation as shown in Figure 4–31(b). The best approximation for bending moment appears at the midpoint of the element.

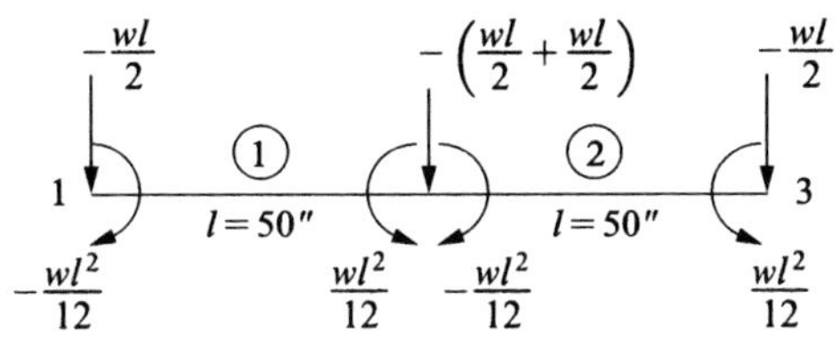

Figure 4–32 Beam discretized into two elements and work-equivalent load replacement for each element

The shear force is derived by taking three derivatives on the displacement function. For the uniformly loaded beam, the resulting shear force shown in Figure 4–31(c) is a constant throughout the single-element model. Again, the best approximation for shear force is at the midpoint of the element.

It should be noted that if we use Eq. (4.4.11), that is, $\underline{f} = \underline{k}\underline{d} - \underline{f}_o$, and subtract off the $\underline{f}_o$ matrix, we also obtain the correct nodal forces and moments in each element. For instance, from the one-element finite element solution we have for the bending moment at node 1

$$m_1^{(1)} = \frac{EI}{L^3}\left[-6L\left(\frac{-wL^4}{8EI}\right) + 2L^2\left(\frac{-wL^3}{6EI}\right)\right] - \left(\frac{-wL^2}{12}\right) = \frac{wL^2}{2}$$

and at node 2

$$m_2^{(1)} = 0$$

To improve the finite element solution we need to use more elements in the model (refine the mesh) or use a higher-order element, such as a fifth-order approximation for the displacement function, that is, $\hat{v}(x) = a_1 + a_2x + a_3x^2 + a_4x^3 + a_5x^4 + a_6x^5$, with three nodes (with an extra node at the middle of the element).

We now present the two-element finite element solution for the cantilever beam subjected to a uniformly distributed load. Figure 4–32 shows the beam discretized into two elements of equal length and the work-equivalent load replacement for each element. Using the beam element stiffness matrix [Eq. (4.1.13)], we obtain the element stiffness matrices as follows:

$$\begin{matrix} & & 1 & & 2 & \\ & & 2 & & 3 & \end{matrix}$$

$$\underline{k}^{(1)} = \underline{k}^{(2)} = \frac{EI}{l^3}\begin{bmatrix} 12 & 6l & -12 & 6l \\ 6l & 4l^2 & -6l & 2l^2 \\ -12 & -6l & 12 & -6l \\ 6l & 2l^2 & -6l & 4l^2 \end{bmatrix} \tag{4.5.21}$$

where $l = 50$ in. is the length of each element and the numbers above the columns indicate the degrees of freedom associated with each element.

Applying the boundary conditions $\hat{d}_{1y} = 0$ and $\hat{\phi}_1 = 0$ to reduce the number of equations for a normal longhand solution, we obtain the global equations for solution as

$$\frac{EI}{l^3}\begin{bmatrix} 24 & 0 & -12 & 6l \\ 0 & 8l^2 & -6l & 2l^2 \\ -12 & -6l & 12 & -6l \\ 6l & 2l^2 & -6l & 4l^2 \end{bmatrix}\begin{Bmatrix} \hat{d}_{2y} \\ \hat{\phi}_2 \\ \hat{d}_{3y} \\ \hat{\phi}_3 \end{Bmatrix} = \begin{Bmatrix} -wl \\ 0 \\ -wl/2 \\ wl^2/12 \end{Bmatrix} \tag{4.5.22}$$

Solving Eq. (4.5.22) for the displacements and slopes, we obtain

$$\hat{d}_{2y} = \frac{-17wl^4}{24EI} \qquad \hat{d}_{3y} = \frac{-2wl^4}{EI} \qquad \hat{\phi}_2 = \frac{-7wl^3}{6EI} \qquad \hat{\phi}_3 = \frac{-4wl^3}{3EI} \tag{4.5.23}$$

Substituting the numerical values $w = 20$ lb/in., $l = 50$ in., $E = 30 \times 10^6$ psi, and $I = 100$ in.^{4} into Eq. (4.5.23), we obtain

$$\hat{d}_{2y} = -0.02951 \text{ in.} \qquad \hat{d}_{3y} = -0.0833 \text{ in.} \qquad \hat{\phi}_2 = -9.722 \times 10^{-4} \text{ rad}$$

$$\hat{\phi}_3 = -11.11 \times 10^{-4} \text{ rad}$$

The two-element solution yields nodal displacements that match the beam theory results exactly [see Eqs. (4.5.9) and (4.5.12)]. A plot of the two-element displacement throughout the length of the beam would be a cubic displacement within each element. Within element 1, the plot would start at a displacement of 0 at node 1 and finish at a displacement of -0.0295 at node 2. A cubic function would connect these values. Similarly, within element 2, the plot would start at a displacement of -0.0295 and finish at a displacement of -0.0833 in. at node 2 [see Figure 4–31(a)]. A cubic function would again connect these values.

4.6 Beam Element with Nodal Hinge

In some beams an internal hinge may be present. In general, this internal hinge causes a discontinuity in the slope of the deflection curve at the hinge.

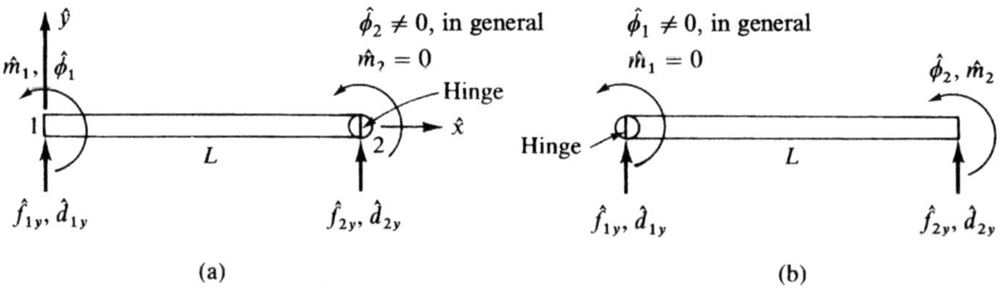

Figure 4–33 Beam element with (a) hinge at right end and (b) hinge at left end

Also, the bending moment is zero at the hinge. We could construct other types of connections that release other generalized end forces; that is, connections can be designed to make the shear force or axial force zero at the connection. These special conditions can be treated by starting with the generalized unreleased beam stiffness matrix [Eq. (4.1.14)] and eliminating the known zero force or moment. This yields a modified stiffness matrix with the desired force or moment equal to zero and the corresponding displacement or slope eliminated.

We now consider the most common cases of a beam element with a nodal hinge at the right end or left end, as shown in Figure 4–33. For the beam element with a hinge at its right end, the moment $\hat{m}_2$ is zero and we partition the $\hat{\underline{k}}$ matrix

[Eq. (4.1.14)] to eliminate the degree of freedom $\hat{\phi}_2$ (which is not zero, in general) associated with $\hat{m}_2 = 0$ as follows:

$$\underline{\hat{k}} = \frac{EI}{L^3}\left[\begin{array}{ccc|c} 12 & 6L & -12 & 6L \\ 6L & 4L^2 & -6L & 2L^2 \\ -12 & -6L & 12 & -6L \\ \hline 6L & 2L^2 & -6L & 4L^2 \end{array}\right] \tag{4.6.1}$$

We condense out the degree of freedom $\hat{\phi}_2$ associated with $\hat{m}_2 = 0$. Partitioning allows us to condense out the degree of freedom $\hat{\phi}_2$ associated with $\hat{m}_2 = 0$. That is, Eq. (4.6.1) is partitioned as shown below:

$$\underline{\hat{k}} = \left[\begin{array}{c|c} \underset{3\times 3}{\underline{K}_{11}} & \underset{3\times 1}{\underline{K}_{12}} \\ \hline \underset{1\times 3}{\underline{K}_{21}} & \underset{1\times 1}{\underline{K}_{22}} \end{array}\right] \tag{4.6.2}$$

The condensed stiffness matrix is then found by using the equation $\underline{\hat{f}} = \underline{\hat{k}}\underline{\hat{d}}$ partitioned as follows:

$$\left\{\begin{array}{c} \underset{3\times 1}{\underline{f}_1} \\ \hline \underset{1\times 1}{\underline{f}_2} \end{array}\right\} = \left[\begin{array}{c|c} \underset{3\times 3}{\underline{K}_{11}} & \underset{3\times 1}{\underline{K}_{12}} \\ \hline \underset{1\times 3}{\underline{K}_{21}} & \underset{1\times 1}{\underline{K}_{22}} \end{array}\right]\left\{\begin{array}{c} \underset{3\times 1}{\underline{d}_1} \\ \hline \underset{1\times 1}{\underline{d}_2} \end{array}\right\} \tag{4.6.3}$$

where

$$\underline{d}_1 = \begin{Bmatrix} \hat{d}_{1y} \\ \hat{\phi}_1 \\ \hat{d}_{2y} \end{Bmatrix} \qquad \underline{d}_2 = \{\hat{\phi}_2\} \tag{4.6.4}$$

Equations (4.6.3) in expanded form are

$$\underline{f}_1 = \underline{K}_{11}\underline{d}_1 + \underline{K}_{12}\underline{d}_2 \tag{4.6.5}$$

$$\underline{f}_2 = \underline{K}_{21}\underline{d}_1 + \underline{K}_{22}\underline{d}_2$$

Solving for $\underline{d}_2$ in the second of Eqs. (4.6.5), we obtain

$$\underline{d}_2 = \underline{K}_{22}^{-1}(\underline{f}_2 - \underline{K}_{21}\underline{d}_1) \tag{4.6.6}$$

Substituting Eq. (4.6.6) into the first of Eqs. (4.6.5), we obtain

$$\underline{f}_1 = (\underline{K}_{11} - \underline{K}_{12}\underline{K}_{22}^{-1}\underline{K}_{21})\underline{d}_1 + \underline{K}_{12}\underline{K}_{22}^{-1}\underline{f}_2 \tag{4.6.7}$$

Combining the second term on the right side of Eq. (4.6.7) with $\underline{f}_1$, we obtain

$$\underline{f}_c = \underline{K}_c\underline{d}_1 \tag{4.6.8}$$

where the condensed stiffness matrix is

$$\underline{K}_c = \underline{K}_{11} - \underline{K}_{12}\underline{K}_{22}^{-1}\underline{K}_{21} \tag{4.6.9}$$

and the condensed force matrix is

$$\underline{f}_c = \underline{f}_1 - \underline{K}_{12}\underline{K}_{22}^{-1}\underline{f}_2 \tag{4.6.10}$$

Substituting the partitioned parts of $\underset{\sim}{\hat{k}}$ from Eq. (4.6.1) into Eq. (4.6.9), we obtain the condensed stiffness matrix as

$$\underset{\sim}{K_c} = [K_{11}] - [K_{12}][K_{22}]^{-1}[K_{21}]$$

$$= \frac{EI}{L^3}\begin{bmatrix} 12 & 6L & -12 \\ 6L & 4L^2 & -6L \\ -12 & -6L & 12 \end{bmatrix} - \frac{EI}{L^3}\begin{Bmatrix} 6L \\ 2L^2 \\ -6L \end{Bmatrix}\frac{1}{4L^2}[6L \quad 2L^2 \quad -6L]$$

$$= \frac{3EI}{L^3}\begin{bmatrix} 1 & L & -1 \\ L & L^2 & -L \\ -1 & -L & 1 \end{bmatrix} \qquad (4.6.11)$$

and the element equations (force/displacement equations) with the hinge at node 2 are

$$\begin{Bmatrix} \hat{f}_{1y} \\ \hat{m}_1 \\ \hat{f}_{2y} \end{Bmatrix} = \frac{3EI}{L^3}\begin{bmatrix} 1 & L & -1 \\ L & L^2 & -L \\ -1 & -L & 1 \end{bmatrix}\begin{Bmatrix} \hat{d}_{1y} \\ \hat{\phi}_1 \\ \hat{d}_{2y} \end{Bmatrix} \qquad (4.6.12)$$

The generalized rotation $\hat{\phi}_2$ has been eliminated from the equation and will not be calculated using this scheme. However, $\hat{\phi}_2$ is not zero in general. We can expand Eq. (4.6.12) to include $\hat{\phi}_2$ by adding zeros in the fourth row and column of the $\underset{\sim}{\hat{k}}$ matrix to maintain $\hat{m}_2 = 0$, as follows:

$$\begin{Bmatrix} \hat{f}_{1y} \\ \hat{m}_1 \\ \hat{f}_{2y} \\ \hat{m}_2 \end{Bmatrix} = \frac{3EI}{L^3}\begin{bmatrix} 1 & L & -1 & 0 \\ L & L^2 & -L & 0 \\ -1 & -L & 1 & 0 \\ 0 & 0 & 0 & 0 \end{bmatrix}\begin{Bmatrix} \hat{d}_{1y} \\ \hat{\phi}_1 \\ \hat{d}_{2y} \\ \hat{\phi}_2 \end{Bmatrix} \qquad (4.6.13)$$

For the beam element with a hinge at its left end, the moment $\hat{m}_1$ is zero, and we partition the $\underset{\sim}{\hat{k}}$ matrix [Eq. (4.1.14)] to eliminate the zero moment $\hat{m}_1$ and its corresponding rotation $\hat{\phi}_1$ to obtain

$$\begin{Bmatrix} \hat{f}_{1y} \\ \hat{f}_{2y} \\ \hat{m}_2 \end{Bmatrix} = \frac{3EI}{L^3}\begin{bmatrix} 1 & -1 & L \\ -1 & 1 & -L \\ L & -L & L^2 \end{bmatrix}\begin{Bmatrix} \hat{d}_{1y} \\ \hat{d}_{2y} \\ \hat{\phi}_2 \end{Bmatrix} \qquad (4.6.14)$$

The expanded form of Eq. (4.6.14) including $\hat{\phi}_1$ is

$$\begin{Bmatrix} \hat{f}_{1y} \\ \hat{m}_1 \\ \hat{f}_{2y} \\ \hat{m}_2 \end{Bmatrix} = \frac{3EI}{L^3}\begin{bmatrix} 1 & 0 & -1 & L \\ 0 & 0 & 0 & 0 \\ -1 & 0 & 1 & -L \\ L & 0 & -L & L^2 \end{bmatrix}\begin{Bmatrix} \hat{d}_{1y} \\ \hat{\phi}_1 \\ \hat{d}_{2y} \\ \hat{\phi}_2 \end{Bmatrix} \qquad (4.6.15)$$

Example 4.10

Determine the displacement and rotation at node 2 and the element forces for the uniform beam with an internal hinge at node 2 shown in Figure 4–34. Let EI be a constant.

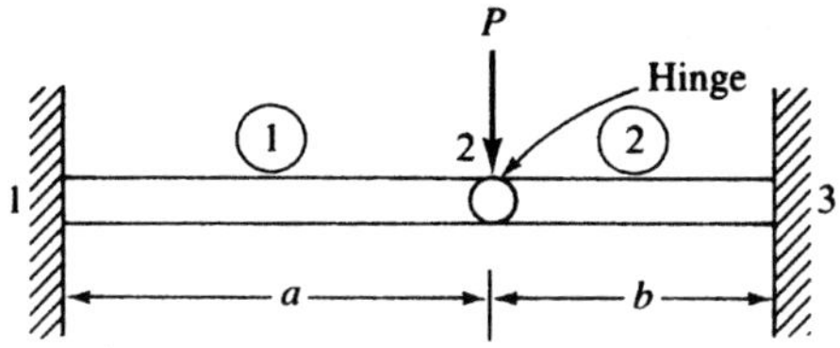

Figure 4–34 Beam with internal hinge

We can assume the hinge is part of element 1. Therefore, using Eq. (4.6.13), the stiffness matrix of element 1 is

$$\underline{k}^{(1)} = \frac{3EI}{a^3} \begin{bmatrix} 1 & a & -1 & 0 \\ a & a^2 & -a & 0 \\ -1 & -a & 1 & 0 \\ 0 & 0 & 0 & 0 \end{bmatrix} \quad \begin{matrix} d_{1y} & \phi_1 & d_{2y} & \phi_2 \end{matrix} \tag{4.6.16}$$

The stiffness matrix of element 2 is obtained from Eq. (4.1.14) as

$$\underline{k}^{(2)} = \frac{EI}{b^3} \begin{bmatrix} 12 & 6b & -12 & 6b \\ 6b & 4b^2 & -6b & 2b^2 \\ -12 & -6b & 12 & -6b \\ 6b & 2b^2 & -6b & 4b^2 \end{bmatrix} \quad \begin{matrix} d_{2y} & \phi_2 & d_{3y} & \phi_3 \end{matrix} \tag{4.6.17}$$

Superimposing Eqs. (4.6.16) and (4.6.17) and applying the boundary conditions

$$d_{1y} = 0, \qquad \phi_1 = 0, \qquad d_{3y} = 0, \qquad \phi_3 = 0$$

we obtain the total stiffness matrix and total set of equations as

$$EI \begin{bmatrix} \dfrac{3}{a^3} + \dfrac{12}{b^3} & \dfrac{6}{b^2} \\ \dfrac{6}{b^2} & \dfrac{4}{b} \end{bmatrix} \begin{Bmatrix} d_{2y} \\ \phi_2 \end{Bmatrix} = \begin{Bmatrix} -P \\ 0 \end{Bmatrix} \tag{4.6.18}$$

Solving Eq. (4.6.18), we obtain

$$d_{2y} = \frac{-a^3 b^3 P}{3(b^3 + a^3)EI}$$
$$\phi_2 = \frac{a^3 b^2 P}{2(b^3 + a^3)EI} \tag{4.6.19}$$

The value ϕ_2 is actually that associated with element 2—that is, ϕ_2 in Eq. (4.6.19) is actually $\phi_2^{(2)}$. The value of ϕ_2 at the right end of element 1 ($\phi_2^{(1)}$) is, in general, not equal to $\phi_2^{(2)}$. If we had chosen to assume the hinge to be part of element 2, then we would have used Eq. (4.1.14) for the stiffness matrix of element 1 and Eq. (4.6.15) for the stiffness matrix of element 2. This would have enabled us to obtain $\phi_2^{(1)}$, which is different from $\phi_2^{(2)}$.

Using Eq. (4.6.12) for element 1, we obtain the element forces as

$$\begin{Bmatrix} \hat{f}_{1y} \\ \hat{m}_1 \\ \hat{f}_{2y} \end{Bmatrix} = \frac{3EI}{a^3} \begin{bmatrix} 1 & a & -1 \\ a & a^2 & -a \\ -1 & -a & 1 \end{bmatrix} \begin{Bmatrix} 0 \\ 0 \\ \dfrac{-a^3b^3P}{3(b^3+a^3)EI} \end{Bmatrix} \tag{4.6.20}$$

Simplifying Eq. (4.6.20), we obtain the forces as

$$\hat{f}_{1y} = \frac{b^3P}{b^3+a^3}$$
$$\hat{m}_1 = \frac{ab^3P}{b^3+a^3} \tag{4.6.21}$$
$$\hat{f}_{2y} = -\frac{b^3P}{b^3+a^3}$$

Using Eq. (4.6.17) and the results from Eq. (4.6.19), we obtain the element 2 forces as

$$\begin{Bmatrix} \hat{f}_{2y} \\ \hat{m}_2 \\ \hat{f}_{3y} \\ \hat{m}_3 \end{Bmatrix} = \frac{EI}{b^3} \begin{bmatrix} 12 & 6b & -12 & 6b \\ 6b & 4b^2 & -6b & 2b^2 \\ -12 & -6b & 12 & -6b \\ 6b & 2b^2 & -6b & 4b^2 \end{bmatrix} \begin{Bmatrix} -\dfrac{a^3b^3P}{3(b^3+a^3)EI} \\ \dfrac{a^3b^2P}{2(b^3+a^3)EI} \\ 0 \\ 0 \end{Bmatrix} \tag{4.6.22}$$

Simplifying Eq. (4.6.22), we obtain the element forces as

$$\hat{f}_{2y} = -\frac{a^3P}{b^3+a^3}$$
$$\hat{m}_2 = 0$$
$$\hat{f}_{3y} = \frac{a^3P}{b^3+a^3} \tag{4.6.23}$$
$$\hat{m}_3 = -\frac{ba^3P}{b^3+a^3}$$

■

It should be noted that another way to solve the nodal hinge of Example 4.10 would be to assume a nodal hinge at the right end of element one and at the left end of element two. Hence, we would use the three-equation stiffness matrix of Eq. (4.6.12) for the left element and the three-equation stiffness matrix of Eq. (4.6.14) for the right element. This results in the hinge rotation being condensed out of the global equations. You can verify that we get the same result for the displacement as given by Eq. (4.6.19). However, we must then go back to Eq. (4.6.6) using it separately for each element to obtain the rotation at node two for each element. We leave this verification to your discretion.

4.7 Potential Energy Approach to Derive Beam Element Equations

We will now derive the beam element equations using the principle of minimum potential energy. The procedure is similar to that used in Section 3.10 in deriving the bar element equations. Again, our primary purpose in applying the principle of minimum potential energy is to enhance your understanding of the principle. It will be used routinely in subsequent chapters to develop element stiffness equations. We use the same notation here as in Section 3.10.

The total potential energy for a beam is

$$\pi_p = U + \Omega \tag{4.7.1}$$

where the general one-dimensional expression for the strain energy U for a beam is given by

$$U = \iiint_V \frac{1}{2}\sigma_x \varepsilon_x \, dV \tag{4.7.2}$$

and for a single beam element subjected to both distributed and concentrated nodal loads, the potential energy of forces is given by

$$\Omega = -\iint_{S_1} \hat{T}_y \hat{v} \, dS - \sum_{i=1}^{2} \hat{P}_{iy}\hat{d}_{iy} - \sum_{i=1}^{2} \hat{m}_i \hat{\phi}_i \tag{4.7.3}$$

where body forces are now neglected. The terms on the right-hand side of Eq. (4.7.3) represent the potential energy of (1) transverse surface loading $\hat{T}_y$ (in units of force per unit surface area, acting over surface S_1 and moving through displacements over which $\hat{T}_y$ act); (2) nodal concentrated force $\hat{P}_{iy}$ moving through displacements $\hat{d}_{iy}$; and (3) moments $\hat{m}_i$ moving through rotations $\hat{\phi}_i$. Again, $\hat{v}$ is the transverse displacement function for the beam element of length L shown in Figure 4–35.

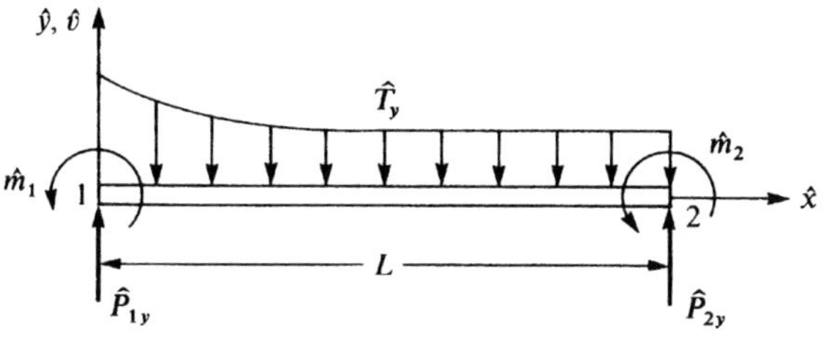

Figure 4–35 Beam element subjected to surface loading and concentrated nodal forces

Consider the beam element to have constant cross-sectional area A. The differential volume for the beam element can then be expressed as

$$dV = dA\,d\hat{x} \tag{4.7.4}$$

and the differential area over which the surface loading acts is

$$dS = b\,d\hat{x} \tag{4.7.5}$$

where b is the constant width. Using Eqs. (4.7.4) and (4.7.5) in Eqs. (4.7.1)–(4.7.3), the total potential energy becomes

$$\pi_p = \iiint_{\hat{x}\ A} \frac{1}{2}\sigma_x\varepsilon_x\,dA\,d\hat{x} - \int_0^L b\hat{T}_y\hat{v}\,d\hat{x} - \sum_{i=1}^{2}(\hat{P}_{iy}\hat{d}_{iy} + \hat{m}_i\hat{\phi}_i) \tag{4.7.6}$$

Substituting Eq. (4.1.4) for $\hat{v}$ into the strain/displacement relationship Eq. (4.1.10), repeated here for convenience as

$$\varepsilon_x = -\hat{y}\frac{d^2\hat{v}}{d\hat{x}^2} \tag{4.7.7}$$

we express the strain in terms of nodal displacements and rotations as

$$\{\varepsilon_x\} = -\hat{y}\left[\frac{12\hat{x} - 6L}{L^3}\quad \frac{6\hat{x}L - 4L^2}{L^3}\quad \frac{-12\hat{x} + 6L}{L^3}\quad \frac{6\hat{x}L - 2L^2}{L^3}\right]\{\hat{d}\} \tag{4.7.8}$$

or

$$\{\varepsilon_x\} = -\hat{y}[B]\{\hat{d}\} \tag{4.7.9}$$

where we define

$$[B] = \left[\frac{12\hat{x} - 6L}{L^3}\quad \frac{6\hat{x}L - 4L^2}{L^3}\quad \frac{-12\hat{x} + 6L}{L^3}\quad \frac{6\hat{x}L - 2L^2}{L^3}\right] \tag{4.7.10}$$

The stress/strain relationship is given by

$$\{\sigma_x\} = [D]\{\varepsilon_x\} \tag{4.7.11}$$

where

$$[D] = [E] \tag{4.7.12}$$

and E is the modulus of elasticity. Using Eq. (4.7.9) in Eq. (4.7.11), we obtain

$$\{\sigma_x\} = -\hat{y}[D][B]\{\hat{d}\} \tag{4.7.13}$$

Next, the total potential energy Eq. (4.7.6) is expressed in matrix notation as

$$\pi_p = \iiint_{\hat{x}\ A} \frac{1}{2}\{\sigma_x\}^T\{\varepsilon_x\}\,dA\,d\hat{x} - \int_0^L b\hat{T}_y[\hat{v}]^T\,d\hat{x} - \{\hat{d}\}^T\{\hat{P}\} \tag{4.7.14}$$

Using Eqs. (4.1.5), (4.7.9), (4.7.12), and (4.7.13), and defining $w = b\hat{T}_y$ as the line load (load per unit length) in the $\hat{y}$ direction, we express the total potential energy, Eq. (4.7.14), in matrix form as

$$\pi_p = \int_0^L \frac{EI}{2}\{\hat{d}\}^T[B]^T[B]\{\hat{d}\}\,d\hat{x} - \int_0^L w\{\hat{d}\}^T[N]^T\,d\hat{x} - \{\hat{d}\}^T\{\hat{P}\} \tag{4.7.15}$$

where we have used the definition of the moment of inertia

$$I = \iint_A y^2\,dA \tag{4.7.16}$$

to obtain the first term on the right-hand side of Eq. (4.7.15). In Eq. (4.7.15), π_p is now expressed as a function of $\{\hat{d}\}$.

Differentiating π_p in Eq. (4.7.15) with respect to $\hat{d}_{1y}, \hat{\phi}_1, \hat{d}_{2y}$, and $\hat{\phi}_2$ and equating each term to zero to minimize π_p, we obtain four element equations, which are written in matrix form as

$$EI \int_0^L [B]^T [B]\, d\hat{x} \{\hat{d}\} - \int_0^L [N]^T w\, d\hat{x} - \{\hat{P}\} = 0 \tag{4.7.17}$$

The derivation of the four element equations is left as an exercise (see Problem 4.45). Representing the nodal force matrix as the sum of those nodal forces resulting from distributed loading and concentrated loading, we have

$$\{\hat{f}\} = \int_0^L [N]^T w\, d\hat{x} + \{\hat{P}\} \tag{4.7.18}$$

Using Eq. (4.7.18), the four element equations given by explicitly evaluating Eq. (4.7.17) are then identical to Eq. (4.1.13). The integral term on the right side of Eq. (4.7.18) also represents the work-equivalent replacement of a distributed load by nodal concentrated loads. For instance, letting $w(\hat{x}) = -w$ (constant), substituting shape functions from Eq. (4.1.7) into the integral, and then performing the integration result in the same nodal equivalent loads as given by Eqs. (4.4.5)–(4.4.7).

Because $\{\hat{f}\} = [\hat{k}]\{\hat{d}\}$, we have, from Eq. (4.7.17),

$$[\hat{k}] = EI \int_0^L [B]^T [B]\, d\hat{x} \tag{4.7.19}$$

Using Eq. (4.7.10) in Eq. (4.7.19) and integrating, $[\hat{k}]$ is evaluated in explicit form as

$$[\hat{k}] = \frac{EI}{L^3} \begin{bmatrix} 12 & 6L & -12 & 6L \\ & 4L^2 & -6L & 2L^2 \\ & & 12 & -6L \\ \text{Symmetry} & & & 4L^2 \end{bmatrix} \tag{4.7.20}$$

Equation (4.7.20) represents the local stiffness matrix for a beam element. As expected, Eq. (4.7.20) is identical to Eq. (4.1.14) developed previously.

4.8 Galerkin's Method for Deriving Beam Element Equations

We will now illustrate Galerkin's method to formulate the beam element stiffness equations. We begin with the basic differential Eq. (4.1.1h) with transverse loading w now included; that is,

$$EI \frac{d^4 \hat{v}}{d\hat{x}^4} + w = 0 \tag{4.8.1}$$

We now define the residual R to be Eq. (4.8.1). Applying Galerkin's criterion [Eq. (3.12.3)] to Eq. (4.8.1), we have

$$\int_0^L \left(EI \frac{d^4 \hat{v}}{d\hat{x}^4} + w \right) N_i\, d\hat{x} = 0 \qquad (i = 1, 2, 3, 4) \tag{4.8.2}$$

where the shape functions N_i are defined by Eqs. (4.1.7).

We now apply integration by parts twice to the first term in Eq. (4.8.2) to yield

$$\int_0^L EI(\hat{v},_{\hat{x}\hat{x}\hat{x}\hat{x}})N_i\,d\hat{x} = \int_0^L EI(\hat{v},_{\hat{x}\hat{x}})(N_i,_{\hat{x}\hat{x}})\,d\hat{x} + EI[N_i(\hat{v},_{\hat{x}\hat{x}\hat{x}}) - (N_i,_{\hat{x}})(\hat{v},_{\hat{x}\hat{x}})]_0^L \quad (4.8.3)$$

where the notation of the comma followed by the subscript $\hat{x}$ indicates differentiation with respect to $\hat{x}$. Again, integration by parts introduces the boundary conditions.

Because $\hat{v} = [N]\{\hat{d}\}$ as given by Eq. (4.1.5), we have

$$\hat{v},_{\hat{x}\hat{x}} = \left[\frac{12\hat{x} - 6L}{L^3} \quad \frac{6\hat{x}L - 4L^2}{L^3} \quad \frac{-12\hat{x} + 6L}{L^3} \quad \frac{6\hat{x}L - 2L^2}{L^3}\right]\{\hat{d}\} \quad (4.8.4)$$

or, using Eq. (4.7.10),

$$\hat{v},_{\hat{x}\hat{x}} = [B]\{\hat{d}\} \quad (4.8.5)$$

Substituting Eq. (4.8.5) into Eq. (4.8.3), and then Eq. (4.8.3) into Eq. (4.8.2), we obtain

$$\int_0^L (N_i,_{\hat{x}\hat{x}})EI[B]\,d\hat{x}\{\hat{d}\} + \int_0^L N_i w\,d\hat{x} + [N_i\hat{V} - (N_i,_{\hat{x}})\hat{m}]|_0^L = 0 \qquad (i = 1, 2, 3, 4) \quad (4.8.6)$$

where Eqs. (4.1.11) have been used in the boundary terms. Equation (4.8.6) is really four equations (one each for $N_i = N_1, N_2, N_3$, and N_4). Instead of directly evaluating Eq. (4.8.6) for each N_i, as was done in Section 3.12, we can express the four equations of Eq. (4.8.6) in matrix form as

$$\int_0^L [B]^T EI[B]\,d\hat{x}\{\hat{d}\} = \int_0^L -[N]^T w\,d\hat{x} + ([N]^T,_{\hat{x}}\hat{m} - [N]^T\hat{V})|_0^L \quad (4.8.7)$$

where we have used the relationship $[N],_{xx} = [B]$ in Eq. (4.8.7).

Observe that the integral term on the left side of Eq. (4.8.7) is identical to the stiffness matrix previously given by Eq. (4.7.19) and that the first term on the right side of Eq. (4.8.7) represents the equivalent nodal forces due to distributed loading [also given in Eq. (4.7.18)]. The two terms in parentheses on the right side of Eq. (4.8.7) are the same as the concentrated force matrix $\{\hat{P}\}$ of Eq. (4.7.18). We explain this by evaluating $[N],_{\hat{x}}$ and $[N]$, where $[N]$ is defined by Eq. (4.1.6), at the ends of the element as follows:

$$\begin{array}{ll} [N],_{\hat{x}}|_0 = [0 \quad 1 \quad 0 \quad 0] & [N],_{\hat{x}}|_L = [0 \quad 0 \quad 0 \quad 1] \\ [N]|_0 = [1 \quad 0 \quad 0 \quad 0] & [N]|_L = [0 \quad 0 \quad 1 \quad 0] \end{array} \quad (4.8.8)$$

Therefore, when we use Eqs. (4.8.8) in Eq. (4.8.7), the following terms result:

$$\begin{Bmatrix} 0 \\ 0 \\ 0 \\ 1 \end{Bmatrix}\hat{m}(L) - \begin{Bmatrix} 0 \\ 1 \\ 0 \\ 0 \end{Bmatrix}\hat{m}(0) - \begin{Bmatrix} 0 \\ 0 \\ 1 \\ 0 \end{Bmatrix}\hat{V}(L) + \begin{Bmatrix} 1 \\ 0 \\ 0 \\ 0 \end{Bmatrix}\hat{V}(0) \quad (4.8.9)$$

These nodal shear forces and moments are illustrated in Figure 4–36.

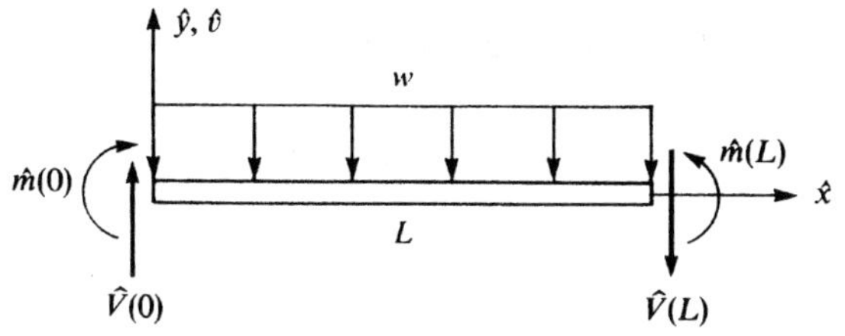

Figure 4–36 Beam element with shear forces, moments, and a distributed load

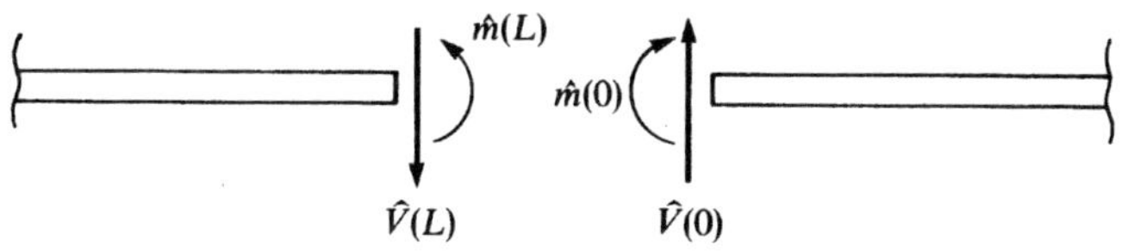

Figure 4–37 Shear forces and moments acting on adjacent elements meeting at a node

Note that when element matrices are assembled, two shear forces and two moments from adjacent elements contribute to the concentrated force and concentrated moment at the node common to the adjacent elements as shown in Figure 4–37. These concentrated shear forces $\hat{V}(0) - \hat{V}(L)$ and moments $\hat{m}(L) - \hat{m}(0)$ are often zero; that is, $\hat{V}(0) = \hat{V}(L)$ and $\hat{m}(L) = \hat{m}(0)$ occur except when a concentrated nodal force or moment exists at the node. In the actual computations, we handle the expressions given by Eq. (4.8.9) by including them as concentrated nodal values making up the matrix $\{P\}$.

References

[1] Gere, J. M., *Mechanics of Materials*, 5th ed., Brooks/Cole Publishers, Pacific Grove, CA, 2001.

[2] Hsieh, Y. Y., *Elementary Theory of Structures*, 2nd ed., Prentice-Hall, Englewood Cliffs, NJ, 1982.

[3] Fraeijes de Veubeke, B., "Upper and Lower Bounds in Matrix Structural Analysis," *Matrix Methods of Structural Analysis*, AGAR Dograph 72, B. Fraeijes de Veubeke, ed., Macmillan, New York, 1964.

[4] Juvinall, R. C., and Marshek, K. M., *Fundamentals of Machine Component Design*, 4th. ed., John Wiley & Sons, New York, 2006.

[5] Przemieneicki, J. S., *Theory of Matrix Structural Analysis*, McGraw-Hill, New York, 1968.

[6] McGuire, W., and Gallagher, R. H., *Matrix Structural Analysis*, John Wiley & Sons, New York, 1979.

[7] Severn, R. T., "Inclusion of Shear Deflection in the Stiffness Matrix for a Beam Element", *Journal of Strain Analysis*, Vol. 5, No. 4, 1970, pp. 239–241.

[8] Narayanaswami, R., and Adelman, H. M., "Inclusion of Transverse Shear Deformation in Finite Element Displacement Formulations", *AIAA Journal*, Vol. 12, No. 11, 1974, 1613–1614.

[9] Timoshenko, S., *Vibration Problems in Engineering,* 3rd. ed., Van Nostrand Reinhold Company, 1955.
[10] Clark, S. K., *Dynamics of Continous Elements,* Prentice Hall, 1972.
[11] Algor Interactive Systems, 260 Alpha Dr., Pittsburgh, PA 15238.

Problems

4.1 Use Eqs. (4.1.7) to plot the shape functions N_1 and N_3 and the derivatives $(dN_2/d\hat{x})$ and $(dN_4/d\hat{x})$, which represent the shapes (variations) of the slopes $\hat{\phi}_1$ and $\hat{\phi}_2$ over the length of the beam element.

4.2 Derive the element stiffness matrix for the beam element in Figure 4–1 if the rotational degrees of freedom are assumed positive clockwise instead of counterclockwise. Compare the two different nodal sign conventions and discuss. Compare the resulting stiffness matrix to Eq. (4.1.14).

Solve all problems using the finite element stiffness method.

4.3 For the beam shown in Figure P4–3, determine the rotation at pin support A and the rotation and displacement under the load P. Determine the reactions. Draw the shear force and bending moment diagrams. Let EI be constant throughout the beam.

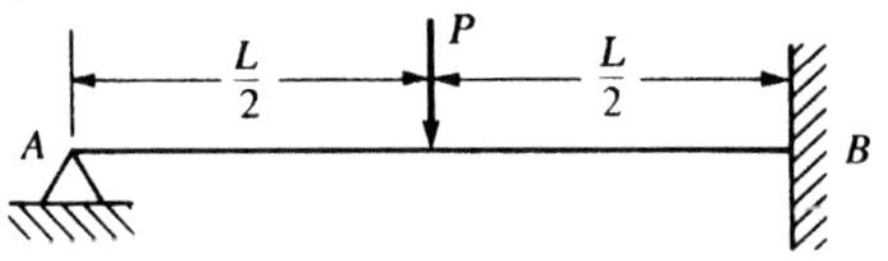

Figure P4–3

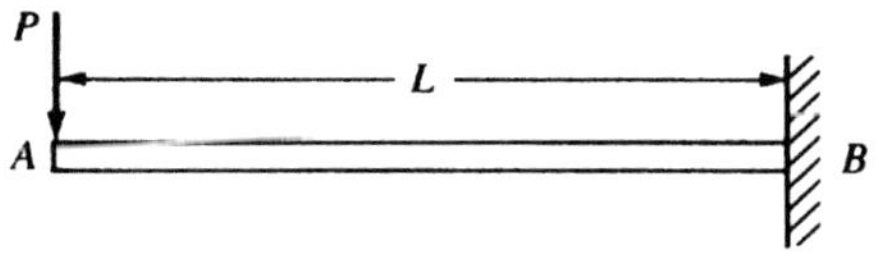

Figure P4–4

4.4 For the cantilever beam subjected to the free-end load P shown in Figure P4–4, determine the maximum deflection and the reactions. Let EI be constant throughout the beam.

4.5–4.11 For the beams shown in Figures P4–5—P4–11, determine the displacements and the slopes at the nodes, the forces in each element, and the reactions. Also, draw the shear force and bending moment diagrams.

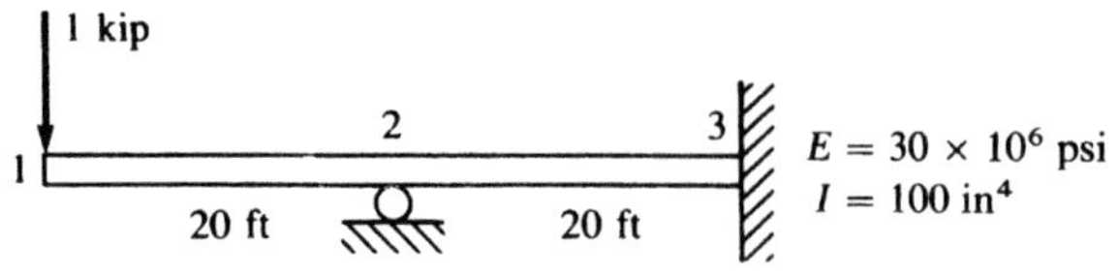

Figure P4–5

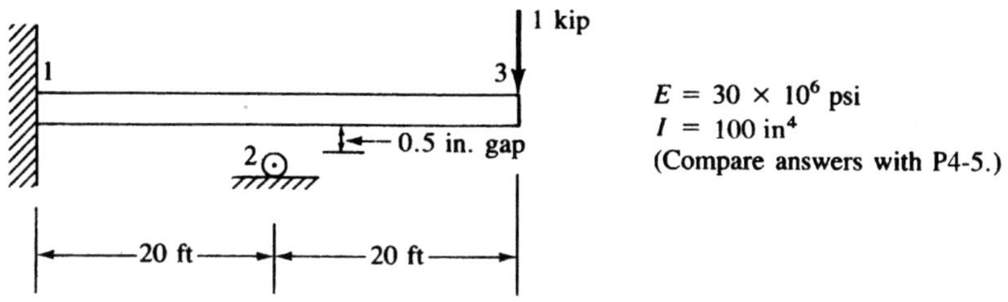

Figure P4–6

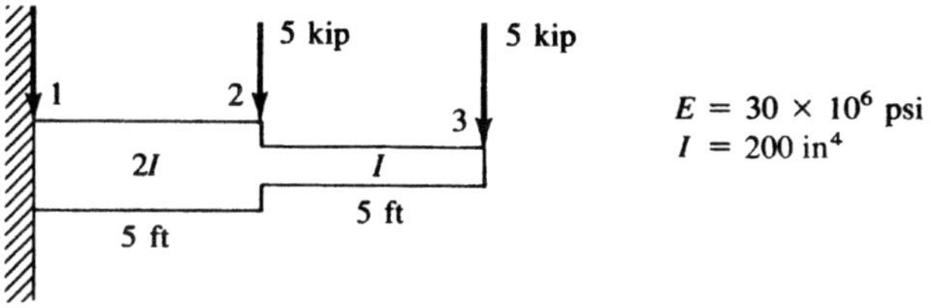

Figure P4–7

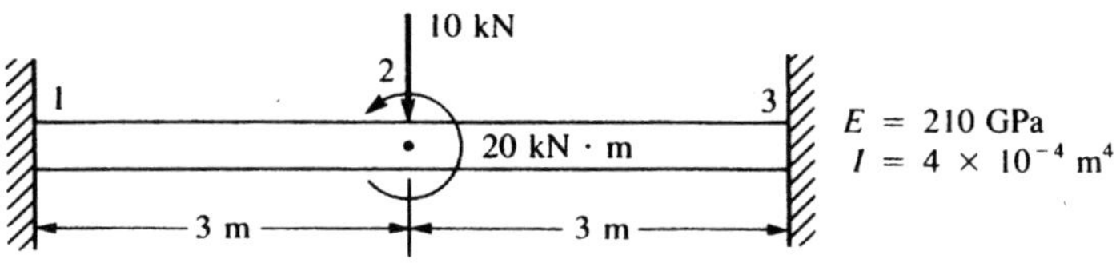

Figure P4–8

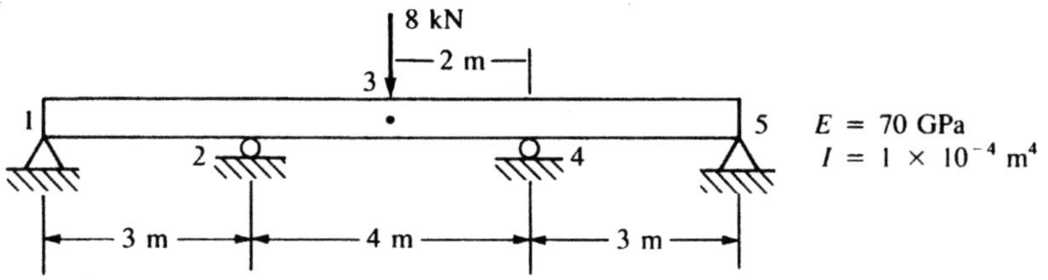

Figure P4–9

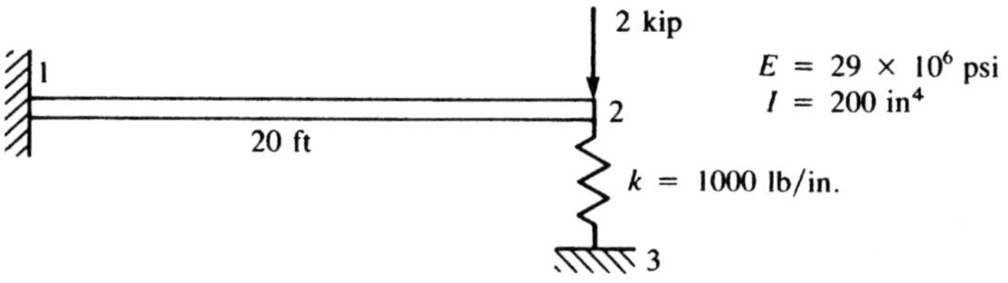

Figure P4–10

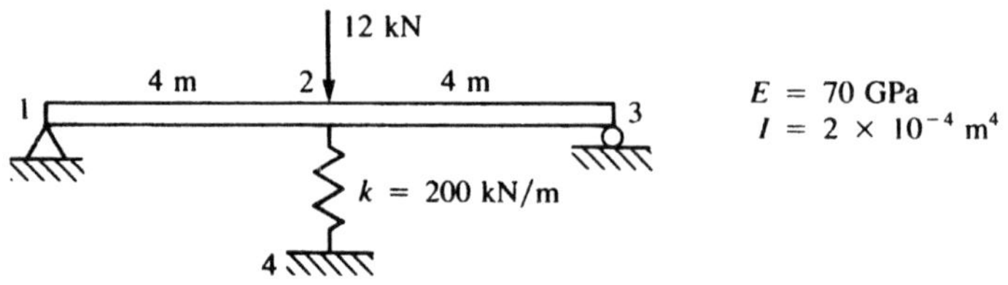

Figure P4–11

4.12 For the fixed-fixed beam subjected to the uniform load w shown in Figure P4–12, determine the midspan deflection and the reactions. Draw the shear force and bending moment diagrams. The middle section of the beam has a bending stiffness of $2EI$; the other sections have bending stiffnesses of EI.

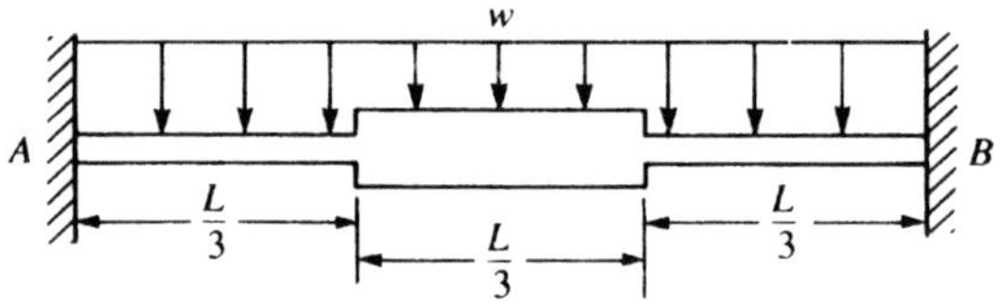

Figure P4–12

4.13 Determine the midspan deflection and the reactions and draw the shear force and bending moment diagrams for the fixed-fixed beam subjected to uniformly distributed load w shown in Figure P4–13. Assume EI constant throughout the beam. Compare your answers with the classical solution (that is, with the appropriate equivalent joint forces given in Appendix D).

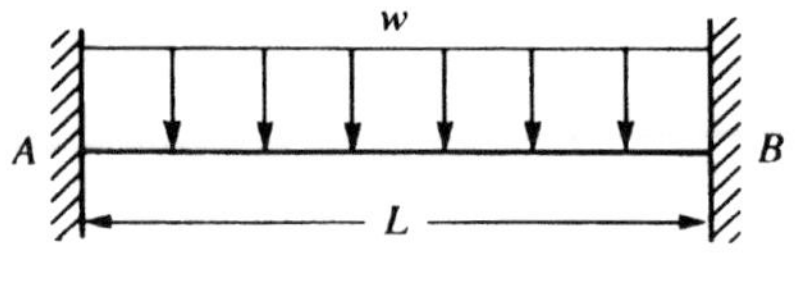

Figure P4–13

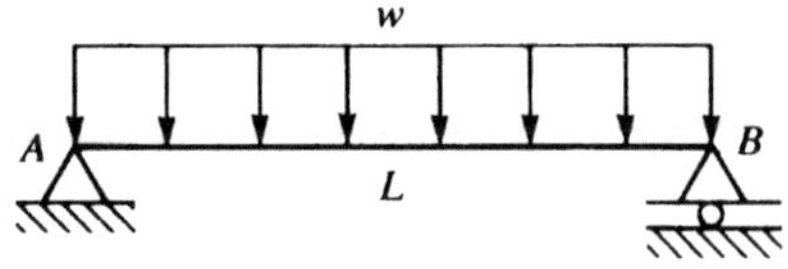

Figure P4–14

4.14 Determine the midspan deflection and the reactions and draw the shear force and bending moment diagrams for the simply supported beam subjected to the uniformly distributed load w shown in Figure P4–14. Assume EI constant throughout the beam.

4.15 For the beam loaded as shown in Figure P4–15, determine the free-end deflection and the reactions and draw the shear force and bending moment diagrams. Assume EI constant throughout the beam.

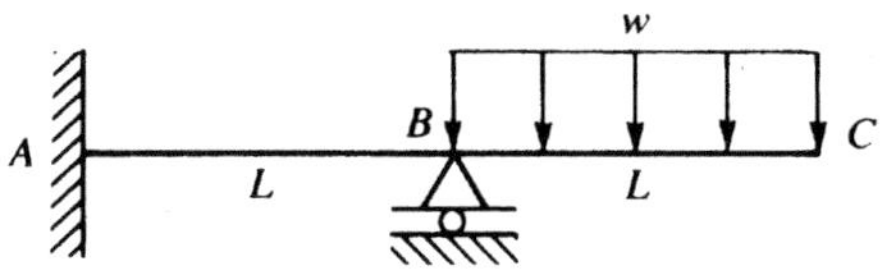

Figure P4–15

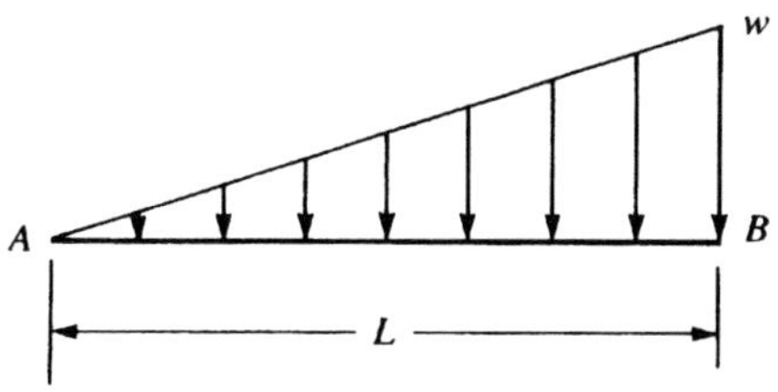

Figure P4–16

4.16 Using the concept of work equivalence, determine the nodal forces and moments (called *equivalent nodal forces*) used to replace the linearly varying distributed load shown in Figure P4–16.

4.17 For the beam shown in Figure 4–17, determine the displacement and slope at the center and the reactions. The load is symmetrical with respect to the center of the beam. Assume EI constant throughout the beam.

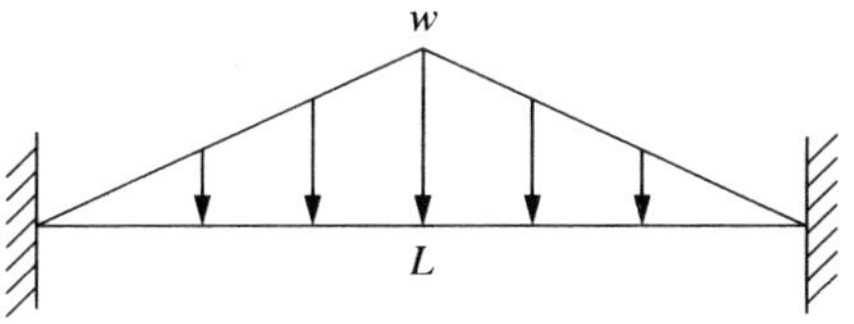

Figure P4–17

4.18 For the beam subjected to the linearly varying line load w shown in Figure P4–18, determine the right-end rotation and the reactions. Assume EI constant throughout the beam.

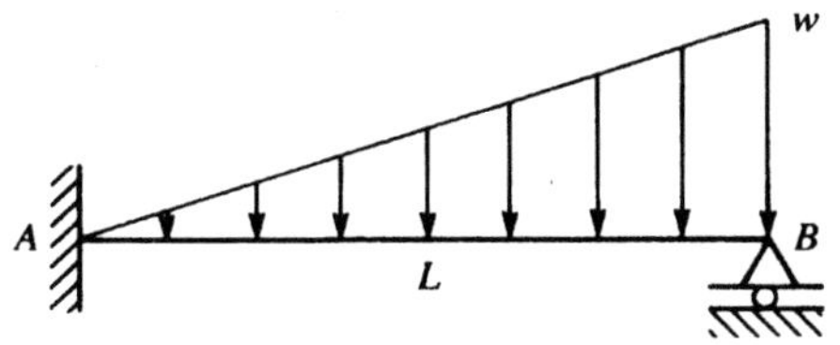

Figure P4–18

4.19–4.24 For the beams shown in Figures P4–19—P4–24, determine the nodal displacements and slopes, the forces in each element, and the reactions.

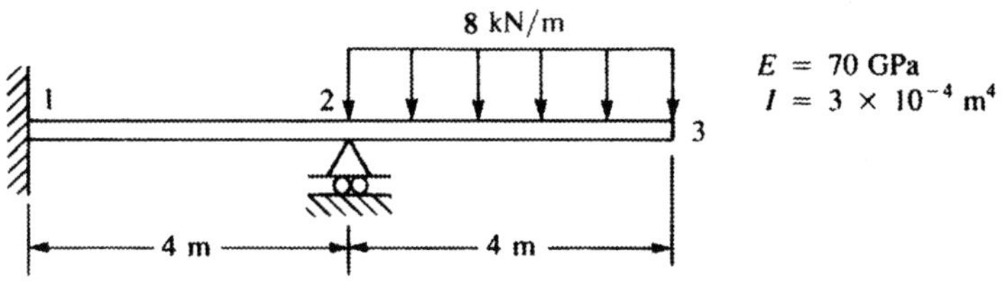

Figure P4–19

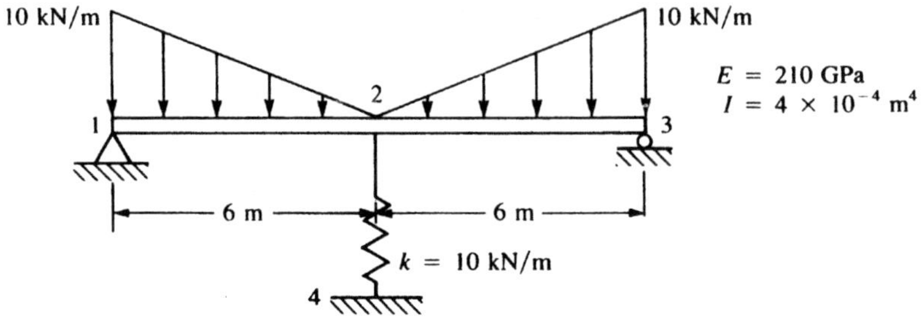

Figure P4–20

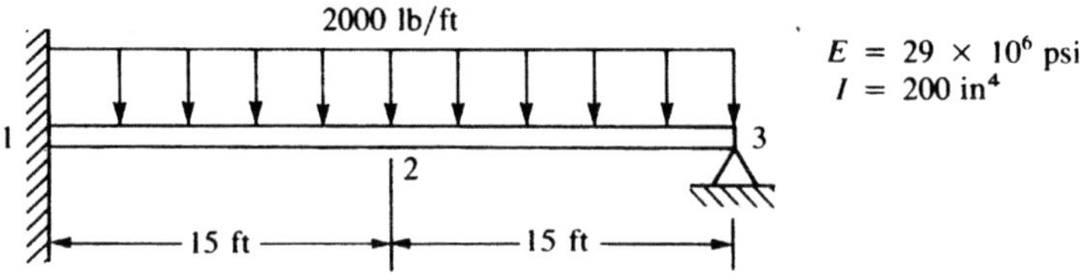

Figure P4–21

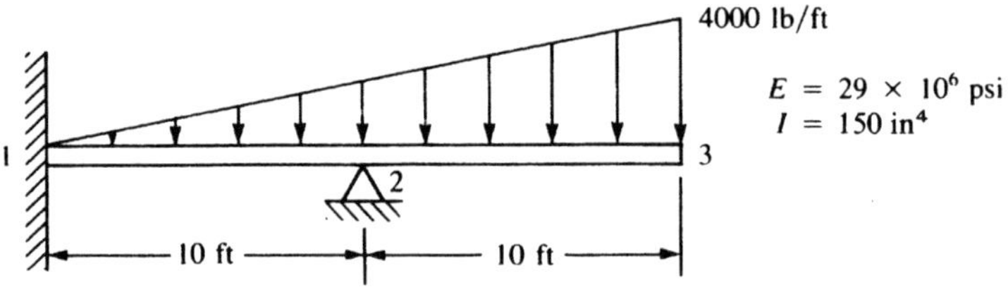

Figure P4–22

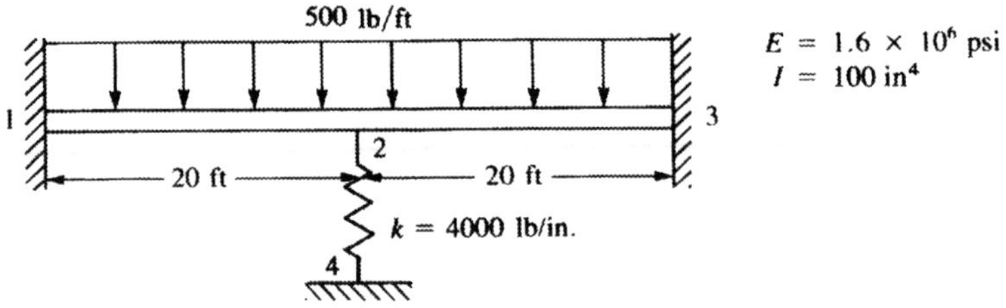

Figure P4–23

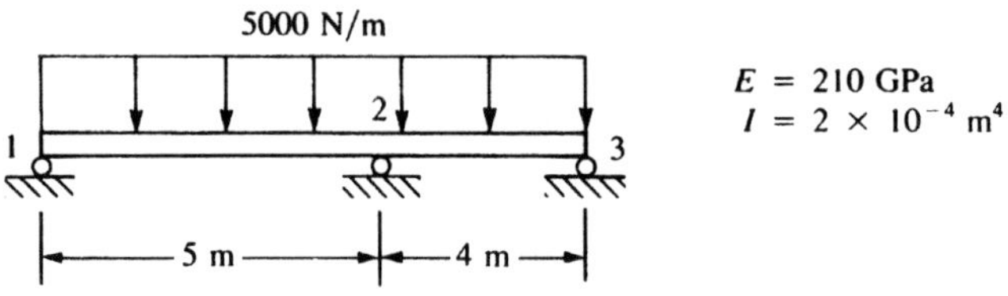

Figure P4–24

4.25–31 For the beams shown in Figures P4–25—P4–30, determine the maximum deflection and maximum bending stress. Let $E = 200$ GPa or 30×10^6 psi for all beams as appropriate for the rest of the units in the problem. Let c be the half-depth of each beam.

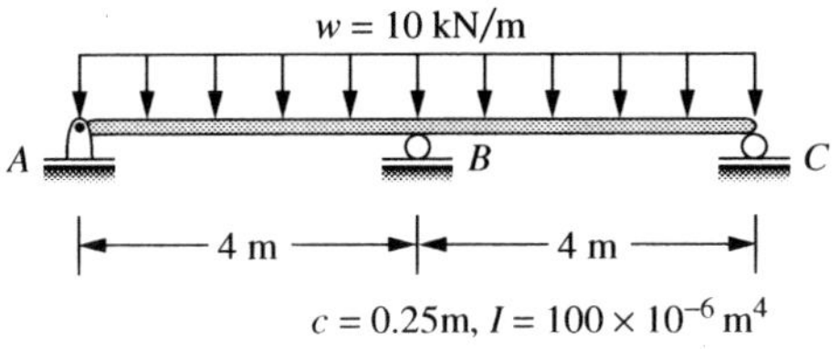

Figure P4–25

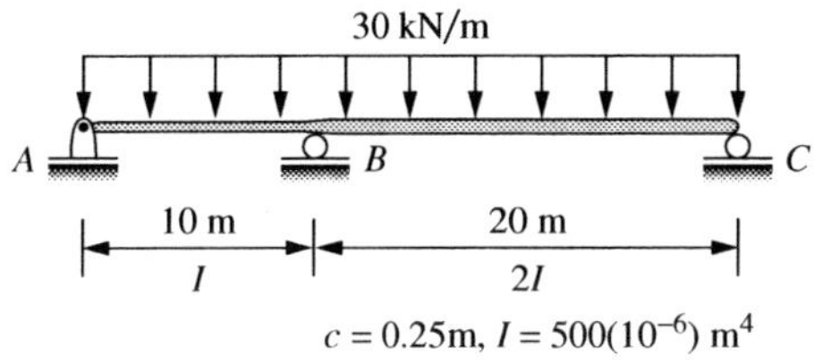

Figure P4–26

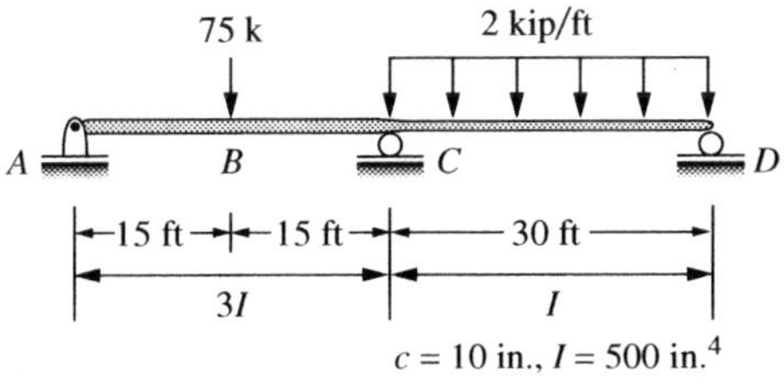

Figure P4–27

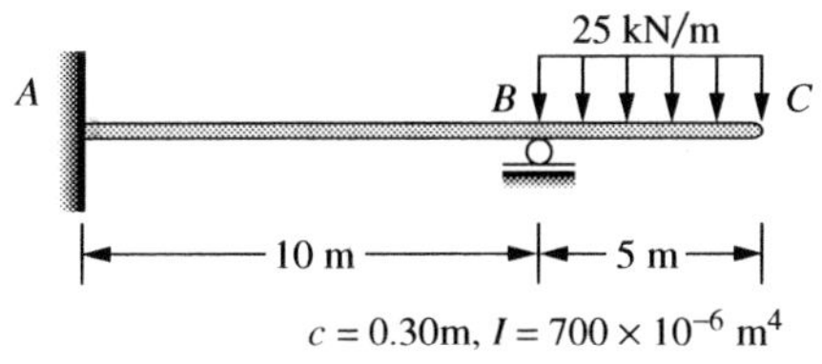

Figure P4–28

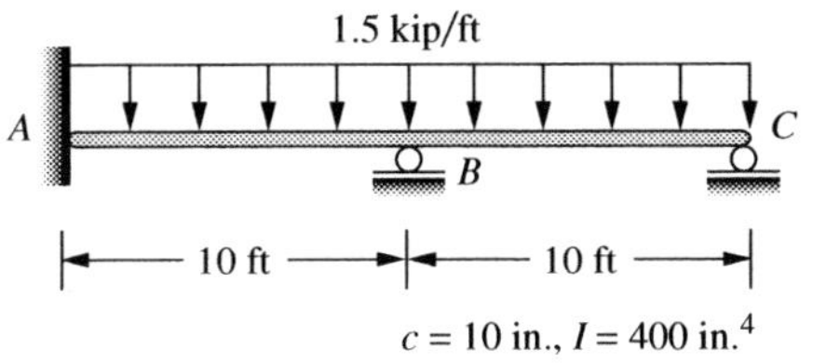

Figure P4–29

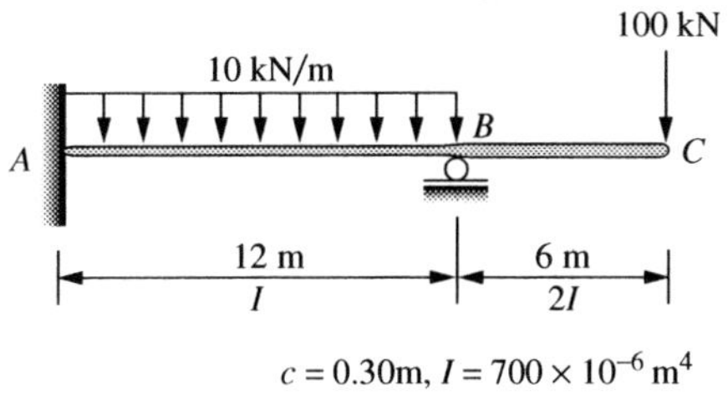

Figure P4–30

For the beam design problems shown in Figures P4–31 through P4–36, determine the size of beam to support the loads shown, based on requirements listed next to each beam.

4.31 Design a beam of ASTM A36 steel with allowable bending stress of 160 MPa to support the load shown in Figure P4–31. Assume a standard wide flange beam from Appendix F or some other source can be used.

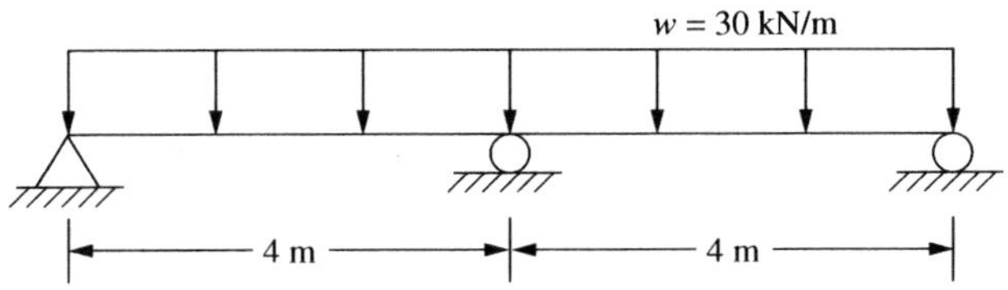

Figure P4–31

4.32 Select a standard steel pipe from Appendix F to support the load shown. The allowable bending stress must not exceed 24 ksi, and the allowable deflection must not exceed L/360 of any span.

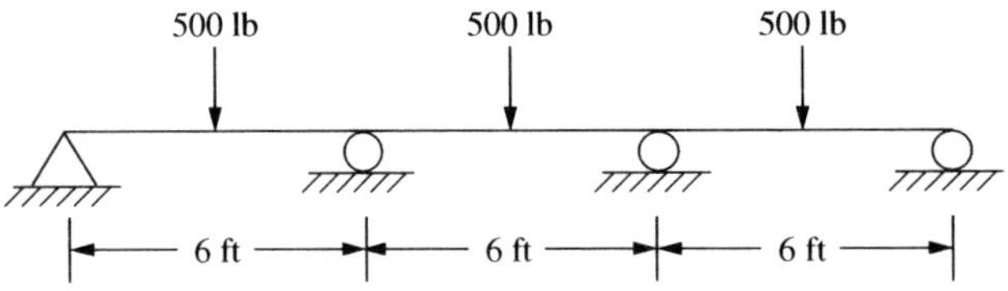

Figure P4–32

4.33 Select a rectangular structural tube from Appendix F to support the loads shown for the beam in Figure P4–33. The allowable bending stress should not exceed 24 ksi.

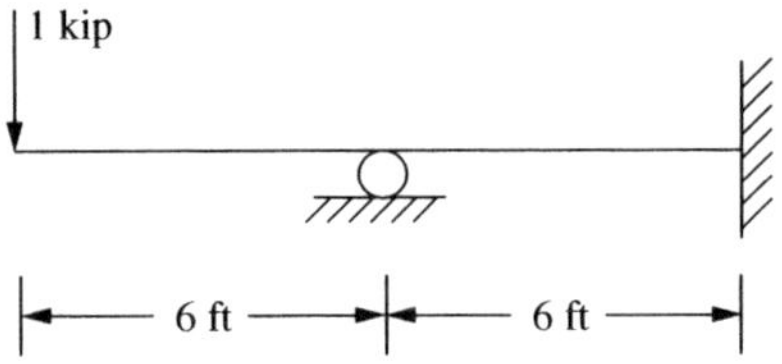

Figure P4–33

4.34 Select a standard W section from Appendix F or some other source to support the loads shown for the beam in Figure P4–34. The bending stress must not exceed 160 MPa.

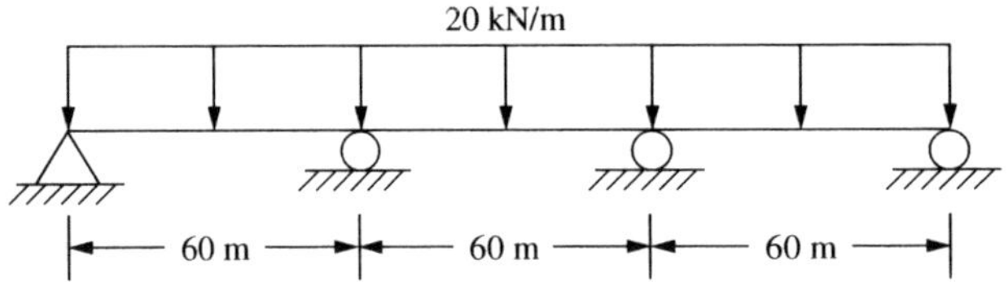

Figure P4–34

4.35 For the beam shown in Figure P4–35, determine a suitable sized W section from Appendix F or from another suitable source such that the bending stress does not exceed 150 MPa and the maximum deflection does not exceed L/360 of any span.

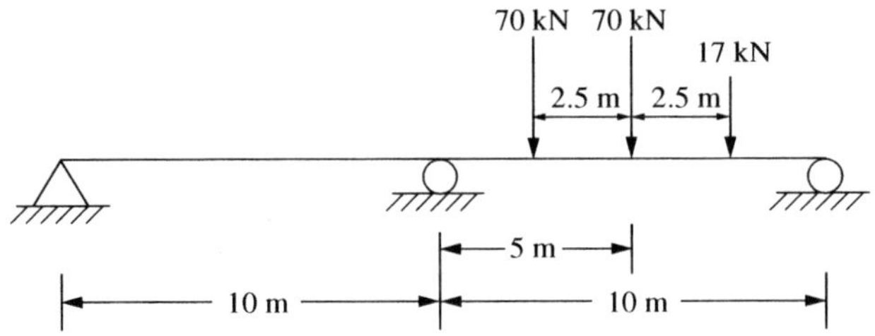

Figure P4–35

4.36 For the stepped shaft shown in Figure P4–36, determine a solid circular cross section for each section shown such that the bending stress does not exceed 160 MPa and the maximum deflection does not exceed L/360 of the span.

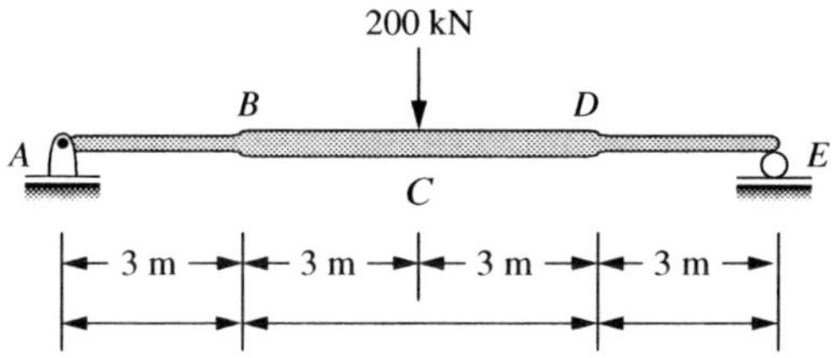

Figure P4–36

4.37 For the beam shown in Figure P4–37 subjected to the concentrated load P and distributed load w, determine the midspan displacement and the reactions. Let EI be constant throughout the beam.

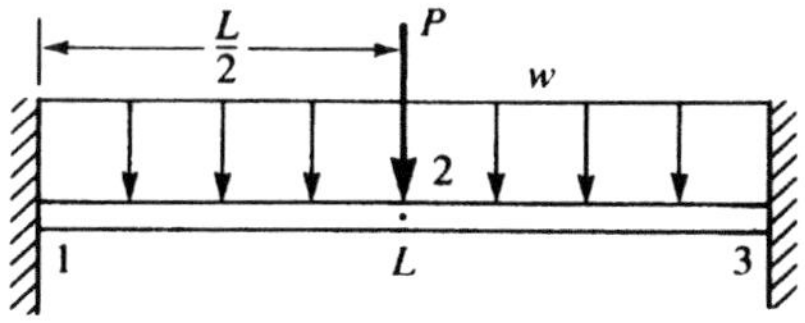

Figure P4–37

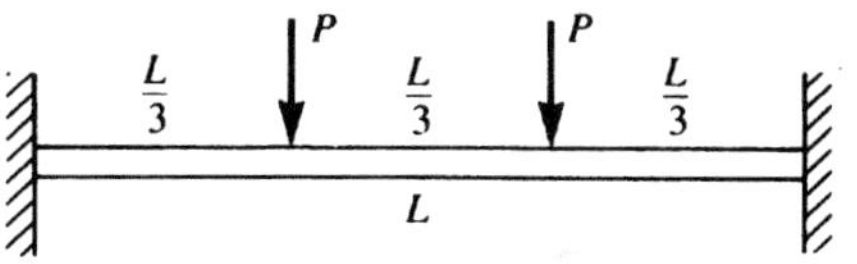

Figure P4–38

4.38 For the beam shown in Figure P4–38 subjected to the two concentrated loads P, determine the deflection at the midspan. Use the equivalent load replacement method. Let EI be constant throughout the beam.

4.39 For the beam shown in Figure P4–39 subjected to the concentrated load P and the linearly varying line load w, determine the free-end deflection and rotation and the reactions. Use the equivalent load replacement method. Let EI be constant throughout the beam.

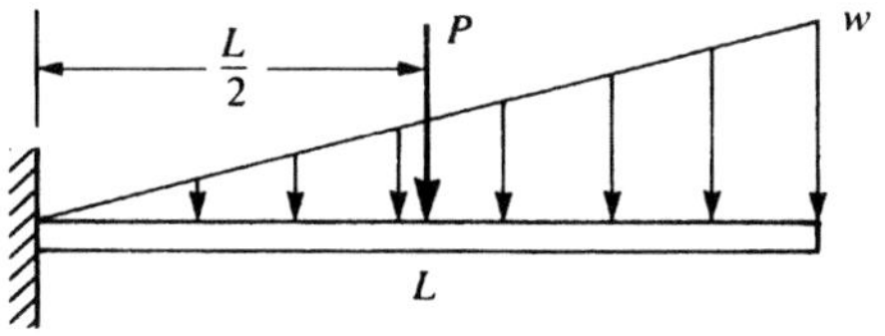

Figure P4–39

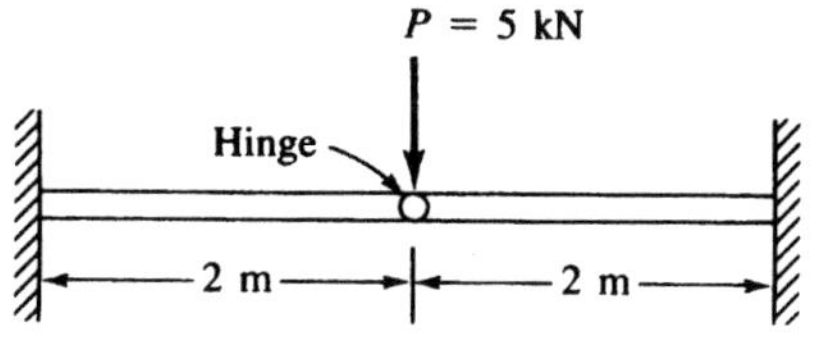

Figure P4–40

4.40–42 For the beams shown in Figures P4–40—P4–42, with internal hinge, determine the deflection at the hinge. Let $E = 210$ GPa and $I = 2 \times 10^{-4}$ m^4.

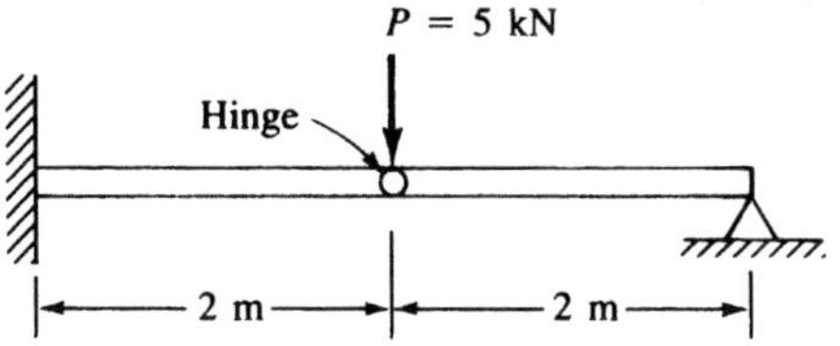

Figure P4–41

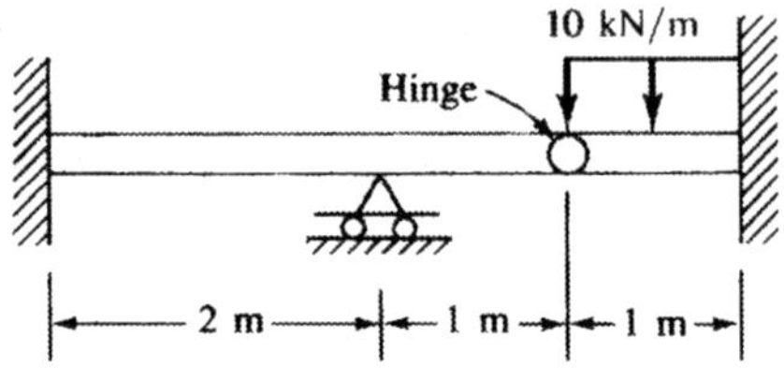

Figure P4–42

4.43 Derive the stiffness matrix for a beam element with a nodal linkage—that is, the shear is 0 at node i, but the usual shear and moment resistance are present at node j (see Figure P4–43).

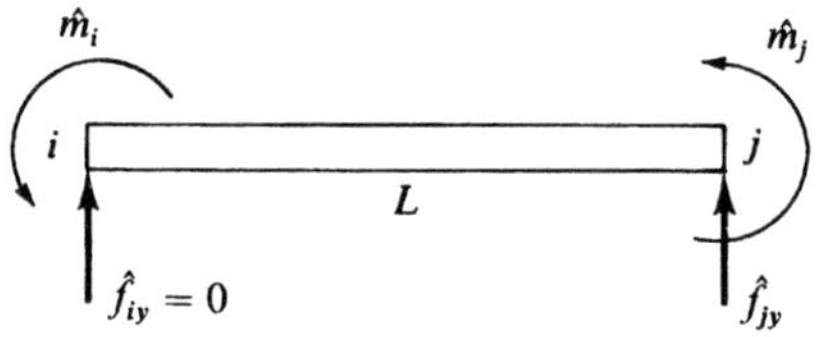

Figure P4–43

4.44 Develop the stiffness matrix for a fictitious pure shear panel element (Figure P4–44) in terms of the shear modulus, G the shear web area, A_W, and the length, L. Notice the Y and v are the shear force and transverse displacement at each node, respectively.

Given 1) $\tau = G_\gamma$, 2) $Y = \tau_{A_w}$, 3) $Y_1 + Y_2 = 0$, 4) $\gamma = \dfrac{v_2 - v_1}{\underline{L}}$

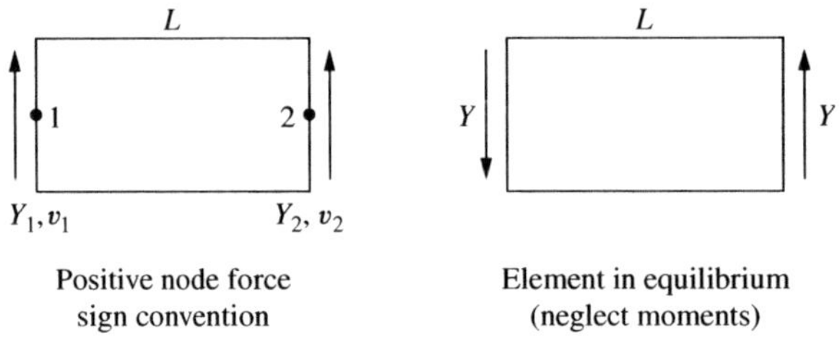

Figure P4–44

4.45 Explicitly evaluate π_p of Eq. (4.7.15); then differentiate π_p with respect to $\hat{d}_{1y}, \hat{\phi}_1, \hat{d}_{2y}$, and $\hat{\phi}_2$ and set each of these equations to zero (that is, minimize π_p) to obtain the four element equations for the beam element. Then express these equations in matrix form.

4.46 Determine the free-end deflection for the tapered beam shown in Figure P4–46. Here $I(x) = I_0(1 + nx/L)$ where I_0 is the moment of inertia at $x = 0$. Compare the exact beam theory solution with a two-element finite element solution for $n = 2$.

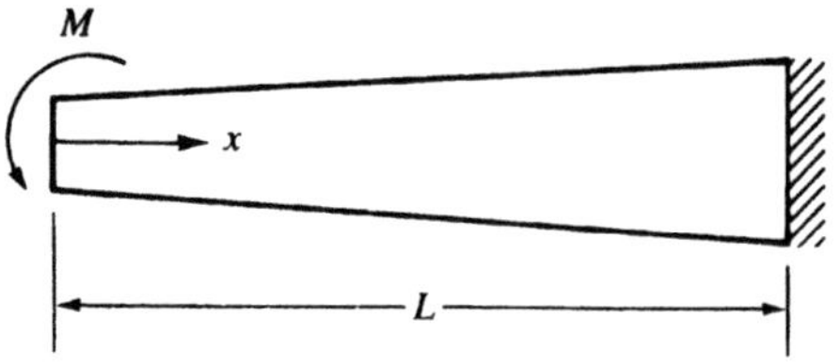

Figure P4–46

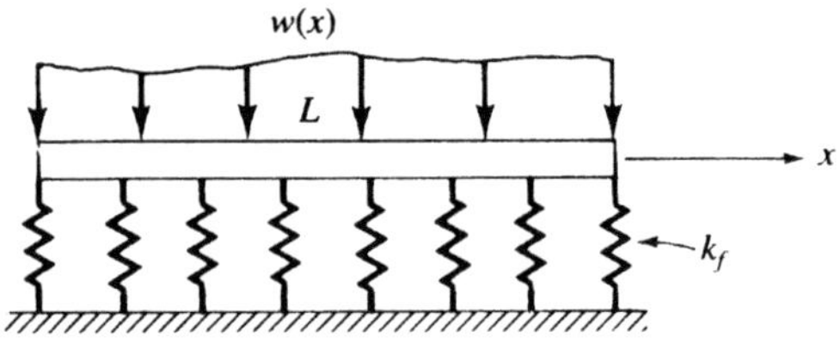

Figure P4–47

4.47 Derive the equations for the beam element on an elastic foundation (Figure P4–47) using the principle of minimum potential energy. Here k_f is the subgrade spring constant per unit length. The potential energy of the beam is

$$\pi_p = \int_0^L \frac{1}{2} EI(v'')^2\,dx + \int_o^L \frac{k_f v^2}{2}\,dx - \int_0^L wv\,dx$$

4.48 Derive the equations for the beam element on an elastic foundation (see Figure P4–47) using Galerkin's method. The basic differential equation for the beam on an elastic foundation is

$$(EIv'')'' = -w + k_f v$$

4.49–76 Solve problems 4.5–4.11, 4.19–4.36, and 4.40–4.42 using a suitable computer program.

4.77 For the beam shown, use a computer program to determine the deflection at the mid-span using four beam elements, making the shear area zero and then making the shear area equal 5/6 times the cross-sectional area (b times h). Then make the beam have decreasing spans of 200 mm, 100 mm, and 50 mm with zero shear area and then 5/6 times the cross-sectional area. Compare the answers. Based on your program answers, can you conclude whether your program includes the effects of transverse shear deformation?

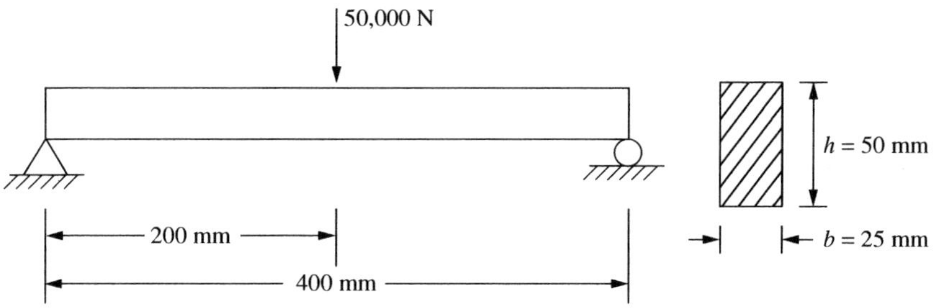

Figure P4–77

4.78 For the beam shown in Figure P4–77, use a longhand solution to solve the problem. Compare answers using the beam stiffness matrix, Eq. (4.1.14), without transverse shear deformation effects and then Eq. (4.1.15o), which includes the transverse shear effects.

CHAPTER 5

Frame and Grid Equations

Introduction

Many structures, such as buildings (Figure 5–1) and bridges, are composed of frames and/or grids. This chapter develops the equations and methods for solution of plane frames and grids.

First, we will develop the stiffness matrix for a beam element arbitrarily oriented in a plane. We will then include the axial nodal displacement degree of freedom in the local beam element stiffness matrix. Then we will combine these results to develop the stiffness matrix, including axial deformation effects, for an arbitrarily oriented beam element, thus making it possible to analyze plane frames. Specific examples of plane frame analysis follow. We will then consider frames with inclined or skewed supports.

Next, we will develop the grid element stiffness matrix. We will present the solution of a grid deck system to illustrate the application of the grid equations. We will then develop the stiffness matrix for a beam element arbitrarily oriented in space. We will also consider the concept of substructure analysis.

5.1 Two-Dimensional Arbitrarily Oriented Beam Element

We can derive the stiffness matrix for an arbitrarily oriented beam element, as shown in Figure 5–2, in a manner similar to that used for the bar element in Chapter 3. The local axes $\hat{x}$ and $\hat{y}$ are located along the beam element and transverse to the beam element, respectively, and the global axes x and y are located to be convenient for the total structure.

Recall that we can relate local displacements to global displacements by using Eq. (3.3.16), repeated here for convenience as

$$\begin{Bmatrix} \hat{d}_x \\ \hat{d}_y \end{Bmatrix} = \begin{bmatrix} C & S \\ -S & C \end{bmatrix} \begin{Bmatrix} d_x \\ d_y \end{Bmatrix} \tag{5.1.1}$$

Figure 5–1 The Arizona Cardinal Football Stadium under construction—a rigid building frame (Courtesy Ed Yack)

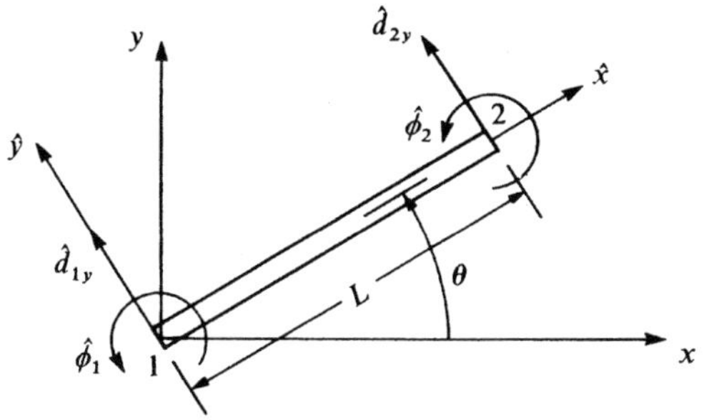

Figure 5–2 Arbitrarily oriented beam element

Using the second equation of Eqs. (5.1.1) for the beam element, we relate local nodal degrees of freedom to global degrees of freedom by

$$\begin{Bmatrix} \hat{d}_{1y} \\ \hat{\phi}_1 \\ \hat{d}_{2y} \\ \hat{\phi}_2 \end{Bmatrix} = \begin{bmatrix} -S & C & 0 & 0 & 0 & 0 \\ 0 & 0 & 1 & 0 & 0 & 0 \\ 0 & 0 & 0 & -S & C & 0 \\ 0 & 0 & 0 & 0 & 0 & 1 \end{bmatrix} \begin{Bmatrix} d_{1x} \\ d_{1y} \\ \phi_1 \\ d_{2x} \\ d_{2y} \\ \phi_2 \end{Bmatrix} \tag{5.1.2}$$

where, for a beam element, we define

$$\underline{T} = \begin{bmatrix} -S & C & 0 & 0 & 0 & 0 \\ 0 & 0 & 1 & 0 & 0 & 0 \\ 0 & 0 & 0 & -S & C & 0 \\ 0 & 0 & 0 & 0 & 0 & 1 \end{bmatrix} \tag{5.1.3}$$

as the *transformation matrix*. The axial effects are not yet included. Equation (5.1.2) indicates that rotation is invariant with respect to either coordinate system. For example, $\hat{\phi}_1 = \phi_1$, and moment $\hat{m}_1 = m_1$ can be considered to be a vector pointing normal to the $\hat{x}$-$\hat{y}$ plane or to the x-y plane by the usual right-hand rule. From either viewpoint, the moment is in the $\hat{z} = z$ direction. Therefore, moment is unaffected as the element changes orientation in the x-y plane.

Substituting Eq. (5.1.3) for $\underline{T}$ and Eq. (4.1.14) for $\hat{\underline{k}}$ into Eq. (3.4.22), $\underline{k} = \underline{T}^T \hat{\underline{k}} \underline{T}$, we obtain the global element stiffness matrix as

$$\underline{k} = \frac{EI}{L^3} \begin{array}{c} \begin{matrix} d_{1x} & d_{1y} & \phi_1 & d_{2x} & d_{2y} & \phi_2 \end{matrix} \\ \begin{bmatrix} 12S^2 & -12SC & -6LS & -12S^2 & 12SC & -6LS \\ & 12C^2 & 6LC & 12SC & -12C^2 & 6LC \\ & & 4L^2 & 6LS & -6LC & 2L^2 \\ & & & 12S^2 & -12SC & 6LS \\ & & & & 12C^2 & -6LC \\ & \text{Symmetry} & & & & 4L^2 \end{bmatrix} \end{array} \tag{5.1.4}$$

where, again, $C = \cos\theta$ and $S = \sin\theta$. It is not necessary here to expand $\underline{T}$ given by Eq. (5.1.3) to make it a square matrix to be able to use Eq. (3.4.22). Because Eq. (3.4.22) is a generally applicable equation, the matrices used must merely be of the correct order for matrix multiplication (see Appendix A for more on matrix multiplication). The stiffness matrix Eq. (5.1.4) is the global element stiffness matrix for a beam element that includes shear and bending resistance. Local axial effects are not yet included. The transformation from local to global stiffness by multiplying matrices $\underline{T}^T \hat{\underline{k}} \underline{T}$, as done in Eq. (5.1.4), is usually done on the computer.

We will now include the axial effects in the element, as shown in Figure 5–3. The element now has three degrees of freedom per node $(\hat{d}_{ix}, \hat{d}_{iy}, \hat{\phi}_i)$. For axial effects, we recall from Eq. (3.1.13),

$$\begin{Bmatrix} \hat{f}_{1x} \\ \hat{f}_{2x} \end{Bmatrix} = \frac{AE}{L} \begin{bmatrix} 1 & -1 \\ -1 & 1 \end{bmatrix} \begin{Bmatrix} \hat{d}_{1x} \\ \hat{d}_{2x} \end{Bmatrix} \tag{5.1.5}$$

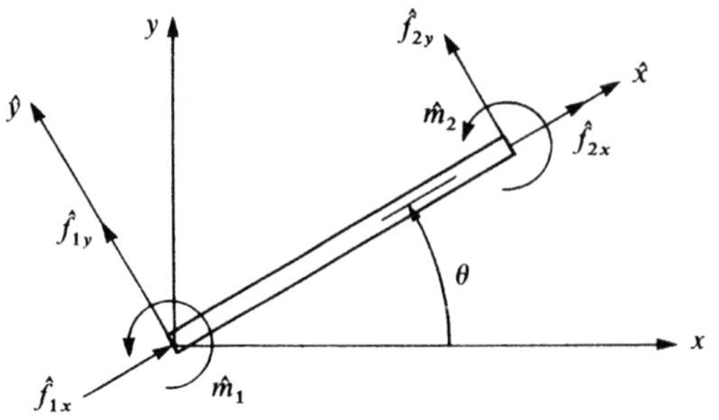

Figure 5–3 Local forces acting on a beam element

Combining the axial effects of Eq. (5.1.5) with the shear and principal bending moment effects of Eq. (4.1.13), we have, in local coordinates,

$$\begin{Bmatrix} \hat{f}_{1x} \\ \hat{f}_{1y} \\ \hat{m}_1 \\ \hat{f}_{2x} \\ \hat{f}_{2y} \\ \hat{m}_2 \end{Bmatrix} = \begin{bmatrix} C_1 & 0 & 0 & -C_1 & 0 & 0 \\ 0 & 12C_2 & 6C_2L & 0 & -12C_2 & 6C_2L \\ 0 & 6C_2L & 4C_2L^2 & 0 & -6C_2L & 2C_2L^2 \\ -C_1 & 0 & 0 & C_1 & 0 & 0 \\ 0 & -12C_2 & -6C_2L & 0 & 12C_2 & -6C_2L \\ 0 & 6C_2L & 2C_2L^2 & 0 & -6C_2L & 4C_2L^2 \end{bmatrix} \begin{Bmatrix} \hat{d}_{1x} \\ \hat{d}_{1y} \\ \hat{\phi}_1 \\ \hat{d}_{2x} \\ \hat{d}_{2y} \\ \hat{\phi}_2 \end{Bmatrix} \tag{5.1.6}$$

where

$$C_1 = \frac{AE}{L} \quad \text{and} \quad C_2 = \frac{EI}{L^3} \tag{5.1.7}$$

and, therefore,

$$\underline{\hat{k}} = \begin{bmatrix} C_1 & 0 & 0 & -C_1 & 0 & 0 \\ 0 & 12C_2 & 6C_2L & 0 & -12C_2 & 6C_2L \\ 0 & 6C_2L & 4C_2L^2 & 0 & -6C_2L & 2C_2L^2 \\ -C_1 & 0 & 0 & C_1 & 0 & 0 \\ 0 & -12C_2 & -6C_2L & 0 & 12C_2 & -6C_2L \\ 0 & 6C_2L & 2C_2L^2 & 0 & -6C_2L & 4C_2L^2 \end{bmatrix} \tag{5.1.8}$$

The $\underline{\hat{k}}$ matrix in Eq. (5.1.8) now has three degrees of freedom per node and now includes axial effects (in the $\hat{x}$ direction), as well as shear force effects (in the $\hat{y}$ direction) and principal bending moment effects (about the $\hat{z} = z$ axis). Using Eqs. (5.1.1) and (5.1.2), we now relate the local to the global displacements by

$$\begin{Bmatrix} \hat{d}_{1x} \\ \hat{d}_{1y} \\ \hat{\phi}_1 \\ \hat{d}_{2x} \\ \hat{d}_{2y} \\ \hat{\phi}_2 \end{Bmatrix} = \begin{bmatrix} C & S & 0 & 0 & 0 & 0 \\ -S & C & 0 & 0 & 0 & 0 \\ 0 & 0 & 1 & 0 & 0 & 0 \\ 0 & 0 & 0 & C & S & 0 \\ 0 & 0 & 0 & -S & C & 0 \\ 0 & 0 & 0 & 0 & 0 & 1 \end{bmatrix} \begin{Bmatrix} d_{1x} \\ d_{1y} \\ \phi_1 \\ d_{2x} \\ d_{2y} \\ \phi_2 \end{Bmatrix} \tag{5.1.9}$$

where $\underline{T}$ has now been expanded to include local axial deformation effects as

$$\underline{T} = \begin{bmatrix} C & S & 0 & 0 & 0 & 0 \\ -S & C & 0 & 0 & 0 & 0 \\ 0 & 0 & 1 & 0 & 0 & 0 \\ 0 & 0 & 0 & C & S & 0 \\ 0 & 0 & 0 & -S & C & 0 \\ 0 & 0 & 0 & 0 & 0 & 1 \end{bmatrix} \tag{5.1.10}$$

Substituting $\underline{T}$ from Eq. (5.1.10) and $\underline{\hat{k}}$ from Eq. (5.1.8) into Eq. (3.4.22), we obtain the general transformed global stiffness matrix for a beam element that includes axial force, shear force, and bending moment effects as follows:

$$
\underline{k} = \frac{E}{L} \times
\begin{bmatrix}
AC^2+\frac{12I}{L^2}S^2 & \left(A-\frac{12I}{L^2}\right)CS & -\frac{6I}{L}S & -\left(AC^2+\frac{12I}{L^2}S^2\right) & -\left(A-\frac{12I}{L^2}\right)CS & -\frac{6I}{L}S \\
 & AS^2+\frac{12I}{L^2}C^2 & \frac{6I}{L}C & -\left(A-\frac{12I}{L^2}\right)CS & -\left(AS^2+\frac{12I}{L^2}C^2\right) & \frac{6I}{L}C \\
 & & 4I & \frac{6I}{L}S & -\frac{6I}{L}C & 2I \\
 & & & AC^2+\frac{12I}{L^2}S^2 & \left(A-\frac{12I}{L^2}\right)CS & \frac{6I}{L}S \\
 & & & & AS^2+\frac{12I}{L^2}C^2 & -\frac{6I}{L}C \\
\text{Symmetry} & & & & & 4I
\end{bmatrix}
\tag{5.1.11}
$$

The analysis of a rigid plane frame can be undertaken by applying stiffness matrix Eq. (5.1.11). *A* **rigid plane frame** *is defined here as a series of beam elements rigidly connected to each other*; that is, the original angles made between elements at their joints remain unchanged after the deformation due to applied loads or applied displacements.

Furthermore, moments are transmitted from one element to another at the joints. Hence, moment continuity exists at the rigid joints. In addition, the element centroids, as well as the applied loads, lie in a common plane (*x*-*y* plane). From Eq. (5.1.11), we observe that the element stiffnesses of a frame are functions of *E*, *A*, *L*, *I*, and the angle of orientation θ of the element with respect to the global-coordinate axes. It should be noted that computer programs often refer to the frame element as a beam element, with the understanding that the program is using the stiffness matrix in Eq. (5.1.11) for plane frame analysis.

5.2 Rigid Plane Frame Examples

To illustrate the use of the equations developed in Section 5.1, we will now perform complete solutions for the following rigid plane frames.

Example 5.1

As the first example of rigid plane frame analysis, solve the simple "bent" shown in Figure 5–4.

The frame is fixed at nodes 1 and 4 and subjected to a positive horizontal force of 10,000 lb applied at node 2 and to a positive moment of 5000 lb-in. applied at node 3. The global-coordinate axes and the element lengths are shown in Figure 5–4.

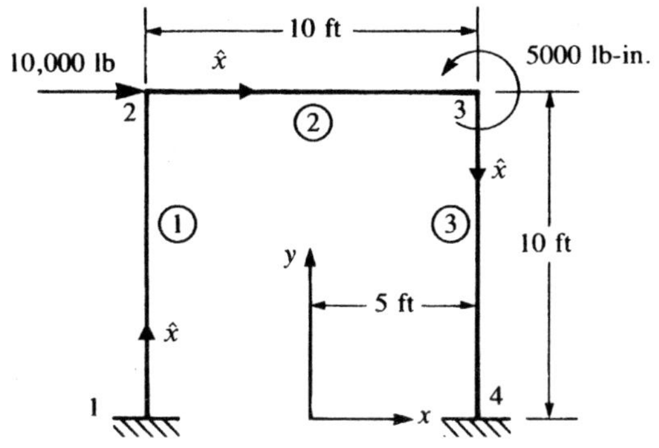

Figure 5–4 Plane frame for analysis, also showing local $\hat{x}$ axis for each element

Let $E = 30 \times 10^6$ psi and $A = 10$ in^2 for all elements, and let $I = 200$ in^4 for elements 1 and 3, and $I = 100$ in^4 for element 2.

Using Eq. (5.1.11), we obtain the global stiffness matrices for each element.

Element 1

For element 1, the angle between the global x and the local $\hat{x}$ axes is 90° (counterclockwise) because $\hat{x}$ is assumed to be directed from node 1 to node 2. Therefore,

$$C = \cos 90^\circ = \frac{x_2 - x_1}{L^{(1)}} = \frac{-60 - (-60)}{120} = 0$$

$$S = \sin 90^\circ = \frac{y_2 - y_1}{L^{(1)}} = \frac{120 - 0}{120} = 1$$

Also,

$$\frac{12I}{L^2} = \frac{12(200)}{(10 \times 12)^2} = 0.167 \text{ in}^2 \qquad (5.2.1)$$

$$\frac{6I}{L} = \frac{6(200)}{10 \times 12} = 10.0 \text{ in}^3$$

$$\frac{E}{L} = \frac{30 \times 10^6}{10 \times 12} = 250{,}000 \text{ lb/in}^3$$

Then, using Eqs. (5.2.1) to help in evaluating Eq. (5.1.11) for element 1, we obtain the element global stiffness matrix as

$$\underline{k}^{(1)} = 250{,}000 \begin{array}{c} \begin{matrix} d_{1x} & d_{1y} & \phi_1 & d_{2x} & d_{2y} & \phi_2 \end{matrix} \\ \begin{bmatrix} 0.167 & 0 & -10 & -0.167 & 0 & -10 \\ 0 & 10 & 0 & 0 & -10 & 0 \\ -10 & 0 & 800 & 10 & 0 & 400 \\ -0.167 & 0 & 10 & 0.167 & 0 & 10 \\ 0 & -10 & 0 & 0 & 10 & 0 \\ -10 & 0 & 400 & 10 & 0 & 800 \end{bmatrix} \end{array} \frac{\text{lb}}{\text{in.}} \qquad (5.2.2)$$

where all diagonal terms are positive.

Element 2

For element 2, the angle between x and $\hat{x}$ is zero because $\hat{x}$ is directed from node 2 to node 3. Therefore,

$$C = 1 \qquad S = 0$$

Also,

$$\frac{12I}{L^2} = \frac{12(100)}{120^2} = 0.0835 \text{ in}^2$$

$$\frac{6I}{L} = \frac{6(100)}{120} = 5.0 \text{ in}^3 \tag{5.2.3}$$

$$\frac{E}{L} = 250{,}000 \text{ lb/in}^3$$

Using the quantities obtained in Eqs. (5.2.3) in evaluating Eq. (5.1.11) for element 2, we obtain

$$\underline{k}^{(2)} = 250{,}000 \begin{array}{c} \begin{array}{cccccc} d_{2x} & d_{2y} & \phi_2 & d_{3x} & d_{3y} & \phi_3 \end{array} \\ \begin{bmatrix} 10 & 0 & 0 & -10 & 0 & 0 \\ 0 & 0.0835 & 5 & 0 & -0.0835 & 5 \\ 0 & 5 & 400 & 0 & -5 & 200 \\ -10 & 0 & 0 & 10 & 0 & 0 \\ 0 & -0.0835 & -5 & 0 & 0.0835 & -5 \\ 0 & 5 & 200 & 0 & -5 & 400 \end{bmatrix} \end{array} \frac{\text{lb}}{\text{in.}} \tag{5.2.4}$$

Element 3

For element 3, the angle between x and $\hat{x}$ is 270° (or −90°) because $\hat{x}$ is directed from node 3 to node 4. Therefore,

$$C = 0 \qquad S = -1$$

Therefore, evaluating Eq. (5.1.11) for element 3, we obtain

$$\underline{k}^{(3)} = 250{,}000 \begin{array}{c} \begin{array}{cccccc} d_{3x} & d_{3y} & \phi_3 & d_{4x} & d_{4y} & \phi_4 \end{array} \\ \begin{bmatrix} 0.167 & 0 & 10 & -0.167 & 0 & 10 \\ 0 & 10 & 0 & 0 & -10 & 0 \\ 10 & 0 & 800 & -10 & 0 & 400 \\ -0.167 & 0 & -10 & 0.167 & 0 & -10 \\ 0 & -10 & 0 & 0 & 10 & 0 \\ 10 & 0 & 400 & -10 & 0 & 800 \end{bmatrix} \end{array} \frac{\text{lb}}{\text{in.}} \tag{5.2.5}$$

Superposition of Eqs. (5.2.2), (5.2.4), and (5.2.5) and application of the boundary conditions $d_{1x} = d_{1y} = \phi_1 = 0$ and $d_{4x} = d_{4y} = \phi_4 = 0$ at nodes 1 and 4 yield the reduced set of equations for a longhand solution as

$$\begin{Bmatrix} 10{,}000 \\ 0 \\ 0 \\ 0 \\ 0 \\ 5000 \end{Bmatrix} = 250{,}000 \begin{bmatrix} 10.167 & 0 & 10 & -10 & 0 & 0 \\ 0 & 10.0835 & 5 & 0 & -0.0835 & 5 \\ 10 & 5 & 1200 & 0 & -5 & 200 \\ -10 & 0 & 0 & 10.167 & 0 & 10 \\ 0 & -0.0835 & -5 & 0 & 10.0835 & -5 \\ 0 & 5 & 200 & 10 & -5 & 1200 \end{bmatrix} \begin{Bmatrix} d_{2x} \\ d_{2y} \\ \phi_2 \\ d_{3x} \\ d_{3y} \\ \phi_3 \end{Bmatrix} \tag{5.2.6}$$

Solving Eq. (5.2.6) for the displacements and rotations, we have

$$\begin{Bmatrix} d_{2x} \\ d_{2y} \\ \phi_2 \\ d_{3x} \\ d_{3y} \\ \phi_3 \end{Bmatrix} = \begin{Bmatrix} 0.211 \text{ in.} \\ 0.00148 \text{ in.} \\ -0.00153 \text{ rad} \\ 0.209 \text{ in.} \\ -0.00148 \text{ in.} \\ -0.00149 \text{ rad} \end{Bmatrix} \tag{5.2.7}$$

The results indicate that the top of the frame moves to the right with negligible vertical displacement and small rotations of elements at nodes 2 and 3.

The element forces can now be obtained using $\hat{\underline{f}} = \hat{\underline{k}}\underline{T}\underline{d}$ for each element, as was previously done in solving truss and beam problems. We will illustrate this procedure only for element 1. For element 1, on using Eq. (5.1.10) for $\underline{T}$ and Eq. (5.2.7) for the displacements at node 2, we have

$$\underline{T}\underline{d} = \begin{bmatrix} 0 & 1 & 0 & 0 & 0 & 0 \\ -1 & 0 & 0 & 0 & 0 & 0 \\ 0 & 0 & 1 & 0 & 0 & 0 \\ 0 & 0 & 0 & 0 & 1 & 0 \\ 0 & 0 & 0 & -1 & 0 & 0 \\ 0 & 0 & 0 & 0 & 0 & 1 \end{bmatrix} \begin{Bmatrix} d_{1x} = 0 \\ d_{1y} = 0 \\ \phi_1 = 0 \\ d_{2x} = 0.211 \\ d_{2y} = 0.00148 \\ \phi_2 = -0.00153 \end{Bmatrix} \tag{5.2.8}$$

On multiplying the matrices in Eq. (5.2.8), we obtain

$$\underline{T}\underline{d} = \begin{Bmatrix} 0 \\ 0 \\ 0 \\ 0.00148 \\ -0.211 \\ -0.00153 \end{Bmatrix} \tag{5.2.9}$$

Then using $\underline{\hat{k}}$ from Eq. (5.1.8), we obtain element 1 local forces as

$$\underline{\hat{f}} = \underline{\hat{k}}\underline{T}\underline{d} = 250{,}000 \begin{bmatrix} 10 & 0 & 0 & -10 & 0 & 0 \\ 0 & 0.167 & 10 & 0 & -0.167 & 10 \\ 0 & 10 & 800 & 0 & -10 & 400 \\ -10 & 0 & 0 & 10 & 0 & 0 \\ 0 & -0.167 & -10 & 0 & 0.167 & -10 \\ 0 & 10 & 400 & 0 & -10 & 800 \end{bmatrix} \begin{Bmatrix} 0 \\ 0 \\ 0 \\ 0.00148 \\ -0.211 \\ -0.00153 \end{Bmatrix} \tag{5.2.10}$$

Simplifying Eq. (5.2.10), we obtain the local forces acting on element 1 as

$$\begin{Bmatrix} \hat{f}_{1x} \\ \hat{f}_{1y} \\ \hat{m}_1 \\ \hat{f}_{2x} \\ \hat{f}_{2y} \\ \hat{m}_2 \end{Bmatrix} = \begin{Bmatrix} -3700 \text{ lb} \\ 4990 \text{ lb} \\ 376{,}000 \text{ lb-in.} \\ 3700 \text{ lb} \\ -4990 \text{ lb} \\ 223{,}000 \text{ lb-in.} \end{Bmatrix} \tag{5.2.11}$$

A free-body diagram of each element is shown in Figure 5–5 along with equilibrium verification. In Figure 5–5, the $\hat{x}$ axis is directed from node 1 to node 2—consistent with the order of the nodal degrees of freedom used in developing the stiffness matrix for the element. Since the x-y plane was initially established as shown in Figure 5–4, the z axis is directed outward—consequently, so is the $\hat{z}$ axis (recall $\hat{z} = z$). The $\hat{y}$ axis is then established such that $\hat{x}$ cross $\hat{y}$ yields the direction of $\hat{z}$. The signs on the resulting element forces in Eq. (5.2.11) are thus consistently shown in Figure 5–5. The forces in elements 2 and 3 can be obtained in a manner similar to that used to obtain Eq. (5.2.11) for the nodal forces in element 1. Here we report only the final results for the forces in elements 2 and 3 and leave it to your discretion to perform the detailed calculations. The element forces (shown in Figure 5–5(b) and (c)) are as follows:

Element 2

$$\begin{aligned} &\hat{f}_{2x} = 5010 \text{ lb} && \hat{f}_{2y} = -3700 \text{ lb} && \hat{m}_2 = -223{,}000 \text{ lb-in.} \\ &\hat{f}_{3x} = -5010 \text{ lb} && \hat{f}_{3y} = 3700 \text{ lb} && \hat{m}_3 = -221{,}000 \text{ lb-in.} \end{aligned} \tag{5.2.12a}$$

Element 3

$$\begin{aligned} &\hat{f}_{3x} = 3700 \text{ lb} && \hat{f}_{3y} = 5010 \text{ lb} && \hat{m}_3 = 226{,}000 \text{ lb-in.} \\ &\hat{f}_{4x} = -3700 \text{ lb} && \hat{f}_{4y} = -5010 \text{ lb} && \hat{m}_4 = 375{,}000 \text{ lb-in.} \end{aligned} \tag{5.2.12b}$$

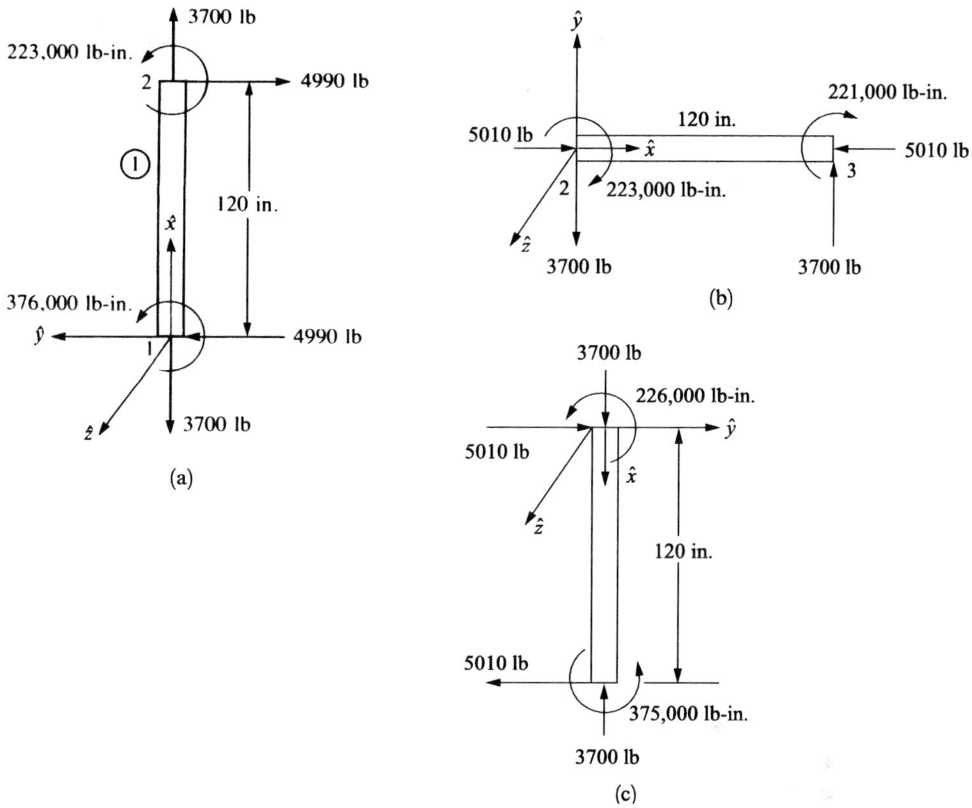

Figure 5–5 Free-body diagrams of (a) element 1, (b) element 2, and (c) element 3

Considering the free body of element 1, the equilibrium equations are

$$\sum F_{\hat{x}}: \; -4990 + 4990 = 0$$

$$\sum F_{\hat{y}}: \; -3700 + 3700 = 0$$

$$\sum M_2: \; 376{,}000 + 223{,}000 - 4990(120 \text{ in.}) \cong 0$$

Considering moment equilibrium at node 2, we see from Eqs. (5.2.12a) and (5.2.12b) that on element 1, $\hat{m}_2 = 223{,}000$ lb-in., and the opposite value, $-223{,}000$ lb-in., occurs on element 2. Similarly, moment equilibrium is satisfied at node 3, as $\hat{m}_3$ from elements 2 and 3 add to the 5000 lb-in. applied moment. That is, from Eqs. (5.2.12a) and (5.2.12b) we have

$$-221{,}000 + 226{,}000 = 5000 \text{ lb-in.}$$ ■

Example 5.2

To illustrate the procedure for solving frames subjected to distributed loads, solve the rigid plane frame shown in Figure 5–6. The frame is fixed at nodes 1 and 3 and subjected to a uniformly distributed load of 1000 lb/ft applied downward over element 2. The global-coordinate axes have been established at node 1. The element lengths are shown in the figure. Let $E = 30 \times 10^6$ psi, $A = 100$ in^2, and $I = 1000$ in^4 for both elements of the frame.

We begin by replacing the distributed load acting on element 2 by nodal forces and moments acting at nodes 2 and 3. Using Eqs. (4.4.5)–(4.4.7) (or Appendix D), the equivalent nodal forces and moments are calculated as

$$\begin{aligned} f_{2y} &= -\frac{wL}{2} = -\frac{(1000)40}{2} = -20{,}000 \text{ lb} = -20 \text{ kip} \\ m_2 &= -\frac{wL^2}{12} = -\frac{(1000)40^2}{12} = -133{,}333 \text{ lb-ft} = -1600 \text{ k-in.} \end{aligned} \tag{5.2.13}$$

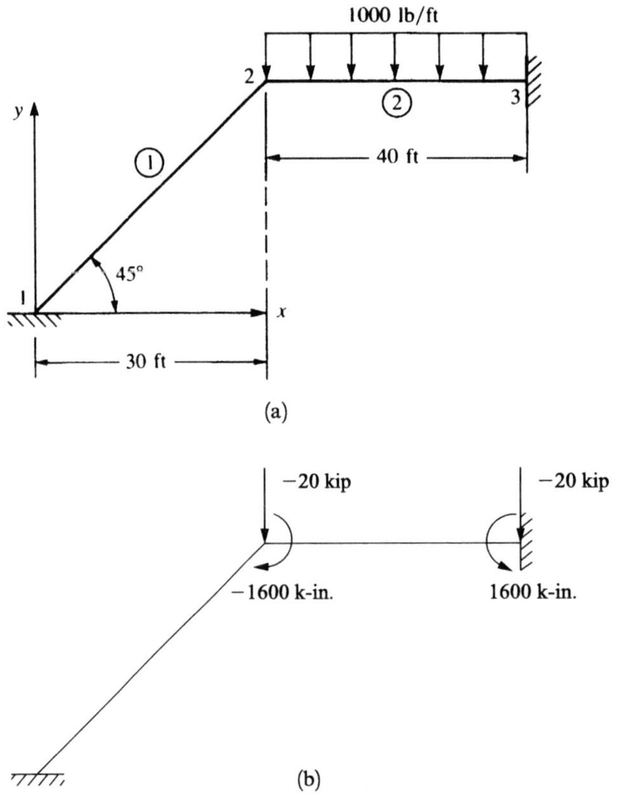

Figure 5–6 (a) Plane frame for analysis and (b) equivalent nodal forces on frame

$$f_{3y} = -\frac{wL}{2} = -\frac{(1000)40}{2} = -20{,}000\text{ lb} = -20\text{ kip}$$

$$m_3 = \frac{wL^2}{12} = \frac{(1000)40^2}{12} = 133{,}333\text{ lb-ft} = 1600\text{ k-in.}$$

We then use Eq. (5.1.11), to determine each element stiffness matrix:

Element 1

$$\theta^{(1)} = 45^\circ \qquad C = 0.707 \qquad S = 0.707 \qquad L^{(1)} = 42.4\text{ ft} = 509.0\text{ in.}$$

$$\frac{E}{L} = \frac{30 \times 10^3}{509} = 58.93$$

$$\underline{k}^{(1)} = 58.93 \begin{bmatrix} 50.02 & 49.98 & 8.33 \\ 49.98 & 50.02 & -8.33 \\ 8.33 & -8.33 & 4000 \end{bmatrix} \frac{\text{kip}}{\text{in.}} \tag{5.2.14}$$

Simplifying Eq. (5.2.14), we obtain

$$\underline{k}^{(1)} = \begin{array}{c} \begin{matrix} d_{2x} & d_{2y} & \phi_2 \end{matrix} \\ \begin{bmatrix} 2948 & 2945 & 491 \\ 2945 & 2948 & -491 \\ 491 & -491 & 235{,}700 \end{bmatrix} \end{array} \frac{\text{kip}}{\text{in.}} \tag{5.2.15}$$

where only the parts of the stiffness matrix associated with degrees of freedom at node 2 are included because node 1 is fixed.

Element 2

$$\theta^{(2)} = 0^\circ \qquad C = 1 \qquad S = 0 \qquad L^{(2)} = 40\text{ ft} = 480\text{ in.}$$

$$\frac{E}{L} = \frac{30 \times 10^3}{480} = 62.50$$

$$\underline{k}^{(2)} = 62.50 \begin{bmatrix} 100 & 0 & 0 \\ 0 & 0.052 & 12.5 \\ 0 & 12.5 & 4000 \end{bmatrix} \frac{\text{kip}}{\text{in.}} \tag{5.2.16}$$

Simplifying Eq. (5.2.16), we obtain

$$\underline{k}^{(2)} = \begin{array}{c} \begin{matrix} d_{2x} & d_{2y} & \phi_2 \end{matrix} \\ \begin{bmatrix} 6250 & 0 & 0 \\ 0 & 3.25 & 781.25 \\ 0 & 781.25 & 250{,}000 \end{bmatrix} \end{array} \frac{\text{kip}}{\text{in.}} \tag{5.2.17}$$

where, again, only the parts of the stiffness matrix associated with degrees of freedom at node 2 are included because node 3 is fixed. On superimposing the stiffness matrices of the elements, using Eqs. (5.2.15) and (5.2.17), and using Eq. (5.2.13) for the nodal forces and moments only at node 2 (because the structure is fixed at node 3), we have

$$\begin{Bmatrix} F_{2x} = 0 \\ F_{2y} = -20 \\ M_2 = -1600 \end{Bmatrix} = \begin{bmatrix} 9198 & 2945 & 491 \\ 2945 & 2951 & 290 \\ 491 & 290 & 485{,}700 \end{bmatrix} \begin{Bmatrix} d_{2x} \\ d_{2y} \\ \phi_2 \end{Bmatrix} \tag{5.2.18}$$

Solving Eq. (5.2.18) for the displacements and the rotation at node 2, we obtain

$$\begin{Bmatrix} d_{2x} \\ d_{2y} \\ \phi_2 \end{Bmatrix} = \begin{Bmatrix} 0.0033 \text{ in.} \\ -0.0097 \text{ in.} \\ -0.0033 \text{ rad} \end{Bmatrix} \tag{5.2.19}$$

The results indicate that node 2 moves to the right ($d_{2x} = 0.0033$ in.) and down ($d_{2y} = -0.0097$ in.) and the rotation of the joint is clockwise ($\phi_2 = -0.0033$ rad).

The local forces in each element can now be determined. The procedure for elements that are subjected to a distributed load must be applied to element 2. Recall that the local forces are given by $\hat{\underline{f}} = \hat{\underline{k}}\underline{T}\underline{d}$. For element 1, we then have

$$\underline{T}\underline{d} = \begin{bmatrix} 0.707 & 0.707 & 0 & 0 & 0 & 0 \\ -0.707 & 0.707 & 0 & 0 & 0 & 0 \\ 0 & 0 & 1 & 0 & 0 & 0 \\ 0 & 0 & 0 & 0.707 & 0.707 & 0 \\ 0 & 0 & 0 & -0.707 & 0.707 & 0 \\ 0 & 0 & 0 & 0 & 0 & 1 \end{bmatrix} \begin{Bmatrix} 0 \\ 0 \\ 0 \\ 0.0033 \\ -0.0097 \\ -0.0033 \end{Bmatrix} \tag{5.2.20}$$

Simplifying Eq. (5.2.20) yields

$$\underline{T}\underline{d} = \begin{Bmatrix} 0 \\ 0 \\ 0 \\ -0.00452 \\ -0.0092 \\ -0.0033 \end{Bmatrix} \tag{5.2.21}$$

Using Eq. (5.2.21) and Eq. (5.1.8) for $\hat{\underline{k}}$, we obtain

$$\begin{Bmatrix} \hat{f}_{1x} \\ \hat{f}_{1y} \\ \hat{m}_1 \\ \hat{f}_{2x} \\ \hat{f}_{2y} \\ \hat{m}_2 \end{Bmatrix} = \begin{bmatrix} 5893 & 0 & 0 & -5893 & 0 & 0 \\ & 2.730 & 694.8 & 0 & -2.730 & 694.8 \\ & & 117{,}900 & 0 & -694.8 & 117{,}900 \\ & & & 5893 & 0 & 0 \\ & & & & 2.730 & -694.8 \\ \text{Symmetry} & & & & & 235{,}800 \end{bmatrix} \begin{Bmatrix} 0 \\ 0 \\ 0 \\ -0.00452 \\ -0.0092 \\ -0.0033 \end{Bmatrix} \tag{5.2.22}$$

Simplifying Eq. (5.2.22) yields the local forces in element 1 as

$$\begin{aligned} \hat{f}_{1x} &= 26.64 \text{ kip} & \hat{f}_{1y} &= -2.268 \text{ kip} & \hat{m}_{1x} &= -389.1 \text{ k-in.} \\ \hat{f}_{2x} &= -26.64 \text{ kip} & \hat{f}_{2y} &= 2.268 \text{ kip} & \hat{m}_{2x} &= -778.2 \text{ k-in.} \end{aligned} \tag{5.2.23}$$

For element 2, the local forces are given by Eq. (4.4.11) because a distributed load is acting on the element. From Eqs. (5.1.10) and (5.2.19), we then have

$$\underline{T}\underline{d} = \begin{bmatrix} 1 & 0 & 0 & 0 & 0 & 0 \\ 0 & 1 & 0 & 0 & 0 & 0 \\ 0 & 0 & 1 & 0 & 0 & 0 \\ 0 & 0 & 0 & 1 & 0 & 0 \\ 0 & 0 & 0 & 0 & 1 & 0 \\ 0 & 0 & 0 & 0 & 0 & 1 \end{bmatrix} \begin{Bmatrix} 0.0033 \\ -0.0097 \\ -0.0033 \\ 0 \\ 0 \\ 0 \end{Bmatrix} \tag{5.2.24}$$

Simplifying Eq. (5.2.24), we obtain

$$\begin{Bmatrix} 0.0033 \\ -0.0097 \\ -0.0033 \\ 0 \\ 0 \\ 0 \end{Bmatrix} \tag{5.2.25}$$

Using Eq. (5.2.25) and Eq. (5.1.8) for $\hat{\underline{k}}$, we have

$$\hat{\underline{k}}\hat{\underline{d}} = \hat{\underline{k}}\underline{T}\underline{d} = \begin{bmatrix} 6250 & 0 & 0 & -6250 & 0 & 0 \\ & 3.25 & 781.1 & 0 & -3.25 & 781.1 \\ & & 250{,}000 & 0 & -781.1 & 125{,}000 \\ & & & 6250 & 0 & 0 \\ & & & & 3.25 & -781.1 \\ \text{Symmetry} & & & & & 250{,}000 \end{bmatrix} \begin{Bmatrix} 0.0033 \\ -0.0097 \\ -0.0033 \\ 0 \\ 0 \\ 0 \end{Bmatrix} \tag{5.2.26}$$

Simplifying Eq. (5.2.26) yields

$$\hat{\underline{k}}\hat{\underline{d}} = \begin{Bmatrix} 20.63 \\ -2.58 \\ -832.57 \\ -20.63 \\ 2.58 \\ -412.50 \end{Bmatrix} \tag{5.2.27}$$

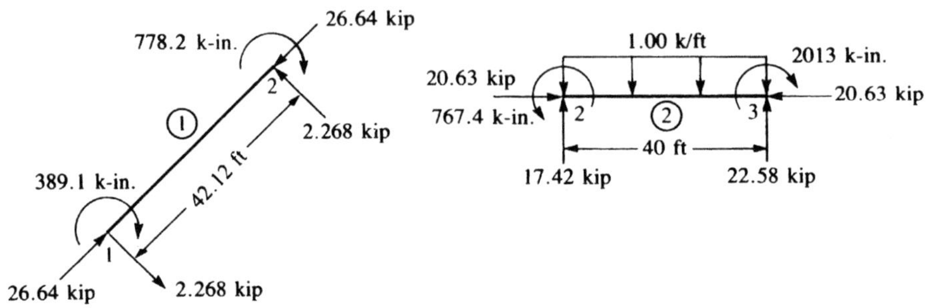

Figure 5–7 Free-body diagrams of elements 1 and 2

To obtain the actual element local nodal forces, we apply Eq. (4.4.11); that is, we must subtract the equivalent nodal forces [Eqs. (5.2.13)] from Eq. (5.2.27) to yield

$$\begin{Bmatrix} \hat{f}_{2x} \\ \hat{f}_{2y} \\ \hat{m}_2 \\ \hat{f}_{3x} \\ \hat{f}_{3y} \\ \hat{m}_3 \end{Bmatrix} = \begin{Bmatrix} 20.63 \\ -2.58 \\ -832.57 \\ -20.63 \\ 2.58 \\ -412.50 \end{Bmatrix} - \begin{Bmatrix} 0 \\ -20 \\ -1600 \\ 0 \\ -20 \\ 1600 \end{Bmatrix} \tag{5.2.28}$$

Simplifying Eq. (5.2.28), we obtain

$$\begin{aligned} \hat{f}_{2x} &= 20.63 \text{ kip} \qquad \hat{f}_{2y} = 17.42 \text{ kip} \qquad \hat{m}_2 = 767.4 \text{ k-in.} \\ \hat{f}_{3x} &= -20.63 \text{ kip} \qquad \hat{f}_{3y} = 22.58 \text{ kip} \qquad \hat{m}_3 = -2013 \text{ k-in.} \end{aligned} \tag{5.2.29}$$

Using Eqs. (5.2.23) and (5.2.29) for the local forces in each element, we can construct the free-body diagram for each element, as shown in Figure 5–7. From the free-body diagrams, one can confirm the equilibrium of each element, the total frame, and joint 2 as desired. ■

In Example 5.3, we will illustrate the equivalent joint force replacement method for a frame subjected to a load acting on an element instead of at one of the joints of the structure. Since no distributed loads are present, the point of application of the concentrated load could be treated as an extra joint in the analysis, and we could solve the problem in the same manner as Example 5.1.

This approach has the disadvantage of increasing the total number of joints, as well as the size of the total structure stiffness matrix $\underline{K}$. For small structures solved by computer, this does not pose a problem. However, for very large structures, this might reduce the maximum size of the structure that could be analyzed. Certainly, this additional node greatly increases the longhand solution time for the structure. Hence, we will illustrate a standard procedure based on the concept of equivalent joint forces applied to the case of concentrated loads. We will again use Appendix D.

Example 5.3

Solve the frame shown in Figure 5–8(a). The frame consists of the three elements shown and is subjected to a 15-kip horizontal load applied at midlength of element 1. Nodes 1, 2, and 3 are fixed, and the dimensions are shown in the figure. Let $E = 30 \times 10^6$ psi, $I = 800$ in^4, and $A = 8$ in^2 for all elements.

1. We first express the applied load in the element 1 local coordinate system (here $\hat{x}$ is directed from node 1 to node 4). This is shown in Figure 5–8(b).

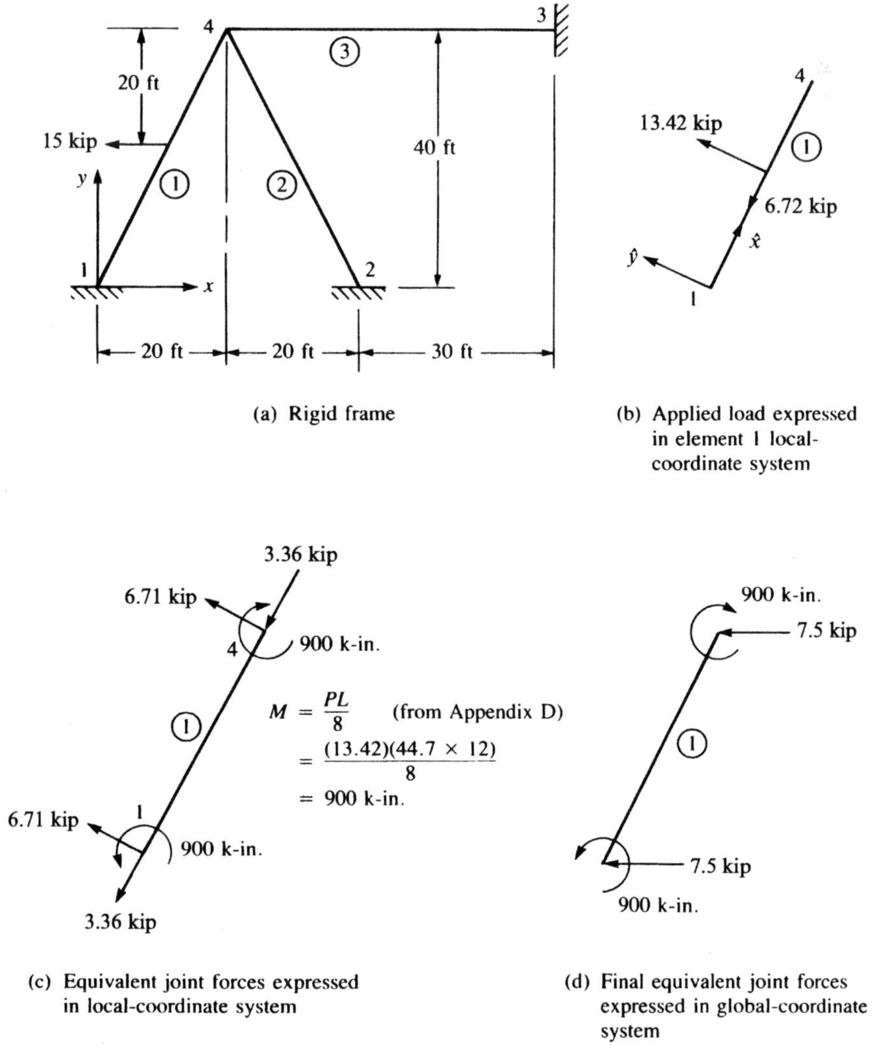

(a) Rigid frame

(b) Applied load expressed in element 1 local-coordinate system

(c) Equivalent joint forces expressed in local-coordinate system

(d) Final equivalent joint forces expressed in global-coordinate system

Figure 5–8 Rigid frame with a load applied on an element

2. Next, we determine the equivalent joint forces at each end of element 1, using the table in Appendix D. (These forces are of opposite sign from what are traditionally known as *fixed-end forces* in classical structural analysis theory [1].) These equivalent forces (and moments) are shown in Figure 5–8(c).
3. We then transform the equivalent joint forces from the present local-coordinate-system forces into the global-coordinate-system forces, using the equation $\underline{f} = \underline{T}^T\underline{\hat{f}}$, where $\underline{T}$ is defined by Eq. (5.1.10). These global joint forces are shown in Figure 5–8(d).
4. Then we analyze the structure in Figure 5–8(d), using the equivalent joint forces (plus actual joint forces, if any) in the usual manner.
5. We obtain the final internal forces developed at the ends of each element that has an applied load (here element 1 only) by subtracting step 2 joint forces from step 4 joint forces; that is, Eq. (4.4.11) is applied locally to all elements that originally had loads acting on them.

The solution of the structure as shown in Figure 5–8(d) now follows. Using Eq. (5.1.11), we obtain the global stiffness matrix for each element.

Element 1

For element 1, the angle between the global x and the local $\hat{x}$ axes is 63.43° because $\hat{x}$ is assumed to be directed from node 1 to node 4. Therefore,

$$C = \cos 63.43^\circ = \frac{x_4 - x_1}{L^{(1)}} = \frac{20 - 0}{44.7} = 0.447$$

$$S = \sin 63.43^\circ = \frac{y_4 - y_1}{L^{(1)}} = \frac{40 - 0}{44.7} = 0.895$$

$$\frac{12I}{L^2} = \frac{12(800)}{(44.7 \times 12)^2} = 0.0334 \qquad \frac{6I}{L} = \frac{6(800)}{44.7 \times 12} = 8.95$$

$$\frac{E}{L} = \frac{30 \times 10^3}{44.7 \times 12} = 55.9$$

Using the preceding results in Eq. (5.1.11) for $\underline{k}$, we obtain

$$\underline{k}^{(1)} = \begin{matrix} d_{4x} & d_{4y} & \phi_4 \end{matrix} \\ \begin{bmatrix} 90.9 & 178 & 448 \\ 178 & 359 & -224 \\ 448 & -224 & 179{,}000 \end{bmatrix} \qquad (5.2.30)$$

where only the parts of the stiffness matrix associated with degrees of freedom at node 4 are included because node 1 is fixed and, hence, not needed in the solution for the nodal displacements.

Element 3

For element 3, the angle between x and $\hat{x}$ is zero because $\hat{x}$ is directed from node 4 to node 3. Therefore,

$$C = 1 \qquad S = 0 \qquad \frac{12I}{L^2} = \frac{12(800)}{(50 \times 12)^2} = 0.0267$$

$$\frac{6I}{L} = \frac{6(800)}{50 \times 12} = 8.00 \qquad \frac{E}{L} = \frac{30 \times 10^3}{50 \times 12} = 50$$

Substituting these results into $\underline{k}$, we obtain

$$\underline{k}^{(3)} = \begin{bmatrix} 400 & 0 & 0 \\ 0 & 1.334 & 400 \\ 0 & 400 & 160{,}000 \end{bmatrix} \quad \begin{matrix} d_{4x} & d_{4y} & \phi_4 \end{matrix} \tag{5.2.31}$$

since node 3 is fixed.

Element 2

For element 2, the angle between x and $\hat{x}$ is 116.57° because $\hat{x}$ is directed from node 2 to node 4. Therefore,

$$C = \frac{20 - 40}{44.7} = -0.447 \qquad S = \frac{40 - 0}{44.7} = 0.895$$

$$\frac{12I}{L^2} = 0.0334 \qquad \frac{6I}{L} = 8.95 \qquad \frac{E}{L} = 55.9$$

since element 2 has the same properties as element 1. Substituting these results into $\underline{k}$, we obtain

$$\underline{k}^{(2)} = \begin{bmatrix} 90.9 & -178 & 448 \\ -178 & 359 & 224 \\ 448 & 224 & 179{,}000 \end{bmatrix} \quad \begin{matrix} d_{4x} & d_{4y} & \phi_4 \end{matrix} \tag{5.2.32}$$

since node 2 is fixed. On superimposing the stiffness matrices given by Eqs. (5.2.30), (5.2.31), and (5.2.32), and using the nodal forces given in Figure 5–8(d) at node 4 only, we have

$$\begin{Bmatrix} -7.50 \text{ kip} \\ 0 \\ -900 \text{ k-in.} \end{Bmatrix} = \begin{bmatrix} 582 & 0 & 896 \\ 0 & 719 & 400 \\ 896 & 400 & 518{,}000 \end{bmatrix} \begin{Bmatrix} d_{4x} \\ d_{4y} \\ \phi_4 \end{Bmatrix} \tag{5.2.33}$$

Simultaneously solving the three equations in Eq. (5.2.33), we obtain

$$\begin{aligned} d_{4x} &= -0.0103 \text{ in.} \\ d_{4y} &= 0.000956 \text{ in.} \\ \phi_4 &= -0.00172 \text{ rad} \end{aligned} \tag{5.2.34}$$

Next, we determine the element forces by again using $\underline{\hat{f}} = \underline{\hat{k}}\underline{T}\underline{d}$. In general, we have

$$\underline{T}\underline{d} = \begin{bmatrix} C & S & 0 & 0 & 0 & 0 \\ -S & C & 0 & 0 & 0 & 0 \\ 0 & 0 & 1 & 0 & 0 & 0 \\ 0 & 0 & 0 & C & S & 0 \\ 0 & 0 & 0 & -S & C & 0 \\ 0 & 0 & 0 & 0 & 0 & 1 \end{bmatrix} \begin{Bmatrix} d_{ix} \\ d_{iy} \\ \phi_i \\ d_{jx} \\ d_{jy} \\ \phi_j \end{Bmatrix}$$

Thus, the preceding matrix multiplication yields

$$\underline{T}\underline{d} = \begin{Bmatrix} Cd_{ix} + Sd_{iy} \\ -Sd_{ix} + Cd_{iy} \\ \phi_i \\ Cd_{jx} + Sd_{jy} \\ -Sd_{jx} + Cd_{jy} \\ \phi_j \end{Bmatrix} \tag{5.2.35}$$

Element 1

$$\underline{T}\underline{d} = \begin{Bmatrix} 0 \\ 0 \\ 0 \\ (0.447)(-0.0103) + (0.895)(0.000956) \\ (-0.895)(-0.0103) + (0.447)(0.000956) \\ -0.00172 \end{Bmatrix} = \begin{Bmatrix} 0 \\ 0 \\ 0 \\ -0.00374 \\ 0.00963 \\ -0.00172 \end{Bmatrix} \tag{5.2.36}$$

Using Eq. (5.1.8) for $\underline{\hat{k}}$ and Eq. (5.2.36), we obtain

$$\underline{\hat{k}}\underline{T}\underline{d} = \begin{bmatrix} 447 & 0 & 0 & -447 & 0 & 0 \\ 0 & 1.868 & 500.5 & 0 & -1.868 & 500.5 \\ 0 & 500.5 & 179{,}000 & 0 & -500.5 & 89{,}490 \\ -447 & 0 & 0 & 447 & 0 & 0 \\ 0 & -1.868 & -500.5 & 0 & 1.868 & -500.5 \\ 0 & 500.5 & 89{,}490 & 0 & -500.5 & 179{,}000 \end{bmatrix} \times \begin{Bmatrix} 0 \\ 0 \\ 0 \\ -0.00374 \\ 0.00963 \\ -0.00172 \end{Bmatrix} \tag{5.2.37}$$

These values are now called *effective nodal forces*. Multiplying the matrices of Eq. (5.2.37) and using Eq. (4.4.11) to subtract the equivalent nodal forces in local coordinates for the element shown in Figure 5–8(c), we obtain the final nodal forces in

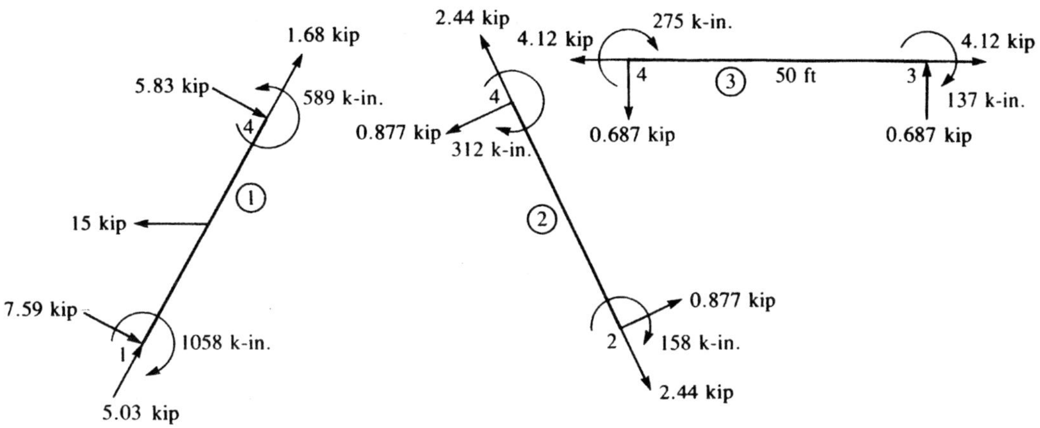

Figure 5–9 Free-body diagrams of all elements of the frame in Figure 5–8(a)

in element 1 as

$$\underline{\hat{f}}^{(1)} = \begin{Bmatrix} 1.67 \\ -0.88 \\ -158 \\ -1.67 \\ 0.88 \\ -311 \end{Bmatrix} - \begin{Bmatrix} -3.36 \\ 6.71 \\ 900 \\ -3.36 \\ 6.71 \\ -900 \end{Bmatrix} = \begin{Bmatrix} 5.03 \text{ kip} \\ -7.59 \text{ kip} \\ -1058 \text{ k-in.} \\ 1.68 \text{ kip} \\ -5.83 \text{ kip} \\ 589 \text{ k-in.} \end{Bmatrix} \tag{5.2.38}$$

Similarly, we can use Eqs. (5.2.35) and (5.1.8) for elements 3 and 2 to obtain the local nodal forces in these elements. Since these elements do not have any applied loads on them, the final nodal forces in local coordinates associated with each element are given by $\underline{\hat{f}} = \underline{\hat{k}}\underline{T}\underline{d}$. These forces have been determined as follows:

Element 3

$$\begin{array}{lll} \hat{f}_{4x} = -4.12 \text{ kip} & \hat{f}_{4y} = -0.687 \text{ kip} & \hat{m}_4 = -275 \text{ k-in.} \\ \hat{f}_{3x} = 4.12 \text{ kip} & \hat{f}_{3y} = 0.687 \text{ kip} & \hat{m}_3 = -137 \text{ k-in.} \end{array} \tag{5.2.39}$$

Element 2

$$\begin{array}{lll} \hat{f}_{2x} = -2.44 \text{ kip} & \hat{f}_{2y} = -0.877 \text{ kip} & \hat{m}_2 = -158 \text{ k-in.} \\ \hat{f}_{4x} = 2.44 \text{ kip} & \hat{f}_{4y} = 0.877 \text{ kip} & \hat{m}_4 = -312 \text{ k-in.} \end{array} \tag{5.2.40}$$

Free-body diagrams of all elements are shown in Figure 5–9. Each element has been determined to be in equilibrium, as often occurs even if errors are made in the longhand calculations. However, equilibrium at node 4 and equilibrium of the whole frame are also satisfied. For instance, using the results of Eqs. (5.2.38)–(5.2.40) to check equilibrium at node 4, which is implicit in the formulation of the global equations, we have

$$\sum M_4 = 589 - 275 - 312 = 2 \text{ k-in.} \qquad \text{(close to zero)}$$

$$\sum F_x = 1.68(0.447) + 5.83(0.895) - 2.44(0.447)$$
$$- 0.877(0.895) - 4.12 = -0.027 \text{ kip} \qquad \text{(close to zero)}$$

$$\sum F_y = 1.68(0.895) - 5.83(0.447) + 2.44(0.895)$$
$$- 0.877(0.447) - 0.687 = 0.004 \text{ kip} \qquad \text{(close to zero)}$$

Thus, the solution has been verified to be correct within the accuracy associated with a longhand solution. ■

To illustrate the solution of a problem involving both bar and frame elements, we will solve the following example.

Example 5.4

The bar element 2 is used to stiffen the cantilever beam element 1, as shown in Figure 5–10. Determine the displacements at node 1 and the element forces. For the bar, let $A = 1.0 \times 10^{-3}$ m^2. For the beam, let $A = 2 \times 10^{-3}$ m^2, $I = 5 \times 10^{-5}$ m^4, and $L = 3$ m. For both the bar and the beam elements, let $E = 210$ GPa. Let the angle between the beam and the bar be 45°. A downward force of 500 kN is applied at node 1.

For brevity's sake, since nodes 2 and 3 are fixed, we keep only the parts of $\underline{k}$ for each element that are needed to obtain the global $\underline{K}$ matrix necessary for solution of the nodal degrees of freedom. Using Eq. (3.4.23), we obtain $\underline{k}$ for the bar as

$$\underline{k}^{(2)} = \frac{(1 \times 10^{-3})(210 \times 10^6)}{(3/\cos 45^\circ)} \begin{bmatrix} 0.5 & 0.5 \\ 0.5 & 0.5 \end{bmatrix}$$

or, simplifying this equation, we obtain

$$\underline{k}^{(2)} = 70 \times 10^3 \begin{bmatrix} \overset{d_{1x}}{0.354} & \overset{d_{1y}}{0.354} \\ 0.354 & 0.354 \end{bmatrix} \frac{\text{kN}}{\text{m}} \tag{5.2.41}$$

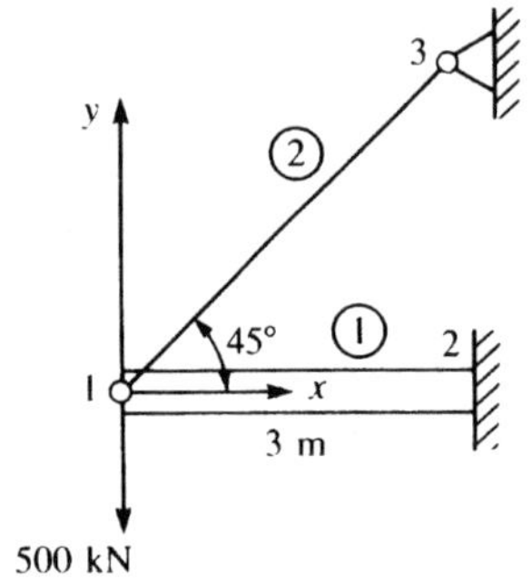

Figure 5–10 Cantilever beam with a bar element support

Using Eq. (5.1.11), we obtain $\underline{k}$ for the beam (including axial effects) as

$$\underline{k}^{(1)} = 70 \times 10^3 \begin{array}{c} \\ \end{array}\overset{\begin{matrix} d_{1x} & d_{1y} & \phi_1 \end{matrix}}{\begin{bmatrix} 2 & 0 & 0 \\ 0 & 0.067 & 0.10 \\ 0 & 0.10 & 0.20 \end{bmatrix}} \frac{\text{kN}}{\text{m}} \tag{5.2.42}$$

where $(E/L) \times 10^{-3}$ has been factored out in evaluating Eq. (5.2.42).

We assemble Eqs. (5.2.41) and (5.2.42) in the usual manner to obtain the global stiffness matrix as

$$\underline{K} = 70 \times 10^3 \begin{bmatrix} 2.354 & 0.354 & 0 \\ 0.354 & 0.421 & 0.10 \\ 0 & 0.10 & 0.20 \end{bmatrix} \frac{\text{kN}}{\text{m}} \tag{5.2.43}$$

The global equations are then written for node 1 as

$$\begin{Bmatrix} F_{1x} \\ F_{1y} \\ M_1 \end{Bmatrix} = \begin{Bmatrix} 0 \\ -500 \\ 0 \end{Bmatrix} = 70 \times 10^3 \begin{bmatrix} 2.354 & 0.354 & 0 \\ 0.354 & 0.421 & 0.10 \\ 0 & 0.10 & 0.20 \end{bmatrix} \begin{Bmatrix} d_{1x} \\ d_{1y} \\ \phi_1 \end{Bmatrix} \tag{5.2.44}$$

Solving Eq. (5.2.44), we obtain

$$d_{1x} = 0.00338 \text{ m} \qquad d_{1y} = -0.0225 \text{ m} \qquad \phi_1 = 0.0113 \text{ rad} \tag{5.2.45}$$

In general, the local element forces are obtained using $\underline{\hat{f}} = \underline{\hat{k}}\underline{T}\underline{d}$. For the bar element, we then have

$$\begin{Bmatrix} \hat{f}_{1x} \\ \hat{f}_{3x} \end{Bmatrix} = \frac{AE}{L} \begin{bmatrix} 1 & -1 \\ -1 & 1 \end{bmatrix} \begin{bmatrix} C & S & 0 & 0 \\ 0 & 0 & C & S \end{bmatrix} \begin{Bmatrix} d_{1x} \\ d_{1y} \\ d_{3x} \\ d_{3y} \end{Bmatrix} \tag{5.2.46}$$

The matrix triple product of Eq. (5.2.46) yields (as one equation)

$$\hat{f}_{1x} = \frac{AE}{L}(Cd_{1x} + Sd_{1y}) \tag{5.2.47}$$

Substituting the numerical values into Eq. (5.2.47), we obtain

$$\hat{f}_{1x} = \frac{(1 \times 10^{-3} \text{ m}^2)(210 \times 10^6 \text{ kN/m}^2)}{4.24 \text{ m}} \left[\frac{\sqrt{2}}{2}(0.00338 - 0.0225) \right] \tag{5.2.48}$$

Simplifying Eq. (5.2.48), we obtain the axial force in the bar (element 2) as

$$\hat{f}_{1x} = -670 \text{ kN} \tag{5.2.49}$$

where the negative sign means $\hat{f}_{1x}$ is in the direction opposite $\hat{x}$ for element 2. Similarly, we obtain

$$\hat{f}_{3x} = 670 \text{ kN} \tag{5.2.50}$$

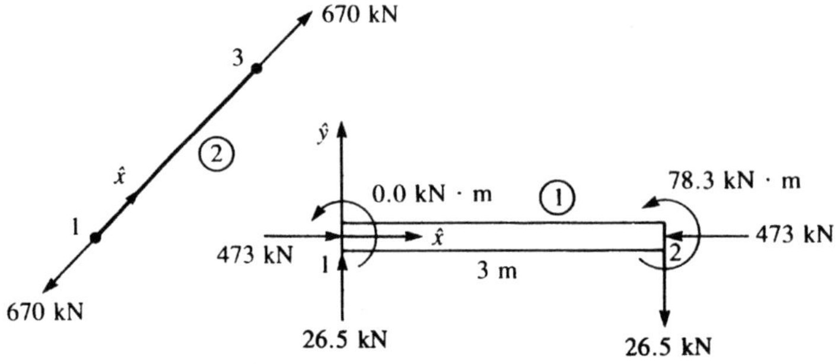

Figure 5–11 Free-body diagrams of the bar (element 2) and beam (element 1) elements of Figure 5–10

which means the bar is in tension as shown in Figure 5–11. Since the local and global axes are coincident for the beam element, we have $\underline{\hat{f}} = \underline{f}$ and $\underline{\hat{d}} = \underline{d}$. Therefore, from Eq. (5.1.6), we have at node 1

$$\begin{Bmatrix} \hat{f}_{1x} \\ \hat{f}_{1y} \\ \hat{m}_1 \end{Bmatrix} = \begin{bmatrix} C_1 & 0 & 0 \\ 0 & 12C_2 & 6C_2L \\ 0 & 6C_2L & 4C_2L^2 \end{bmatrix} \begin{Bmatrix} d_{1x} \\ d_{1y} \\ \phi_1 \end{Bmatrix} \tag{5.2.51}$$

where only the upper part of the stiffness matrix is needed because the displacements at node 2 are equal to zero. Substituting numerical values into Eq. (5.2.51), we obtain

$$\begin{Bmatrix} \hat{f}_{1x} \\ \hat{f}_{1y} \\ \hat{m}_1 \end{Bmatrix} = 70 \times 10^3 \begin{bmatrix} 2 & 0 & 0 \\ 0 & 0.067 & 0.10 \\ 0 & 0.10 & 0.20 \end{bmatrix} \begin{Bmatrix} 0.00338 \\ -0.0225 \\ 0.0113 \end{Bmatrix}$$

The matrix product then yields

$$\hat{f}_{1x} = 473 \text{ kN} \qquad \hat{f}_{1y} = -26.5 \text{ kN} \qquad \hat{m}_1 = 0.0 \text{ kN} \cdot \text{m} \tag{5.2.52}$$

Similarly, using Eq. (5.1.6), we have at node 2,

$$\begin{Bmatrix} \hat{f}_{2x} \\ \hat{f}_{2y} \\ \hat{m}_2 \end{Bmatrix} = 70 \times 10^3 \begin{bmatrix} -2 & 0 & 0 \\ 0 & -0.067 & -0.10 \\ 0 & 0.10 & 0.10 \end{bmatrix} \begin{Bmatrix} 0.00338 \\ -0.0225 \\ 0.0113 \end{Bmatrix}$$

The matrix product then yields

$$\hat{f}_{2x} = -473 \text{ kN} \qquad \hat{f}_{2y} = 26.5 \text{ kN} \qquad \hat{m}_2 = -78.3 \text{ kN} \cdot \text{m} \tag{5.2.53}$$

To help interpret the results of Eqs. (5.2.49), (5.2.50), (5.2.52), and (5.2.53), free-body diagrams of the bar and beam elements are shown in Figure 5–11. To further verify the results, we can show a check on equilibrium of node 1 to be satisfied. You should also verify that moment equilibrium is satisfied in the beam. ■

5.3 Inclined or Skewed Supports—Frame Element

For the frame element with inclined support at node 3 in Figure 5–12, the transformation matrix $\underline{T}$ used to transform global to local nodal displacements is given by Eq. (5.1.10).

In the example shown in Figure 5–12, we use $\underline{T}$ applied to node 3 as follows:

$$\begin{Bmatrix} d'_{3x} \\ d'_{3y} \\ \phi'_3 \end{Bmatrix} = \begin{bmatrix} \cos\alpha & \sin\alpha & 0 \\ -\sin\alpha & \cos\alpha & 0 \\ 0 & 0 & 1 \end{bmatrix} \begin{Bmatrix} d_{3x} \\ d_{3y} \\ \phi_3 \end{Bmatrix}$$

The same steps as given in Section 3.9 then follow for the plane frame. The resulting equations for the plane frame in Figure 5–12 are (see also Eq. (3.9.13))

$$[T_i]\{f\} = [T_i][K][T_i]^T\{d\}$$

or

$$\begin{Bmatrix} F_{1x} \\ F_{1y} \\ M_1 \\ F_{2x} \\ F_{2y} \\ M_2 \\ F'_{3x} \\ F'_{3y} \\ M_3 \end{Bmatrix} = [T_i][K][T_i]^T \begin{Bmatrix} d_{1x} = 0 \\ d_{1y} = 0 \\ \phi_1 = 0 \\ d_{2x} \\ d_{2y} \\ \phi_2 \\ d'_{3x} \\ d'_{3y} = 0 \\ \phi'_3 = \phi_3 \end{Bmatrix}$$

where

$$[T_i] = \begin{bmatrix} [I] & [0] & [0] \\ [0] & [I] & [0] \\ [0] & [0] & [t_3] \end{bmatrix}$$

and

$$[t_3] = \begin{bmatrix} \cos\alpha & \sin\alpha & 0 \\ -\sin\alpha & \cos\alpha & 0 \\ 0 & 0 & 1 \end{bmatrix}$$

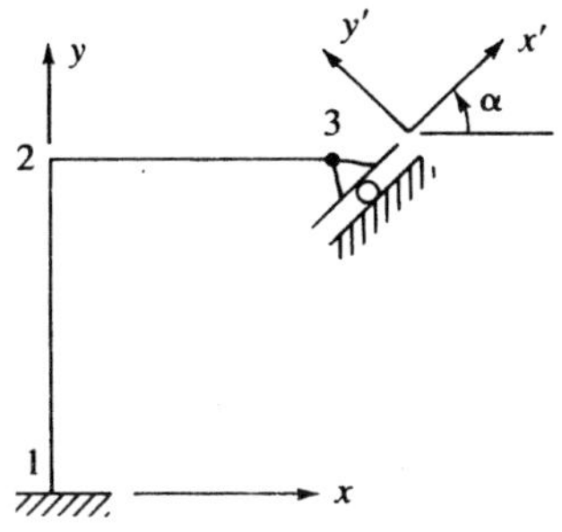

Figure 5–12 Frame with inclined support

5.4 Grid Equations

A **grid** *is a structure on which loads are applied perpendicular to the plane of the structure, as opposed to a plane frame, where loads are applied in the plane of the structure.* We will now develop the grid element stiffness matrix. The elements of a grid are assumed to be rigidly connected, so that the original angles between elements connected together at a node remain unchanged. Both torsional and bending moment continuity then exist at the node point of a grid. Examples of grids include floor and bridge deck systems. A typical grid structure subjected to loads F_1, F_2, F_3, and F_4 is shown in Figure 5–13.

We will now consider the development of the grid element stiffness matrix and element equations. A representative grid element with the nodal degrees of freedom and nodal forces is shown in Figure 5–14. The degrees of freedom at each node for a grid are a vertical deflection $\hat{d}_{iy}$ (normal to the grid), a torsional rotation $\hat{\phi}_{ix}$ about the $\hat{x}$ axis, and a bending rotation $\hat{\phi}_{iz}$ about the $\hat{z}$ axis. Any effect of axial displacement is ignored; that is, $\hat{d}_{ix} = 0$. The nodal forces consist of a transverse force $\hat{f}_{iy}$, a torsional moment $\hat{m}_{ix}$ about the $\hat{x}$ axis, and a bending moment $\hat{m}_{iz}$ about the $\hat{z}$ axis. Grid elements do not resist axial loading; that is $\hat{f}_{ix} = 0$.

To develop the local stiffness matrix for a grid element, we need to include the torsional effects in the basic beam element stiffness matrix Eq. (4.1.14). Recall that Eq. (4.1.14) already accounts for the bending and shear effects.

We can derive the torsional bar element stiffness matrix in a manner analogous to that used for the axial bar element stiffness matrix in Chapter 3. In the derivation, we simply replace $\hat{f}_{ix}$ with $\hat{m}_{ix}$, $\hat{d}_{ix}$ with $\hat{\phi}_{ix}$, E with G (the shear modulus), A with J (the torsional constant, or stiffness factor), σ with τ (shear stress), and ε with γ (shear strain).

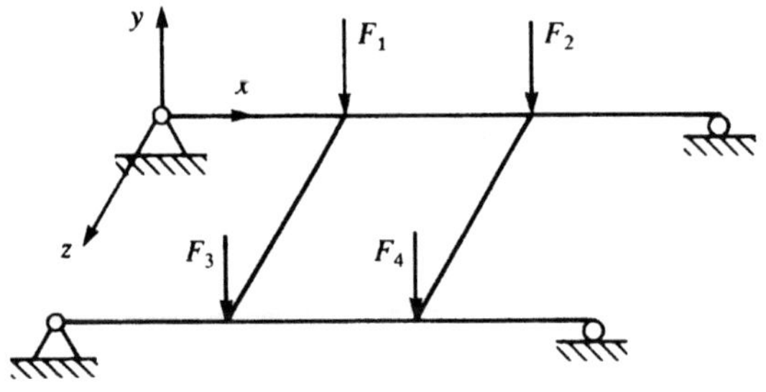

Figure 5–13 Typical grid structure

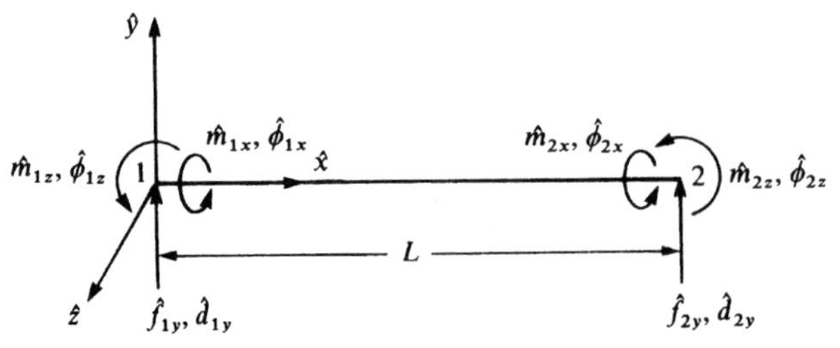

Figure 5–14 Grid element with nodal degrees of freedom and nodal forces

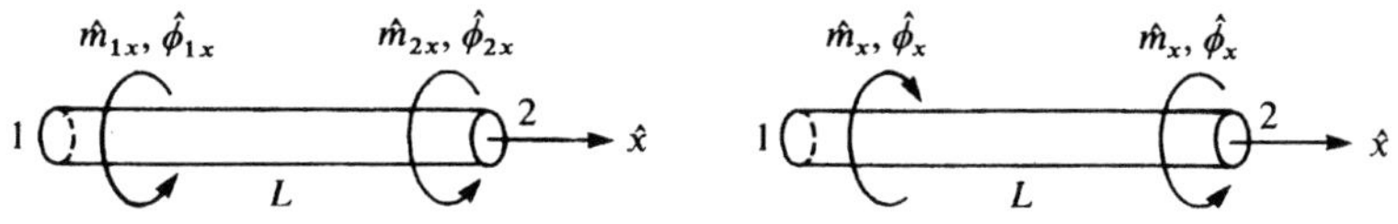

Figure 5–15 Nodal and element torque sign conventions

The actual derivation is briefly presented as follows. We assume a circular cross section with radius R for simplicity but without loss of generalization.

Step 1

Figure 5–15 shows the sign conventions for nodal torque and angle of twist and for element torque.

Step 2

We assume a linear angle-of-twist variation along the $\hat{x}$ axis of the bar such that

$$\hat{\phi} = a_1 + a_2\hat{x} \tag{5.4.1}$$

Using the usual procedure of expressing a_1 and a_2 in terms of unknown nodal angles of twist $\hat{\phi}_{1x}$ and $\hat{\phi}_{2x}$, we obtain

$$\hat{\phi} = \left(\frac{\hat{\phi}_{2x} - \hat{\phi}_{1x}}{L}\right)\hat{x} + \hat{\phi}_{1x} \tag{5.4.2}$$

or, in matrix form, Eq. (5.4.2) becomes

$$\hat{\phi} = [N_1 \quad N_2]\begin{Bmatrix} \hat{\phi}_{1x} \\ \hat{\phi}_{2x} \end{Bmatrix} \tag{5.4.3}$$

with the shape functions given by

$$N_1 = 1 - \frac{\hat{x}}{L} \qquad N_2 = \frac{\hat{x}}{L} \tag{5.4.4}$$

Step 3

We obtain the shear strain γ/angle of twist $\hat{\phi}$ relationship by considering the torsional deformation of the bar segment shown in Figure 5–16. Assuming that all radial lines, such as OA, remain straight during twisting or torsional deformation, we observe that the arc length $\widehat{AB}$ is given by

$$\widehat{AB} = \gamma_{\max}\, d\hat{x} = R\, d\hat{\phi}$$

Solving for the maximum shear strain $\gamma_{\max}$, we obtain

$$\gamma_{\max} = \frac{R\, d\hat{\phi}}{d\hat{x}}$$

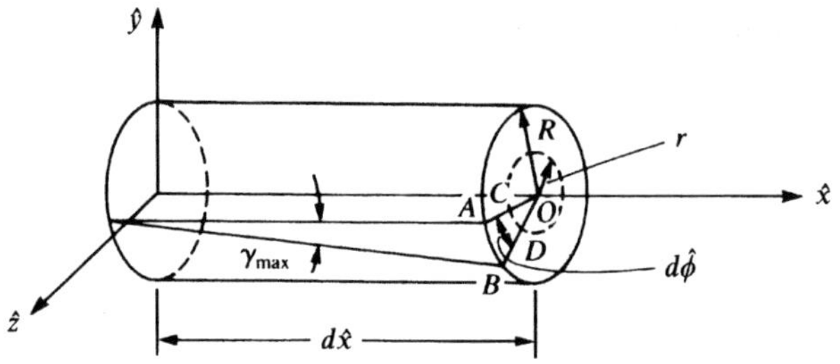

Figure 5–16 Torsional deformation of a bar segment

Similarly, at any radial position r, we then have, from similar triangles OAB and OCD,

$$\gamma = r\frac{d\hat{\phi}}{d\hat{x}} = \frac{r}{L}(\hat{\phi}_{2x} - \hat{\phi}_{1x}) \tag{5.4.5}$$

where we have used Eq. (5.4.2) to derive the final expression in Eq. (5.4.5).

The shear stress τ/shear strain γ relationship for linear-elastic isotropic materials is given by

$$\tau = G\gamma \tag{5.4.6}$$

where G is the shear modulus of the material.

Step 4

We derive the element stiffness matrix in the following manner. From elementary mechanics, we have the shear stress related to the applied torque by

$$\hat{m}_x = \frac{\tau J}{R} \tag{5.4.7}$$

where J is called the *polar moment of inertia* for the circular cross section or, generally, the *torsional constant* for noncircular cross sections. Using Eqs. (5.4.5) and (5.4.6) in Eq. (5.4.7), we obtain

$$\hat{m}_x = \frac{GJ}{L}(\hat{\phi}_{2x} - \hat{\phi}_{1x}) \tag{5.4.8}$$

By the nodal torque sign convention of Figure 5–15,

$$\hat{m}_{1x} = -\hat{m}_x \tag{5.4.9}$$

or, by using Eq. (5.4.8) in Eq. (5.4.9), we obtain

$$\hat{m}_{1x} = \frac{GJ}{L}(\hat{\phi}_{1x} - \hat{\phi}_{2x}) \tag{5.4.10}$$

Similarly,

$$\hat{m}_{2x} = \hat{m}_x \tag{5.4.11}$$

or

$$\hat{m}_{2x} = \frac{GJ}{L}(\hat{\phi}_{2x} - \hat{\phi}_{1x}) \tag{5.4.12}$$

Expressing Eqs. (5.4.10) and (5.4.12) together in matrix form, we have the resulting torsion bar stiffness matrix equation:

$$\begin{Bmatrix} \hat{m}_{1x} \\ \hat{m}_{2x} \end{Bmatrix} = \frac{GJ}{L}\begin{bmatrix} 1 & -1 \\ -1 & 1 \end{bmatrix}\begin{Bmatrix} \hat{\phi}_{1x} \\ \hat{\phi}_{2x} \end{Bmatrix} \tag{5.4.13}$$

Hence, the stiffness matrix for the torsion bar is

$$\underline{\hat{k}} = \frac{GJ}{L}\begin{bmatrix} 1 & -1 \\ -1 & 1 \end{bmatrix} \tag{5.4.14}$$

The cross sections of various structures, such as bridge decks, are often not circular. However, Eqs. (5.4.13) and (5.4.14) are still general; to apply them to other cross sections, we simply evaluate the torsional constant J for the particular cross section. For instance, for cross sections made up of thin rectangular shapes such as channels, angles, or I shapes, we approximate J by

$$J = \sum \frac{1}{3} b_i t_i^3 \tag{5.4.15}$$

where b_i is the length of any element of the cross section and t_i is the thickness of any element of the cross section. In Table 5–1, we list values of J for various common cross sections. The first four cross sections are called *open sections.* Equation (5.4.15) applies only to these open cross sections. (For more information on the J concept, consult References [2] and [3], and for an extensive table of torsional constants for various cross-sectional shapes, consult Reference [4].) We assume the loading to go through the shear center of these open cross sections in order to prevent twisting of the cross section. For more on the shear center consult References [2] and [5].

On combining the torsional effects of Eq. (5.4.13) with the shear and bending effects of Eq. (4.1.13), we obtain the local stiffness matrix equation for a grid element as

$$\begin{Bmatrix} \hat{f}_{1y} \\ \hat{m}_{1x} \\ \hat{m}_{1z} \\ \hat{f}_{2y} \\ \hat{m}_{2x} \\ \hat{m}_{2z} \end{Bmatrix} = \begin{bmatrix} \frac{12EI}{L^3} & 0 & \frac{6EI}{L^2} & \frac{-12EI}{L^3} & 0 & \frac{6EI}{L^2} \\ & \frac{GJ}{L} & 0 & 0 & \frac{-GJ}{L} & 0 \\ & & \frac{4EI}{L} & \frac{-6EI}{L^2} & 0 & \frac{2EI}{L} \\ & & & \frac{12EI}{L^3} & 0 & \frac{-6EI}{L^2} \\ & & & & \frac{GJ}{L} & 0 \\ \text{Symmetry} & & & & & \frac{4EI}{L} \end{bmatrix} \begin{Bmatrix} \hat{d}_{1y} \\ \hat{\phi}_{1x} \\ \hat{\phi}_{1z} \\ \hat{d}_{2y} \\ \hat{\phi}_{2x} \\ \hat{\phi}_{2z} \end{Bmatrix} \tag{5.4.16}$$

Table 5–1 Torsional constants J and shear centers SC for various cross sections

Cross Section	Torsional Constant
1. Channel 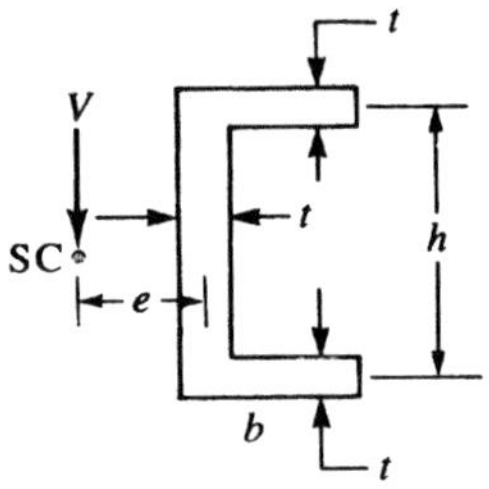	$J = \frac{t^3}{3}(h + 2b)$ $e = \frac{h^2 b^2 t}{4I}$
2. Angle 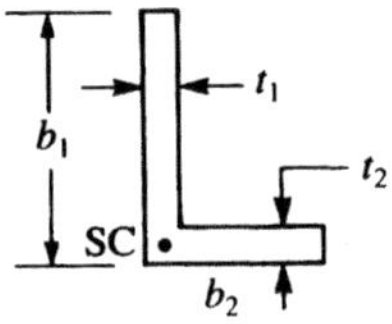	$J = \frac{1}{3}(b_1 t_1^3 + b_2 t_2^3)$
3. Z section 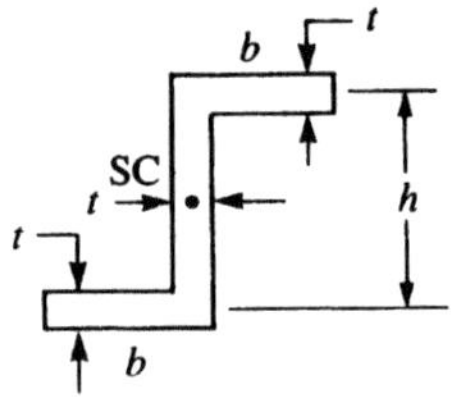	$J = \frac{t^3}{3}(2b + h)$
4. Wide-flanged beam with unequal flanges 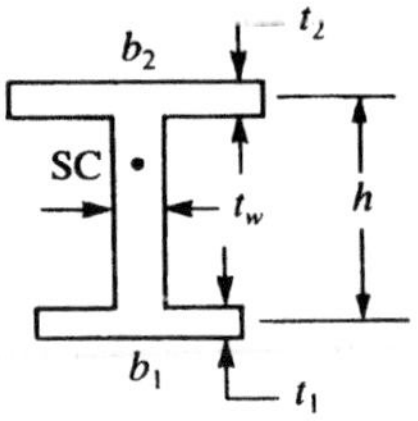	$J = \frac{1}{3}(b_1 t_1^3 + b_2 t_2^3 + h t_w^3)$
5. Solid circular 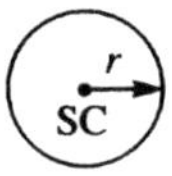	$J = \frac{\pi}{2} r^4$
6. Closed hollow rectangular 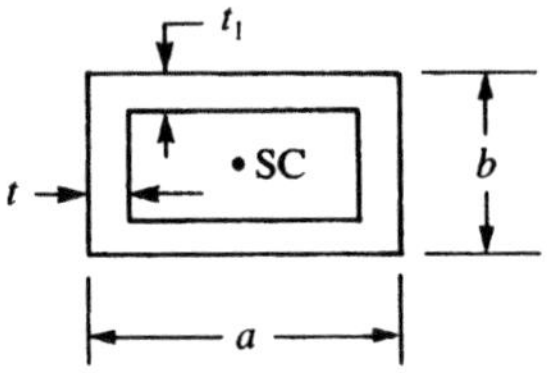	$J = \frac{2 t t_1 (a - t)^2 (b - t_1)^2}{at + bt_1 - t^2 - t_1^2}$

where, from Eq. (5.4.16), the local stiffness matrix for a grid element is

$$\underline{\hat{k}}_G = \begin{array}{c} \begin{array}{cccccc} \hat{d}_{1y} & \hat{\phi}_{1x} & \hat{\phi}_{1z} & \hat{d}_{2y} & \hat{\phi}_{2x} & \hat{\phi}_{2z} \end{array} \\ \begin{bmatrix} \frac{12EI}{L^3} & 0 & \frac{6EI}{L^2} & \frac{-12EI}{L^3} & 0 & \frac{6EI}{L^2} \\ 0 & \frac{GJ}{L} & 0 & 0 & \frac{-GJ}{L} & 0 \\ \frac{6EI}{L^2} & 0 & \frac{4EI}{L} & \frac{-6EI}{L^2} & 0 & \frac{2EI}{L} \\ \frac{-12EI}{L^3} & 0 & \frac{-6EI}{L^2} & \frac{12EI}{L^3} & 0 & \frac{-6EI}{L^2} \\ 0 & \frac{-GJ}{L} & 0 & 0 & \frac{GJ}{L} & 0 \\ \frac{6EI}{L^2} & 0 & \frac{2EI}{L} & \frac{-6EI}{L^2} & 0 & \frac{4EI}{L} \end{bmatrix} \end{array} \tag{5.4.17}$$

and the degrees of freedom are in the order (1) vertical deflection, (2) torsional rotation, and (3) bending rotation, as indicated by the notation used above the columns of Eq. (5.4.17).

The transformation matrix relating local to global degrees of freedom for a grid is given by

$$\underline{T}_G = \begin{bmatrix} 1 & 0 & 0 & 0 & 0 & 0 \\ 0 & C & S & 0 & 0 & 0 \\ 0 & -S & C & 0 & 0 & 0 \\ 0 & 0 & 0 & 1 & 0 & 0 \\ 0 & 0 & 0 & 0 & C & S \\ 0 & 0 & 0 & 0 & -S & C \end{bmatrix} \tag{5.4.18}$$

where θ is now positive, taken counterclockwise from x to $\hat{x}$ in the x-z plane (Figure 5–17) and

$$C = \cos\theta = \frac{x_j - x_i}{L} \qquad S = \sin\theta = \frac{z_j - z_i}{L}$$

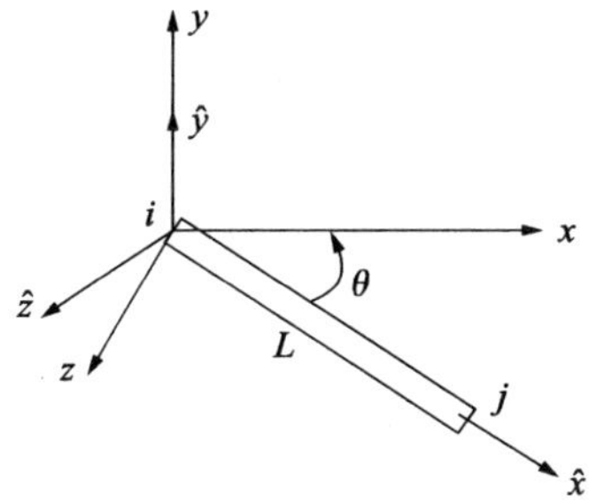

Figure 5–17 Grid element arbitrarily oriented in the *x-z* plane

where L is the length of the element from node i to node j. As indicated by Eq. (5.4.18) for a grid, the vertical deflection $\hat{d}_y$ is invariant with respect to a coordinate transformation (that is, $y = \hat{y}$) (Figure 5–17).

The global stiffness matrix for a grid element arbitrarily oriented in the x-z plane is then given by using Eqs. (5.4.17) and (5.4.18) in

$$\underline{k}_G = \underline{T}_G^T \hat{\underline{k}}_G \underline{T}_G \tag{5.4.19}$$

Now that we have formulated the global stiffness matrix for the grid element, the procedure for solution then follows in the same manner as that for the plane frame.

To illustrate the use of the equations developed in Section 5.4, we will now solve the following grid structures.

Example 5.5

Analyze the grid shown in Figure 5–18. The grid consists of three elements, is fixed at nodes 2, 3, and 4, and is subjected to a downward vertical force (perpendicular to the x-z plane passing through the grid elements) of 100 kip. The global-coordinate axes have been established at node 3, and the element lengths are shown in the figure. Let $E = 30 \times 10^3$ ksi, $G = 12 \times 10^3$ ksi, $I = 400$ in^4, and $J = 110$ in^4 for all elements of the grid.

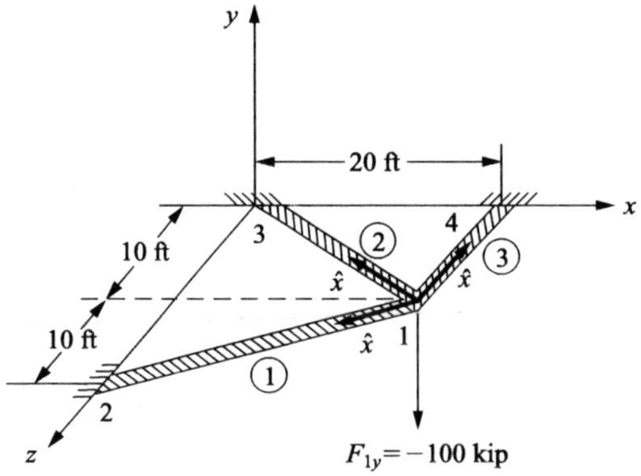

Figure 5–18 Grid for analysis showing local $\hat{x}$ axis for each element

Substituting Eq. (5.4.17) for the local stiffness matrix and Eq. (5.4.18) for the transformation matrix into Eq. (5.4.19), we can obtain each element global stiffness matrix. To expedite the longhand solution, the boundary conditions at nodes 2, 3, and 4,

$$d_{2y} = \phi_{2x} = \phi_{2z} = 0 \qquad d_{3y} = \phi_{3x} = \phi_{3z} = 0 \qquad d_{4y} = \phi_{4x} = \phi_{4z} = 0 \tag{5.4.20}$$

make it possible to use only the upper left-hand 3 × 3 partitioned part of the local stiffness and transformation matrices associated with the degrees of freedom at node 1. Therefore, the global stiffness matrices for each element are as follows:

Element 1

For element 1, we assume the local $\hat{x}$ axis to be directed from node 1 to node 2 for the formulation of the element stiffness matrix. We need the following expressions to evaluate the element stiffness matrix:

$$
\begin{aligned}
C &= \cos\theta = \frac{x_2 - x_1}{L^{(1)}} = \frac{-20 - 0}{22.36} = -0.894 \\
S &= \sin\theta = \frac{z_2 - z_1}{L^{(1)}} = \frac{10 - 0}{22.36} = 0.447 \\
\frac{12EI}{L^3} &= \frac{12(30 \times 10^3)(400)}{(22.36 \times 12)^3} = 7.45 \\
\frac{6EI}{L^2} &= \frac{6(30 \times 10^3)(400)}{(22.36 \times 12)^2} = 1000 \\
\frac{GJ}{L} &= \frac{(12 \times 10^3)(110)}{(22.36 \times 12)} = 4920 \\
\frac{4EI}{L} &= \frac{4(30 \times 10^3)(400)}{(22.36 \times 12)} = 179{,}000
\end{aligned}
\tag{5.4.21}
$$

Considering the boundary condition Eqs. (5.4.20), using the results of Eqs. (5.4.21) in Eq. (5.4.17) for $\hat{\underline{k}}_G$ and Eq. (5.4.18) for $\underline{T}_G$, and then applying Eq. (5.4.19), we obtain the upper left-hand 3 × 3 partitioned part of the global stiffness matrix for element 1 as

$$
\underline{k}^{(1)} = \begin{bmatrix} 1 & 0 & 0 \\ 0 & -0.894 & -0.447 \\ 0 & 0.447 & -0.894 \end{bmatrix} \begin{bmatrix} 7.45 & 0 & 1000 \\ 0 & 4920 & 0 \\ 1000 & 0 & 179{,}000 \end{bmatrix} \begin{bmatrix} 1 & 0 & 0 \\ 0 & -0.894 & 0.447 \\ 0 & -0.447 & -0.894 \end{bmatrix}
$$

Performing the matrix multiplications, we obtain the global element grid stiffness matrix

$$
\underline{k}^{(1)} = \begin{array}{c} \begin{matrix} d_{1y} & \phi_1 & \phi_2 \end{matrix} \\ \begin{bmatrix} 7.45 & -447 & -894 \\ -447 & 39{,}700 & 69{,}600 \\ -894 & 69{,}600 & 144{,}000 \end{bmatrix} \end{array} \frac{\text{kip}}{\text{in.}}
\tag{5.4.22}
$$

where the labels next to the columns indicate the degrees of freedom.

Element 2

For element 2, we assume the local $\hat{x}$ axis to be directed from node 1 to node 3 for the formulation of the element stiffness matrix. We need the following expressions to evaluate the element stiffness matrix:

$$C = \frac{x_3 - x_1}{L^{(2)}} = \frac{-20 - 0}{22.36} = -0.894$$
$$S = \frac{z_3 - z_1}{L^{(2)}} = \frac{-10 - 0}{22.36} = -0.447 \tag{5.4.23}$$

Other expressions used in Eq. (5.4.17) are identical to those in Eqs. (5.4.21) for element 1 because E, G, I, J, and L are identical. Evaluating Eq. (5.4.19) for the global stiffness matrix for element 2, we obtain

$$\underline{k}^{(2)} = \begin{bmatrix} 1 & 0 & 0 \\ 0 & -0.894 & 0.447 \\ 0 & -0.447 & -0.894 \end{bmatrix} \begin{bmatrix} 7.45 & 0 & 1000 \\ 0 & 4920 & 0 \\ 1000 & 0 & 179{,}000 \end{bmatrix} \begin{bmatrix} 1 & 0 & 0 \\ 0 & -0.894 & -0.447 \\ 0 & 0.447 & -0.894 \end{bmatrix}$$

Simplifying, we obtain

$$\underline{k}^{(2)} = \begin{array}{c} \begin{array}{ccc} d_{1y} & \phi_{1x} & \phi_{1z} \end{array} \\ \begin{bmatrix} 7.45 & 447 & -894 \\ 447 & 39{,}700 & -69{,}600 \\ -894 & -69{,}600 & 144{,}000 \end{bmatrix} \end{array} \frac{\text{kip}}{\text{in.}} \tag{5.4.24}$$

Element 3

For element 3, we assume the local $\hat{x}$ axis to be directed from node 1 to node 4. We need the following expressions to evaluate the element stiffness matrix:

$$C = \frac{x_4 - x_1}{L^{(3)}} = \frac{20 - 20}{10} = 0$$
$$S = \frac{z_4 - z_1}{L^{(3)}} = \frac{0 - 10}{10} = -1$$
$$\frac{12EI}{L^3} = \frac{12(30 \times 10^3)(400)}{(10 \times 12)^3} = 83.3 \tag{5.4.25}$$
$$\frac{6EI}{L^2} = \frac{6(30 \times 10^3)(400)}{(10 \times 12)^2} = 5000$$

$$\frac{GJ}{L} = \frac{(12 \times 10^3)(110)}{(10 \times 12)} = 11{,}000$$

$$\frac{4EI}{L} = \frac{4(30 \times 10^3)(400)}{(10 \times 12)} = 400{,}000$$

Using Eqs. (5.4.25), we can obtain the upper part of the global stiffness matrix for element 3 as

$$\underline{k}^{(3)} = \begin{bmatrix} 83.3 & 5000 & 0 \\ 5000 & 400{,}000 & 0 \\ 0 & 0 & 11{,}000 \end{bmatrix} \frac{\text{kip}}{\text{in.}} \quad \begin{matrix} \text{columns: } d_{1y} & \phi_{1x} & \phi_{1z} \end{matrix} \tag{5.4.26}$$

Superimposing the global stiffness matrices from Eqs. (5.4.22), (5.4.24), and (5.4.26), we obtain the total stiffness matrix of the grid (with boundary conditions applied) as

$$\underline{K}_G = \begin{bmatrix} 98.2 & 5000 & -1790 \\ 5000 & 479{,}000 & 0 \\ -1790 & 0 & 299{,}000 \end{bmatrix} \frac{\text{kip}}{\text{in.}} \quad \begin{matrix} \text{columns: } d_{1y} & \phi_{1x} & \phi_{1z} \end{matrix} \tag{5.4.27}$$

The grid matrix equation then becomes

$$\begin{Bmatrix} F_{1y} = -100 \\ M_{1x} = 0 \\ M_{1z} = 0 \end{Bmatrix} = \begin{bmatrix} 98.2 & 5000 & -1790 \\ 5000 & 479{,}000 & 0 \\ -1790 & 0 & 299{,}000 \end{bmatrix} \begin{Bmatrix} d_{1y} \\ \phi_{1x} \\ \phi_{1z} \end{Bmatrix} \tag{5.4.28}$$

The force F_{1y} is negative because the load is applied in the negative y direction. Solving for the displacement and the rotations in Eq. (5.4.28), we obtain

$$\begin{aligned} d_{1y} &= -2.83 \text{ in.} \\ \phi_{1x} &= 0.0295 \text{ rad} \\ \phi_{1z} &= -0.0169 \text{ rad} \end{aligned} \tag{5.4.29}$$

The results indicate that the y displacement at node 1 is downward as indicated by the minus sign, the rotation about the x axis is positive, and the rotation about the z axis is negative. Based on the downward loading location with respect to the supports, these results are expected.

Having solved for the unknown displacement and the rotations, we can obtain the local element forces on formulating the element equations in a manner similar to that for the beam and the plane frame. The local forces (which are needed in the design/analysis stage) are found by applying the equation $\underline{\hat{f}} = \underline{\hat{k}}_G \underline{T}_G \underline{d}$ for each element as follows:

Element 1

Using Eqs. (5.4.17) and (5.4.18) for $\hat{\underline{k}}_G$ and $\underline{T}_G$ and Eq. (5.4.29), we obtain

$$\underline{T}_G\underline{d} = \begin{bmatrix} 1 & 0 & 0 & 0 & 0 & 0 \\ 0 & -0.894 & 0.447 & 0 & 0 & 0 \\ 0 & -0.447 & -0.894 & 0 & 0 & 0 \\ 0 & 0 & 0 & 1 & 0 & 0 \\ 0 & 0 & 0 & 0 & -0.894 & 0.447 \\ 0 & 0 & 0 & 0 & -0.447 & -0.894 \end{bmatrix} \begin{Bmatrix} -2.83 \\ 0.0295 \\ -0.0169 \\ 0 \\ 0 \\ 0 \end{Bmatrix}$$

Multiplying the matrices, we obtain

$$\underline{T}_G\underline{d} = \begin{Bmatrix} -2.83 \\ -0.0339 \\ 0.00192 \\ 0 \\ 0 \\ 0 \end{Bmatrix} \tag{5.4.30}$$

Then $\hat{\underline{f}} = \hat{\underline{k}}_G \underline{T}_G \underline{d}$ becomes

$$\begin{Bmatrix} \hat{f}_{1y} \\ \hat{m}_{1x} \\ \hat{m}_{1z} \\ \hat{f}_{2y} \\ \hat{m}_{2x} \\ \hat{m}_{2z} \end{Bmatrix} = \begin{bmatrix} 7.45 & 0 & 1000 & -7.45 & 0 & 1000 \\ 0 & 4920 & 0 & 0 & -4920 & 0 \\ 1000 & 0 & 179{,}000 & -1000 & 0 & 89{,}500 \\ -7.45 & 0 & -1000 & 7.45 & 0 & -1000 \\ 0 & -4920 & 0 & 0 & 4920 & 0 \\ 1000 & 0 & 89{,}500 & -1000 & 0 & 179{,}000 \end{bmatrix} \begin{Bmatrix} -2.83 \\ -0.0339 \\ 0.00192 \\ 0 \\ 0 \\ 0 \end{Bmatrix} \tag{5.4.31}$$

Multiplying the matrices in Eq. (5.4.31), we obtain the local element forces as

$$\begin{Bmatrix} \hat{f}_{1y} \\ \hat{m}_{1x} \\ \hat{m}_{1z} \\ \hat{f}_{2y} \\ \hat{m}_{2x} \\ \hat{m}_{2z} \end{Bmatrix} = \begin{Bmatrix} -19.2 \text{ kip} \\ -167 \text{ k-in.} \\ -2480 \text{ k-in.} \\ 19.2 \text{ kip} \\ 167 \text{ k-in.} \\ -2660 \text{ k-in.} \end{Bmatrix} \tag{5.4.32}$$

The directions of the forces acting on element 1 are shown in the free-body diagram of element 1 in Figure 5–19.

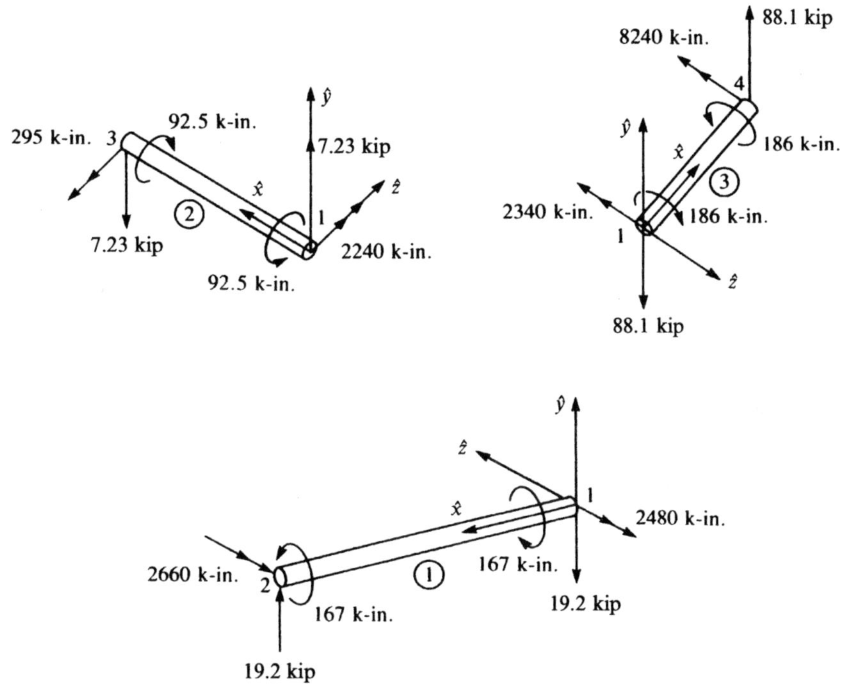

Figure 5–19 Free-body diagrams of the elements of Figure 5–18 showing local-coordinate systems for each

Element 2

Similarly, using $\hat{\underline{f}} = \hat{\underline{k}}_G \underline{T}_G \underline{d}$ for element 2, with the direction cosines in Eqs. (5.4.23), we obtain

$$\begin{Bmatrix} \hat{f}_{1y} \\ \hat{m}_{1x} \\ \hat{m}_{1z} \\ \hat{f}_{3y} \\ \hat{m}_{3x} \\ \hat{m}_{3z} \end{Bmatrix} = \begin{bmatrix} 7.45 & 0 & 1000 & -7.45 & 0 & 1000 \\ 0 & 4920 & 0 & 0 & -4920 & 0 \\ 1000 & 0 & 179{,}000 & -1000 & 0 & 89{,}500 \\ -7.45 & 0 & -1000 & 7.45 & 0 & -1000 \\ 0 & -4920 & 0 & 0 & 4920 & 0 \\ 1000 & 0 & 89{,}500 & -1000 & 0 & 179{,}000 \end{bmatrix}$$

$$\times \begin{bmatrix} 1 & 0 & 0 & 0 & 0 & 0 \\ 0 & -0.894 & -0.447 & 0 & 0 & 0 \\ 0 & 0.447 & -0.894 & 0 & 0 & 0 \\ 0 & 0 & 0 & 1 & 0 & 0 \\ 0 & 0 & 0 & 0 & -0.894 & -0.447 \\ 0 & 0 & 0 & 0 & 0.447 & -0.894 \end{bmatrix} \begin{Bmatrix} -2.83 \\ 0.0295 \\ -0.0169 \\ 0 \\ 0 \\ 0 \end{Bmatrix} \tag{5.4.33}$$

Multiplying the matrices in Eq. (5.4.33), we obtain the local element forces as

$$\begin{aligned}
\hat{f}_{1y} &= 7.23 \text{ kip} \\
\hat{m}_{1x} &= -92.5 \text{ k-in.} \\
\hat{m}_{1z} &= 2240 \text{ k-in.} \\
\hat{f}_{3y} &= -7.23 \text{ kip} \\
\hat{m}_{3x} &= 92.5 \text{ k-in.} \\
\hat{m}_{3z} &= -295 \text{ k-in.}
\end{aligned} \tag{5.4.34}$$

Element 3

Finally, using the direction cosines in Eqs. (5.4.25), we obtain the local element forces as

$$\begin{Bmatrix} \hat{f}_{1y} \\ \hat{m}_{1x} \\ \hat{m}_{1z} \\ \hat{f}_{3y} \\ \hat{m}_{3x} \\ \hat{m}_{3z} \end{Bmatrix} = \begin{bmatrix} 83.3 & 0 & 5000 & -83.3 & 0 & 5000 \\ 0 & 11{,}000 & 0 & 0 & -11{,}000 & 0 \\ 5000 & 0 & 400{,}000 & -5000 & 0 & 200{,}000 \\ -83.3 & 0 & -5000 & 83.33 & 0 & -5000 \\ 0 & -11{,}000 & 0 & 0 & 11{,}000 & 0 \\ 5000 & 0 & 200{,}000 & -5000 & 0 & 400{,}000 \end{bmatrix}$$

$$\times \begin{bmatrix} 1 & 0 & 0 & 0 & 0 & 0 \\ 0 & 0 & -1 & 0 & 0 & 0 \\ 0 & 1 & 0 & 0 & 0 & 0 \\ 0 & 0 & 0 & 1 & 0 & 0 \\ 0 & 0 & 0 & 0 & 0 & -1 \\ 0 & 0 & 0 & 0 & 1 & 0 \end{bmatrix} \begin{Bmatrix} -2.83 \\ 0.0295 \\ -0.0169 \\ 0 \\ 0 \\ 0 \end{Bmatrix} \tag{5.4.35}$$

Multiplying the matrices in Eq. (5.4.35), we obtain the local element forces as

$$\begin{aligned}
\hat{f}_{1y} &= -88.1 \text{ kip} \\
\hat{m}_{1x} &= 186 \text{ k-in.} \\
\hat{m}_{1z} &= -2340 \text{ k-in.} \\
\hat{f}_{4y} &= 88.1 \text{ kip} \\
\hat{m}_{4x} &= -186 \text{ k-in.} \\
\hat{m}_{4z} &= -8240 \text{ k-in.}
\end{aligned} \tag{5.4.36}$$

Free-body diagrams for all elements are shown in Figure 5–19. Each element is in equilibrium. For each element, the $\hat{x}$ axis is shown directed from the first node to the

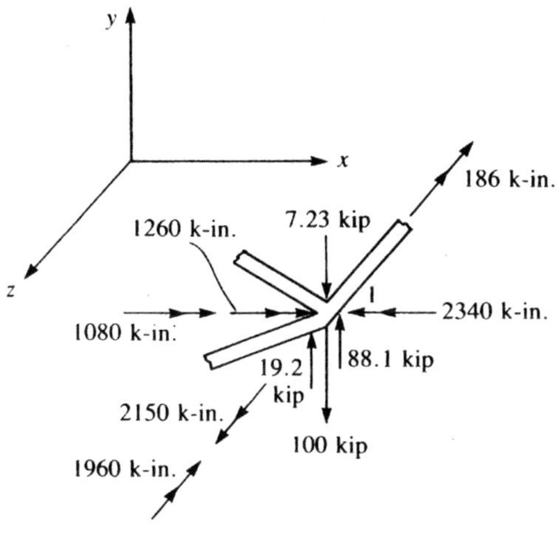

Figure 5–20 Free-body diagram of node 1 of Figure 5–18

second node, the $\hat{y}$ axis coincides with the global y axis, and the $\hat{z}$ axis is perpendicular to the $\hat{x}$-$\hat{y}$ plane with its direction given by the right-hand rule.

To verify equilibrium of node 1, we draw a free-body diagram of the node showing all forces and moments transferred from node 1 of each element, as in Figure 5–20. In Figure 5–20, the local forces and moments from each element have been transformed to global components, and any applied nodal forces have been included. To perform this transformation, recall that, in general, $\underline{\hat{f}} = \underline{T}\underline{f}$, and therefore $\underline{f} = \underline{T}^T\underline{\hat{f}}$ because $\underline{T}^T = \underline{T}^{-1}$. Since we are transforming forces at node 1 of each element, only the upper 3×3 part of Eq. (5.4.18) for $\underline{T}_G$ need be applied. Therefore, by premultiplying the local element forces and moments at node 1 by the transpose of the transformation matrix for each element, we obtain the global nodal forces and moments as follows:

Element 1

$$\begin{Bmatrix} f_{1y} \\ m_{1x} \\ m_{1z} \end{Bmatrix} = \begin{bmatrix} 1 & 0 & 0 \\ 0 & -0.894 & -0.447 \\ 0 & 0.447 & -0.894 \end{bmatrix} \begin{Bmatrix} -19.2 \\ -167 \\ -2480 \end{Bmatrix}$$

Simplifying, we obtain the global-coordinate force and moments as

$$f_{1y} = -19.2 \text{ kip} \qquad m_{1x} = 1260 \text{ k-in.} \qquad m_{1z} = 2150 \text{ k-in.} \tag{5.4.37}$$

where $f_{1y} = \hat{f}_{1y}$ because $y = \hat{y}$.

Element 2

$$\begin{Bmatrix} f_{1y} \\ m_{1x} \\ m_{1z} \end{Bmatrix} = \begin{bmatrix} 1 & 0 & 0 \\ 0 & -0.894 & 0.447 \\ 0 & -0.447 & -0.894 \end{bmatrix} \begin{Bmatrix} 7.23 \\ -92.5 \\ 2240 \end{Bmatrix}$$

Simplifying, we obtain the global-coordinate force and moments as

$$f_{1y} = 7.23 \text{ kip} \qquad m_{1x} = 1080 \text{ k-in.} \qquad m_{1z} = -1960 \text{ k-in.} \tag{5.4.38}$$

Element 3

$$\begin{Bmatrix} f_{1y} \\ m_{1x} \\ m_{1z} \end{Bmatrix} = \begin{bmatrix} 1 & 0 & 0 \\ 0 & 0 & 1 \\ 0 & -1 & 0 \end{bmatrix} \begin{Bmatrix} -88.1 \\ 186 \\ -2340 \end{Bmatrix}$$

Simplifying, we obtain the global-coordinate force and moments as

$$f_{1y} = -88.1 \text{ kip} \qquad m_{1x} = -2340 \text{ k-in.} \qquad m_{1z} = -186 \text{ k-in.} \tag{5.4.39}$$

Then forces and moments from each element that are equal in magnitude but opposite in sign will be applied to node 1. Hence, the free-body diagram of node 1 is shown in Figure 5–20. Force and moment equilibrium are verified as follows:

$$\sum F_{1y} = -100 - 7.23 + 19.2 + 88.1 = 0.07 \text{ kip} \qquad \text{(close to zero)}$$

$$\sum M_{1x} = -1260 - 1080 + 2340 = 0.0 \text{ k-in.}$$

$$\sum M_{1z} = -2150 + 1960 + 186 = -4.00 \text{ k-in.} \qquad \text{(close to zero)}$$

Thus, we have verified the solution to be correct within the accuracy associated with a longhand solution. ■

Example 5.6

Analyze the grid shown in Figure 5–21. The grid consists of two elements, is fixed at nodes 1 and 3, and is subjected to a downward vertical load of 22 kN. The global-coordinate axes and element lengths are shown in the figure. Let $E = 210$ GPa, $G = 84$ GPa, $I = 16.6 \times 10^{-5}$ m^4, and $J = 4.6 \times 10^{-5}$ m^4.

As in Example 5.5, we use the boundary conditions and express only the part of the stiffness matrix associated with the degrees of freedom at node 2. The boundary conditions at nodes 1 and 3 are

$$d_{1y} = \phi_{1x} = \phi_{1z} = 0 \qquad d_{3y} = \phi_{3x} = \phi_{3z} = 0 \tag{5.4.40}$$

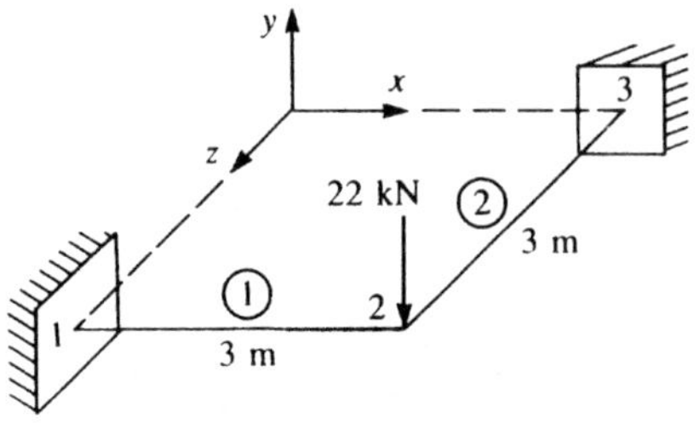

Figure 5–21 Grid example

The global stiffness matrices for each element are obtained as follows:

Element 1

For element 1, we have the local $\hat{x}$ axis coincident with the global x axis. Therefore, we obtain

$$C = \frac{x_2 - x_1}{L^{(1)}} = \frac{3}{3} = 1 \qquad S = \frac{z_2 - z_1}{L^{(1)}} = \frac{3-3}{3} = 0$$

Other expressions needed to evaluate the stiffness matrix are

$$\begin{aligned}
\frac{12EI}{L^3} &= \frac{12(210 \times 10^6 \text{ kN/m}^2)(16.6 \times 10^{-5} \text{ m}^4)}{(3 \text{ m})^3} = 1.55 \times 10^4 \\
\frac{6EI}{L^2} &= \frac{6(210 \times 10^6)(16.6 \times 10^{-5})}{(3)^2} = 2.32 \times 10^4 \\
\frac{GJ}{L} &= \frac{(84 \times 10^6)(4.6 \times 10^{-5})}{3} = 1.28 \times 10^3 \\
\frac{4EI}{L} &= \frac{4(210 \times 10^6)(16.6 \times 10^{-5})}{3} = 4.65 \times 10^4
\end{aligned} \tag{5.4.41}$$

Considering the boundary condition Eqs. (5.4.40), using the results of Eqs. (5.4.41) in Eq. (5.4.17) for $\hat{\underline{k}}_G$ and Eq. (5.4.18) for $\underline{T}_G$, and then applying Eq. (5.4.19), we obtain the reduced part of the global stiffness matrix associated only with the degrees of freedom at node 2 as

$$\underline{k}^{(1)} = \begin{bmatrix} 1 & 0 & 0 \\ 0 & 1 & 0 \\ 0 & 0 & 1 \end{bmatrix} \begin{bmatrix} 1.55 & 0 & -2.32 \\ 0 & 0.128 & 0 \\ -2.32 & 0 & 4.65 \end{bmatrix} (10^4) \begin{bmatrix} 1 & 0 & 0 \\ 0 & 1 & 0 \\ 0 & 0 & 1 \end{bmatrix}$$

Since the local axes associated with element 1 are parallel to the global axes, we observe that $\underline{T}_G$ is merely the identity matrix; therefore, $\underline{k}_G = \hat{\underline{k}}_G$. Performing the matrix multiplications, we obtain

$$\underline{k}^{(1)} = \begin{bmatrix} 1.55 & 0 & -2.32 \\ 0 & 0.128 & 0 \\ -2.32 & 0 & 4.65 \end{bmatrix} (10^4) \, \frac{\text{kN}}{\text{m}} \tag{5.4.42}$$

Element 2

For element 2, we assume the local $\hat{x}$ axis to be directed from node 2 to node 3 for the formulation of $\underline{k}$. Therefore,

$$C = \frac{x_3 - x_2}{L^{(2)}} = \frac{0-0}{3} = 0 \qquad S = \frac{z_3 - z_2}{L^{(2)}} = \frac{0-3}{3} = -1 \tag{5.4.43}$$

Other expressions used in Eq. (5.4.17) are identical to those obtained in Eqs. (5.4.41) for element 1. Evaluating Eq. (5.4.19) for the global stiffness matrix, we obtain

$$\underline{k}^{(2)} = \begin{bmatrix} 1 & 0 & 0 \\ 0 & 0 & 1 \\ 0 & -1 & 0 \end{bmatrix} \begin{bmatrix} 1.55 & 0 & 2.32 \\ 0 & 0.128 & 0 \\ 2.32 & 0 & 4.65 \end{bmatrix} (10^4) \begin{bmatrix} 1 & 0 & 0 \\ 0 & 0 & -1 \\ 0 & 1 & 0 \end{bmatrix}$$

where the reduced part of $\underline{k}$ is now associated with node 2 for element 2. Again performing the matrix multiplications, we have

$$\underline{k}^{(2)} = \begin{bmatrix} 1.55 & 2.32 & 0 \\ 2.32 & 4.65 & 0 \\ 0 & 0 & 0.128 \end{bmatrix} (10^4)\ \frac{\text{kN}}{\text{m}} \tag{5.4.44}$$

Superimposing the global stiffness matrices from Eqs. (5.4.42) and (5.4.44), we obtain the total global stiffness matrix (with boundary conditions applied) as

$$\underline{K}_G = \begin{bmatrix} 3.10 & 2.32 & -2.32 \\ 2.32 & 4.78 & 0 \\ -2.32 & 0 & 4.78 \end{bmatrix} (10^4)\ \frac{\text{kN}}{\text{m}} \tag{5.4.45}$$

The grid matrix equation becomes

$$\begin{Bmatrix} F_{2y} = -22 \\ M_{2x} = 0 \\ M_{2z} = 0 \end{Bmatrix} = \begin{bmatrix} 3.10 & 2.32 & -2.32 \\ 2.32 & 4.78 & 0 \\ -2.32 & 0 & 4.78 \end{bmatrix} \begin{Bmatrix} d_{2y} \\ \phi_{2x} \\ \phi_{2z} \end{Bmatrix} (10^4) \tag{5.4.46}$$

Solving for the displacement and the rotations in Eq. (5.4.46), we obtain

$$\begin{aligned} d_{2y} &= -0.259 \times 10^{-2}\ \text{m} \\ \phi_{2x} &= 0.126 \times 10^{-2}\ \text{rad} \\ \phi_{2z} &= -0.126 \times 10^{-2}\ \text{rad} \end{aligned} \tag{5.4.47}$$

We determine the local element forces by applying the local equation $\underline{\hat{f}} = \underline{\hat{k}}_G \underline{T}_G \underline{d}$ for each element as follows:

Element 1

Using Eq. (5.4.17) for $\underline{\hat{k}}_G$, Eq. (5.4.18) for $\underline{T}_G$, and Eqs. (5.4.47), we obtain

$$\underline{T}_G \underline{d} = \begin{bmatrix} 1 & 0 & 0 & 0 & 0 & 0 \\ 0 & 1 & 0 & 0 & 0 & 0 \\ 0 & 0 & 1 & 0 & 0 & 0 \\ 0 & 0 & 0 & 1 & 0 & 0 \\ 0 & 0 & 0 & 0 & 1 & 0 \\ 0 & 0 & 0 & 0 & 0 & 1 \end{bmatrix} \begin{Bmatrix} 0 \\ 0 \\ 0 \\ -0.259 \times 10^{-2} \\ 0.126 \times 10^{-2} \\ -0.126 \times 10^{-2} \end{Bmatrix}$$

Multiplying the matrices, we have

$$\underline{T}_G\underline{d} = \begin{Bmatrix} 0 \\ 0 \\ 0 \\ -0.259 \times 10^{-2} \\ 0.126 \times 10^{-2} \\ -0.126 \times 10^{-2} \end{Bmatrix} \tag{5.4.48}$$

Using Eqs. (5.4.17), (5.4.41), and (5.4.48), we obtain the local element forces as

$$\begin{Bmatrix} \hat{f}_{1y} \\ \hat{m}_{1x} \\ \hat{m}_{1z} \\ \hat{f}_{2y} \\ \hat{m}_{2x} \\ \hat{m}_{2z} \end{Bmatrix} = (10^4) \begin{bmatrix} 1.55 & 0 & 2.32 & -1.55 & 0 & 2.32 \\ & 0.128 & 0 & 0 & -0.128 & 0 \\ & & 4.65 & -2.32 & 0 & 2.33 \\ & & & 1.55 & 0 & -2.32 \\ & & & & 0.128 & 0 \\ \text{Symmetry} & & & & & 4.65 \end{bmatrix} \begin{Bmatrix} 0 \\ 0 \\ 0 \\ -0.259 \times 10^{-2} \\ 0.126 \times 10^{-2} \\ -0.126 \times 10^{-2} \end{Bmatrix} \tag{5.4.49}$$

Multiplying the matrices in Eq. (5.4.49), we obtain

$$\begin{array}{lll} \hat{f}_{1y} = 11.0 \text{ kN} & \hat{m}_{1x} = -1.50 \text{ kN} \cdot \text{m} & \hat{m}_{1z} = 31.0 \text{ kN} \cdot \text{m} \\ \hat{f}_{2y} = -11.0 \text{ kN} & \hat{m}_{2x} = 1.50 \text{ kN} \cdot \text{m} & \hat{m}_{2z} = 1.50 \text{ kN} \cdot \text{m} \end{array} \tag{5.4.50}$$

Element 2

We can obtain the local element forces for element 2 in a similar manner. Because the procedure is the same as that used to obtain the element 1 local forces, we will not show the details but will only list the final results:

$$\begin{array}{lll} \hat{f}_{2y} = -11.0 \text{ kN} & \hat{m}_{2x} = 1.50 \text{ kN} \cdot \text{m} & \hat{m}_{2z} = -1.50 \text{ kN} \cdot \text{m} \\ \hat{f}_{3y} = 11.0 \text{ kN} & \hat{m}_{3x} = -1.50 \text{ kN} \cdot \text{m} & \hat{m}_{3z} = -31.0 \text{ kN} \cdot \text{m} \end{array} \tag{5.4.51}$$

Free-body diagrams showing the local element forces are shown in Figure 5–22. ■

5.5 Beam Element Arbitrarily Oriented in Space

In this section, we develop the stiffness matrix for the beam element arbitrarily oriented in space, or three dimensions. This element can then be used to analyze frames in three-dimensional space.

First we consider bending about two axes, as shown in Figure 5–23.

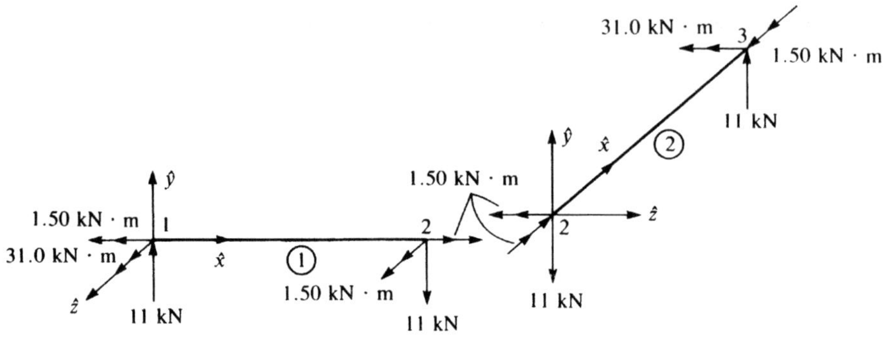

Figure 5–22 Free-body diagram of each element of Figure 5–21

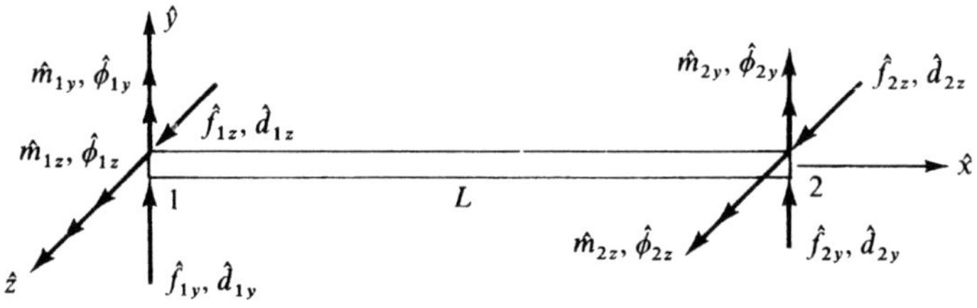

Figure 5–23 Bending about two axes $\hat{y}$ and $\hat{z}$

We establish the following sign convention for the axes. Now we choose positive $\hat{x}$ from node 1 to 2. Then $\hat{y}$ is the principal axis for which the moment of inertia is minimum, I_y. By the right-hand rule we establish $\hat{z}$, and the maximum moment of inertia is I_z.

Bending in $\hat{x}$-$\hat{z}$ Plane

First consider bending in the $\hat{x}$=$\hat{z}$ plane due to $\hat{m}_y$. Then clockwise rotation $\hat{\phi}_y$ is in the same sense as before for single bending. The stiffness matrix due to bending in the $\hat{x}$-$\hat{z}$ plane is then

$$\underline{\hat{k}}_y = \frac{EI_y}{L^4}\begin{bmatrix} 12L & -6L^2 & -12L & -6L^2 \\ & 4L^3 & 6L^2 & 2L^3 \\ & & 12L & 6L^2 \\ \text{Symmetry} & & & 4L^3 \end{bmatrix} \tag{5.5.1}$$

where I_y is the moment of inertia of the cross section about the principal axis $\hat{y}$, the weak axis; that is, $I_y < I_z$.

Bending in the $\hat{z}$-$\hat{y}$ Plane

Now we consider bending in the $\hat{x}$-$\hat{y}$ plane due to $\hat{m}_z$. Now positive rotation $\hat{\phi}_z$ is counterclockwise instead of clockwise. Therefore, some signs change in the stiffness matrix for bending in the $\hat{x}$-$\hat{y}$ plane. The resulting stiffness matrix is

$$\underline{\hat{k}}_z = \frac{EI_z}{L^4}\begin{bmatrix} 12L & 6L^2 & -12L & 6L^2 \\ & 4L^3 & -6L^2 & 2L^3 \\ & & 12L & -6L^2 \\ \text{Symmetry} & & & 4L^3 \end{bmatrix} \tag{5.5.2}$$

Direct superposition of Eqs. (5.5.1) and (5.5.2) with the axial stiffness matrix Eq. (3.1.14) and the torsional stiffness matrix Eq. (5.4.14) yields the element stiffness matrix for the beam or frame element in three-dimensional space as

$$\underline{\hat{k}} = \begin{array}{c} \begin{matrix} \hat{d}_{1x} & \hat{d}_{1y} & \hat{d}_{1z} & \hat{\phi}_{1x} & \hat{\phi}_{1y} & \hat{\phi}_{1z} & \hat{d}_{2x} & \hat{d}_{2y} & \hat{d}_{2z} & \hat{\phi}_{2x} & \hat{\phi}_{2y} & \hat{\phi}_{2z} \end{matrix} \\ \left[\begin{array}{cccccc|cccccc} \frac{AE}{L} & 0 & 0 & 0 & 0 & 0 & -\frac{AE}{L} & 0 & 0 & 0 & 0 & 0 \\ 0 & \frac{12EI_z}{L^3} & 0 & 0 & 0 & \frac{6EI_z}{L^2} & 0 & -\frac{12EI_z}{L^3} & 0 & 0 & 0 & \frac{6EI_z}{L^2} \\ 0 & 0 & \frac{12EI_y}{L^3} & 0 & -\frac{6EI_y}{L^2} & 0 & 0 & 0 & -\frac{12EI_y}{L^3} & 0 & -\frac{6EI_y}{L^2} & 0 \\ 0 & 0 & 0 & \frac{GJ}{L} & 0 & 0 & 0 & 0 & 0 & -\frac{GJ}{L} & 0 & 0 \\ 0 & 0 & -\frac{6EI_y}{L^2} & 0 & \frac{4EI_y}{L} & 0 & 0 & 0 & \frac{6EI_y}{L^2} & 0 & \frac{2EI_y}{L} & 0 \\ 0 & \frac{6EI_z}{L^2} & 0 & 0 & 0 & \frac{4EI_z}{L} & 0 & -\frac{6EI_z}{L^2} & 0 & 0 & 0 & \frac{2EI_z}{L} \\ \hline -\frac{AE}{L} & 0 & 0 & 0 & 0 & 0 & \frac{AE}{L} & 0 & 0 & 0 & 0 & 0 \\ 0 & -\frac{12EI_z}{L^3} & 0 & 0 & 0 & -\frac{6EI_z}{L^2} & 0 & \frac{12EI_z}{L^3} & 0 & 0 & 0 & -\frac{6EI_z}{L^2} \\ 0 & 0 & -\frac{12EI_y}{L^3} & 0 & \frac{6EI_y}{L^2} & 0 & 0 & 0 & \frac{12EI_y}{L^3} & 0 & \frac{6EI_y}{L^2} & 0 \\ 0 & 0 & 0 & -\frac{GJ}{L} & 0 & 0 & 0 & 0 & 0 & \frac{GJ}{L} & 0 & 0 \\ 0 & 0 & -\frac{6EI_y}{L^2} & 0 & \frac{2EI_y}{L} & 0 & 0 & 0 & \frac{6EI_y}{L^2} & 0 & \frac{4EI_y}{L} & 0 \\ 0 & \frac{6EI_z}{L^2} & 0 & 0 & 0 & \frac{2EI_z}{L} & 0 & -\frac{6EI_z}{L^2} & 0 & 0 & 0 & \frac{4EI_z}{L} \end{array}\right] \end{array} \tag{5.5.3}$$

The transformation from local to global axis system is accomplished as follows:

$$\underline{k} = \underline{T}^T \underline{\hat{k}} \underline{T} \tag{5.5.4}$$

where $\underline{\hat{k}}$ is given by Eq. (5.5.3) and $\underline{T}$ is given by

$$\underline{T} = \begin{bmatrix} \underline{\lambda}_{3\times 3} & & & \\ & \underline{\lambda}_{3\times 3} & & \\ & & \underline{\lambda}_{3\times 3} & \\ & & & \underline{\lambda}_{3\times 3} \end{bmatrix} \tag{5.5.5}$$

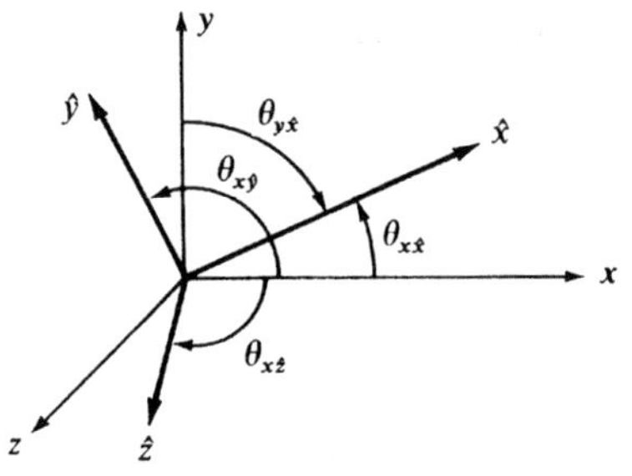

Figure 5–24 Direction cosines associated with the x axis

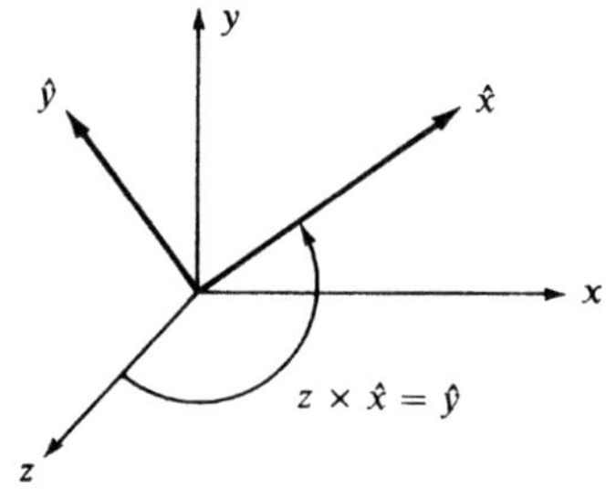

Figure 5–25 Illustration showing how local ŷ axis is determined

where
$$\underline{\lambda} = \begin{bmatrix} C_{x\hat{x}} & C_{y\hat{x}} & C_{z\hat{x}} \\ C_{x\hat{y}} & C_{y\hat{y}} & C_{z\hat{y}} \\ C_{x\hat{z}} & C_{y\hat{z}} & C_{z\hat{z}} \end{bmatrix} \tag{5.5.6}$$

Here $C_{y\hat{x}}$ and $C_{x\hat{y}}$ are not necessarily equal. The direction cosines are shown in part in Figure 5–24.

Remember that direction cosines of the $\hat{x}$ axis member are

$$\hat{x} = \cos\theta_{x\hat{x}}\bar{i} + \cos\theta_{y\hat{x}}\bar{j} + \cos\theta_{z\hat{x}}\bar{k} \tag{5.5.7}$$

where
$$\cos\theta_{x\hat{x}} = \frac{x_2 - x_1}{L} = l$$
$$\cos\theta_{y\hat{x}} = \frac{y_2 - y_1}{L} = m \tag{5.5.8}$$
$$\cos\theta_{z\hat{x}} = \frac{z_2 - z_1}{L} = n$$

The $\hat{y}$ axis is selected to be perpendicular to the $\hat{x}$ and z axes in such a way that the cross product of global z with $\hat{x}$ results in the $\hat{y}$ axis, as shown in Figure 5–25. Therefore,

$$z \times \hat{x} = \hat{y} = \frac{1}{D}\begin{vmatrix} \bar{i} & \bar{j} & \bar{k} \\ 0 & 0 & 1 \\ l & m & n \end{vmatrix} \tag{5.5.9}$$

$$\hat{y} = -\frac{m}{D}\bar{i} + \frac{l}{D}\bar{j} \tag{5.5.10}$$

and
$$D = (l^2 + m^2)^{1/2}$$

The $\hat{z}$ axis will be determined by the orthogonality condition $\hat{z} = \hat{x} \times \hat{y}$ as follows:

$$\hat{z} = \hat{x} \times \hat{y} = \frac{1}{D}\begin{vmatrix} \bar{i} & \bar{j} & \bar{k} \\ l & m & n \\ -m & l & 0 \end{vmatrix} \tag{5.5.11}$$

or

$$\hat{z} = -\frac{ln}{D}\bar{i} - \frac{mn}{D}\bar{j} + D\bar{k} \tag{5.5.12}$$

Combining Eqs. (5.5.7), (5.5.10), and (5.5.12), the 3×3 transformation matrix becomes

$$\underline{\lambda}_{3\times 3} = \begin{bmatrix} l & m & n \\ -\frac{m}{D} & \frac{l}{D} & 0 \\ -\frac{ln}{D} & -\frac{mn}{D} & D \end{bmatrix} \tag{5.5.13}$$

This vector $\underline{\lambda}$ rotates a vector from the local coordinate system into the global one. This is the $\underline{\lambda}$ used in the $\underline{T}$ matrix. In summary, we have

$$\begin{aligned} \cos\theta_{x\hat{y}} &= -\frac{m}{D} \\ \cos\theta_{y\hat{y}} &= \frac{l}{D} \\ \cos\theta_{z\hat{y}} &= 0 \\ \cos\theta_{x\hat{z}} &= -\frac{ln}{D} \\ \cos\theta_{y\hat{z}} &= -\frac{mn}{D} \\ \cos\theta_{z\hat{z}} &= D \end{aligned} \tag{5.5.14}$$

Two exceptions arise when local and global axes have special orientations with respect to each other. If the local $\hat{x}$ axis coincides with the global z axis, then the member is parallel to the global z axis and the $\hat{y}$ axis becomes uncertain, as shown in Figure 5–26(a). In this case the local $\hat{y}$ axis is selected as the global y axis. Then, for

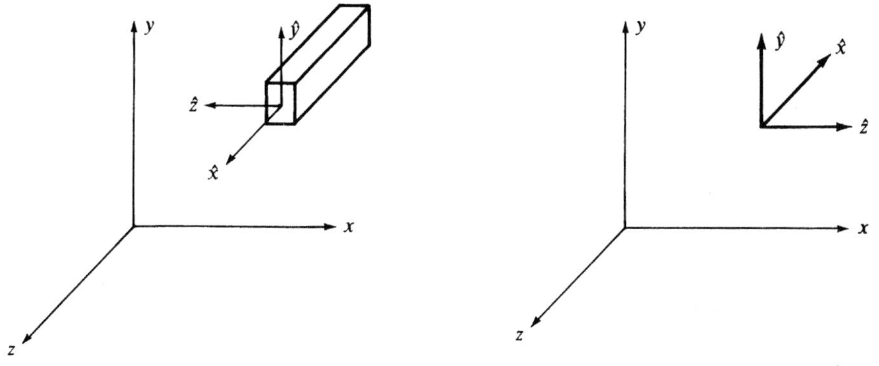

(a) $\hat{x}$ in same direction as z (b) $\hat{x}$ in opposite direction of z

Figure 5–26 Special cases of transformation matrices

the positive $\hat{x}$ axis in the same direction as the global z, $\underline{\lambda}$ becomes

$$\underline{\lambda} = \begin{bmatrix} 0 & 0 & 1 \\ 0 & 1 & 0 \\ -1 & 0 & 0 \end{bmatrix} \tag{5.5.15}$$

For the positive $\hat{x}$ axis opposite the global z [Figure 5–26(b)], $\underline{\lambda}$ becomes

$$\underline{\lambda} = \begin{bmatrix} 0 & 0 & -1 \\ 0 & 1 & 0 \\ 1 & 0 & 0 \end{bmatrix} \tag{5.5.16}$$

Example 5.7

Determine the direction cosines and the rotation matrix of the local $\hat{x}, \hat{y}, \hat{z}$ axes in reference to the global x, y, z axes for the beam element oriented in space with end nodal coordinates of 1 (0, 0, 0) and 2 (3, 4, 12), as shown in Figure 5–27.

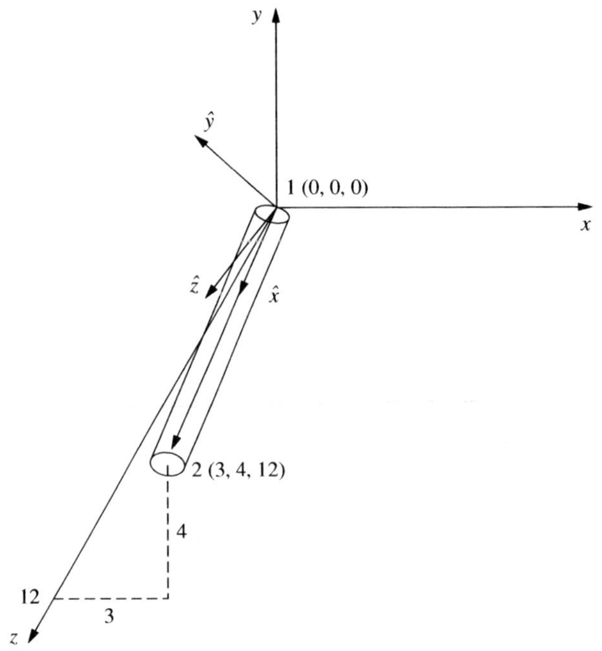

Figure 5–27 Beam element oriented in space

First we determine the length of the element as

$$L = \sqrt{3^2 + 4^2 + 12^2} = 13$$

Now using Eq. (5.5.8), we obtain the direction cosines of the $\hat{x}$ axis as follows:

$$l_x = \frac{x_2 - x_1}{L} = \frac{3 - 0}{13} = \frac{3}{13}$$
$$m_x = \frac{y_2 - y_1}{L} = \frac{4 - 0}{13} = \frac{4}{13} \qquad (5.5.17)$$
$$n_x = \frac{z_2 - z_1}{L} = \frac{12 - 0}{13} = \frac{12}{13}$$

By Eq. (5.5.10) or (5.5.14), we obtain the direction cosines of the $\hat{y}$ axis as follows:

$$D = (l^2 + m^2)^{1/2} = \left[\left(\frac{3}{13}\right)^2 + \left(\frac{4}{13}\right)^2\right]^{1/2} = \frac{5}{13} \qquad (5.5.18)$$

Define the direction cosines of the $\hat{y}$ axis as l_y, m_y, and n_y, where

$$l_y = -\frac{m}{D} = -\frac{4}{5}$$
$$m_y = \frac{l}{D} = \frac{3}{5} \qquad (5.5.19)$$
$$n_y = 0$$

For the $\hat{z}$ axis, define the direction cosines as l_z, m_z, n_z and again use Eq. (5.5.12) or (5.5.14) as follows:

$$l_z = -\frac{ln}{D} = \frac{\left(-\frac{3}{13}\right)\left(\frac{12}{13}\right)}{\frac{5}{13}} = -\frac{36}{65}$$
$$m_z = -\frac{mn}{D} = \frac{\left(-\frac{4}{13}\right)\left(\frac{12}{13}\right)}{\frac{5}{13}} = -\frac{48}{65} \qquad (5.5.20)$$
$$n_z = D = \frac{5}{13}$$

Now check that $l^2 + m^2 + n^2 = 1$.

$$\text{For } \hat{x} : \frac{3^2 + 4^2 + 12^2}{13^2} = 1$$
$$\text{For } \hat{y} : \frac{(-4)^2 + 3^2}{5^2} = 1 \qquad (5.5.21)$$
$$\text{For } \hat{z} : \left(-\frac{36}{65}\right)^2 + \left(-\frac{48}{65}\right)^2 + \left(\frac{25}{65}\right)^2 = 1$$

By Eq. (5.5.13), the rotation matrix is

$$\lambda_{3\times 3} = \begin{bmatrix} \frac{3}{13} & \frac{4}{13} & \frac{12}{13} \\ -\frac{4}{5} & \frac{3}{5} & 0 \\ -\frac{36}{65} & -\frac{48}{65} & \frac{5}{13} \end{bmatrix} \qquad (5.5.22)$$

Based on the resulting direction cosines from Eqs. (5.5.17), (5.5.19), and (5.5.20), the local axes are also shown in Figure 5–27. ■

Example 5.8

Determine the displacements and rotations at the free node (node 1) and the element local forces and moments for the space frame shown in Figure 5–28. Also verify equilibrium at node 1. Let $E = 30{,}000$ ksi, $G = 10{,}000$ ksi, $J = 50$ in.^{4}, $I_y = 100$ in.^{4}, $I_z = 100$ in.^{4}, A = 10 in.^{2}, and $L = 100$ in. for all three beam elements.

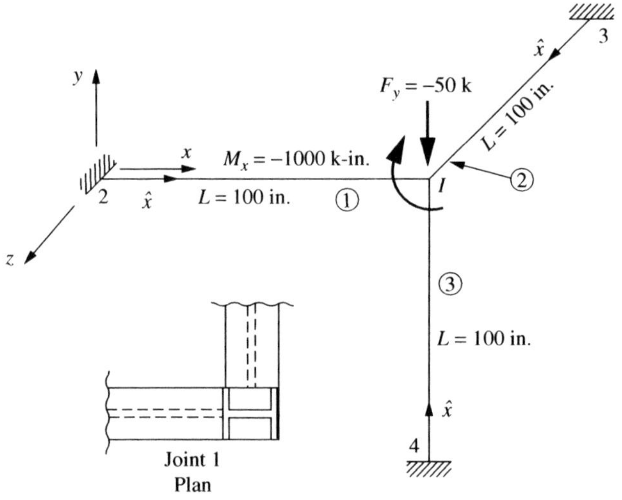

Figure 5–28 Space frame for analysis

Use Eq. (5.5.4) to obtain the global stiffness matrix for each element. This requires us to first use Eq. (5.5.3) to obtain each local stiffness matrix, Eq. (5.5.5) to obtain the transformation matrix for each element, and Eqs. (5.5.6) and (5.5.14) to obtain the direction cosine matrix for each element.

Element 1

We establish the local $\hat{x}$ axis to go from node 2 to node 1 as shown in Figure 5–28. Therefore, using Eq. (5.5.8), we obtain the direction cosines of the $\hat{x}$ axis as follows:

$$l = 1, \quad m = 0, \quad n = 0 \tag{5.5.23}$$

Also,

$$D = (l^2 + m^2)^{1/2} = 1.$$

Using Eqs. (5.5.10) and (5.5.14), we obtain the direction cosines of the $\hat{y}$ axis as follows:

$$l_y = -\frac{m}{D} = 0 \qquad m_y = \frac{l}{D} = 1 \qquad n_y = 0 \tag{5.5.24}$$

Using Eqs. (5.5.12) and (5.5.14), we obtain the direction cosines of the $\hat{z}$ axis as follows:

$$l_z = -\frac{ln}{D} = 0 \qquad m_z = -\frac{mn}{D} = 0 \qquad n_z = D = 1 \tag{5.5.25}$$

Using Eqs. (5.5.23) through (5.5.25) in Eq. (5.5.13), we have

$$\underline{\lambda} = \begin{bmatrix} 1 & 0 & 0 \\ 0 & 1 & 0 \\ 0 & 0 & 1 \end{bmatrix} \tag{5.5.26}$$

Using Eq. (5.5.3), we obtain the local stiffness matrix for element one as

$$\hat{k}^{(1)} = \begin{array}{c} \begin{matrix} \hat{d}_{2x} & \hat{d}_{2y} & \hat{d}_{2z} & \hat{\phi}_{2x} & \hat{\phi}_{2y} & \hat{\phi}_{2z} & \hat{d}_{1x} & \hat{d}_{1y} & \hat{d}_{1z} & \hat{\phi}_{1x} & \hat{\phi}_{1y} & \hat{\phi}_{1z} \end{matrix} \\ \begin{bmatrix} 3\cdot10^3 & 0 & 0 & 0 & 0 & 0 & -3\cdot10^3 & 0 & 0 & 0 & 0 & 0 \\ 0 & 36 & 0 & 0 & 0 & 1.8\cdot10^3 & 0 & -36 & 0 & 0 & 0 & 1.8\cdot10^3 \\ 0 & 0 & 36 & 0 & -1.8\cdot10^3 & 0 & 0 & 0 & -36 & 0 & -1.8\cdot10^3 & 0 \\ 0 & 0 & 0 & 5\cdot10^3 & 0 & 0 & 0 & 0 & 0 & -5\cdot10^3 & 0 & 0 \\ 0 & 0 & -1.8\cdot10^3 & 0 & 1.2\cdot10^5 & 0 & 0 & 0 & 1.8\cdot10^3 & 0 & 6\cdot10^4 & 0 \\ 0 & 1.8\cdot10^3 & 0 & 0 & 0 & 1.2\cdot10^5 & 0 & -1.8\cdot10^3 & 0 & 0 & 0 & 6\cdot10^4 \\ -3\cdot10^3 & 0 & 0 & 0 & 0 & 0 & 3\cdot10^3 & 0 & 0 & 0 & 0 & 0 \\ 0 & -36 & 0 & 0 & 0 & -1.8\cdot10^3 & 0 & 36 & 0 & 0 & 0 & -1.8\cdot10^3 \\ 0 & 0 & -36 & 0 & 1.8\cdot10^3 & 0 & 0 & 0 & 36 & 0 & 1.8\cdot10^3 & 0 \\ 0 & 0 & 0 & -5\cdot10^3 & 0 & 0 & 0 & 0 & 0 & 5\cdot10^3 & 0 & 0 \\ 0 & 0 & -1.8\cdot10^3 & 0 & 6\cdot10^4 & 0 & 0 & 0 & 1.8\cdot10^3 & 0 & 1.2\cdot10^5 & 0 \\ 0 & 1.8\cdot10^3 & 0 & 0 & 0 & 6\cdot10^4 & 0 & -1.8\cdot10^3 & 0 & 0 & 0 & 1.2\cdot10^5 \end{bmatrix} \end{array} \tag{5.5.27}$$

Using Eq. (5.5.26) in Eq. (5.5.5), we obtain the transformation matrix from local to global axis system as

$$\underline{T} = \begin{bmatrix} 1 & 0 & 0 & 0 & 0 & 0 & 0 & 0 & 0 & 0 & 0 & 0 \\ 0 & 1 & 0 & 0 & 0 & 0 & 0 & 0 & 0 & 0 & 0 & 0 \\ 0 & 0 & 1 & 0 & 0 & 0 & 0 & 0 & 0 & 0 & 0 & 0 \\ 0 & 0 & 0 & 1 & 0 & 0 & 0 & 0 & 0 & 0 & 0 & 0 \\ 0 & 0 & 0 & 0 & 1 & 0 & 0 & 0 & 0 & 0 & 0 & 0 \\ 0 & 0 & 0 & 0 & 0 & 1 & 0 & 0 & 0 & 0 & 0 & 0 \\ 0 & 0 & 0 & 0 & 0 & 0 & 1 & 0 & 0 & 0 & 0 & 0 \\ 0 & 0 & 0 & 0 & 0 & 0 & 0 & 1 & 0 & 0 & 0 & 0 \\ 0 & 0 & 0 & 0 & 0 & 0 & 0 & 0 & 1 & 0 & 0 & 0 \\ 0 & 0 & 0 & 0 & 0 & 0 & 0 & 0 & 0 & 1 & 0 & 0 \\ 0 & 0 & 0 & 0 & 0 & 0 & 0 & 0 & 0 & 0 & 1 & 0 \\ 0 & 0 & 0 & 0 & 0 & 0 & 0 & 0 & 0 & 0 & 0 & 1 \end{bmatrix} \tag{5.5.28}$$

Finally, using Eq. (5.5.4), we obtain the global stiffness matrix for element 1 as

$$\underline{k}^{(1)} = \underline{T}^T \hat{\underline{k}}^{(1)} \underline{T} = \begin{matrix} d_{2x} & d_{2y} & d_{2z} & \phi_{2x} & \phi_{2y} & \phi_{2z} & d_{1x} & d_{1y} & d_{1z} & \phi_{1x} & \phi_{1y} & \phi_{1z} \end{matrix}$$
$$\begin{bmatrix}
3\cdot10^3 & 0 & 0 & 0 & 0 & 0 & -3\cdot10^3 & 0 & 0 & 0 & 0 & 0 \\
0 & 36 & 0 & 0 & 0 & 1.8\cdot10^3 & 0 & -36 & 0 & 0 & 0 & 1.8\cdot10^3 \\
0 & 0 & 36 & 0 & -1.8\cdot10^3 & 0 & 0 & 0 & -36 & 0 & -1.8\cdot10^3 & 0 \\
0 & 0 & 0 & 5\cdot10^3 & 0 & 0 & 0 & 0 & 0 & -5\cdot10^3 & 0 & 0 \\
0 & 0 & -1.8\cdot10^3 & 0 & 1.2\cdot10^5 & 0 & 0 & 0 & 1.8\cdot10^3 & 0 & 6\cdot10^4 & 0 \\
0 & 1.8\cdot10^3 & 0 & 0 & 0 & 1.2\cdot10^5 & 0 & -1.8\cdot10^3 & 0 & 0 & 0 & 6\cdot10^4 \\
-3\cdot10^3 & 0 & 0 & 0 & 0 & 0 & 3\cdot10^3 & 0 & 0 & 0 & 0 & 0 \\
0 & -36 & 0 & 0 & 0 & -1\cdot8.10^3 & 0 & 36 & 0 & 0 & 0 & -1.8\cdot10^3 \\
0 & 0 & -36 & 0 & 1.8\cdot10^3 & 0 & 0 & 0 & 36 & 0 & 1.8\cdot10^3 & 0 \\
0 & 0 & 0 & -5\cdot10^3 & 0 & 0 & 0 & 0 & 0 & 5\cdot10^3 & 0 & 0 \\
0 & 0 & -1.8\cdot10^3 & 0 & 6\cdot10^4 & 0 & 0 & 0 & 1.8\cdot10^3 & 0 & 1.2\cdot10^5 & 0 \\
0 & 1.8\cdot10^3 & 0 & 0 & 0 & 6\cdot10^4 & 0 & -1.8\cdot10^3 & 0 & 0 & 0 & 1.2\cdot10^5
\end{bmatrix} \tag{5.5.29}$$

Element 2

We establish the local $\hat{x}$ axis from node 3 to node 1 as shown in Figure 5–28. We note that the local $\hat{x}$ axis coincides with the global z axis. Therefore, by Eq. (5.5.15), we obtain

$$\underline{\lambda} = \begin{bmatrix} 0 & 0 & 1 \\ 0 & 1 & 0 \\ -1 & 0 & 0 \end{bmatrix} \tag{5.5.30}$$

The local stiffness matrix is the same as the one in Eq. (5.5.27) as all properties are the same as for element one. However, we must remember that the degrees of freedom are for node 3 and then node 1.

Using Eq. (5.5.30) in Eq. (5.5.5), we obtain the transformation matrix as follows:

$$\underline{T} = \begin{bmatrix}
0 & 0 & 1 & 0 & 0 & 0 & 0 & 0 & 0 & 0 & 0 & 0 \\
0 & 1 & 0 & 0 & 0 & 0 & 0 & 0 & 0 & 0 & 0 & 0 \\
-1 & 0 & 0 & 0 & 0 & 0 & 0 & 0 & 0 & 0 & 0 & 0 \\
0 & 0 & 0 & 0 & 0 & 1 & 0 & 0 & 0 & 0 & 0 & 0 \\
0 & 0 & 0 & 0 & 1 & 0 & 0 & 0 & 0 & 0 & 0 & 0 \\
0 & 0 & 0 & -1 & 0 & 0 & 0 & 0 & 0 & 0 & 0 & 0 \\
0 & 0 & 0 & 0 & 0 & 0 & 0 & 0 & 1 & 0 & 0 & 0 \\
0 & 0 & 0 & 0 & 0 & 0 & 0 & 1 & 0 & 0 & 0 & 0 \\
0 & 0 & 0 & 0 & 0 & 0 & -1 & 0 & 0 & 0 & 0 & 0 \\
0 & 0 & 0 & 0 & 0 & 0 & 0 & 0 & 0 & 0 & 0 & 1 \\
0 & 0 & 0 & 0 & 0 & 0 & 0 & 0 & 0 & 0 & 1 & 0 \\
0 & 0 & 0 & 0 & 0 & 0 & 0 & 0 & 0 & -1 & 0 & 0
\end{bmatrix} \tag{5.5.31}$$

Finally, using Eq. (5.5.31) in Eq. (5.5.4), we obtain the global stiffness matrix for element two as

$$
\begin{matrix} d_{3x} & d_{3y} & d_{3z} & \phi_{3x} & \phi_{3y} & \phi_{3z} & d_{1x} & d_{1y} & d_{1z} & \phi_{1x} & \phi_{1y} & \phi_{1z} \end{matrix}
$$

$$
k^{(2)} = \begin{bmatrix}
36 & 0 & 0 & 0 & 1.8 \cdot 10^3 & 0 & -36 & 0 & 0 & 0 & 1.8 \cdot 10^3 & 0 \\
0 & 36 & 0 & -1.8 \cdot 10^3 & 0 & 0 & 0 & -36 & 0 & -1.8 \cdot 10^3 & 0 & 0 \\
0 & 0 & 3 \cdot 10^3 & 0 & 0 & 0 & 0 & 0 & -3 \cdot 10^3 & 0 & 0 & 0 \\
0 & -1.8 \cdot 10^3 & 0 & 1.2 \cdot 10^5 & 0 & 0 & 0 & 1.8 \cdot 10^3 & 0 & 6 \cdot 10^4 & 0 & 0 \\
1.8 \cdot 10^3 & 0 & 0 & 0 & 1.2 \cdot 10^5 & 0 & -1.8 \cdot 10^3 & 0 & 0 & 0 & 6 \cdot 10^4 & 0 \\
0 & 0 & 0 & 0 & 0 & 5 \cdot 10^3 & 0 & 0 & 0 & 0 & 0 & -5 \cdot 10^3 \\
-36 & 0 & 0 & 0 & -1.8 \cdot 10^3 & 0 & 36 & 0 & 0 & 0 & -1.8 \cdot 10^3 & 0 \\
0 & -36 & 0 & 1.8 \cdot 10^3 & 0 & 0 & 0 & 36 & 0 & 1.8 \cdot 10^3 & 0 & 0 \\
0 & 0 & -3 \cdot 10^3 & 0 & 0 & 0 & 0 & 0 & 3 \cdot 10^3 & 0 & 0 & 0 \\
0 & -1.8 \cdot 10^3 & 0 & 6 \cdot 10^4 & 0 & 0 & 0 & 1.8 \cdot 10^3 & 0 & 1.2 \cdot 10^5 & 0 & 0 \\
1.8 \cdot 10^3 & 0 & 0 & 0 & 6 \cdot 10^4 & 0 & -1.8 \cdot 10^3 & 0 & 0 & 0 & 1.2 \cdot 10^5 & 0 \\
0 & 0 & 0 & 0 & 0 & -5 \cdot 10^3 & 0 & 0 & 0 & 0 & 0 & 5 \cdot 10^3
\end{bmatrix} \tag{5.5.32}
$$

Element 3

We establish the local $\hat{x}$ axis from node 4 to node 1 for element 3 as shown in Figure 5–28. The direction cosines are now

$$
l = \frac{0-0}{100} = 0 \qquad m = \frac{0-(-100)}{100} = 1 \qquad n = \frac{0-0}{100} = 0 \tag{5.5.33}
$$

Also $D = 1$.

Using Eq. (5.5.14), we obtain the rest of the direction cosines as

$$
l_y = -\frac{m}{D} = -1 \qquad m_y = \frac{L}{D} = 0 \qquad n_y = 0 \tag{5.5.34}
$$

and

$$
l_z = -\frac{ln}{D} = 0 \qquad m_z = -\frac{mn}{D} = 0 \qquad n_z = D = 1 \tag{5.5.35}
$$

Using Eqs. (5.5.33) through (5.5.35), we obtain

$$
\underset{\sim}{\lambda} = \begin{bmatrix} 0 & 1 & 0 \\ -1 & 0 & 0 \\ 0 & 0 & 1 \end{bmatrix} \tag{5.5.36}
$$

The transformation matrix for element three is then obtained by using Eq. (5.5.5) as:

$$\underline{T} = \begin{bmatrix} 0 & 1 & 0 & 0 & 0 & 0 & 0 & 0 & 0 & 0 & 0 & 0 \\ -1 & 0 & 0 & 0 & 0 & 0 & 0 & 0 & 0 & 0 & 0 & 0 \\ 0 & 0 & 1 & 0 & 0 & 0 & 0 & 0 & 0 & 0 & 0 & 0 \\ 0 & 0 & 0 & 0 & 1 & 0 & 0 & 0 & 0 & 0 & 0 & 0 \\ 0 & 0 & 0 & -1 & 0 & 0 & 0 & 0 & 0 & 0 & 0 & 0 \\ 0 & 0 & 0 & 0 & 0 & 1 & 0 & 0 & 0 & 0 & 0 & 0 \\ 0 & 0 & 0 & 0 & 0 & 0 & 0 & 1 & 0 & 0 & 0 & 0 \\ 0 & 0 & 0 & 0 & 0 & 0 & -1 & 0 & 0 & 0 & 0 & 0 \\ 0 & 0 & 0 & 0 & 0 & 0 & 0 & 0 & 1 & 0 & 0 & 0 \\ 0 & 0 & 0 & 0 & 0 & 0 & 0 & 0 & 0 & 0 & 1 & 0 \\ 0 & 0 & 0 & 0 & 0 & 0 & 0 & 0 & 0 & -1 & 0 & 0 \\ 0 & 0 & 0 & 0 & 0 & 0 & 0 & 0 & 0 & 0 & 0 & 1 \end{bmatrix} \tag{5.5.37}$$

The element three properties are identical to the element one properties; therefore, the local stiffness matrix is identical to the one in Eq. (5.5.27). We must remember that the degrees of freedom are now in the order node 4 and then node 1.

Using Eq. (5.5.37) in Eq. (5.5.4), we obtain the global stiffness matrix for element three as

$$k^{(3)} = \begin{bmatrix} 36 & 0 & 0 & 0 & 0 & -1.8 \cdot 10^3 & -36 & 0 & 0 & 0 & 0 & -1.8 \cdot 10^3 \\ 0 & 3 \cdot 10^3 & 0 & 0 & 0 & 0 & 0 & -3 \cdot 10^3 & 0 & 0 & 0 & 0 \\ 0 & 0 & 36 & 1.8 \cdot 10^3 & 0 & 0 & 0 & 0 & -36 & 1.8 \cdot 10^3 & 0 & 0 \\ 0 & 0 & 1.8 \cdot 10^3 & 1.2 \cdot 10^5 & 0 & 0 & 0 & 0 & -1.8 \cdot 10^3 & 6 \cdot 10^4 & 0 & 0 \\ 0 & 0 & 0 & 0 & 5 \cdot 10^3 & 0 & 0 & 0 & 0 & 0 & -5 \cdot 10^3 & 0 \\ -1.8 \cdot 10^3 & 0 & 0 & 0 & 0 & 1.2 \cdot 10^5 & 1.8 \cdot 10^3 & 0 & 0 & 0 & 0 & 6 \cdot 10^4 \\ -36 & 0 & 0 & 0 & 0 & 1.8 \cdot 10^3 & 36 & 0 & 0 & 0 & 0 & 1.8 \cdot 10^3 \\ 0 & -3 \cdot 10^3 & 0 & 0 & 0 & 0 & 0 & 3 \cdot 10^3 & 0 & 0 & 0 & 0 \\ 0 & 0 & -36 & -1.8 \cdot 10^3 & 0 & 0 & 0 & 0 & 36 & -1.8 \cdot 10^3 & 0 & 0 \\ 0 & 0 & 1.8 \cdot 10^3 & 6 \cdot 10^4 & 0 & 0 & 0 & 0 & -1.8 \cdot 10^3 & 1.2 \cdot 10^5 & 0 & 0 \\ 0 & 0 & 0 & 0 & -5 \cdot 10^3 & 0 & 0 & 0 & 0 & 0 & 5 \cdot 10^3 & 0 \\ -1.8 \cdot 10^3 & 0 & 0 & 0 & 0 & 6 \cdot 10^4 & 1.8 \cdot 10^3 & 0 & 0 & 0 & 0 & 1.2 \cdot 10^5 \end{bmatrix}$$

(columns labelled d_{4x}, d_{4y}, d_{4z}, ϕ_{4x}, ϕ_{4y}, ϕ_{4z}, d_{1x}, d_{1y}, d_{1z}, ϕ_{1x}, ϕ_{1y}, ϕ_{1z})

(5.5.38)

Applying the boundary conditions that displacements in the x, y, and z directions are all zero at nodes two, three, and four, and rotations about the x, y, and z axes are all zero at nodes two, three, and four, we obtain the reduced global stiffness matrix. Also, the applied global force is directed in the negative y direction at node one and so expressed as $F_{1y} = -50$ kips, and the global moment about the x axis at node 1 is

$M_{1x} = -1000$ k-in. With these considerations, the final global equations are

$$\begin{Bmatrix} 0 \\ -50 \\ 0 \\ -1000 \\ 0 \\ 0 \end{Bmatrix} = \begin{bmatrix} 3.072\times10^3 & 0 & 0 & 0 & -1.8\times10^3 & 1.8\times10^3 \\ 0 & 3.072\times10^3 & 0 & 1.8\times10^3 & 0 & -1.8\times10^3 \\ 0 & 0 & 3.072\times10^3 & -1.8\times10^3 & 1.8\times10^3 & 0 \\ 0 & 1.8\times10^3 & -1.8\times10^3 & 2.45\times10^5 & 0 & 0 \\ -1.8\times10^3 & 0 & 1.8\times10^3 & 0 & 2.45\times10^5 & 0 \\ 1.8\times10^3 & -1.8\times10^3 & 0 & 0 & 0 & 2.45\times10^5 \end{bmatrix} \begin{Bmatrix} d_{1x} \\ d_{1y} \\ d_{1z} \\ \phi_{1x} \\ \phi_{1y} \\ \phi_{1z} \end{Bmatrix} \quad (5.5.39)$$

Finally, solving simultaneously for the displacements and rotations at node one, we obtain

$$\underline{d} = \begin{bmatrix} 7.098\times10^{-5}\text{ in.} \\ -0.014\text{ in.} \\ -2.352\times10^{-3}\text{ in.} \\ -3.996\times10^{-3}\text{ rad} \\ 1.78\times10^{-5}\text{ rad} \\ -1.033\times10^{-4}\text{ rad} \end{bmatrix} \quad (5.5.40)$$

We now determine the element local forces and moments using the equation $\underline{\hat{f}} = \underline{\hat{k}}\,\underline{T}\,\underline{d}$ for each element as previously done for plane frames and trusses. As we are dealing with space frame elements, these element local forces and moments are now the normal force, two shear forces, torsional moment, and two bending moments at each end of each element.

Element 1

Using Eq. (5.5.27) for the local stiffness matrix, Eq. (5.5.28) for the transformation matrix, $\underline{T}$, and Eq. (5.5.40) for the displacements, we obtain the local element forces and moments as

$$\underline{\hat{f}}^{(1)} = \begin{Bmatrix} -0.213\text{ Kip} \\ 0.318\text{ Kip} \\ 0.053\text{ Kip} \\ 19.98\text{ Kip}\cdot\text{in.} \\ -3.165\text{ Kip}\cdot\text{in.} \\ 18.991\text{ Kip}\cdot\text{in.} \\ 0.213\text{ Kip} \\ -0.318\text{ Kip} \\ -0.053\text{ Kip} \\ -19.98\text{ Kip}\cdot\text{in} \\ -2.097\text{ Kip}\cdot\text{in} \\ 12.79\text{ Kip}\cdot\text{in} \end{Bmatrix} \quad (5.5.41)$$

Element 2

Using Eq. (5.5.27) for the local stiffness matrix, Eq. (5.5.28) for the transformation matrix and Eq. (5.5.40) for the displacements, we obtain the local forces and moments as

$$\underline{\hat{f}}^{(2)} = \begin{Bmatrix} 7.056\ \text{Kip} \\ 7.697\ \text{Kip} \\ -0.029\ \text{Kip} \\ 0.517\ \text{Kip}\cdot\text{in} \\ 0.94\ \text{Kip}\cdot\text{in} \\ 264.957\ \text{Kip}\cdot\text{in} \\ -7.056\ \text{Kip} \\ -7.697\ \text{Kip} \\ 0.029\ \text{Kip} \\ -0.517\ \text{Kip}\cdot\text{in} \\ 2.008\ \text{Kip}\cdot\text{in} \\ 504.722\ \text{Kip}\cdot\text{in} \end{Bmatrix} \tag{5.5.42}$$

Element 3

Similarly, using Eqs. (5.5.27), (5.5.37), and (5.5.40), we obtain the local forces and moments as

$$\underline{\hat{f}}^{(3)} = \begin{Bmatrix} 41.985\ \text{Kip} \\ -0.183\ \text{Kip} \\ -7.108\ \text{Kip} \\ -0.089\ \text{Kip}\cdot\text{in} \\ 235.532\ \text{Kip}\cdot\text{in} \\ -6.073\ \text{Kip}\cdot\text{in} \\ -41.985\ \text{Kip} \\ 0.183\ \text{Kip} \\ 7.108\ \text{Kip} \\ 0.089\ \text{Kip}\cdot\text{in} \\ 475.297\ \text{Kip}\cdot\text{in} \\ -12.273\ \text{Kip}\cdot\text{in} \end{Bmatrix} \tag{5.5.43}$$

We can verify equilibrium of node 1 by considering the node one forces and moments from each element that transfer to the node. We use the results from Eqs. (5.5.41), (5.5.42), and (5.5.43) to establish the proper forces and moments transferred to node 1. (Note that based on Newton's third law, the opposite forces and moments from each element are sent to node 1.) For instance, we observe from summing forces in the global y direction (shown in the diagram that follows)

$$0.318\ \text{kip} + 7.697\ \text{kip} + 41.985\ \text{kip} - 50\ \text{kip} = 0 \tag{5.5.44}$$

In Eq. (5.5.44), 0.318 kip is from element one local $\hat{y}$ force that is coincident with the global y direction; 7.697 kip is from element two local $\hat{y}$ force that is coincident with the global y direction, while 41.985 kip from element three is from the local $\hat{x}$direction that is coincident with the global y direction. We observe these axes from Figure 5–28. Verification of the other equilibrium equations is left to your discretion.

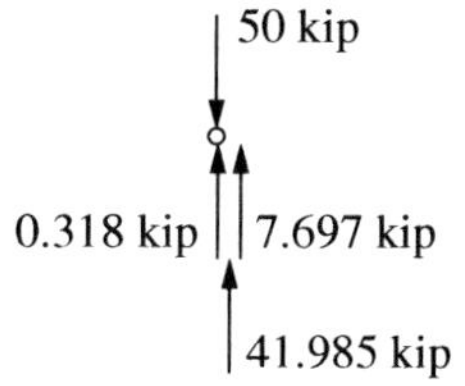

Global y force equilibrium ■

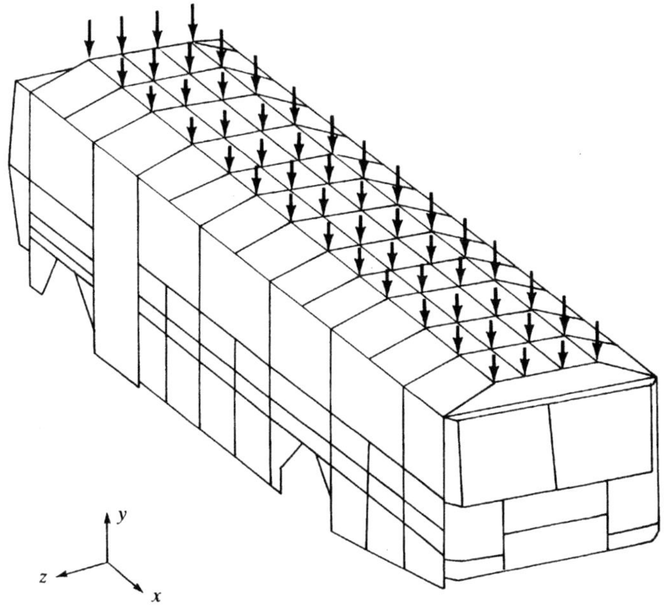

Figure 5–29 Finite element model of bus frame subjected to roof load [6]

An example using the frame element in three-dimensional space is shown in Figure 5–29. Figure 5–29 shows a bus frame subjected to a static roof-crush analysis. In this model, 599 frame elements and 357 nodes were used. A total downward load of 100 kN was uniformly spread over the 56 nodes of the roof portion of the frame. Figure 5–30 shows the rear of the frame and the displaced view of the rear frame. Other frame models with additional loads simulating rollover and front-end collisions were studied in Reference [6].

5.6 Concept of Substructure Analysis

The problem of exceeding memory capacity on todays personal computers has decreased significantly for most applications. However, for those structures that are too large to be analyzed as a single system or treated as a whole; that is, the final

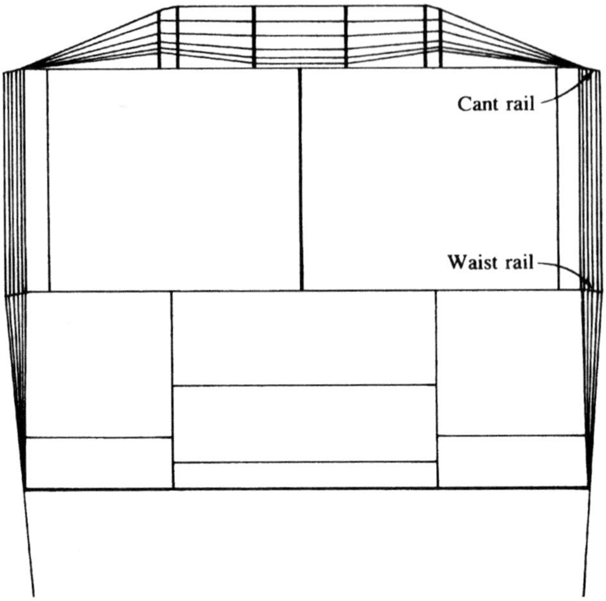

Figure 5–30 Displaced view of the frame of Figure 5–29 made of square section members

stiffness matrix and equations for solution exceed the memory capacity of the computer, the concept of **substructure** analysis can be used. The procedure to overcome this problem is to separate the whole structure into smaller units called **substructures**. For example, the space frame of an airplane, as shown in Figure 5–31(a), may require thousands of nodes and elements to model and describe completely the response of the whole structure. If we separate the aircraft into substructures, such as parts of the fuselage or body, wing sections, and so on, as shown in Figure 5–31(b), then we can solve the problem more readily and on computers with limited memory.

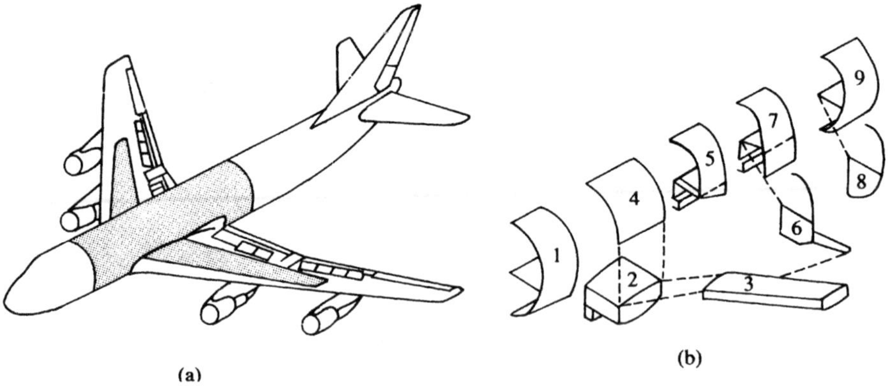

Figure 5–31 Airplane frame showing substructuring. (a) Boeing 747 aircraft (shaded area indicates portion of the airframe analyzed by finite element method). (b) Substructures for finite element analysis of shaded region

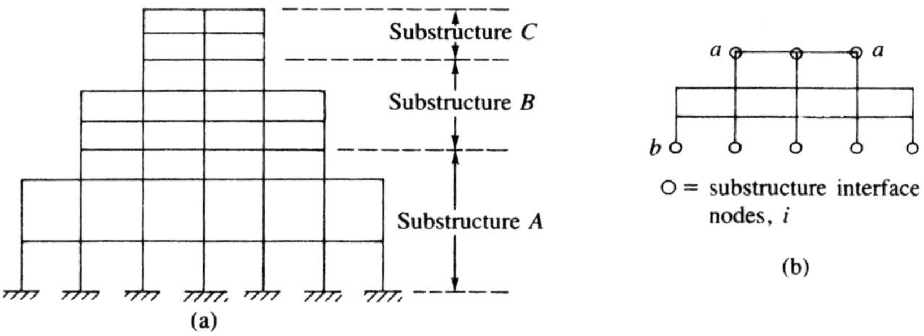

Figure 5–32 (a) Rigid frame for substructure analysis and (b) substructure B

The analysis of the airplane frame is performed by treating each substructure separately while ensuring force and displacement compatibility at the intersections where partitioning occurs.

To describe the procedure of substructuring, consider the rigid frame shown in Figure 5–32 (even though this frame could be analyzed as a whole). First we define individual separate substructures. Normally, we make these substructures of similar size, and to reduce computations, we make as few cuts as possible. We then separate the frame into three parts, A, B, and C.

We now analyze a typical substructure B shown in Figure 5–32(b). This substructure includes the beams at the top (a-a), but the beams at the bottom (b-b) are included in substructure A, although the beams at top could be included in substructure C and the beams at the bottom could be included in substructure B.

The force/displacement equations for substructure B are partitioned with the interface displacements and forces separated from the interior ones as follows:

$$\begin{Bmatrix} \underline{F}_i^B \\ \hline \underline{F}_e^B \end{Bmatrix} = \left[\begin{array}{c|c} \underline{K}_{ii}^B & \underline{K}_{ie}^B \\ \hline \underline{K}_{ei}^B & \underline{K}_{ee}^B \end{array}\right] \begin{Bmatrix} \underline{d}_i^B \\ \hline \underline{d}_e^B \end{Bmatrix} \tag{5.6.1}$$

where the superscript B denotes the substructure B, subscript i denotes the interface nodal forces and displacements, and subscript e denotes the interior nodal forces and displacements to be eliminated by static condensation. Using static condensation, Eq. (5.6.1) becomes

$$\underline{F}_i^B = \underline{K}_{ii}^B \underline{d}_i^B + \underline{K}_{ie}^B \underline{d}_e^B \tag{5.6.2}$$

$$\underline{F}_e^B = \underline{K}_{ei}^B \underline{d}_i^B + \underline{K}_{ee}^B \underline{d}_e^B \tag{5.6.3}$$

We eliminate the interior displacements $\underline{d}_e$ by solving Eq. (5.6.3) for $\underline{d}_e^B$, as follows:

$$\underline{d}_e^B = [\underline{K}_{ee}^B]^{-1}[\underline{F}_e^B - \underline{K}_{ei}^B \underline{d}_i^B] \tag{5.6.4}$$

Then we substitute Eq. (5.6.4) for $\underline{d}_e^B$ into Eq. (5.6.2) to obtain

$$\underline{F}_i^B - \underline{K}_{ie}^B[\underline{K}_{ee}^B]^{-1}\underline{F}_e^B = (\underline{K}_{ii}^B - \underline{K}_{ie}^B[\underline{K}_{ee}^B]^{-1}\underline{K}_{ei}^B)\underline{d}_i^B \tag{5.6.5}$$

We define

$$\underline{\bar{F}}_i^B = \underline{K}_{ie}^B[\underline{K}_{ee}^B]^{-1}\underline{F}_e^B \qquad \text{and} \qquad \underline{\bar{K}}_{ii}^B = \underline{K}_{ii}^B - \underline{K}_{ie}^B[\underline{K}_{ee}^B]^{-1}\underline{K}_{ei}^B \tag{5.6.6}$$

Substituting Eq. (5.6.6) into (5.6.5), we obtain

$$\underline{F}_i^B - \underline{\bar{F}}_i^B = \underline{\bar{K}}_{ii}^B \underline{d}_i^B \tag{5.6.7}$$

Similarly, we can write force/displacement equations for substructures A and C. These equations can be partitioned in a manner similar to Eq. (5.6.1) to obtain

$$\begin{Bmatrix} \underline{F}_i^A \\ --- \\ \underline{F}_e^A \end{Bmatrix} = \left[\begin{array}{c|c} \underline{K}_{ii}^A & \underline{K}_{ie}^A \\ \hline \underline{K}_{ei}^A & \underline{K}_{ee}^A \end{array}\right] \begin{Bmatrix} \underline{d}_i^A \\ --- \\ \underline{d}_e^A \end{Bmatrix} \tag{5.6.8}$$

Eliminating $\underline{d}_e^A$, we obtain

$$\underline{F}_i^A - \underline{\bar{F}}_i^A = \underline{\bar{K}}_{ii}^A \underline{d}_i^A \tag{5.6.9}$$

Similarly, for substructure C, we have

$$\underline{F}_i^C - \underline{\bar{F}}_i^C = \underline{\bar{K}}_{ii}^C \underline{d}_i^C \tag{5.6.10}$$

The whole frame is now considered to be made of superelements A, B, and C connected at interface nodal points (each superelement being made up of a collection of individual smaller elements). Using compatibility, we have

$$\underline{d}_{i\,\text{top}}^A = \underline{d}_{i\,\text{bottom}}^B \qquad \text{and} \qquad \underline{d}_{i\,\text{top}}^B = \underline{d}_{i\,\text{bottom}}^C \tag{5.6.11}$$

That is, the interface displacements at the common locations where cuts were made must be the same.

The response of the whole structure can now be obtained by direct superposition of Eqs. (5.6.7), (5.6.9), and (5.6.10), where now the final equations are expressed in terms of the interface displacements at the eight interface nodes only [Figure 5–32(b)] as

$$\underline{F}_i - \underline{\bar{F}}_i = \underline{\bar{K}}_{ii} \underline{d}_i \tag{5.6.12}$$

The solution of Eq. (5.6.12) gives the displacements at the interface nodes. To obtain the displacements within each substructure, we use the force-displacement Eqs. (5.6.4) for $\underline{d}_e^B$ with similar equations for substructures A and C. Example 5.9 illustrates the concept of substructure analysis. In order to solve by hand, a relatively simple structure is used.

Example 5.9

Solve for the displacement and rotation at node 3 for the beam in Figure 5–33 by using substructuring. Let $E = 29 \times 10^3$ ksi and $I = 1000$ in^4.

To illustrate the substructuring concept, we divide the beam into two substructures, labeled 1 and 2 in Figure 5–34. The 10-kip force has been assigned to node 3 of substructure 2, although it could have been assigned to either substructure or a fraction of it assigned to each substructure.

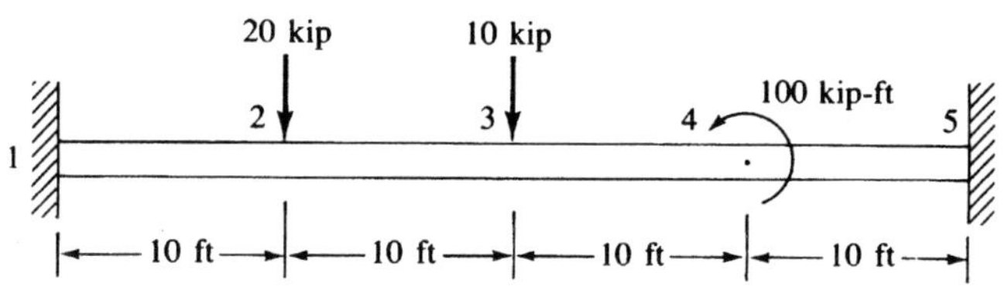

Figure 5–33 Beam analyzed by substructuring

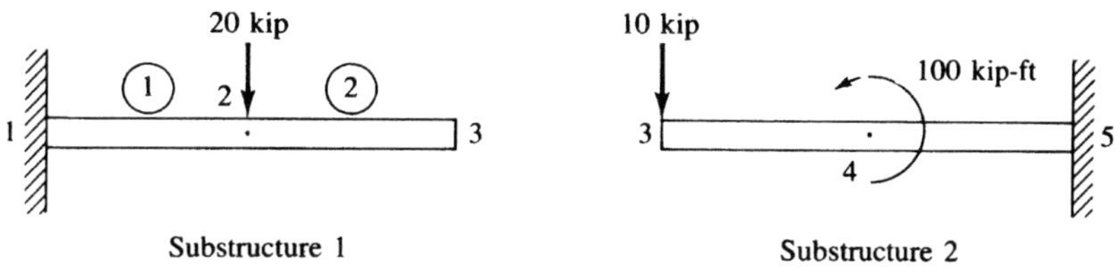

Figure 5–34 Beam of Figure 5–33 separated into substructures

The stiffness matrix for each beam element is given by Eq. (4.1.14) as

$$
\begin{array}{cc}
1 & 2 \\
2 & 3 \\
3 & 4 \\
4 & 5
\end{array}
$$

$$
\underline{k}^{(1)} = \underline{k}^{(2)} = \underline{k}^{(3)} = \underline{k}^{(4)} = \frac{29 \times 10^6}{(120)^3} \begin{bmatrix} 12 & 6(120) & -12 & 6(120) \\ 6(120) & 4(120)^2 & -6(120) & 2(120)^2 \\ -12 & -6(120) & 12 & -6(120) \\ 6(120) & 2(120)^2 & -6(120) & 4(120)^2 \end{bmatrix} \tag{5.6.13}
$$

$$
= 16.78 \begin{bmatrix} 12 & 720 & -12 & 720 \\ 720 & 57{,}600 & -720 & 28{,}800 \\ -12 & -720 & 12 & -720 \\ 720 & 28{,}800 & -720 & 57{,}600 \end{bmatrix} \tag{5.6.14}
$$

For substructure 1, we add the stiffness matrices of elements 1 and 2 together. The equations are

$$
16.78 \left[\begin{array}{cc|cc} 12+12 & -720+720 & -12 & 720 \\ -720+720 & 57{,}600+57{,}600 & -720 & 28{,}800 \\ \hline -12 & -720 & 12 & -720 \\ 720 & 28{,}800 & -720 & 57{,}600 \end{array} \right] \left\{ \begin{array}{c} d_{2y} \\ \phi_2 \\ \hline d_{3y} \\ \phi_3 \end{array} \right\} = \left\{ \begin{array}{c} -20 \\ 0 \\ \hline 0 \\ 0 \end{array} \right\} \tag{5.6.15}
$$

where the boundary conditions $d_{1y} = \phi_1 = 0$ were used to reduce the equations.

Rewriting Eq. (5.6.15) with the interface displacements first allows us to use Eq. (5.6.6) to condense out, or eliminate, the interior degrees of freedom, d_{2y} and ϕ_2. These reordered equations are

$$\begin{aligned}
&16.78(12d_{3y} - 720\phi_3 - 12d_{2y} - 720\phi_2) &&= 0\\
&16.78(-720d_{3y} + 57{,}600\phi_3 + 720d_{2y} + 28{,}800\phi_2) &&= 0\\
&16.78(-12d_{3y} + 720\phi_3 + 24d_{2y} + \phi_2) &&= -20\\
&16.78(-720d_{2y} + 28{,}800\phi_3 + 0d_{2y} + 115{,}200\phi_2) &&= 0
\end{aligned} \tag{5.6.16}$$

Using Eq. (5.6.6), we obtain equations for the interface degrees of freedom as

$$16.78\left\{\begin{bmatrix} 12 & -720 \\ -720 & 57{,}600 \end{bmatrix} - \begin{bmatrix} -12 & -720 \\ 720 & 28{,}800 \end{bmatrix}\begin{bmatrix} 24 & 0 \\ 0 & 115{,}200 \end{bmatrix}^{-1}\begin{bmatrix} -12 & 720 \\ -720 & 28{,}800 \end{bmatrix}\right\}\begin{Bmatrix} d_{3y} \\ \phi_3 \end{Bmatrix}$$
$$= \begin{Bmatrix} 0 \\ 0 \end{Bmatrix} - \begin{bmatrix} -12 & -720 \\ 720 & 28{,}800 \end{bmatrix}\begin{bmatrix} 24 & 0 \\ 0 & 115{,}200 \end{bmatrix}^{-1}\begin{Bmatrix} -20 \\ 0 \end{Bmatrix} \tag{5.6.17}$$

Simplifying Eq. (5.6.17), we obtain

$$\begin{bmatrix} 25.17 & -3020 \\ -3020 & 483{,}264 \end{bmatrix}\begin{Bmatrix} d_{3y} \\ \phi_3 \end{Bmatrix} = \begin{Bmatrix} -10 \\ 600 \end{Bmatrix} \tag{5.6.18}$$

For substructure 2, we add the stiffness matrices of elements 3 and 4 together. The equations are

$$16.78\begin{bmatrix} 12 & 720 & -12 & 720 \\ 720 & 57{,}600 & -720 & 28{,}800 \\ -12 & -720 & 12+12 & -720+720 \\ 720 & 28{,}800 & -720+720 & 57{,}600+57{,}600 \end{bmatrix}\begin{Bmatrix} d_{3y} \\ \phi_3 \\ d_{4y} \\ \phi_4 \end{Bmatrix} = \begin{Bmatrix} -10 \\ 0 \\ 0 \\ 1200 \end{Bmatrix} \tag{5.6.19}$$

where boundary conditions $d_{5y} = \phi_5 = 0$ were used to reduce the equations.

Using static condensation, Eq. (5.6.6), we obtain equations with only the interface displacements d_{3y} and ϕ_3. These equations are

$$16.78\left\{\begin{bmatrix} 12 & 720 \\ 720 & 57{,}600 \end{bmatrix} - \begin{bmatrix} -12 & 720 \\ -720 & 28{,}800 \end{bmatrix}\begin{bmatrix} 24 & 0 \\ 0 & 115{,}200 \end{bmatrix}^{-1}\begin{bmatrix} -12 & -720 \\ 720 & 28{,}800 \end{bmatrix}\right\}\begin{Bmatrix} d_{3y} \\ \phi_3 \end{Bmatrix}$$
$$= \begin{Bmatrix} -10 \\ 0 \end{Bmatrix} - \begin{bmatrix} -12 & 720 \\ -720 & 28{,}800 \end{bmatrix}\begin{bmatrix} 24 & 0 \\ 0 & 115{,}200 \end{bmatrix}^{-1}\begin{Bmatrix} 0 \\ 1200 \end{Bmatrix} \tag{5.6.20}$$

Simplifying Eq. (5.6.20), we obtain

$$\begin{bmatrix} 25.17 & 3020 \\ 3020 & 483{,}264 \end{bmatrix}\begin{Bmatrix} d_{3y} \\ \phi_3 \end{Bmatrix} = \begin{Bmatrix} -17.5 \\ -300 \end{Bmatrix} \tag{5.6.21}$$

Adding Eqs. (5.6.18) and (5.6.21), we obtain the final nodal equilibrium equations at the interface degrees of freedom as

$$\begin{bmatrix} 50.34 & 0 \\ 0 & 966{,}528 \end{bmatrix} \begin{Bmatrix} d_{3y} \\ \phi_3 \end{Bmatrix} = \begin{Bmatrix} -27.5 \\ 300 \end{Bmatrix} \tag{5.6.22}$$

Solving Eq. (5.6.22) for the displacement and rotation at node 3, we obtain

$$\begin{aligned} d_{3y} &= -0.5463 \text{ in.} \\ \phi_3 &= 0.0003104 \text{ rad} \end{aligned} \tag{5.6.23}$$

We could now return to Eq. (5.6.15) or Eq. (5.6.16) to obtain d_{2y} and ϕ_2 and to Eq. (5.6.19) to obtain d_{4y} and ϕ_4. ■

We emphasize that this example is used as a simple illustration of substructuring and is not typical of the size of problems where substructuring is normally performed. Generally, substructuring is used when the number of degrees of freedom is very large, as might occur, for instance, for very large structures such as the airframe in Figure 5–31.

References

[1] Kassimali, A., *Structural Analysis,* 2nd ed., Brooks/Cole Publishers, Pacific Grove, CA, 1999.

[2] Budynas, R. G., *Advanced Strength and Applied Stress Analysis,* 2nd ed., McGraw-Hill, New York, 1999.

[3] Allen, H. G., and Bulson, P. S., *Background to Buckling,* McGraw-Hill, London, 1980.

[4] Roark, R. J., and Young, W. C., *Formulas for Stress and Strain,* 6th ed., McGraw-Hill, New York, 1989.

[5] Gere, J. M., *Mechanics of Materials,* 5th ed., Brooks/Cole Publishers, Pacific Grove, CA, 2001.

[6] Parakh, Z. K., *Finite Element Analysis of Bus Frames under Simulated Crash Loadings,* M.S. Thesis, Rose-Hulman Institute of Technology, Terre Haute, Indiana, May 1989.

[7] Martin, H. C., *Introduction to Matrix Methods of Structural Analysis,* McGraw-Hill, New York, 1966.

[8] Juvinall, R. C., and Marshek, K. M., *Fundamentals of Machine Component Design,* 4th ed., p. 198, Wiley, 2005.

Problems

Solve all problems using the finite element stiffness method.

5.1 For the rigid frame shown in Figure P5–1, determine (1) the displacement components and the rotation at node 2, (2) the support reactions, and (3) the forces in each element. Then check equilibrium at node 2. Let $E = 30 \times 10^6$ psi, $A = 10$ in^2, and $I = 500$ in^4 for both elements.

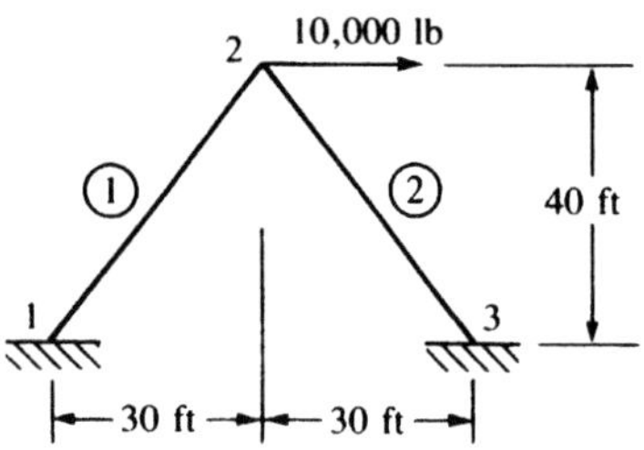

Figure P5–1

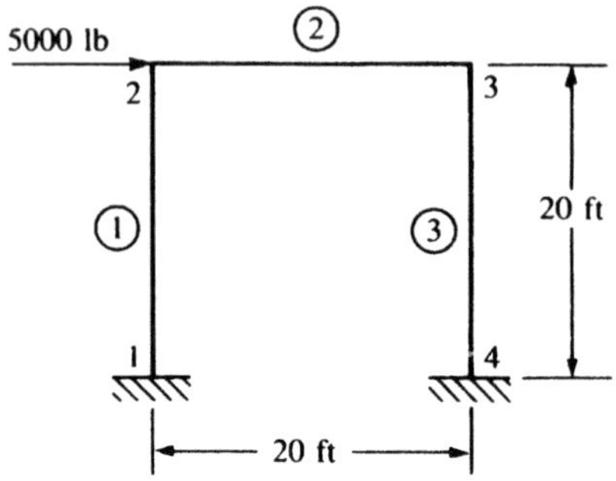

Figure P5–2

5.2 For the rigid frame shown in Figure P5–2, determine (1) the nodal displacement components and rotations, (2) the support reactions, and (3) the forces in each element. Let $E = 30 \times 10^6$ psi, $A = 10$ in^2, and $I = 200$ in^4 for all elements.

5.3 For the rigid stairway frame shown in Figure P5–3, determine (1) the displacements at node 2, (2) the support reactions, and (3) the local nodal forces acting on each element. Draw the bending moment diagram for the whole frame. Remember that the angle between elements 1 and 2 is preserved as deformation takes place; similarly for the angle between elements 2 and 3. Furthermore, owing to symmetry, $d_{2x} = -d_{3x}$, $d_{2y} = d_{3y}$, and $\phi_2 = -\phi_3$. What size A36 steel channel section would be needed to keep the allowable bending stress less than two-thirds of the yield stress? (For A36 steel, the yield stress is 36,000 psi.)

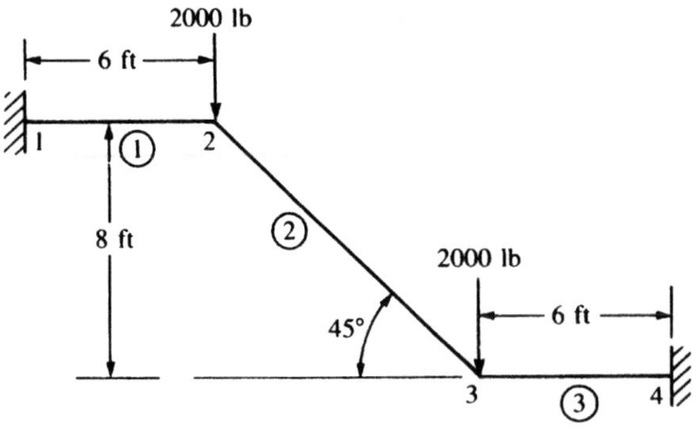

Figure P5–3

5.4 For the rigid frame shown in Figure P5–4, determine (1) the nodal displacements and rotation at node 4, (2) the reactions, and (3) the forces in each element. Then check equilibrium at node 4. Finally, draw the shear force and bending moment diagrams for each element. Let $E = 30 \times 10^3$ ksi, $A = 8$ in^2, and $I = 800$ in^4 for all elements.

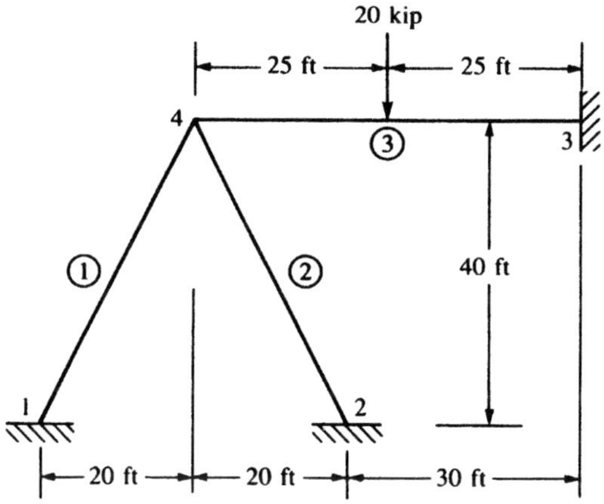

Figure P5–4

5.5–5.15 For the rigid frames shown in Figures P5–5—P5–15, determine the displacements and rotations of the nodes, the element forces, and the reactions. The values of E, A, and I to be used are listed next to each figure.

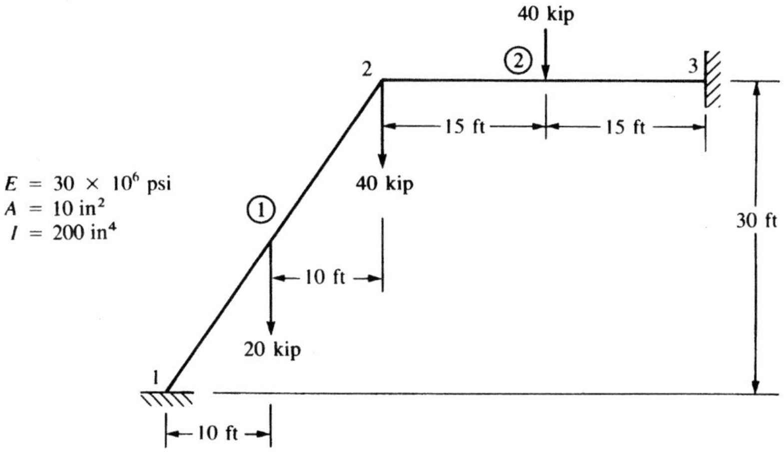

Figure P5–5

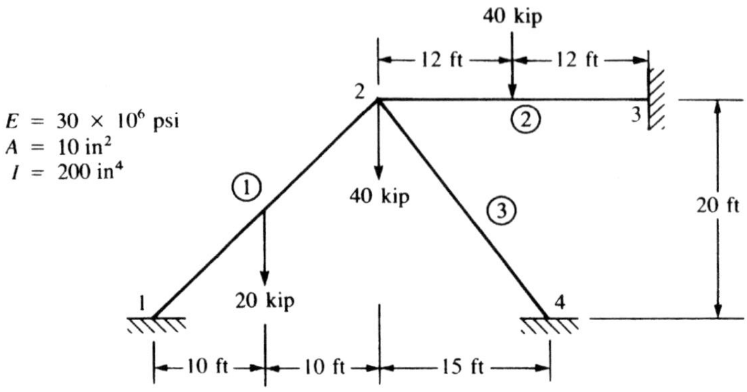

Figure P5–6

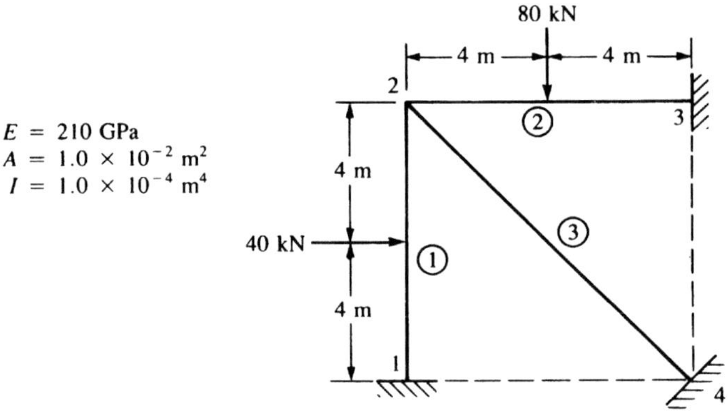

Figure P5–7

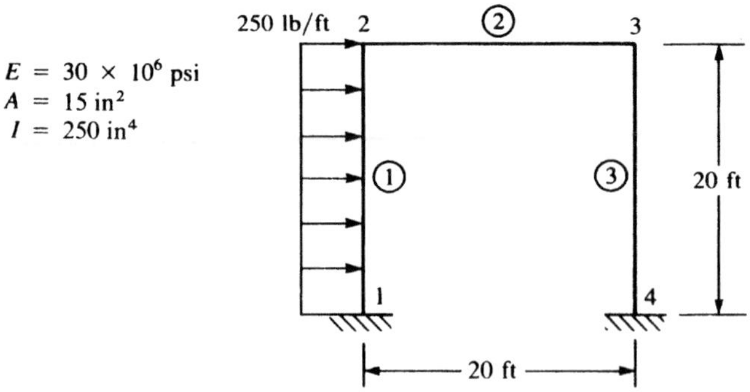

Figure P5–8

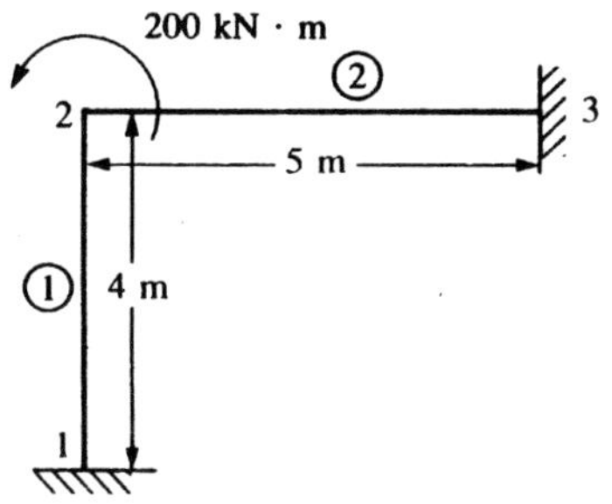

$E = 210$ GPa
$A = 2 \times 10^{-2}$ m^2
$I = 2 \times 10^{-4}$ m^4

Figure P5–9

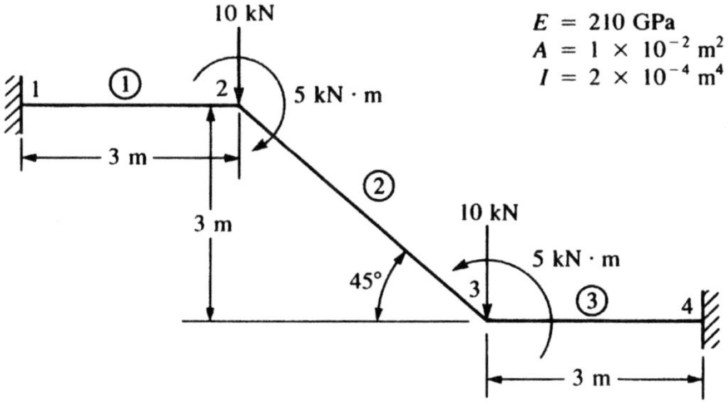

$E = 210$ GPa
$A = 1 \times 10^{-2}$ m^2
$I = 2 \times 10^{-4}$ m^4

Figure P5–10

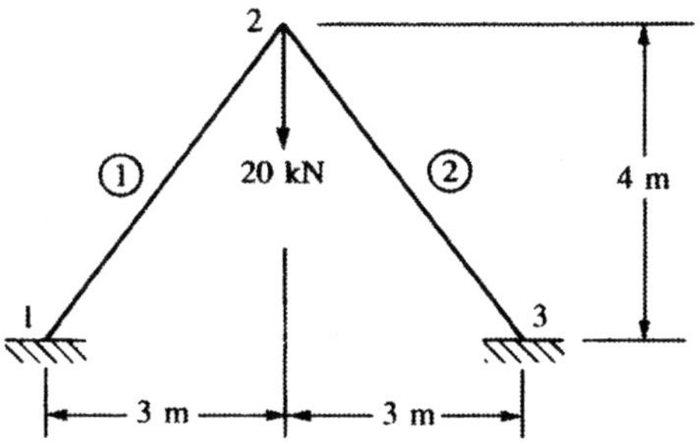

$E = 70$ GPa
$A = 3 \times 10^{-2}$ m^2
$I = 3 \times 10^{-4}$ m^4

Figure P5–11

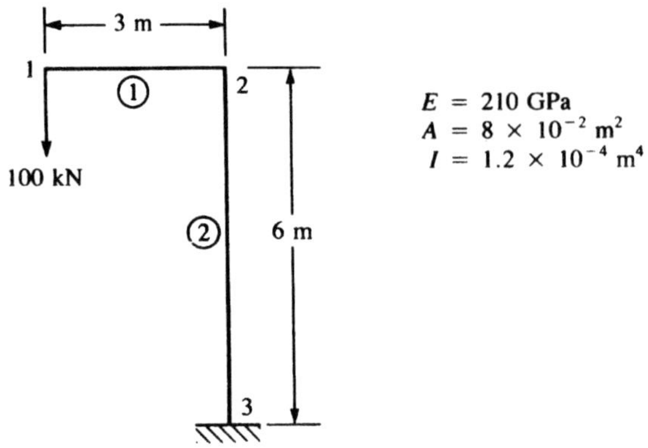

Figure P5–12

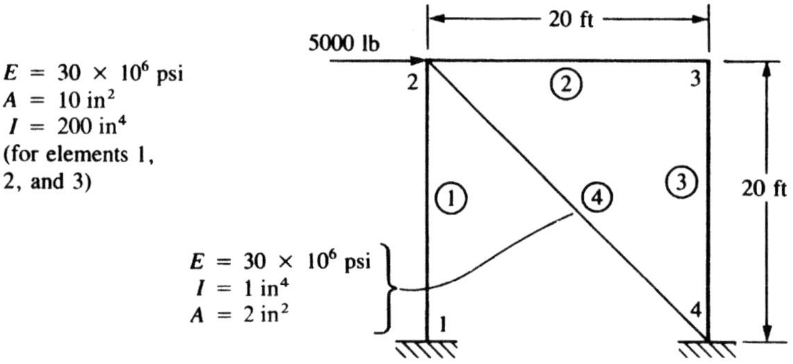

Figure P5–13

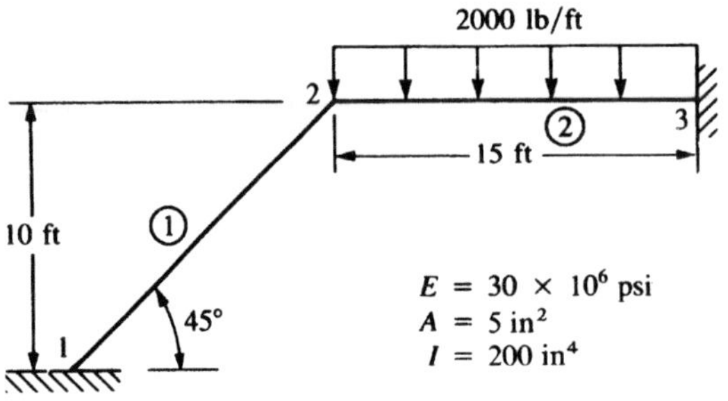

Figure P5–14

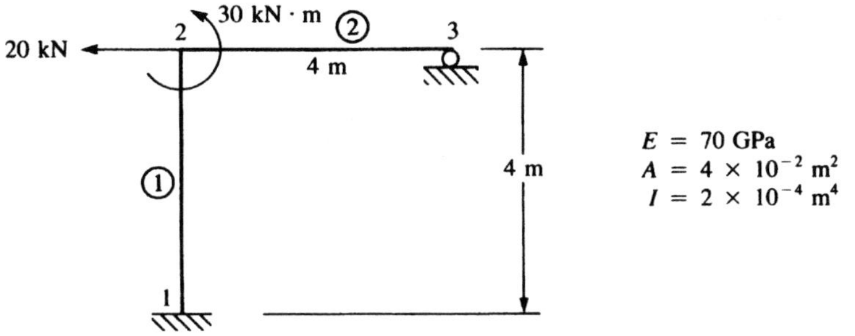

Figure P5–15

5.16–5.18 Solve the structures in Figures P5–16—P5–18 by using substructuring.

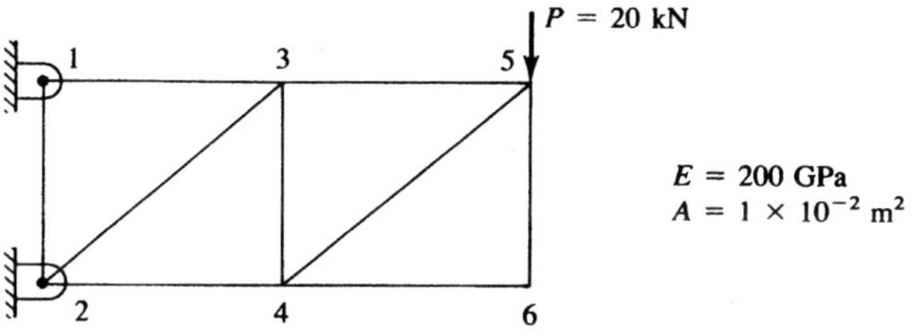

Figure P5–16 (Substructure the truss at nodes 3 and 4)

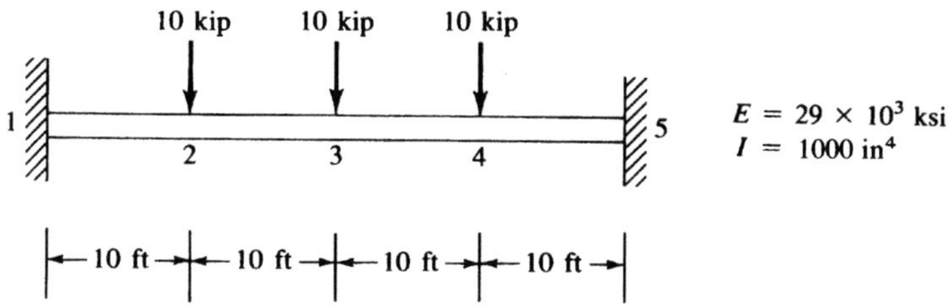

Figure P5–17 (Substructure the beam at node 3)

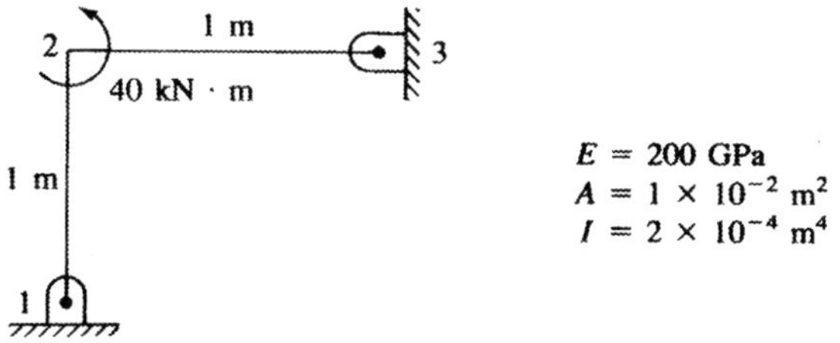

Figure P5–18 (Substructure the beam at node 2)

Solve Problems 5.19–5.39 by using a computer program.

5.19 For the rigid frame shown in Figure P5–19, determine (1) the nodal displacement components and (2) the support reactions. (3) Draw the shear force and bending moment diagrams. For all elements, let $E = 30 \times 10^6$ psi, $I = 200$ in^4, and $A = 10$ in^2.

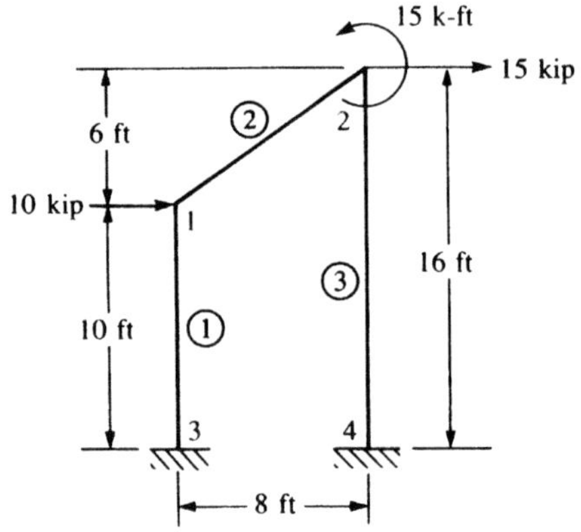

Figure P5–19

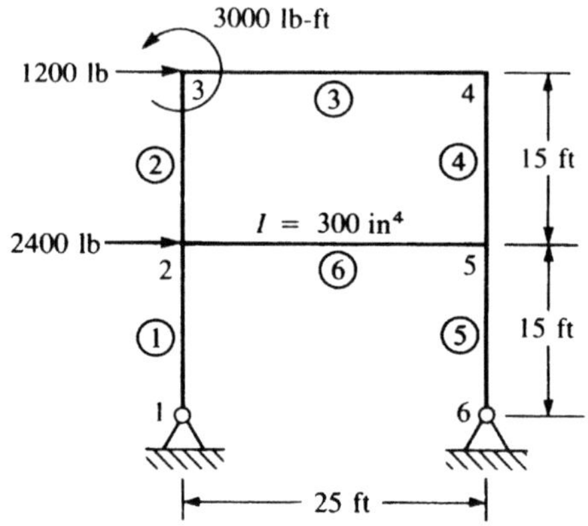

Figure P5–20

5.20 For the rigid frame shown in Figure P5–20, determine (1) the nodal displacement components and (2) the support reactions. (3) Draw the shear force and bending moment diagrams. Let $E = 30 \times 10^6$ psi, $I = 200$ in^4, and $A = 10$ in^2 for all elements, except as noted in the figure.

5.21 For the slant-legged rigid frame shown in Figure P5–21, size the structure for minimum weight based on a maximum bending stress of 20 ksi in the horizontal beam elements and a maximum compressive stress (due to bending and direct axial load) of 15 ksi in the slant-legged elements. Use the same element size for the two slant-legged elements and the same element size for the two 10-foot sections of the horizontal element. Assume A36 steel is used.

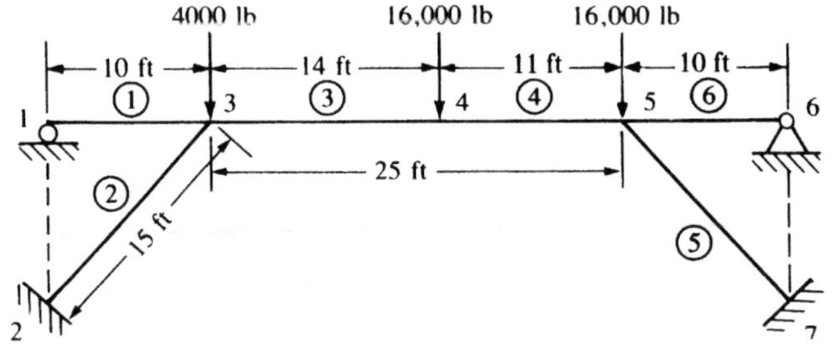

Figure P5–21

5.22 For the rigid building frame shown in Figure P5–22, determine the forces in each element and calculate the bending stresses. Assume all the vertical elements have $A = 10$ in^2 and $I = 100$ in^4 and all horizontal elements have $A = 15$ in^2 and $I = 150$ in^4. Let $E = 29 \times 10^6$ psi for all elements. Let $c = 5$ in. for the vertical elements and $c = 6$ in. for the horizontal elements, where c denotes the distance from the neutral axis to the top or bottom of the beam cross section, as used in the bending stress formula $\sigma = (Mc/I)$.

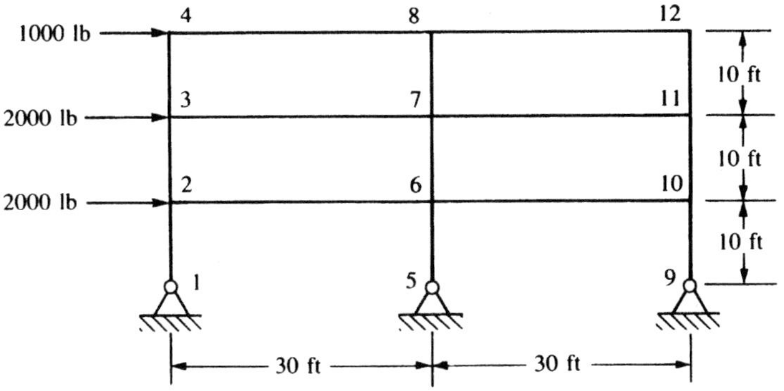

Figure P5–22

5.23–5.38 For the rigid frames or beams shown in Figures P5–23—P5–38, determine the displacements and rotations at the nodes, the element forces, and the reactions.

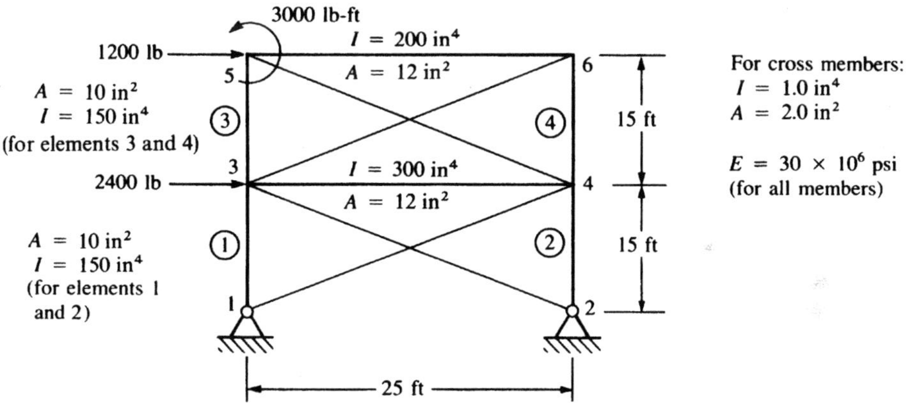

Figure P5–23

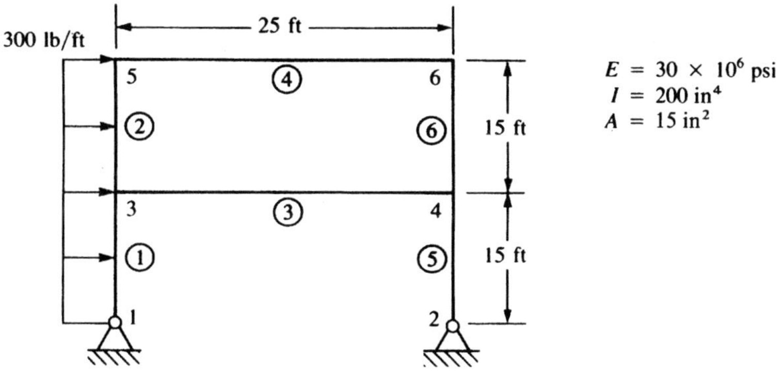

Figure P5–24

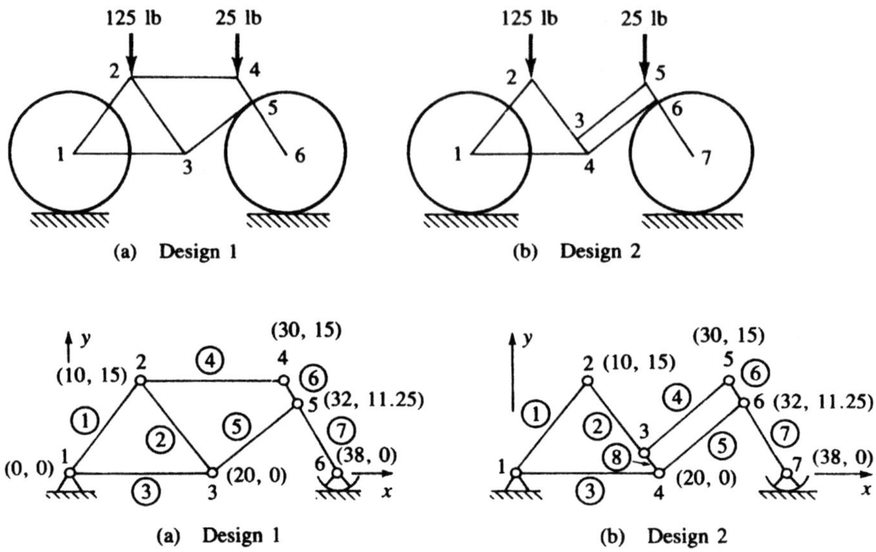

Case 1
$E = 30 \times 10^6$ psi
Case 2
$E = 10 \times 10^6$ psi

$A_1 = 0.1$ in^2
$A_2 = A_3 = A_4 = A_5 = 0.15$ in^2
$A_6 = A_7 = A_8 = 0.3$ in^2
$I_1 = 0.01$ in^4
$I_2 = I_3 = I_4 = I_5 = 0.02$ in^4
$I_6 = I_7 = I_8 = 0.1$ in^4

Figure P5–25 Two bicycle frame models (coordinates shown in inches)

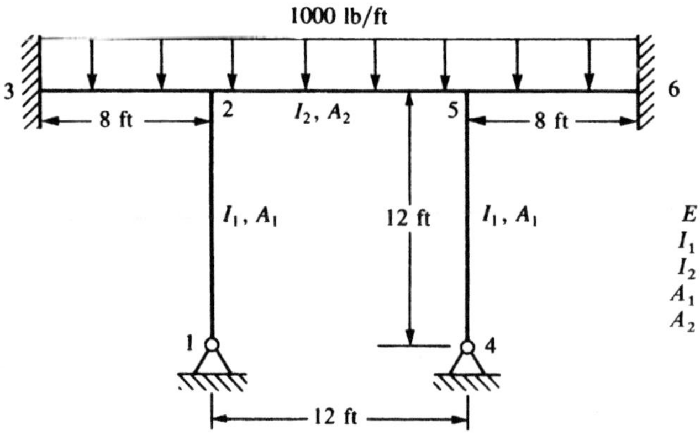

$E = 30 \times 10^6$ psi
$I_1 = 300$ in^4
$I_2 = 600$ in^4
$A_1 = 15$ in^2
$A_2 = 30$ in^2

Figure P5–26

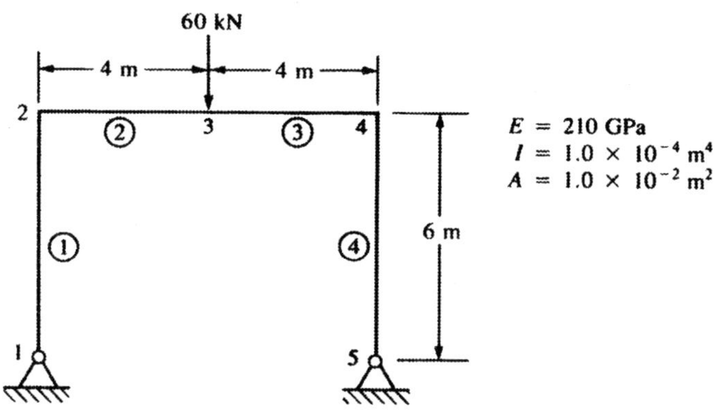

Figure P5–27

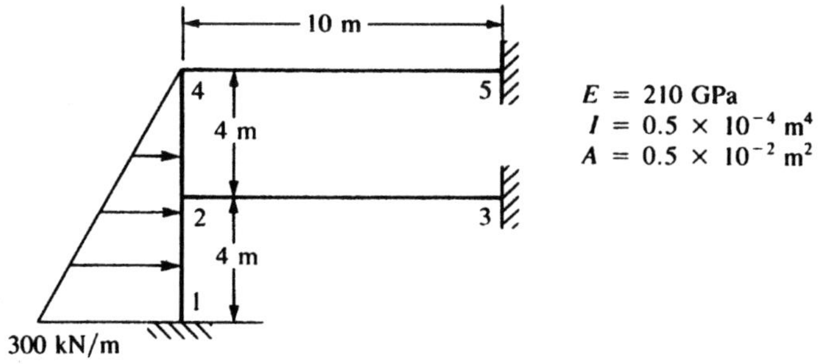

Figure P5–28

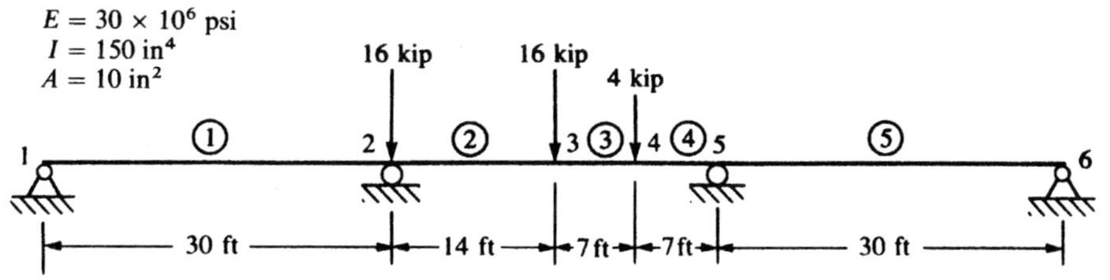

Figure P5–29

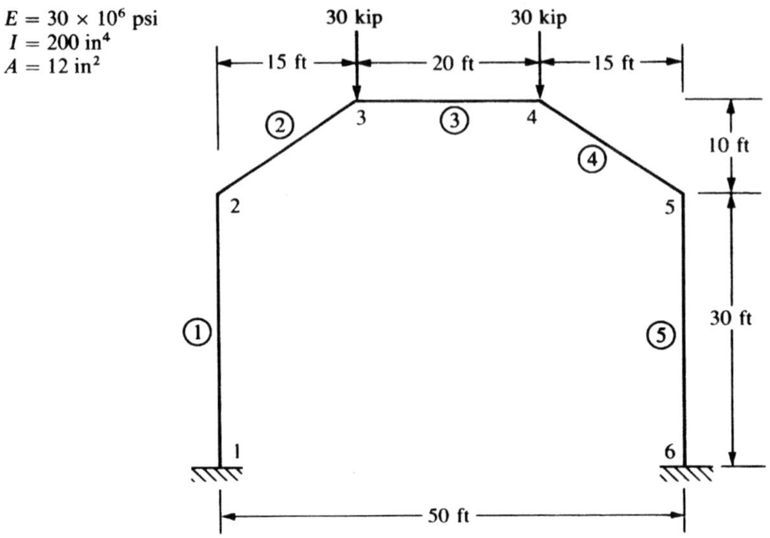

Figure P5–30

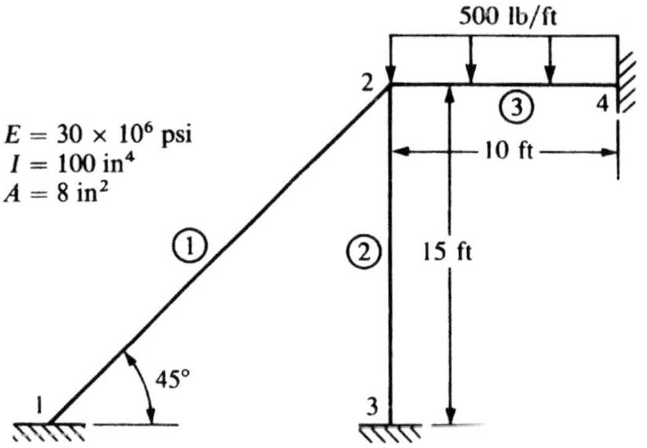

Figure P5–31

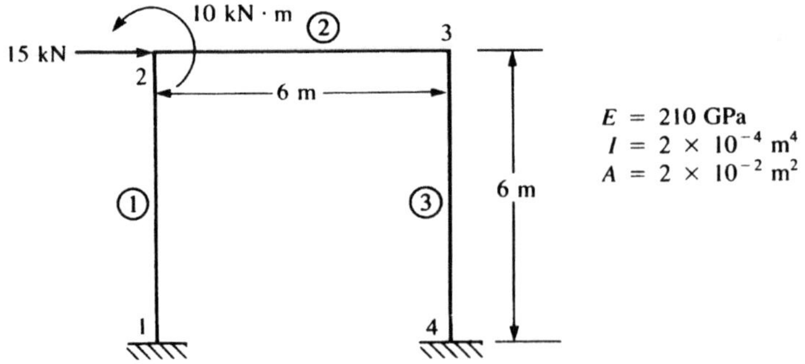

Figure P5–32

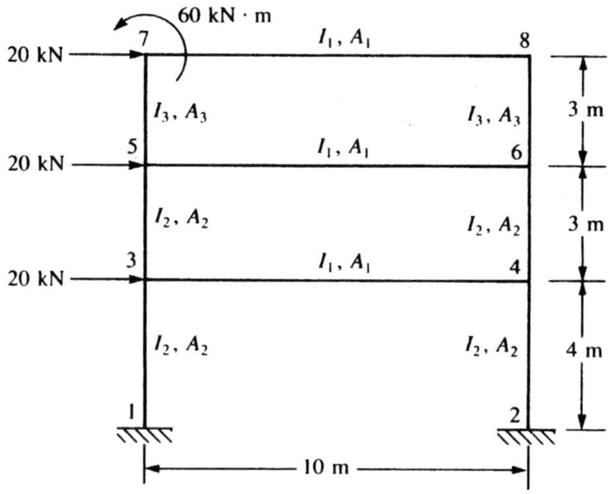

$E = 210$ GPa
$I_1 = 2 \times 10^{-4}$ m^4
$A_1 = 2 \times 10^{-2}$ m^2
$I_2 = 1 \times 10^{-4}$ m^4
$A_2 = 1 \times 10^{-2}$ m^2
$I_3 = 0.5 \times 10^{-4}$ m^4
$A_3 = 0.5 \times 10^{-2}$ m^2

Figure P5–33

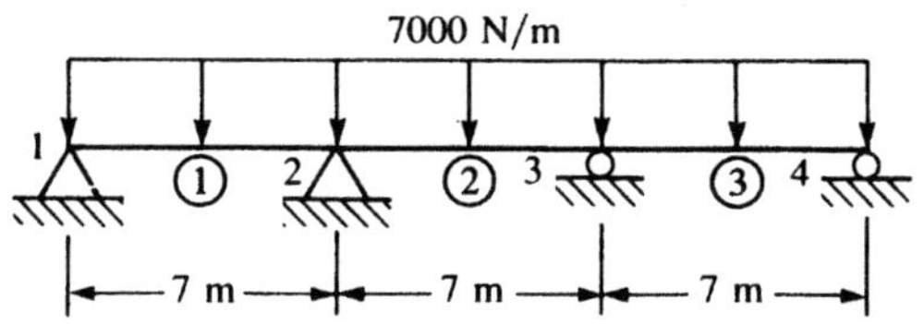

$E = 210$ GPa
$I = 1 \times 10^{-4}$ m^4
$A = 1 \times 10^{-2}$ m^2

Figure P5–34

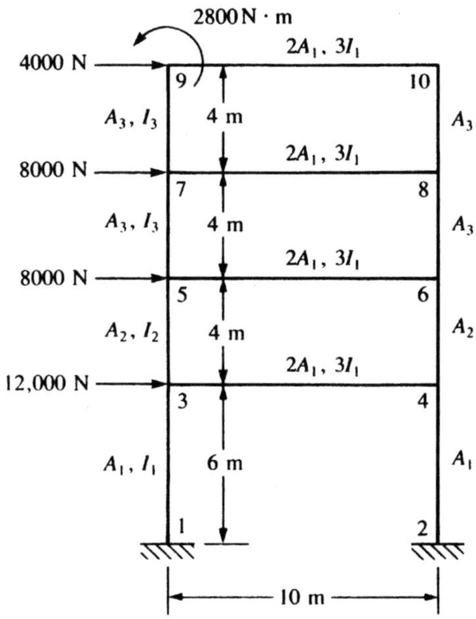

$E = 210$ GPa
$A_1 = 2.0 \times 10^{-2}$ m^2
$I_1 = 2.0 \times 10^{-4}$ m^4
$A_2 = 1.5 \times 10^{-2}$ m^2
$I_2 = 1.5 \times 10^{-4}$ m^4
$A_3 = 1.0 \times 10^{-2}$ m^2
$I_3 = 1.0 \times 10^{-4}$ m^4

Figure P5–35

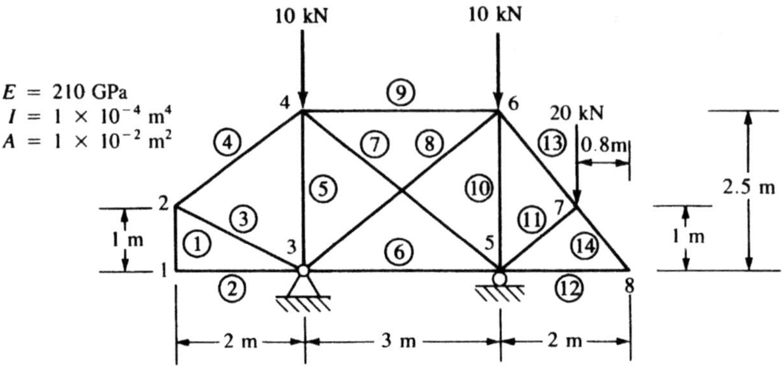

Figure P5–36

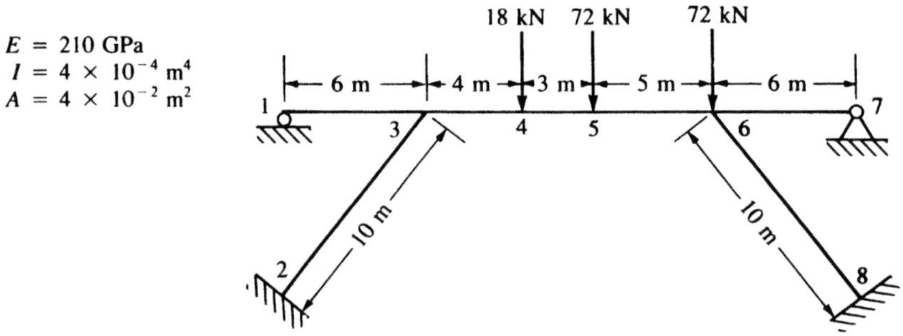

Figure P5–37

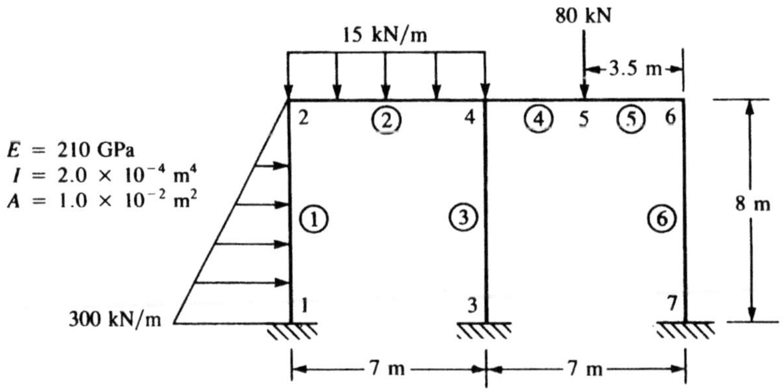

Figure P5–38

5.39 Consider the plane structure shown in Figure P5–39. First assume the structure to be a plane frame with rigid joints, and analyze using a frame element. Then assume the structure to be pin-jointed and analyze as a plane truss, using a truss element. If the structure is actually a truss, is it appropriate to model it as a rigid frame? How

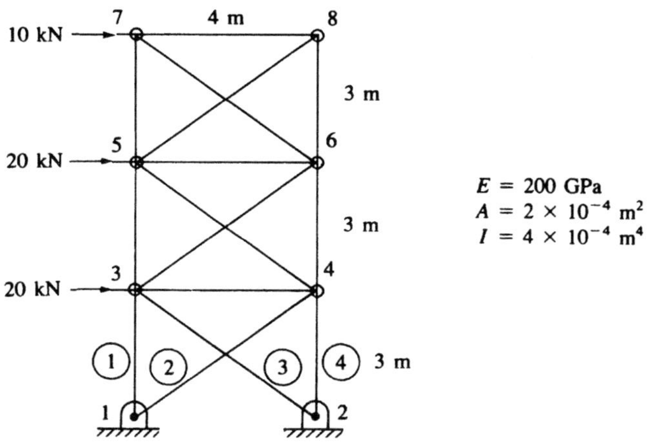

Figure P5–39

can you model the truss using the frame (or beam) element? In other words, what idealization could you make in your model to use the beam element to approximate a truss?

5.40 For the two-story, two-bay rigid frame shown, determine (1) the nodal displacement components and (2) the shear force and bending moments in each member. Let $E = 200$ GPa, $I = 2 \times 10^{-4}\,\text{m}^4$ for each horizontal member and $I = 1.5 \times 10^{-4}\,\text{m}^4$ for each vertical member.

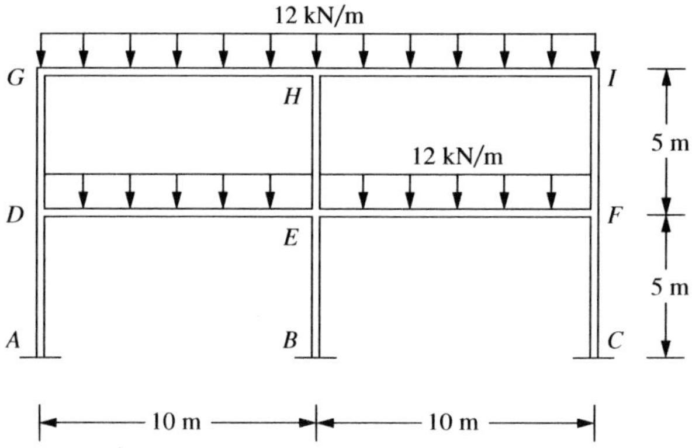

Figure P5–40

5.41 For the two-story, three-bay rigid frame shown, determine (1) the nodal displacements and (2) the member end shear forces and bending moments. (3) Draw the shear force and bending moment diagrams for each member. Let $E = 200$ GPa, $I = 1.29 \times 10^{-4}\,\text{m}^4$ for the beams and $I = 0.462 \times 10^{-4}\,\text{m}^4$ for the columns.

The properties for I correspond to a W 610 × 155 and a W 410 × 114 wide-flange section, respectively, in metric units.

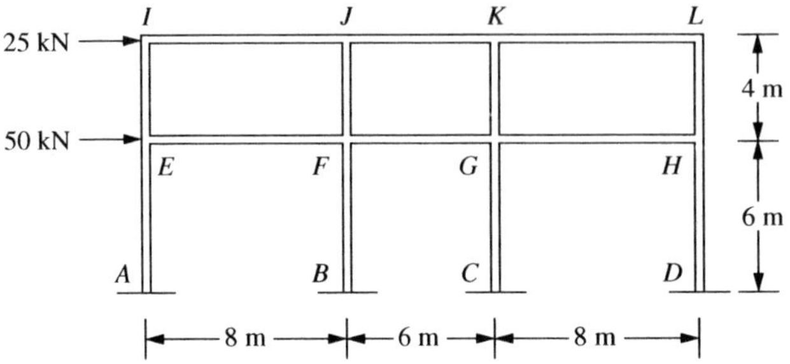

Figure P5–41

5.42 For the rigid frame shown, determine (1) the nodal displacements and rotations and (2) the member shear forces and bending moments. Let $E = 200$ GPa, $I = 0.795 \times 10^{-4}\ \text{m}^4$ for the horizontal members and $I = 0.316 \times 10^{-4}\ \text{m}^4$ for the vertical members. These I values correspond to a W 460 × 158 and a W 410 × 85 wide-flange section, respectively.

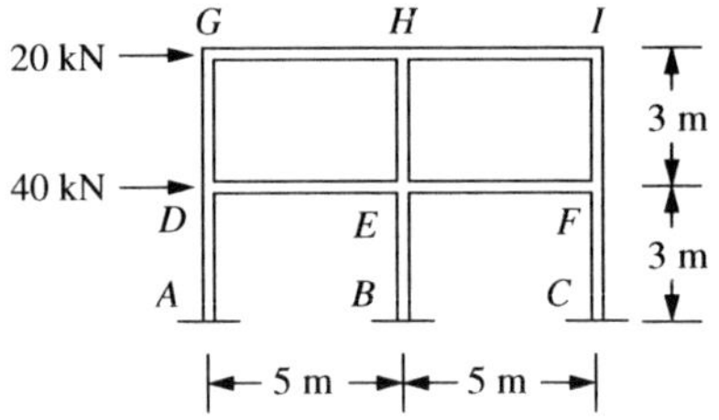

Figure P5–42

5.43 For the rigid frame shown, determine (1) the nodal displacements and rotations and (2) the shear force and bending moments in each member. Let $E = 29 \times 10^6$ psi, $I = 3100\ \text{in.}^4$ for the horizontal members and $I = 1110\ \text{in.}^4$ for the vertical members. The I values correspond to a W 24 × 104 and a W 16 × 77.

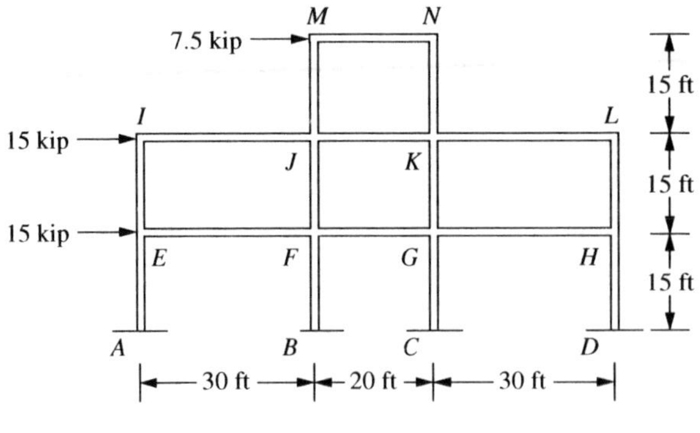

Figure P5–43

5.44 A structure is fabricated by welding together three lengths of I-shaped members as shown in Figure P5–44. The yield strength of the members is 36 ksi, $E = 29e6$ psi, and Poisson's ratio is 0.3. The members all have cross-section properties corresponding to a W18 by 76. That is, $A = 22.3\ \text{in}^2$, depth of section is $d = 18.21$ in., $I_x = 1330\ \text{in}^4$, $S_x = 146\ \text{in}^3$, $I_y = 152\ \text{in}^4$, and $S_y = 27.6\ \text{in}^3$. Determine whether a load of $Q = 10{,}000$ lb downward is safe against general yielding of the material. The factor of safety against general yielding is to be 2.0. Also, determine the maximum vertical and horizontal deflections of the structure.

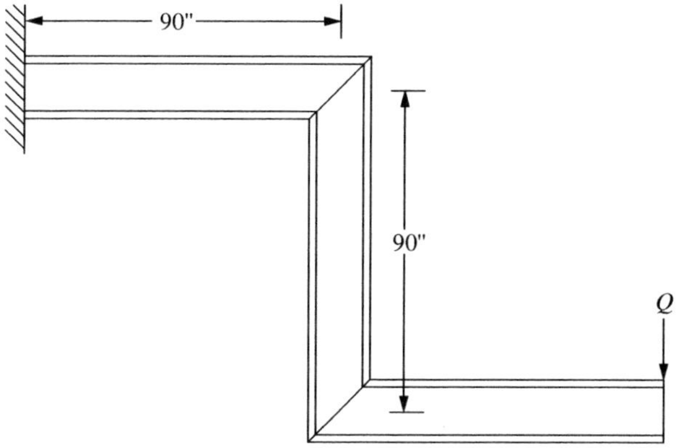

Figure P5–44

5.45 For the tapered beam shown in Figure P5–45, determine the maximum deflection using one, two, four, and eight elements. Calculate the moment of inertia at the mid-length station for each element. Let $E = 30 \times 10^6$ psi, $I_0 = 100\ \text{in}^4$, and $L = 100$ in. Run cases where $n = 1, 3,$ and 7. Use a beam element. The analytical solution for $n = 7$ is given by Reference [7]:

$$v_1 = \frac{PL^3}{49EI_0}(1/7 \ln 8 + 2.5) = \frac{1}{17.55}\frac{PL^3}{EI_0}$$

$$\theta_1 = \frac{PL^2}{49EI_0}(\ln 8 - 7) = -\frac{1}{9.95}\frac{PL^2}{EI_0}$$

$$I(x) = I_0\left(1 + n\frac{x}{L}\right)$$

where n = arbitrary numerical factor and I_0 = moment of inertia of section at $x = 0$.

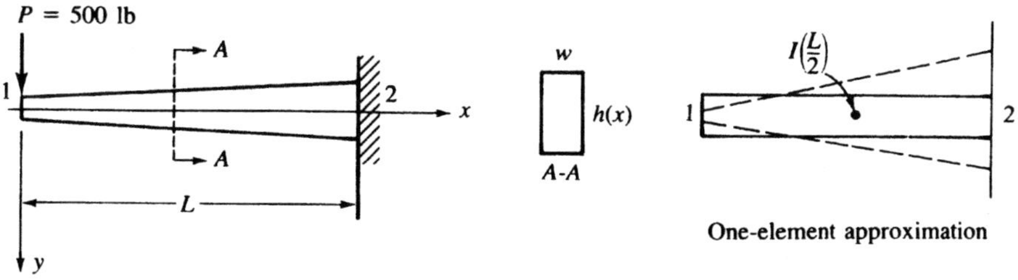

Figure P5–45 Tapered cantilever beam

5.46 Derive the stiffness matrix for the nonprismatic torsion bar shown in Figure P5–46. The radius of the shaft is given by

$$r = r_0 + (x/L)r_0, \text{ where } r_0 \text{ is the radius at } x = 0.$$

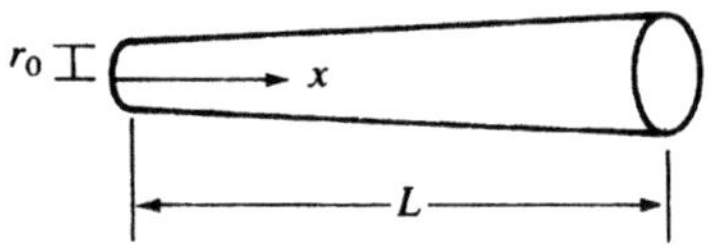

Figure P5–46

5.47 Derive the total potential energy for the prismatic circular cross-section torsion bar shown in Figure P5–47. Also determine the equivalent nodal torques for the bar subjected to uniform torque per unit length (lb-in./in.). Let G be the shear modulus and J be the polar moment of inertia of the bar.

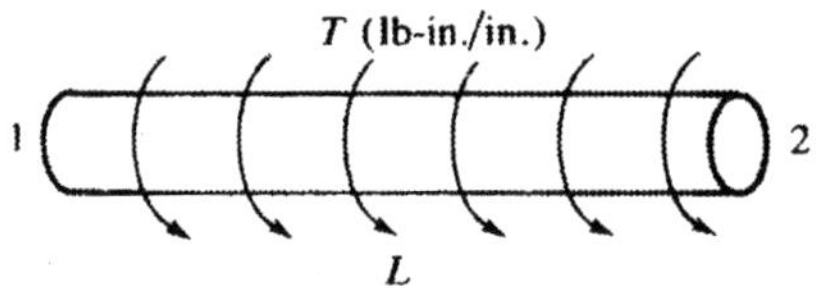

Figure P5–47

5.48 For the grid shown in Figure P5–48, determine the nodal displacements and the local element forces. Let $E = 30 \times 10^6$ psi, $G = 12 \times 10^6$ psi, $I = 200$ in^4, and $J = 100$ in^4 for both elements.

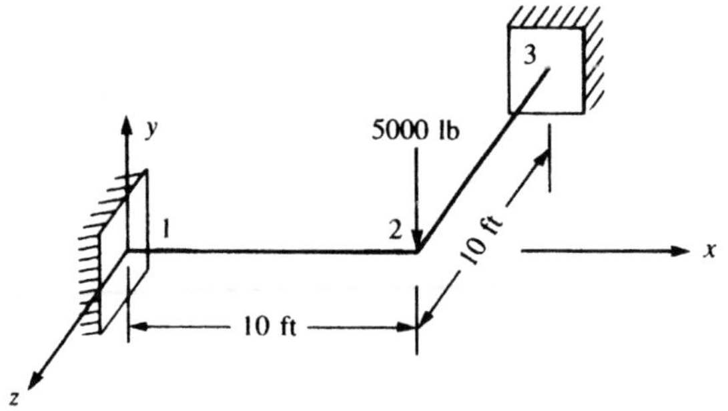

Figure P5–48

5.49 Resolve Problem 5–48 with an additional nodal moment of 1000 k-in. applied about the x axis at node 2.

5.50–5.51 For the grids shown in Figures P5–50 and P5–51, determine the nodal displacements and the local element forces. Let $E = 210$ GPa, $G = 84$ GPa, $I = 2 \times 10^{-4}$ m^4, $J = 1 \times 10^{-4}$ m^4, and $A = 1 \times 10^{-2}$ m^2.

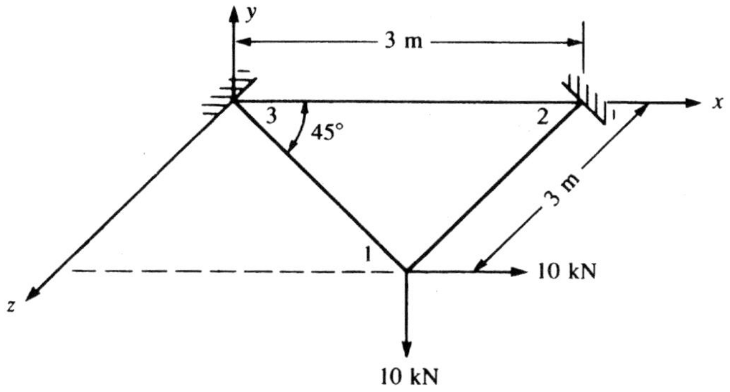

Figure P5–50

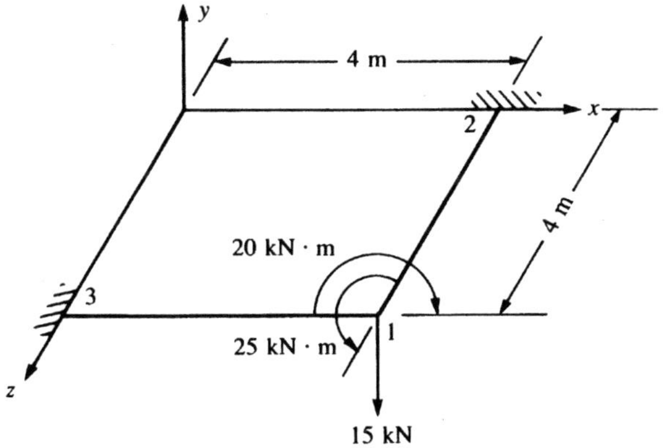

Figure P5–51

5.52–5.57 Solve the grid structures shown in Figures P5–52—P5–57 using a computer program. For grids P5–52—P5–54, let $E = 30 \times 10^6$ psi, $G = 12 \times 10^6$ psi, $I = 200$ in^4, and $J = 100$ in^4, except as noted in the figures. In Figure P5–54, let the cross elements

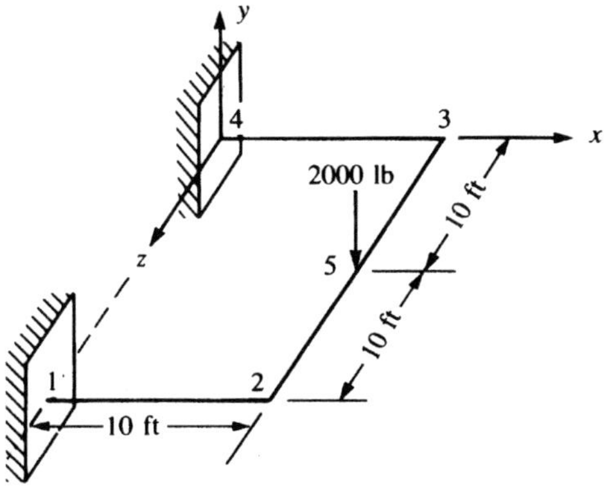

Figure P5–52

have $I = 50$ in^4 and $J = 20$ in^4, with dimensions and loads as in Figure P5–53. For grids P5–55—P5–57, let $E = 210$ GPa, $G = 84$ GPa, $I = 2 \times 10^{-4}$ m^4, $J = 1 \times 10^{-4}$ m^4, and $A = 1 \times 10^{-2}$ m^2.

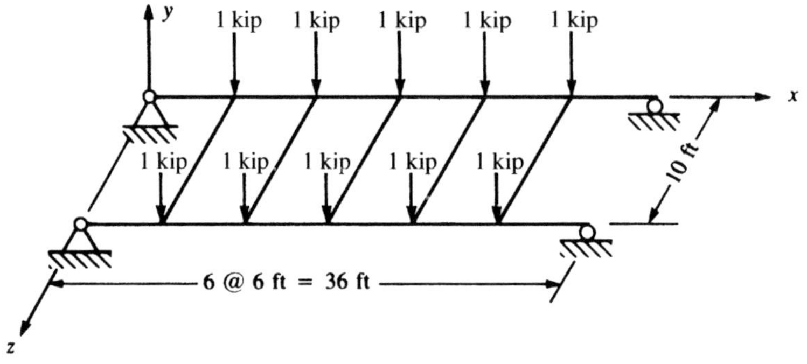

Figure P5–53

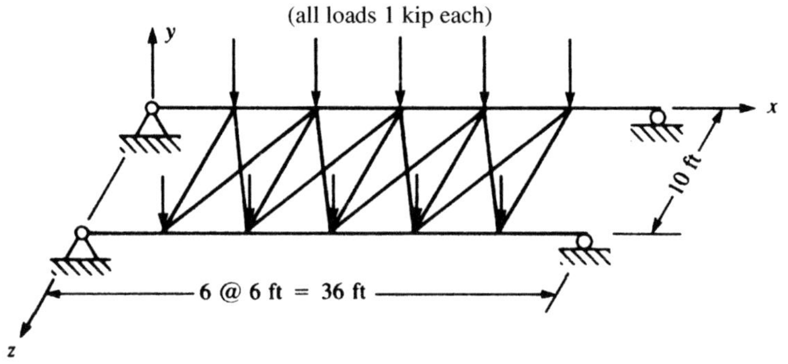

Figure P5–54

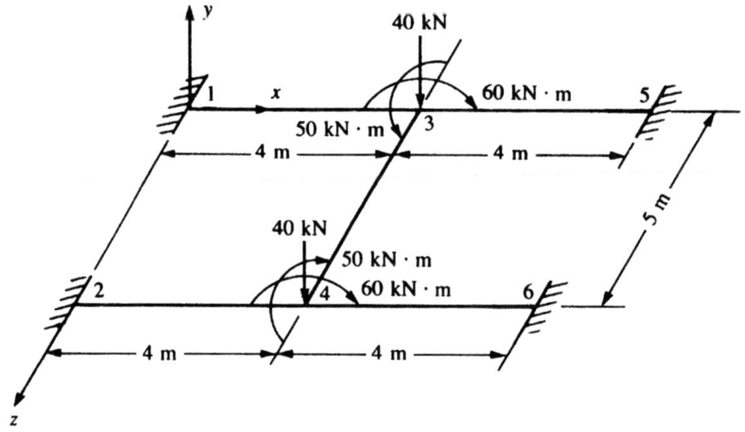

Figure P5–55

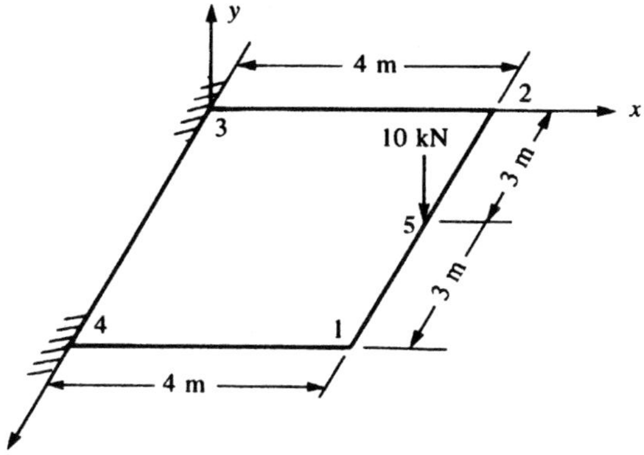

Figure P5–56

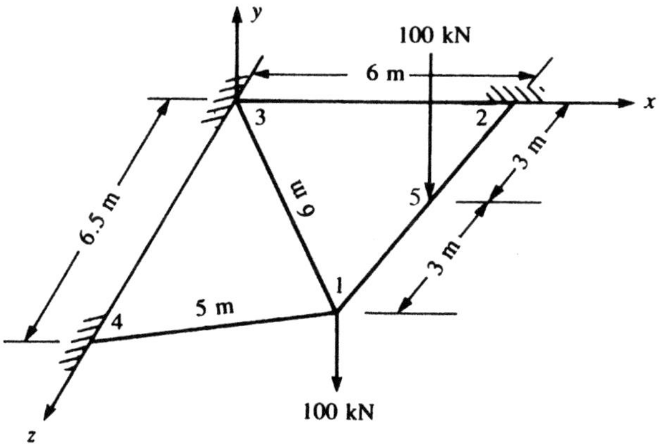

Figure P5–57

5.58–5.59 Determine the displacements and reactions for the space frames shown in Figures P5–58 and P5–59. Let $I_x = 100$ in^4, $I_y = 200$ in^4, $I_z = 1000$ in^4, $E = 30{,}000$ ksi, $G = 10{,}000$ ksi, and $A = 100$ in^2 for both frames.

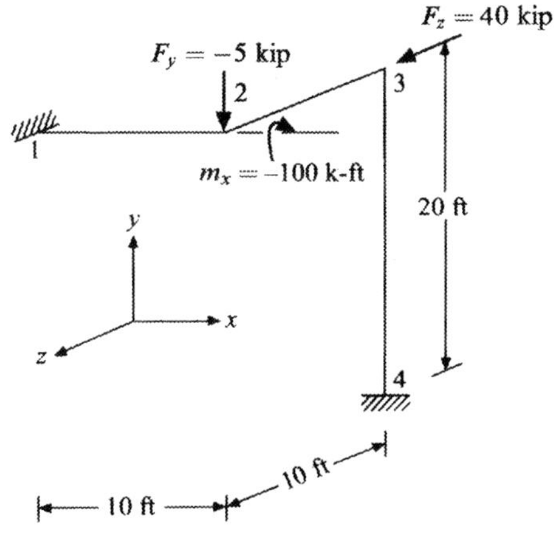

Figure P5–58

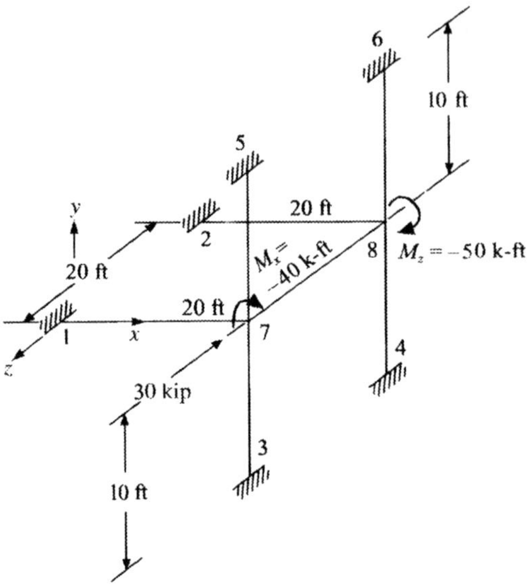

Figure P5–59

Use a computer program to assist in the design problems in Problems 5.60–5.72.

5.60 Design a jib crane as shown in Figure P5–60 that will support a downward load of 6000 lb. Choose a common structural steel shape for all members. Use allowable stresses of $0.66S_y$ (S_y is the yield strength of the material) in bending, and $0.60S_y$ in

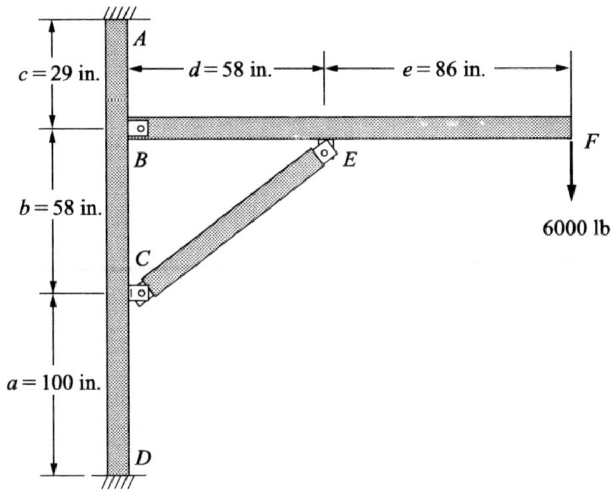

Figure P5–60

tension on gross areas. The maximum deflection should not exceed 1/360 of the length of the horizontal beam. Buckling should be checked using Euler's or Johnson's method as applicable.

5.61 Design the support members, AB and CD, for the platform lift shown in Figure P5–61. Select a mild steel and choose suitable cross-sectional shapes with no more than a 4 : 1 ratio of moments of inertia between the two principal directions of the cross section. You may choose two different cross sections to make up each arm to reduce weight. The actual structure has four support arms, but the loads shown are for one side of the platform with the two arms shown. The loads shown are under operating conditions. Use a factor of safety of 2 for human safety. In developing the finite element model, remove the platform and replace it with statically equivalent loads at the joints at B and D. Use truss elements or beam elements with low bending stiffness to model the arms from B to D, the intermediate connection, E to F, and the hydraulic actuator. The allowable stresses are $0.66S_y$ in bending and $0.60S_y$ in tension. Check buckling using either Euler's method or Johnson's method as appropriate. Also check maximum deflections. Any deflection greater than 1/360 of the length of member AB is considered too large.

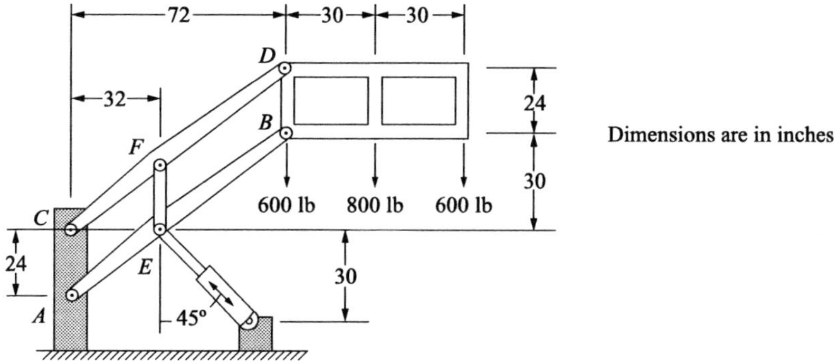

Figure P5–61

5.62 A two-story building frame is to be designed as shown in Figure P5–62. The members are all to be I-beams with rigid connections. We would like the floor joists beams to have a 15-in. depth and the columns to have a 10 in. width. The material is to be A36 structural steel. Two horizontal loads and vertical loads are shown. Select members such that the allowable bending in the beams is 24,000 psi. Check buckling in the columns using Euler's or Johnson's method as appropriate. The allowable deflection in the beams should not exceed 1/360 of each beam span. The overall sway of the frame should not exceed 0.5 in.

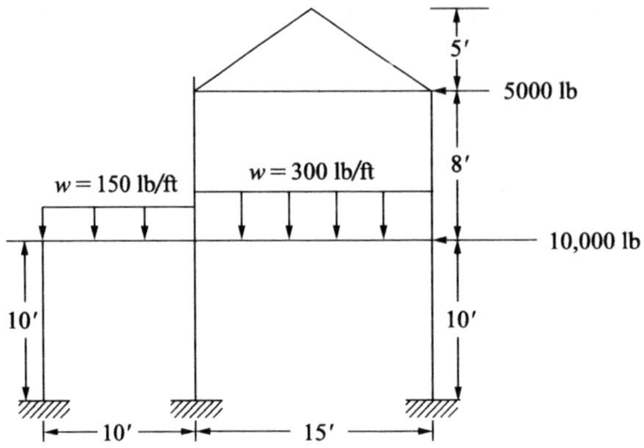

Figure P5–62

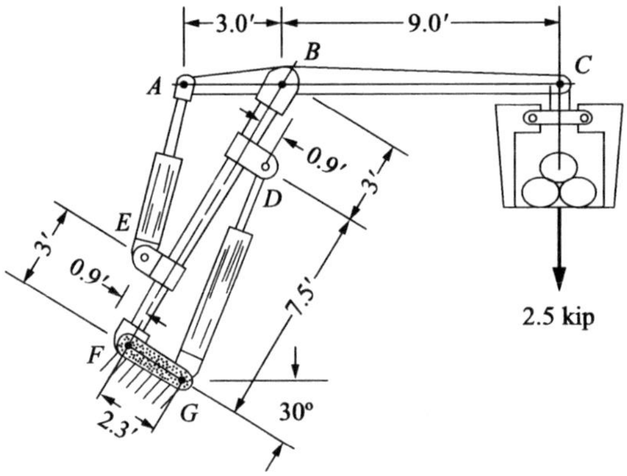

Figure P5–63

5.63 A pulpwood loader as shown in Figure P5–63 is to be designed to lift 2.5 kip. Select a steel and determine a suitable tubular cross section for the main upright member *BF* that has attachments for the hydraulic cylinder actuators *AE* and *DG*. Select a steel and determine a suitable box section for the horizontal load arm *AC*. The horizontal load arm may have two different cross sections *AB* and *BC* to reduce weight. The finite element model should use beam elements for all members except the hydraulic cylinders, which should be truss elements. The pinned joint at *B* between the upright and the horizontal beam is best modeled with end release of the end node of the top element on the upright member. The allowable bending stress is $0.66S_y$ in members *AB* and *BC*. Member *BF* should be checked for buckling. The allowable deflection at *C* should be less than $1/360$ of the length of *BC*. As a bonus, the client would like you to select the size of the hydraulic cylinders *AE* and *DG*.

5.64 A piston ring (with a split as shown in Figure P5–64) is to be expanded by a tool to facilitate its installation. The ring is sufficiently thin (0.2 in. depth) to justify using conventional straight-beam bending formulas. The ring requires a displacement of 0.1 in. at its separation for installation. Determine the force required to produce this separation. In addition, determine the largest stress in the ring. Let $E = 18 \times 10^6$ psi, $G = 7 \times 10^6$ psi, cross-sectional area $A = 0.06$ in.^{2}, and principal moment of inertia $I = 4.5 \times 10^{-4}$ in.^{4}. The inner radius is 1.85 in., and the outer radius is 2.15 in. Use models with 4, 6, 8, 10, and 20 elements in a symmetric model until convergence to the same results occurs. Plot the displacement versus the number of elements for a constant force F predicted by the conventional beam theory equation of Reference [8].

$$\delta = \frac{3\pi F R^3}{EI} + \frac{\pi F R}{EA} + \frac{6\pi F R}{5GA} \qquad \text{where } R = 2.0 \text{ in. and } \delta = 0.1 \text{ in.}$$

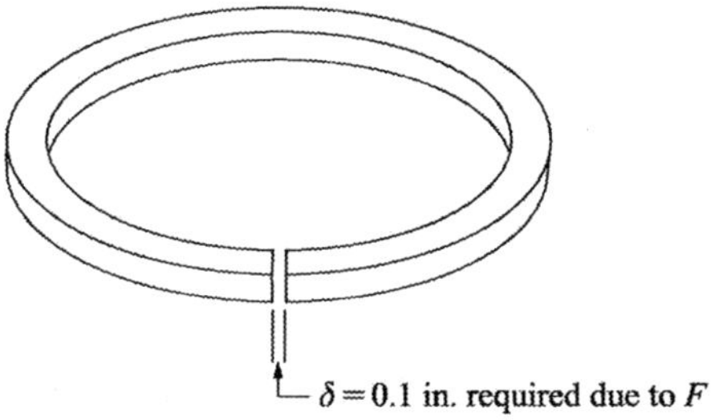

Figure P5–64

5.65 A small hydraulic floor crane as shown in Figure P5–65 carries a 5000-lb load. Determine the size of the beam and column needed. Select either a standard box section or a wide-flange section. Assume a rigid connection between the beam and column. The column is rigidly connected to the floor. The allowable bending stress in the beam is $0.60S_y$. The allowable deflection is $1/360$ of the beam length. Check the column for buckling.

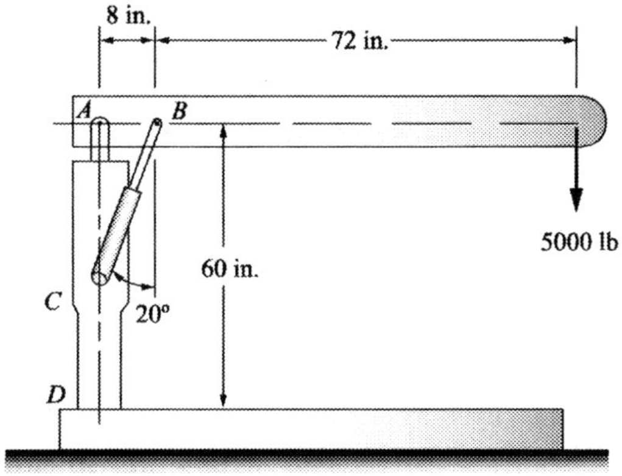

Figure P5–65

5.66 Determine the size of a solid round shaft such that the maximum angle of twist between C and B is 0.26 degrees per meter of length and the deflection of the beam is less than 0.005 inches under the pulley C for the loads shown. Assume simple supports at bearings A and B. Assume the shaft is made from cold-rolled AISI 1020 steel. (Recommended angles of twist in driven shafts can be found in *Machinery's Handbook*, Oberg, E., et. al., 26th ed., Industrial Press, N.Y., 2000.)

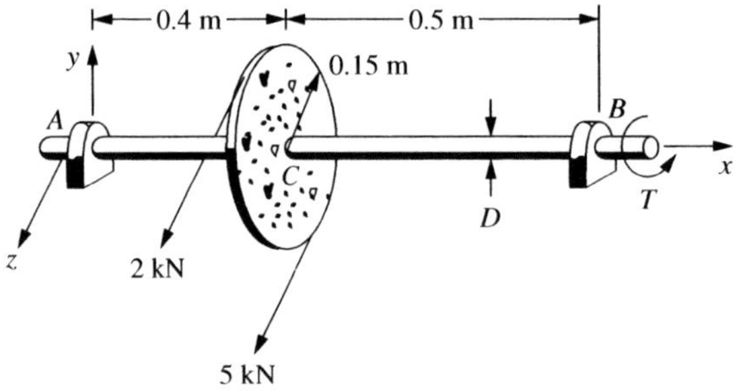

Figure P5–66

5.67 The shaft shown supports a winch load of 780 lb and a torsional moment of 7800 lb-in. at F (26 inches from the center of the bearing at A). In addition, a radial load of 500 lb and an axial load of 400 lb act at point E from a worm gearset. Assume the maximum stress in the shaft cannot be larger than that obtained from the maximum distortional energy theory with a factor of safety of 2.5. Also make sure the angle of twist is less than 1.5 deg between A and D. In your model, assume the bearing at A to be frozen when calculating the angle of twist. Bearings at B, C, and D can be assumed as simple supports. Determine the required shaft diameter.

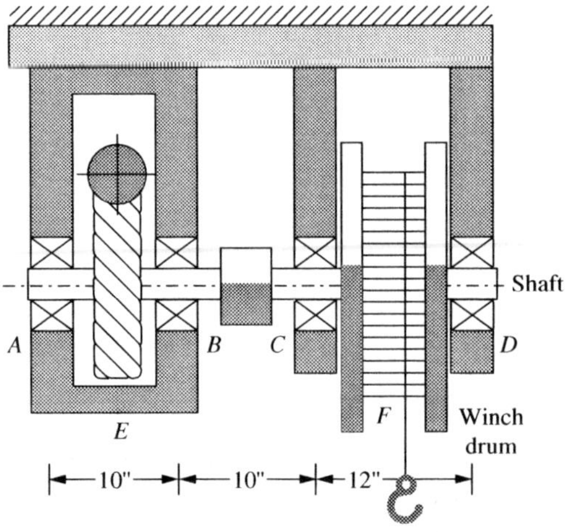

Figure P5–67

5.68 Design the gabled frame subjected to the external wind load shown (comparable to an 80 mph wind speed) for an industrial building. Assume this is one of a typical frame spaced every 20 feet. Select a wide flange section based on allowable bending stress of 20 ksi and an allowable compressive stress of 10 ksi in any member. Neglect the possibility of buckling in any members. Use ASTM A36 steel.

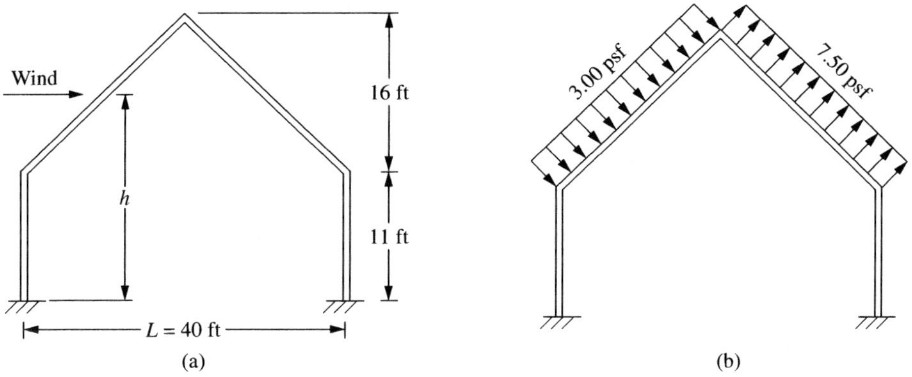

Figure P5–68

5.69 Design the gabled frame shown for a balanced snow load shown (typical of the Midwest) for an apartment building. Select a wide flange section for the frame. Assume the allowable bending stress not to exceed 140 MPa. Use ASTM A36 steel.

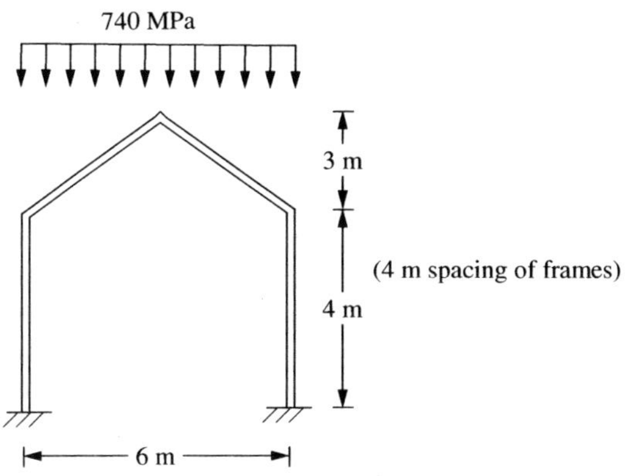

Figure P5–69

5.70 Design a gantry crane that must be able to lift 10 tons as it must lift compressors, motors, heat exchangers, and controls. This load should be placed at the center of one of the main 12-foot-long beams as shown in Figure P5–70 by the hoisting device location. Note that this beam is on one side of the crane. Assume you are using

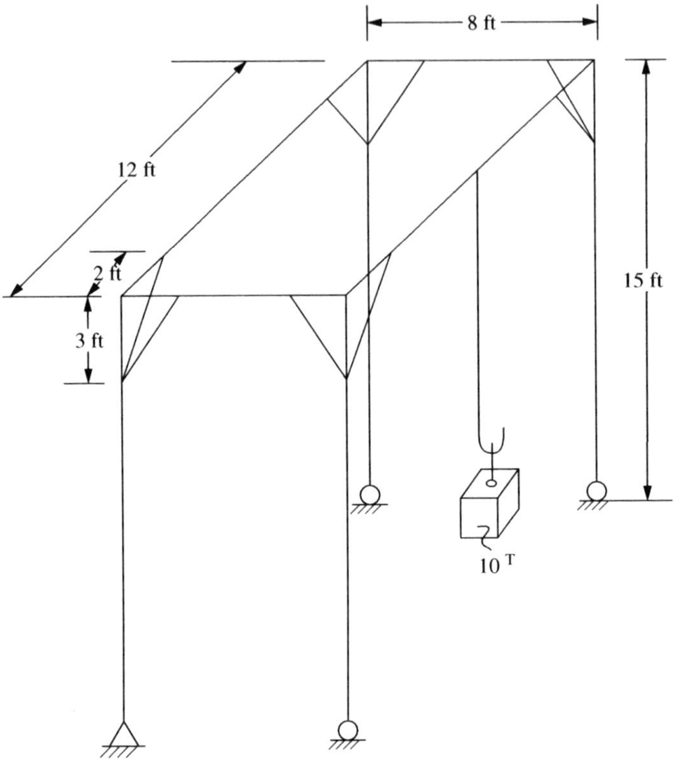

Figure P5–70

ASTM A36 structural steel. The crane must be 12 feet long, 8 feet wide, and 15 feet high. The beams should all be the same size, the columns all the same size, and the bracing all the same size. The corner bracing can be wide flange sections or some other common shape. You must verify that the structure is safe by checking the beam's bending strength and allowable deflection, the column's buckling strength, and the bracing's buckling strength. Use a factor of safety against material yielding of the beams of 5. Verify that the beam deflection is less than L/360, where L is the span of the beam. Check Euler buckling of the long columns and the bracing. Use a factor of safety against buckling of 5. Assume the column-to-beam joints to be rigid while the bracing (a total of eight braces) is pinned to the column and beam at each of the four corners. Also assume the gantry crane is on rollers with one roller locked down to behave as a pin support as shown.

5.71 Design the rigid highway bridge frame structure shown in Figure P5–71 for a moving truck load (shown below) simulating a truck moving across the bridge. Use the load shown and place it along the top girder at various locations. Use the allowable stresses in bending and compression and allowable deflection given in the *Standard Specifications for Highway Bridges*, American Association of State Highway and Transportation Officials (AASHTO), Washington, D.C. or use some other reasonable values.

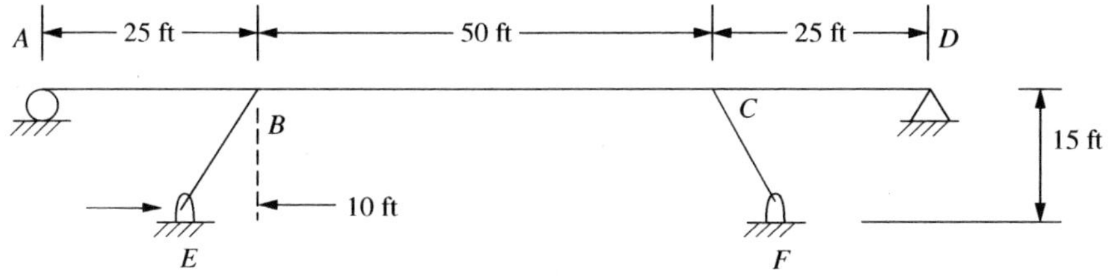

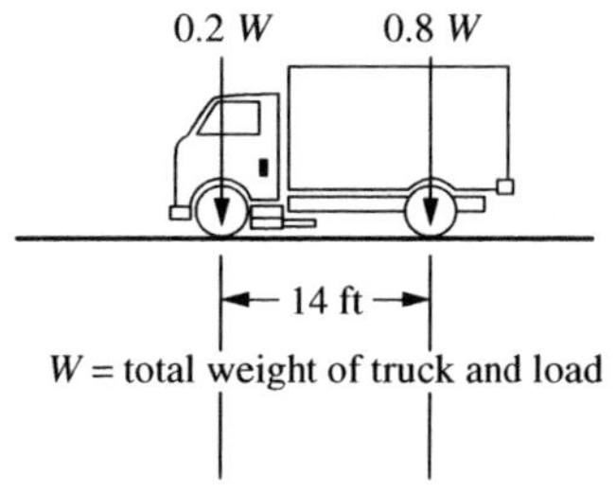

Figure P5–71

5.72 For the tripod space frame shown in Figure P5–72, determine standard steel pipe sections such that the maximum bending stress must not exceed 20 ksi, the compressive stress to prevent buckling must not exceed that given by the Euler buckling formula with a factor of safety of 2 and the maximum deflection will not exceed L/360 in any span, L. Assume the three bottom supports to be fixed. All coordinates shown in units of inches.

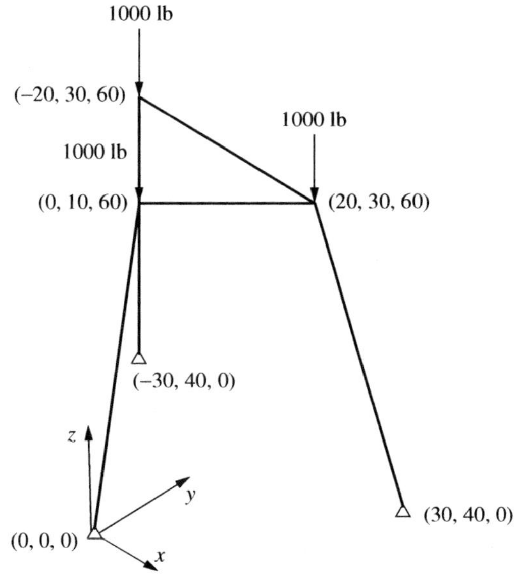

Figure P5–72

CHAPTER 6

Development of the Plane Stress and Plane Strain Stiffness Equations

Introduction

In Chapters 2–5, we considered only line elements. Two or more line elements are connected only at common nodes, forming framed or articulated structures such as trusses, frames, and grids. Line elements have geometric properties such as cross-sectional area and moment of inertia associated with their cross sections. However, only one local coordinate $\hat{x}$ along the length of the element is required to describe a position along the element (hence, they are called *line elements* or one-dimensional elements). Nodal compatibility is then enforced during the formulation of the nodal equilibrium equations for a line element.

This chapter considers the two-dimensional finite element. Two-dimensional (planar) elements are defined by three or more nodes in a two-dimensional plane (that is, x-y). The elements are connected at common nodes and/or along common edges to form continuous structures such as those shown in Figures 1–3, 1–4, 1–6, and 6–6(b). Nodal displacement compatibility is then enforced during the formulation of the nodal equilibrium equations for two-dimensional elements. If proper displacement functions are chosen, compatibility along common edges is also obtained. The two-dimensional element is extremely important for (1) plane stress analysis, which includes problems such as plates with holes, fillets, or other changes in geometry that are loaded in their plane resulting in local stress concentrations, such as illustrated in Figure 6–1; and (2) plane strain analysis, which includes problems such as a long underground box culvert subjected to a uniform load acting constantly over its length, as illustrated in Figure 1–3, a long, cylindrical control rod subjected to a load that remains constant over the rod length (or depth), as illustrated in Figure 1–4, and dams and pipes subjected to loads that remain constant over their lengths as shown in Figure 6–2.

We begin this chapter with the development of the stiffness matrix for a basic two-dimensional or plane finite element, called the *constant-strain triangular element*. We consider the constant-strain triangle (CST) stiffness matrix because its derivation is the simplest among the available two-dimensional elements. The element is called a CST because it has a constant strain throughout it.

We will derive the CST stiffness matrix by using the principle of minimum potential energy because the energy formulation is the most feasible for the development of the equations for both two- and three-dimensional finite elements.

We will then present a simple, thin-plate plane stress example problem to illustrate the assemblage of the plane element stiffness matrices using the direct stiffness method as presented in Chapter 2. We will present the total solution, including the stresses within the plate.

6.1 Basic Concepts of Plane Stress and Plane Strain

In this section, we will describe the concepts of plane stress and plane strain. These concepts are important because the developments in this chapter are directly applicable only to systems assumed to behave in a plane stress or plane strain manner. Therefore, we will now describe these concepts in detail.

Plane Stress

Plane stress *is defined to be a state of stress in which the normal stress and the shear stresses directed perpendicular to the plane are assumed to be zero.* For instance, in Figures 6–1(a) and 6–1(b), the plates in the x-y plane shown subjected to surface tractions T (pressure acting on the surface edge or face of a member in units of force/area) in the plane are under a state of plane stress; that is, the normal stress σ_z and the shear stresses τ_{xz} and τ_{yz} are assumed to be zero. Generally, members that are thin (those with a small z dimension compared to the in-plane x and y dimensions) and whose loads act only in the x-y plane can be considered to be under plane stress.

Plane Strain

Plane strain *is defined to be a state of strain in which the strain normal to the* x-y *plane* ε_z *and the shear strains* γ_{xz} *and* γ_{yz} *are assumed to be zero.* The assumptions of plane strain are realistic for long bodies (say, in the z direction) with constant cross-sectional area subjected to loads that act only in the x and/or y directions and do not vary in the

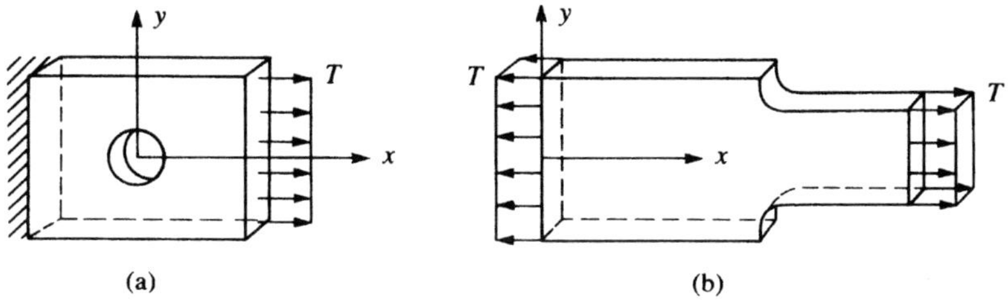

Figure 6–1 Plane stress problems: (a) plate with hole; (b) plate with fillet

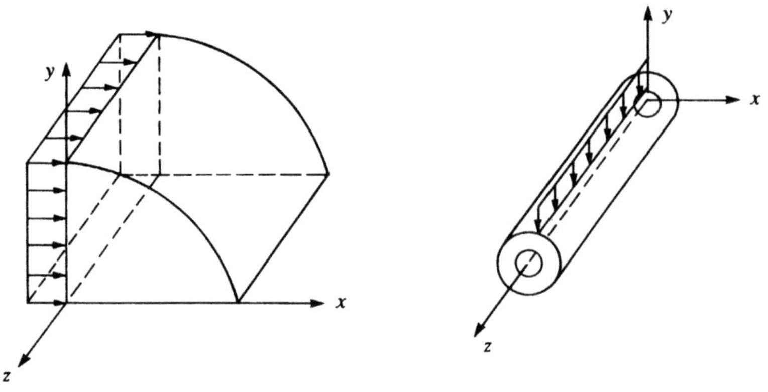

Figure 6–2 Plane strain problems: (a) dam subjected to horizontal loading; (b) pipe subjected to a vertical load

z direction. Some plane strain examples are shown in Figure 6–2 [and in Figures 1–3 (a long underground box culvert) and 1–4 (a hydraulic cylinder rod end)]. In these examples, only a unit thickness (1 in. or 1 ft) of the structure is considered because each unit thickness behaves identically (except near the ends). The finite element models of the structures in Figure 6–2 consist of appropriately discretized cross sections in the x-y plane with the loads acting over unit thicknesses in the x and/or y directions only.

Two-Dimensional State of Stress and Strain

The concept of a two-dimensional state of stress and strain and the stress/strain relationships for plane stress and plane strain are necessary to understand fully the development and applicability of the stiffness matrix for the plane stress/plane strain triangular element. Therefore, we briefly outline the essential concepts of two-dimensional stress and strain (see References [1] and [2] and Appendix C for more details on this subject).

First, we illustrate the two-dimensional state of stress using Figure 6–3. The infinitesimal element with sides dx and dy has normal stresses σ_x and σ_y acting in the x and y directions (here on the vertical and horizontal faces), respectively. The shear stress τ_{xy} acts on the x edge (vertical face) in the y direction. The shear stress τ_{yx} acts on the y edge (horizontal face) in the x direction. Moment equilibrium of the element results in τ_{xy} being equal in magnitude to τ_{yx}. See Appendix C.1 for proof of this equality. Hence, three independent stresses exist and are represented by the vector column matrix

$$\{\sigma\} = \begin{Bmatrix} \sigma_x \\ \sigma_y \\ \tau_{xy} \end{Bmatrix} \tag{6.1.1}$$

The element equilibrium equations are derived in Appendix C.1.

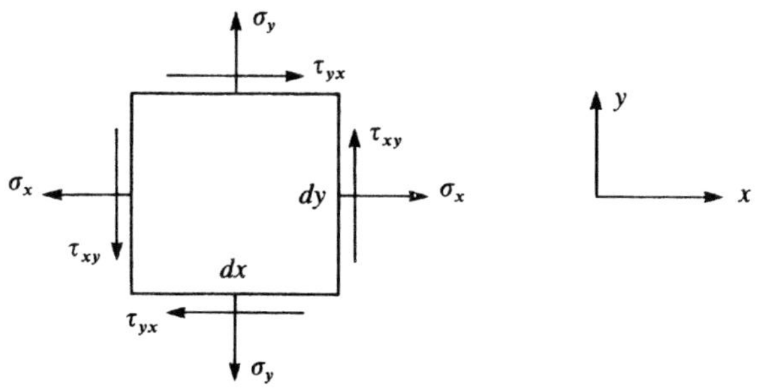

Figure 6–3 Two-dimensional state of stress

The stresses given by Eq. (6.1.1) will be expressed in terms of the nodal displacement degrees of freedom. Hence, once the nodal displacements are determined, these stresses can be evaluated directly.

Recall from strength of materials [2] that the **principal stresses**, which are the maximum and minimum normal stresses in the two-dimensional plane, can be obtained from the following expressions:

$$\sigma_1 = \frac{\sigma_x + \sigma_y}{2} + \sqrt{\left(\frac{\sigma_x - \sigma_y}{2}\right)^2 + \tau_{xy}^2} = \sigma_{\max}$$

$$\sigma_2 = \frac{\sigma_x + \sigma_y}{2} - \sqrt{\left(\frac{\sigma_x - \sigma_y}{2}\right)^2 + \tau_{xy}^2} = \sigma_{\min} \tag{6.1.2}$$

Also, the **principal angle** θ_p, which defines the normal whose direction is perpendicular to the plane on which the maximum or minimum principal stress acts, is defined by

$$\tan 2\theta_p = \frac{2\tau_{xy}}{\sigma_x - \sigma_y} \tag{6.1.3}$$

Figure 6–4 shows the principal stresses σ_1 and σ_2 and the angle θ_p. Recall (as Figure 6–4 indicates) that the shear stress is zero on the planes having principal (maximum and minimum) normal stresses.

In Figure 6–5, we show an infinitesimal element used to represent the general two-dimensional state of strain at some point in a structure. The element is shown to be displaced by amounts u and v in the x and y directions at point A, and to displace or extend an additional (incremental) amount $(\partial u/\partial x)\,dx$ along line AB, and $(\partial v/\partial y)\,dy$ along line AC in the x and y directions, respectively. Furthermore, observing lines AB and AC, we see that point B moves upward an amount $(\partial v/\partial x)\,dx$ with respect to A, and point C moves to the right an amount $(\partial u/\partial y)\,dy$ with respect to A.

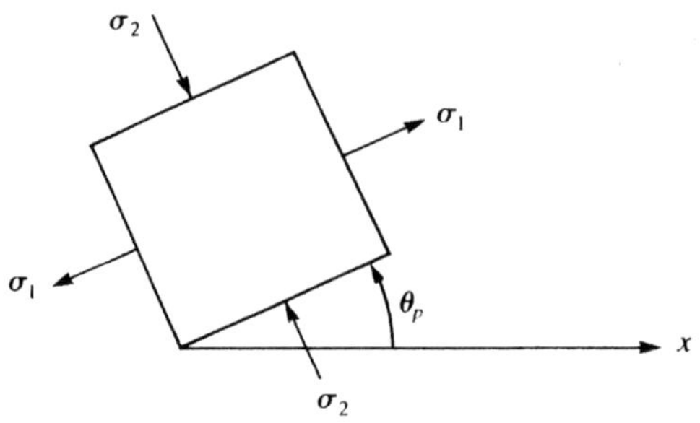

Figure 6–4 Principal stresses and their directions

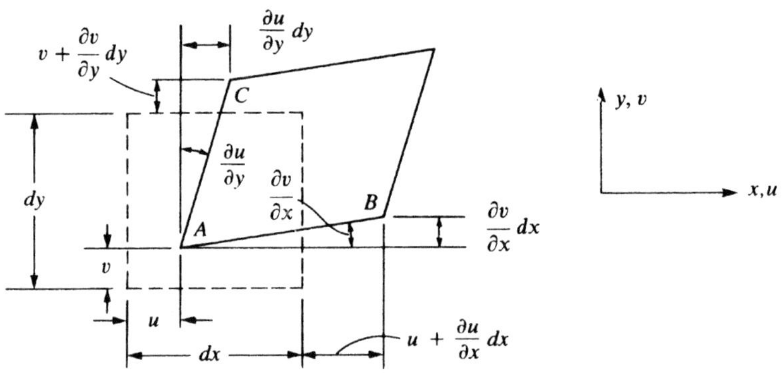

Figure 6–5 Displacements and rotations of lines of an element in the *x*-*y* plane

From the general definitions of normal and shear strains and the use of Figure 6–5, we obtain

$$\varepsilon_x = \frac{\partial u}{\partial x} \qquad \varepsilon_y = \frac{\partial v}{\partial y} \qquad \gamma_{xy} = \frac{\partial u}{\partial y} + \frac{\partial v}{\partial x} \tag{6.1.4}$$

Appendix C.2 shows a detailed derivation of Eqs. (6.1.4). Hence, recall that the strains ε_x and ε_y are the changes in length per unit length of material fibers originally parallel to the x and y axes, respectively, when the element undergoes deformation. These strains are then called *normal* (or *extensional* or *longitudinal*) *strains*. The strain γ_{xy} is the change in the original right angle made between dx and dy when the element undergoes deformation. The strain γ_{xy} is then called a *shear strain*.

The strains given by Eqs. (6.1.4) are generally represented by the vector column matrix

$$\{\varepsilon\} = \begin{Bmatrix} \varepsilon_x \\ \varepsilon_y \\ \gamma_{xy} \end{Bmatrix} \tag{6.1.5}$$

The relationships between strains and displacements referred to the x and y directions given by Eqs. (6.1.4) are sufficient for your understanding of subsequent material in this chapter.

We now present the stress/strain relationships for isotropic materials for both plane stress and plane strain. For plane stress, we assume the following stresses to be zero:

$$\sigma_z = \tau_{xz} = \tau_{yz} = 0 \tag{6.1.6}$$

Applying Eq. (6.1.6) to the three-dimensional stress/strain relationship [see Appendix C, Eq. (C.3.10)], the shear strains $\gamma_{xz} = \gamma_{yz} = 0$, but $\varepsilon_z \neq 0$. For plane stress conditions, we then have

$$\{\sigma\} = [D]\{\varepsilon\} \tag{6.1.7}$$

where

$$[D] = \frac{E}{1-\nu^2}\begin{bmatrix} 1 & \nu & 0 \\ \nu & 1 & 0 \\ 0 & 0 & \dfrac{1-\nu}{2} \end{bmatrix} \tag{6.1.8}$$

is called the *stress*/strain matrix (or *constitutive matrix*), E is the modulus of elasticity, and ν is Poisson's ratio. In Eq. (6.1.7), $\{\sigma\}$ and $\{\varepsilon\}$ are defined by Eqs. (6.1.1) and (6.1.5), respectively.

For plane strain, we assume the following strains to be zero:

$$\varepsilon_z = \gamma_{xz} = \gamma_{yz} = 0 \tag{6.1.9}$$

Applying Eq. (6.1.9) to the three-dimensional stress/strain relationship [Eq. (C.3.10)], the shear stresses $\tau_{xz} = \tau_{yz} = 0$, but $\sigma_z \neq 0$. The stress/strain matrix then becomes

$$[D] = \frac{E}{(1+\nu)(1-2\nu)}\begin{bmatrix} 1-\nu & \nu & 0 \\ \nu & 1-\nu & 0 \\ 0 & 0 & \dfrac{1-2\nu}{2} \end{bmatrix} \tag{6.1.10}$$

The $\{\sigma\}$ and $\{\varepsilon\}$ matrices remain the same as for the plane stress case. The basic partial differential equations for plane stress, as derived in Reference [1], are

$$\begin{aligned} \frac{\partial^2 u}{\partial x^2} + \frac{\partial^2 u}{\partial y^2} &= \frac{1+\nu}{2}\left(\frac{\partial^2 u}{\partial y^2} - \frac{\partial^2 v}{\partial x \partial y}\right) \\ \frac{\partial^2 v}{\partial x^2} + \frac{\partial^2 v}{\partial y^2} &= \frac{1+\nu}{2}\left(\frac{\partial^2 v}{\partial x^2} - \frac{\partial^2 u}{\partial x \partial y}\right) \end{aligned} \tag{6.1.11}$$

6.2 Derivation of the Constant-Strain Triangular Element Stiffness Matrix and Equations

To illustrate the steps and introduce the basic equations necessary for the plane triangular element, consider the thin plate subjected to tensile surface traction loads T_S in Figure 6–6(a).

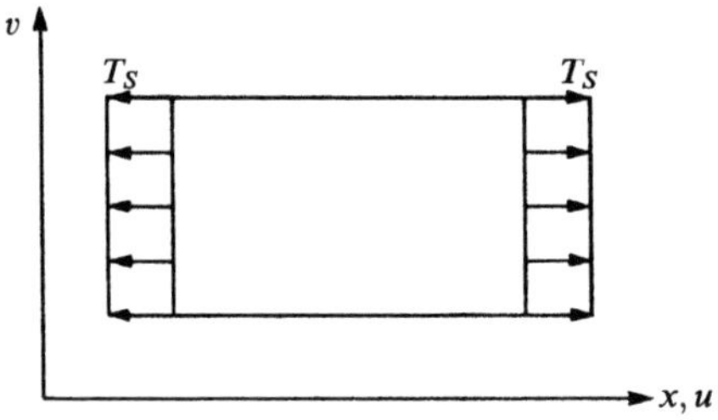

Figure 6–6(a) Thin plate in tension

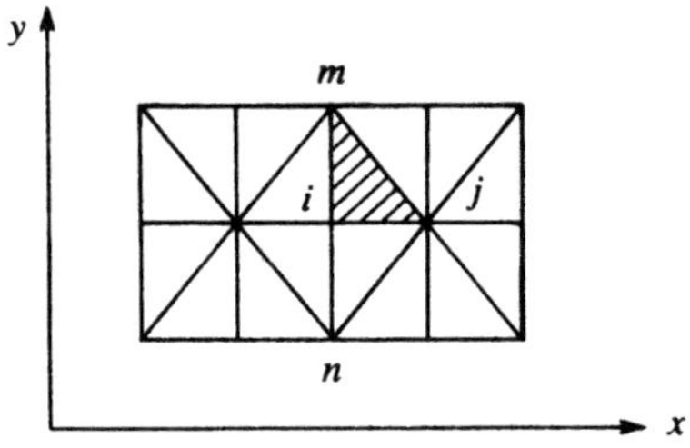

Figure 6–6(b) Discretized plate of Figure 6–6(a) using triangular elements

Step 1 Select Element Type

To analyze the plate, we consider the basic triangular element in Figure 6–7 taken from the discretized plate, as shown in Figure 6–6(b). The discretized plate has been divided into triangular elements, each with nodes such as i, j, and m. We use triangular elements because boundaries of irregularly shaped bodies can be closely approximated in this way, and because the expressions related to the triangular element are comparatively simple. This discretization is called a *coarse-mesh generation* if a few large elements are used. Each node has two degrees of freedom—an x and a y displacement. We will let u_i and v_i represent the node i displacement components in the x and y directions, respectively.

Here all formulations are based on this counterclockwise system of labeling of nodes, although a formulation based on a clockwise system of labeling could be used. Remember that a consistent labeling procedure for the whole body is necessary

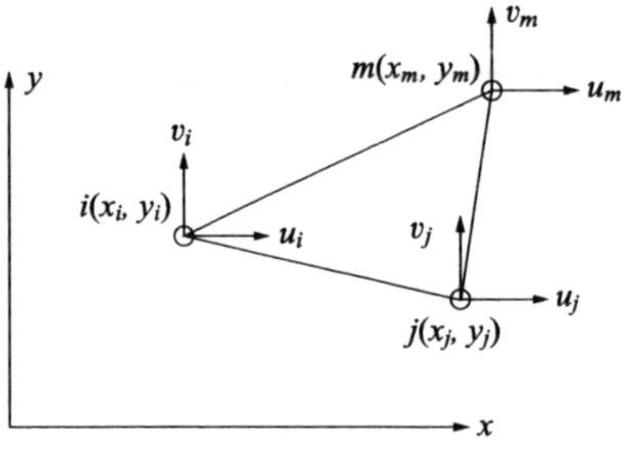

Figure 6–7 Basic triangular element showing degrees of freedom

to avoid problems in the calculations such as negative element areas. Here (x_i, y_i), (x_j, y_j), and (x_m, y_m) are the known nodal coordinates of nodes i, j, and m, respectively.

The nodal displacement matrix is given by

$$\{d\} = \begin{Bmatrix} \underline{d}_i \\ \underline{d}_j \\ \underline{d}_m \end{Bmatrix} = \begin{Bmatrix} u_i \\ v_i \\ u_j \\ v_j \\ u_m \\ v_m \end{Bmatrix} \tag{6.2.1}$$

Step 2 Select Displacement Functions

We select a linear displacement function for each element as

$$\begin{aligned} u(x, y) &= a_1 + a_2 x + a_3 y \\ v(x, y) &= a_4 + a_5 x + a_6 y \end{aligned} \tag{6.2.2}$$

where $u(x, y)$ and $v(x, y)$ describe displacements at any interior point (x_i, y_i) of the element.

The linear function ensures that compatibility will be satisfied. A linear function with specified endpoints has only one path through which to pass—that is, through the two points. Hence, the linear function ensures that the displacements along the edge and at the nodes shared by adjacent elements, such as edge i-j of the two elements shown in Figure 6–6(b), are equal. Using Eqs. (6.2.2), the general displacement function $\{\psi\}$, which stores the functions u and v, can be expressed as

$$\{\psi\} = \begin{Bmatrix} a_1 + a_2 x + a_3 y \\ a_4 + a_5 x + a_6 y \end{Bmatrix} = \begin{bmatrix} 1 & x & y & 0 & 0 & 0 \\ 0 & 0 & 0 & 1 & x & y \end{bmatrix} \begin{Bmatrix} a_1 \\ a_2 \\ a_3 \\ a_4 \\ a_5 \\ a_6 \end{Bmatrix} \tag{6.2.3}$$

To obtain the a's in Eqs. (6.2.2), we begin by substituting the coordinates of the nodal points into Eqs. (6.2.2) to yield

$$\begin{aligned} u_i &= u(x_i, y_i) = a_1 + a_2 x_i + a_3 y_i \\ u_j &= u(x_j, y_j) = a_1 + a_2 x_j + a_3 y_j \\ u_m &= u(x_m, y_m) = a_1 + a_2 x_m + a_3 y_m \\ v_i &= v(x_i, y_i) = a_4 + a_5 x_i + a_6 y_i \\ v_j &= v(x_j, y_j) = a_4 + a_5 x_j + a_6 y_j \\ v_m &= v(x_m, y_m) = a_4 + a_5 x_m + a_6 y_m \end{aligned} \tag{6.2.4}$$

We can solve for the a's beginning with the first three of Eqs. (6.2.4) expressed in matrix form as

$$\begin{Bmatrix} u_i \\ u_j \\ u_m \end{Bmatrix} = \begin{bmatrix} 1 & x_i & y_i \\ 1 & x_j & y_j \\ 1 & x_m & y_m \end{bmatrix} \begin{Bmatrix} a_1 \\ a_2 \\ a_3 \end{Bmatrix} \tag{6.2.5}$$

or, solving for the a's, we have

$$\{a\} = [x]^{-1}\{u\} \tag{6.2.6}$$

where $[x]$ is the 3×3 matrix on the right side of Eq. (6.2.5). The method of cofactors (Appendix A) is one possible method for finding the inverse of $[x]$. Thus,

$$[x]^{-1} = \frac{1}{2A} \begin{bmatrix} \alpha_i & \alpha_j & \alpha_m \\ \beta_i & \beta_j & \beta_m \\ \gamma_i & \gamma_j & \gamma_m \end{bmatrix} \tag{6.2.7}$$

where

$$2A = \begin{vmatrix} 1 & x_i & y_i \\ 1 & x_j & y_j \\ 1 & x_m & y_m \end{vmatrix} \tag{6.2.8}$$

is the determinant of $[x]$, which on evaluation is

$$2A = x_i(y_j - y_m) + x_j(y_m - y_i) + x_m(y_i - y_j) \tag{6.2.9}$$

Here A is the area of the triangle, and

$$\begin{aligned} &\alpha_i = x_j y_m - y_j x_m \qquad &&\alpha_j = y_i x_m - x_i y_m \qquad &&\alpha_m = x_i y_j - y_i x_j \\ &\beta_i = y_j - y_m &&\beta_j = y_m - y_i &&\beta_m = y_i - y_j \\ &\gamma_i = x_m - x_j &&\gamma_j = x_i - x_m &&\gamma_m = x_j - x_i \end{aligned} \tag{6.2.10}$$

Having determined $[x]^{-1}$, we can now express Eq. (6.2.6) in expanded matrix form as

$$\begin{Bmatrix} a_1 \\ a_2 \\ a_3 \end{Bmatrix} = \frac{1}{2A} \begin{bmatrix} \alpha_i & \alpha_j & \alpha_m \\ \beta_i & \beta_j & \beta_m \\ \gamma_i & \gamma_j & \gamma_m \end{bmatrix} \begin{Bmatrix} u_i \\ u_j \\ u_m \end{Bmatrix} \tag{6.2.11}$$

Similarly, using the last three of Eqs. (6.2.4), we can obtain

$$\begin{Bmatrix} a_4 \\ a_5 \\ a_6 \end{Bmatrix} = \frac{1}{2A} \begin{bmatrix} \alpha_i & \alpha_j & \alpha_m \\ \beta_i & \beta_j & \beta_m \\ \gamma_i & \gamma_j & \gamma_m \end{bmatrix} \begin{Bmatrix} v_i \\ v_j \\ v_m \end{Bmatrix} \tag{6.2.12}$$

We will derive the general x displacement function $u(x, y)$ of $\{\psi\}$ (v will follow analogously) in terms of the coordinate variables x and y, known coordinate variables

$\alpha_i, \alpha_j, \ldots, \gamma_m$, and unknown nodal displacements u_i, u_j, and u_m. Beginning with Eqs. (6.2.2) expressed in matrix form, we have

$$\{u\} = [1 \quad x \quad y]\begin{Bmatrix} a_1 \\ a_2 \\ a_3 \end{Bmatrix} \tag{6.2.13}$$

Substituting Eq. (6.2.11) into Eq. (6.2.13), we obtain

$$\{u\} = \frac{1}{2A}[1 \quad x \quad y]\begin{bmatrix} \alpha_i & \alpha_j & \alpha_m \\ \beta_i & \beta_j & \beta_m \\ \gamma_i & \gamma_j & \gamma_m \end{bmatrix}\begin{Bmatrix} u_i \\ u_j \\ u_m \end{Bmatrix} \tag{6.2.14}$$

Expanding Eq. (6.2.14), we have

$$\{u\} = \frac{1}{2A}[1 \quad x \quad y]\begin{Bmatrix} \alpha_i u_i + \alpha_j u_j + \alpha_m u_m \\ \beta_i u_i + \beta_j u_j + \beta_m u_m \\ \gamma_i u_i + \gamma_j u_j + \gamma_m u_m \end{Bmatrix} \tag{6.2.15}$$

Multiplying the two matrices in Eq. (6.2.15) and rearranging, we obtain

$$u(x, y) = \frac{1}{2A}\{(\alpha_i + \beta_i x + \gamma_i y)u_i + (\alpha_j + \beta_j x + \gamma_j y)u_j + (\alpha_m + \beta_m x + \gamma_m y)u_m\} \tag{6.2.16}$$

Similarly, replacing u_i by v_i, u_j by v_j, and u_m by v_m in Eq. (6.2.16), we have the y displacement given by

$$v(x, y) = \frac{1}{2A}\{(\alpha_i + \beta_i x + \gamma_i y)v_i + (\alpha_j + \beta_j x + \gamma_j y)v_j + (\alpha_m + \beta_m x + \gamma_m y)v_m\} \tag{6.2.17}$$

To express Eqs. (6.2.16) and (6.2.17) for u and v in simpler form, we define

$$\begin{aligned} N_i &= \frac{1}{2A}(\alpha_i + \beta_i x + \gamma_i y) \\ N_j &= \frac{1}{2A}(\alpha_j + \beta_j x + \gamma_j y) \\ N_m &= \frac{1}{2A}(\alpha_m + \beta_m x + \gamma_m y) \end{aligned} \tag{6.2.18}$$

Thus, using Eqs. (6.2.18), we can rewrite Eqs. (6.2.16) and (6.2.17) as

$$\begin{aligned} u(x, y) &= N_i u_i + N_j u_j + N_m u_m \\ v(x, y) &= N_i v_i + N_j v_j + N_m v_m \end{aligned} \tag{6.2.19}$$

Expressing Eqs. (6.2.19) in matrix form, we obtain

$$\{\psi\} = \begin{Bmatrix} u(x,y) \\ v(x,y) \end{Bmatrix} = \begin{Bmatrix} N_i u_i + N_j u_j + N_m u_m \\ N_i v_i + N_j v_j + N_m v_m \end{Bmatrix}$$

or

$$\{\psi\} = \begin{bmatrix} N_i & 0 & N_j & 0 & N_m & 0 \\ 0 & N_i & 0 & N_j & 0 & N_m \end{bmatrix} \begin{Bmatrix} u_i \\ v_i \\ u_j \\ v_j \\ u_m \\ v_m \end{Bmatrix} \tag{6.2.20}$$

Finally, expressing Eq. (6.2.20) in abbreviated matrix form, we have

$$\{\psi\} = [N]\{d\} \tag{6.2.21}$$

where $[N]$ is given by

$$[N] = \begin{bmatrix} N_i & 0 & N_j & 0 & N_m & 0 \\ 0 & N_i & 0 & N_j & 0 & N_m \end{bmatrix} \tag{6.2.22}$$

We have now expressed the general displacements as functions of $\{d\}$, in terms of the shape functions N_i, N_j, and N_m. The shape functions represent the shape of $\{\psi\}$ when plotted over the surface of a typical element. For instance, N_i represents the shape of the variable u when plotted over the surface of the element for $u_i = 1$ and all other degrees of freedom equal to zero; that is, $u_j = u_m = v_i = v_j = v_m = 0$. In addition, $u(x_i, y_i)$ must be equal to u_i. Therefore, we must have $N_i = 1$, $N_j = 0$, and $N_m = 0$ at (x_i, y_i). Similarly, $u(x_j, y_j) = u_j$. Therefore, $N_i = 0$, $N_j = 1$, and $N_m = 0$ at (x_j, y_j). Figure 6–8 shows the shape variation of N_i plotted over the surface of a typical element. Note that N_i does not equal zero except along a line connecting and including nodes j and m.

Finally, $N_i + N_j + N_m = 1$ for all x and y locations on the surface of the element so that u and v will yield a constant value when rigid-body displacement occurs. The proof of this relationship follows that given for the bar element in Section 3.2 and is left as an exercise (Problem 6.1). The shape functions are also used to determine the body and surface forces at element nodes, as described in Section 6.3.

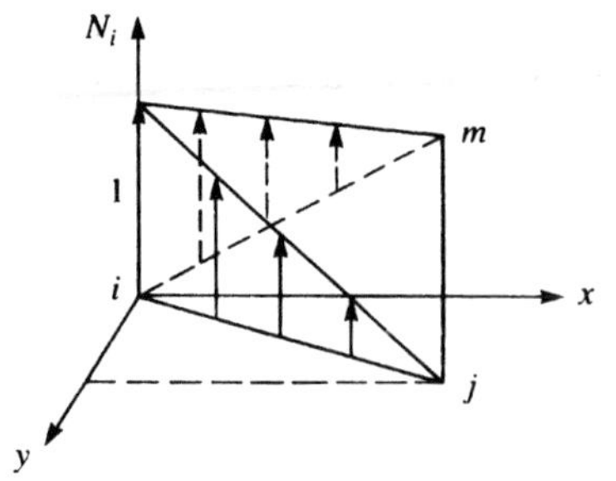

Figure 6–8 Variation of N_i over the x-y surface of a typical element

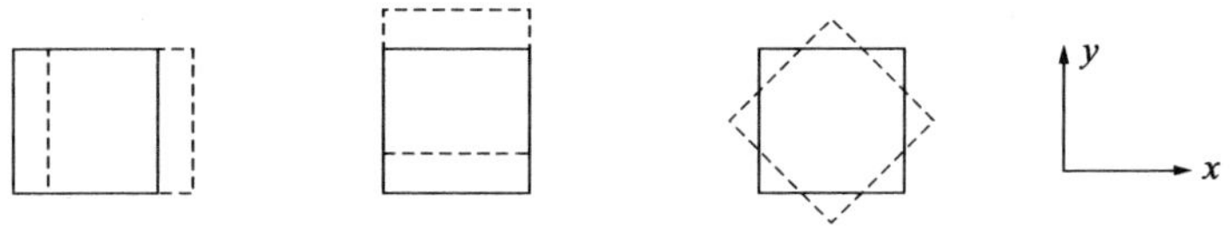

(a) Rigid-body modes of a plane stress element (from left to right, pure translation in x and y directions and pure rotation)

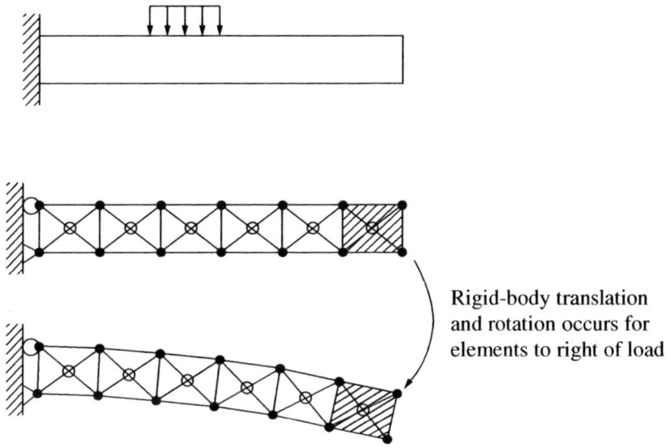

(b) Cantilever beam modeled using constant-strain triangle elements; elements to the right of the loading are stress-free

Figure 6–9 Unstressed elements in a cantilever beam modeled with CST

The requirement of completeness for the constant-strain triangle element used in a two-dimensional plane stress element is illustrated in Figure 6–9. The element must be able to translate uniformly in either the x or y direction in the plane and to rotate without straining as shown in Figure 6–9(a). The reason that the element must be able to translate as a rigid body and to rotate stress-free is illustrated in the example of a cantilever beam modeled with plane stress elements as shown in Figure 6–9(b). By simple statics, the beam elements beyond the loading are stress-free. Hence these elements must be free to translate and rotate without stretching or changing shape.

Step 3 Define the Strain/ Displacement and Stress/Strain Relationships

We express the element strains and stresses in terms of the unknown nodal displacements.

Element Strains

The strains associated with the two-dimensional element are given by

$$\{\varepsilon\} = \begin{Bmatrix} \varepsilon_x \\ \varepsilon_y \\ \gamma_{xy} \end{Bmatrix} = \begin{Bmatrix} \dfrac{\partial u}{\partial x} \\ \dfrac{\partial v}{\partial y} \\ \dfrac{\partial u}{\partial y} + \dfrac{\partial v}{\partial x} \end{Bmatrix} \tag{6.2.23}$$

Using Eqs. (6.2.19) for the displacements, we have

$$\frac{\partial u}{\partial x} = u_{,x} = \frac{\partial}{\partial x}(N_i u_i + N_j u_j + N_m u_m) \tag{6.2.24}$$

or

$$u_{,x} = N_{i,x} u_i + N_{j,x} u_j + N_{m,x} u_m \tag{6.2.25}$$

where the comma followed by a variable indicates differentiation with respect to that variable. We have used $u_{i,x} = 0$ because $u_i = u(x_i, y_i)$ is a constant value; similarly, $u_{j,x} = 0$ and $u_{m,x} = 0$.

Using Eqs. (6.2.18), we can evaluate the expressions for the derivatives of the shape functions in Eq. (6.2.25) as follows:

$$N_{i,x} = \frac{1}{2A}\frac{\partial}{\partial x}(\alpha_i + \beta_i x + \gamma_i y) = \frac{\beta_i}{2A} \tag{6.2.26}$$

Similarly,

$$N_{j,x} = \frac{\beta_j}{2A} \quad \text{and} \quad N_{m,x} = \frac{\beta_m}{2A} \tag{6.2.27}$$

Therefore, using Eqs. (6.2.26) and (6.2.27) in Eq. (6.2.25), we have

$$\frac{\partial u}{\partial x} = \frac{1}{2A}(\beta_i u_i + \beta_j u_j + \beta_m u_m) \tag{6.2.28}$$

Similarly, we can obtain

$$\begin{aligned} \frac{\partial v}{\partial y} &= \frac{1}{2A}(\gamma_i v_i + \gamma_j v_j + \gamma_m v_m) \\ \frac{\partial u}{\partial y} + \frac{\partial v}{\partial x} &= \frac{1}{2A}(\gamma_i u_i + \beta_i v_i + \gamma_j u_j + \beta_j v_j + \gamma_m u_m + \beta_m v_m) \end{aligned} \tag{6.2.29}$$

Using Eqs. (6.2.28) and (6.2.29) in Eq. (6.2.23), we obtain

$$\{\varepsilon\} = \frac{1}{2A}\begin{bmatrix} \beta_i & 0 & \beta_j & 0 & \beta_m & 0 \\ 0 & \gamma_i & 0 & \gamma_j & 0 & \gamma_m \\ \gamma_i & \beta_i & \gamma_j & \beta_j & \gamma_m & \beta_m \end{bmatrix} \begin{Bmatrix} u_i \\ v_i \\ u_j \\ v_j \\ u_m \\ v_m \end{Bmatrix} \tag{6.2.30}$$

or
$$\{\varepsilon\} = [\underline{B}_i \quad \underline{B}_j \quad \underline{B}_m]\begin{Bmatrix} \underline{d}_i \\ \underline{d}_j \\ \underline{d}_m \end{Bmatrix} \tag{6.2.31}$$

where

$$[B_i] = \frac{1}{2A}\begin{bmatrix} \beta_i & 0 \\ 0 & \gamma_i \\ \gamma_i & \beta_i \end{bmatrix} \qquad [B_j] = \frac{1}{2A}\begin{bmatrix} \beta_j & 0 \\ 0 & \gamma_j \\ \gamma_j & \beta_j \end{bmatrix} \qquad [B_m] = \frac{1}{2A}\begin{bmatrix} \beta_m & 0 \\ 0 & \gamma_m \\ \gamma_m & \beta_m \end{bmatrix} \tag{6.2.32}$$

Finally, in simplified matrix form, Eq. (6.2.31) can be written as

$$\{\varepsilon\} = [B]\{d\} \tag{6.2.33}$$

where
$$[B] = [\underline{B}_i \quad \underline{B}_j \quad \underline{B}_m] \tag{6.2.34}$$

The $\underline{B}$ matrix is independent of the x and y coordinates. It depends solely on the element nodal coordinates, as seen from Eqs. (6.2.32) and (6.2.10). The strains in Eq. (6.2.33) will be constant; hence, the element is called a *constant-strain triangle* (CST).

Stress/Strain Relationship

In general, the in-plane stress/strain relationship is given by

$$\begin{Bmatrix} \sigma_x \\ \sigma_y \\ \tau_{xy} \end{Bmatrix} = [D]\begin{Bmatrix} \varepsilon_x \\ \varepsilon_y \\ \gamma_{xy} \end{Bmatrix} \tag{6.2.35}$$

where $[D]$ is given by Eq. (6.1.8) for plane stress problems and by Eq. (6.1.10) for plane strain problems. Using Eq. (6.2.33) in Eq. (6.2.35), we obtain the in-plane stresses in terms of the unknown nodal degrees of freedom as

$$\{\sigma\} = [D][B]\{d\} \tag{6.2.36}$$

where the stresses $\{\sigma\}$ are also constant everywhere within the element.

Step 4 Derive the Element Stiffness Matrix and Equations

Using the principle of minimum potential energy, we can generate the equations for a typical constant-strain triangular element. Keep in mind that for the basic plane stress element, the total potential energy is now a function of the nodal displacements $u_i, v_i, u_j, \ldots, v_m$ (that is, $\{d\}$) such that

$$\pi_p = \pi_p(u_i, v_i, u_j, \ldots, v_m) \tag{6.2.37}$$

Here the total potential energy is given by

$$\pi_p = U + \Omega_b + \Omega_p + \Omega_s \tag{6.2.38}$$

where the strain energy is given by

$$U = \frac{1}{2}\iiint_V \{\varepsilon\}^T\{\sigma\}\,dV \tag{6.2.39}$$

or, using Eq. (6.2.35), we have

$$U = \frac{1}{2}\iiint_V \{\varepsilon\}^T[D]\{\varepsilon\}\,dV \tag{6.2.40}$$

where we have used $[D]^T = [D]$ in Eq. (6.2.40).

The potential energy of the body forces is given by

$$\Omega_b = -\iiint_V \{\psi\}^T\{X\}\,dV \tag{6.2.41}$$

where $\{\psi\}$ is again the general displacement function, and $\{X\}$ is the body weight/unit volume or weight density matrix (typically, in units of pounds per cubic inch or kilonewtons per cubic meter).

The potential energy of concentrated loads is given by

$$\Omega_p = -\{d\}^T\{P\} \tag{6.2.42}$$

where $\{d\}$ represents the usual nodal displacements, and $\{P\}$ now represents the concentrated external loads.

The potential energy of distributed loads (or surface tractions) moving through respective surface displacements is given by

$$\Omega_s = -\iint_S \{\psi_S\}^T\{T_S\}\,dS \tag{6.2.43}$$

where $\{T_S\}$ represents the surface tractions (typically in units of pounds per square inch or kilonewtons per square meter), $\{\psi_S\}$ represents the field of surface displacements through which the surface tractions act, and S represents the surfaces over which the tractions $\{T_S\}$ act. Similar to Eq. (6.2.21), we express $\{\psi_S\}$ as $\{\psi_S\} = [N_S]\{d\}$, where $[N_S]$ represents the shape function matrix evaluated along the surface where the surface traction acts.

Using Eq. (6.2.21) for $\{\psi\}$ and Eq. (6.2.33) for the strains in Eqs. (6.2.40)–(6.2.43), we have

$$\pi_p = \frac{1}{2}\iiint_V \{d\}^T[B]^T[D][B]\{d\}\,dV - \iiint_V \{d\}^T[N]^T\{X\}\,dV$$
$$- \{d\}^T\{P\} - \iint_S \{d\}^T[N_S]^T\{T_S\}\,dS \tag{6.2.44}$$

The nodal displacements $\{d\}$ are independent of the general x-y coordinates, so $\{d\}$ can be taken out of the integrals of Eq. (6.2.44). Therefore,

$$\pi_p = \frac{1}{2}\{d\}^T \iiint\limits_V [B]^T[D][B]\,dV\{d\} - \{d\}^T \iiint\limits_V [N]^T\{X\}\,dV - \{d\}^T\{P\} - \{d\}^T \iint\limits_S [N_S]^T\{T_S\}\,dS \tag{6.2.45}$$

From Eqs. (6.2.41)–(6.2.43) we can see that the last three terms of Eq. (6.2.45) represent the total load system $\{f\}$ on an element; that is,

$$\{f\} = \iiint\limits_V [N]^T\{X\}\,dV + \{P\} + \iint\limits_S [N_S]^T\{T_S\}\,dS \tag{6.2.46}$$

where the first, second, and third terms on the right side of Eq. (6.2.46) represent the body forces, the concentrated nodal forces, and the surface tractions, respectively. Using Eq. (6.2.46) in Eq. (6.2.45), we obtain

$$\pi_p = \frac{1}{2}\{d\}^T \iiint\limits_V [B]^T[D][B]\,dV\{d\} - \{d\}^T\{f\} \tag{6.2.47}$$

Taking the first variation, or equivalently, as shown in Chapters 2 and 3, the partial derivative of π_p with respect to the nodal displacements since $\pi_p = \pi_p(\underline{d})$ (as was previously done for the bar and beam elements in Chapters 3 and 4, respectively), we obtain

$$\frac{\partial \pi_p}{\partial\{d\}} = \left[\iiint\limits_V [B]^T[D][B]\,dV\right]\{d\} - \{f\} = 0 \tag{6.2.48}$$

Rewriting Eq. (6.2.48), we have

$$\iiint\limits_V [B]^T[D][B]\,dV\{d\} = \{f\} \tag{6.2.49}$$

where the partial derivative with respect to matrix $\{d\}$ was previously defined by Eq. (2.6.12). From Eq. (6.2.49) we can see that

$$[k] = \iiint\limits_V [B]^T[D][B]\,dV \tag{6.2.50}$$

For an element with constant thickness, t, Eq. (6.2.50) becomes

$$[k] = t\iint\limits_A [B]^T[D][B]\,dx\,dy \tag{6.2.51}$$

where the integrand is not a function of x or y for the constant-strain triangular element and thus can be taken out of the integral to yield

$$[k] = tA[B]^T[D][B] \tag{6.2.52}$$

where A is given by Eq. (6.2.9), $[B]$ is given by Eq. (6.2.34), and $[D]$ is given by Eq. (6.1.8) or Eq. (6.1.10). We will assume elements of constant thickness. (This assumption is convergent to the actual situation as the element size is decreased.)

From Eq. (6.2.52) we see that $[k]$ is a function of the nodal coordinates (because $[B]$ and A are defined in terms of them) and of the mechanical properties E and ν (of which $[D]$ is a function). The expansion of Eq. (6.2.52) for an element is

$$[k] = \begin{bmatrix} [k_{ii}] & [k_{ij}] & [k_{im}] \\ [k_{ji}] & [k_{jj}] & [k_{jm}] \\ [k_{mi}] & [k_{mj}] & [k_{mm}] \end{bmatrix} \tag{6.2.53}$$

where the 2×2 submatrices are given by

$$\begin{aligned} [k_{ii}] &= [B_i]^T [D][B_i]tA \\ [k_{ij}] &= [B_i]^T [D][B_j]tA \\ [k_{im}] &= [B_i]^T [D][B_m]tA \end{aligned} \tag{6.2.54}$$

and so forth. In Eqs. (6.2.54), $[B_i]$, $[B_j]$, and $[B_m]$ are defined by Eqs. (6.2.32). The $[k]$ matrix is seen to be a 6×6 matrix (equal in order to the number of degrees of freedom per node, two, times the total number of nodes per element, three).

In general, Eq. (6.2.46) must be used to evaluate the surface and body forces. When Eq. (6.2.46) is used to evaluate the surface and body forces, these forces are called *consistent loads* because they are derived from the consistent (energy) approach. For higher-order elements, typically with quadratic or cubic displacement functions, Eq. (6.2.46) should be used. However, for the CST element, the body and surface forces can be lumped at the nodes with equivalent results (this is illustrated in Section 6.3) and added to any concentrated nodal forces to obtain the element force matrix. The element equations are then given by

$$\begin{Bmatrix} f_{1x} \\ f_{1y} \\ f_{2x} \\ f_{2y} \\ f_{3x} \\ f_{3y} \end{Bmatrix} = \begin{bmatrix} k_{11} & k_{12} & \dots & k_{16} \\ k_{21} & k_{22} & \dots & k_{26} \\ \vdots & \vdots & & \vdots \\ k_{61} & k_{62} & \dots & k_{66} \end{bmatrix} \begin{Bmatrix} u_1 \\ v_1 \\ u_2 \\ v_2 \\ u_3 \\ v_3 \end{Bmatrix} \tag{6.2.55}$$

Step 5 Assemble the Element Equations to Obtain the Global Equations and Introduce Boundary Conditions

We obtain the global structure stiffness matrix and equations by using the direct stiffness method as

$$[K] = \sum_{e=1}^{N} [k^{(e)}] \tag{6.2.56}$$

and
$$\{F\} = [K]\{d\} \tag{6.2.57}$$

where, in Eq. (6.2.56), all element stiffness matrices are defined in terms of the global x-y coordinate system, $\{d\}$ is now the total structure displacement matrix, and

$$\{F\} = \sum_{e=1}^{N} \{f^{(e)}\} \tag{6.2.58}$$

is the column of equivalent global nodal loads obtained by lumping body forces and distributed loads at the proper nodes (as well as including concentrated nodal loads) or by consistently using Eq. (6.2.46). (Further details regarding the treatment of body forces and surface tractions will be given in Section 6.3.)

In the formulation of the element stiffness matrix Eq. (6.2.52), the matrix has been derived for a general orientation in global coordinates. Equation (6.2.52) then applies for all elements. All element matrices are expressed in the global-coordinate orientation. Therefore, no transformation from local to global equations is necessary. However, for completeness, we will now describe the method to use if the local axes for the constant-strain triangular element are not parallel to the global axes for the whole structure.

If the local axes for the constant-strain triangular element are not parallel to the global axes for the whole structure, we must apply rotation-of-axes transformations similar to those introduced in Chapter 3 by Eq. (3.3.16) to the element stiffness matrix, as well as to the element nodal force and displacement matrices. We illustrate the transformation of axes for the triangular element shown in Figure 6–10, considering the element to have local axes $\hat{x}$-$\hat{y}$ not parallel to global axes x-y. Local nodal forces are shown in the figure. The transformation from local to global equations follows the procedure outlined in Section 3.4. We have the same general expressions, Eqs. (3.4.14), (3.4.16), and (3.4.22), to relate local to global displacements, forces, and stiffness matrices, respectively; that is,

$$\underline{\hat{d}} = \underline{T}\underline{d} \qquad \underline{\hat{f}} = \underline{T}\underline{f} \qquad \underline{k} = \underline{T}^T\underline{\hat{k}}\underline{T} \tag{6.2.59}$$

where Eq. (3.4.15) for the transformation matrix $\underline{T}$ used in Eqs. (6.2.59) must be expanded because two additional degrees of freedom are present in the constant-strain

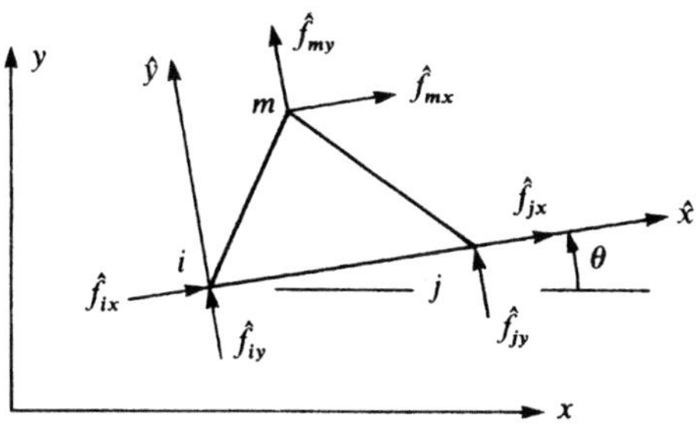

Figure 6–10 Triangular element with local axes not parallel to global axes

triangular element. Thus, Eq. (3.4.15) is expanded to

$$\underline{T} = \begin{bmatrix} C & S & 0 & 0 & 0 & 0 \\ -S & C & 0 & 0 & 0 & 0 \\ 0 & 0 & C & S & 0 & 0 \\ 0 & 0 & -S & C & 0 & 0 \\ 0 & 0 & 0 & 0 & C & S \\ 0 & 0 & 0 & 0 & -S & C \end{bmatrix} \begin{matrix} u_i \\ v_i \\ u_j \\ v_j \\ u_m \\ v_m \end{matrix} \qquad (6.2.60)$$

where $C = \cos\theta$, $S = \sin\theta$, and θ is shown in Figure 6–10.

Step 6 Solve for the Nodal Displacements

We determine the unknown global structure nodal displacements by solving the system of algebraic equations given by Eq. (6.2.57).

Step 7 Solve for the Element Forces (Stresses)

Having solved for the nodal displacements, we obtain the strains and stresses in the global x and y directions in the elements by using Eqs. (6.2.33) and (6.2.36). Finally, we determine the maximum and minimum in-plane principal stresses σ_1 and σ_2 by using the transformation Eqs. (6.1.2), where these stresses are usually assumed to act at the centroid of the element. The angle that one of the principal stresses makes with the x axis is given by Eq. (6.1.3).

Example 6.1

Evaluate the stiffness matrix for the element shown in Figure 6–11. The coordinates are shown in units of inches. Assume plane stress conditions. Let $E = 30 \times 10^6$ psi, $\nu = 0.25$, and thickness $t = 1$ in. Assume the element nodal displacements have been determined to be $u_1 = 0.0$, $v_1 = 0.0025$ in., $u_2 = 0.0012$ in., $v_2 = 0.0$, $u_3 = 0.0$, and $v_3 = 0.0025$ in. Determine the element stresses.

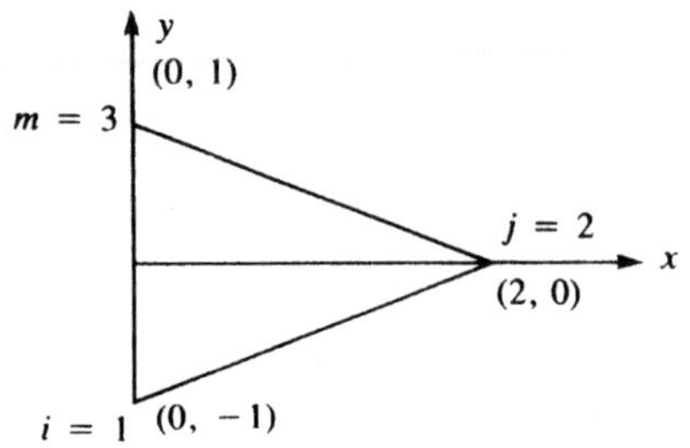

Figure 6–11 Plane stress element for stiffness matrix evaluation

We use Eq. (6.2.52) to obtain the element stiffness matrix. To evaluate $\underline{k}$, we first use Eqs. (6.2.10) to obtain the β's and γ's as follows:

$$\begin{aligned} \beta_i &= y_j - y_m = 0 - 1 = -1 & \gamma_i &= x_m - x_j = 0 - 2 = -2 \\ \beta_j &= y_m - y_i = 1 - (-1) = 2 & \gamma_j &= x_i - x_m = 0 - 0 = 0 \\ \beta_m &= y_i - y_j = -1 - 0 = -1 & \gamma_m &= x_j - x_i = 2 - 0 = 2 \end{aligned} \tag{6.2.61}$$

Using Eqs. (6.2.32) and (6.2.34), we obtain matrix $\underline{B}$ as

$$\underline{B} = \frac{1}{2(2)} \begin{bmatrix} -1 & 0 & 2 & 0 & -1 & 0 \\ 0 & -2 & 0 & 0 & 0 & 2 \\ -2 & -1 & 0 & 2 & 2 & -1 \end{bmatrix} \tag{6.2.62}$$

where we have used $A = 2$ in.^{2} in Eq. (6.2.62).

Using Eq. (6.1.8) for plane stress conditions,

$$\underline{D} = \frac{30 \times 10^6}{1 - (0.25)^2} \begin{bmatrix} 1 & 0.25 & 0 \\ 0.25 & 1 & 0 \\ 0 & 0 & \dfrac{1 - 0.25}{2} \end{bmatrix} \text{psi} \tag{6.2.63}$$

Substituting Eqs. (6.2.62) and (6.2.63) into Eq. (6.2.52), we obtain

$$\underline{k} = \frac{(2)30 \times 10^6}{4(0.9375)} \begin{bmatrix} -1 & 0 & -2 \\ 0 & -2 & -1 \\ 2 & 0 & 0 \\ 0 & 0 & 2 \\ -1 & 0 & 2 \\ 0 & 2 & -1 \end{bmatrix}$$

$$\times \begin{bmatrix} 1 & 0.25 & 0 \\ 0.25 & 1 & 0 \\ 0 & 0 & 0.375 \end{bmatrix} \frac{1}{2(2)} \begin{bmatrix} -1 & 0 & 2 & 0 & -1 & 0 \\ 0 & -2 & 0 & 0 & 0 & 2 \\ -2 & -1 & 0 & 2 & 2 & -1 \end{bmatrix}$$

Performing the matrix triple product, we have

$$\underline{k} = 4.0 \times 10^6 \begin{bmatrix} 2.5 & 1.25 & -2 & -1.5 & -0.5 & 0.25 \\ 1.25 & 4.375 & -1 & -0.75 & -0.25 & -3.625 \\ -2 & -1 & 4 & 0 & -2 & 1 \\ -1.5 & -0.75 & 0 & 1.5 & 1.5 & -0.75 \\ -0.5 & -0.25 & -2 & 1.5 & 2.5 & -1.25 \\ 0.25 & -3.625 & 1 & -0.75 & -1.25 & 4.375 \end{bmatrix} \frac{\text{lb}}{\text{in.}} \tag{6.2.64}$$

To evaluate the stresses, we use Eq. (6.2.36). Substituting Eqs. (6.2.62) and (6.2.63), along with the given nodal displacements, into Eq. (6.2.36), we obtain

$$\begin{Bmatrix} \sigma_x \\ \sigma_y \\ \tau_{xy} \end{Bmatrix} = \frac{30 \times 10^6}{1-(0.25)^2} \begin{bmatrix} 1 & 0.25 & 0 \\ 0.25 & 1 & 0 \\ 0 & 0 & 0.375 \end{bmatrix}$$

$$\times \frac{1}{2(2)} \begin{bmatrix} -1 & 0 & 2 & 0 & -1 & 0 \\ 0 & -2 & 0 & 0 & 0 & 2 \\ -2 & -1 & 0 & 2 & 2 & -1 \end{bmatrix} \begin{Bmatrix} 0.0 \\ 0.0025 \\ 0.0012 \\ 0.0 \\ 0.0 \\ 0.0025 \end{Bmatrix} \tag{6.2.65}$$

Performing the matrix triple product in Eq. (6.2.65), we have

$$\sigma_x = 19{,}200 \text{ psi} \qquad \sigma_y = 4800 \text{ psi} \qquad \tau_{xy} = -15{,}000 \text{ psi} \tag{6.2.66}$$

Finally, the principal stresses and principal angle are obtained by substituting the results from Eqs. (6.2.66) into Eqs. (6.1.2) and (6.1.3) as follows:

$$\sigma_1 = \frac{19{,}200 + 4800}{2} + \left[\left(\frac{19{,}200 - 4800}{2}\right)^2 + (-15{,}000)^2\right]^{1/2}$$

$$= 28{,}639 \text{ psi}$$

$$\sigma_2 = \frac{19{,}200 + 4800}{2} - \left[\left(\frac{19{,}200 - 4800}{2}\right)^2 + (-15{,}000)^2\right]^{1/2} \tag{6.2.67}$$

$$= -4639 \text{ psi}$$

$$\theta_p = \frac{1}{2}\tan^{-1}\left[\frac{2(-15{,}000)}{19{,}200 - 4800}\right] = -32.2^\circ$$

■

6.3 Treatment of Body and Surface Forces

Body Forces

Using the first term on the right side of Eq. (6.2.46), we can evaluate the body forces at the nodes as

$$\{f_b\} = \iiint_V [N]^T \{X\}\, dV \tag{6.3.1}$$

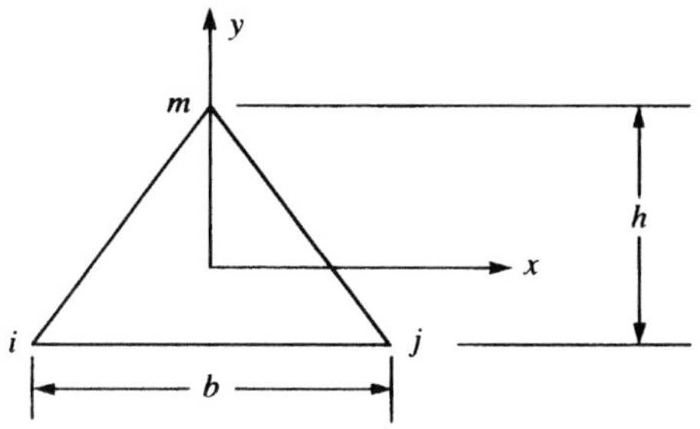

Figure 6–12 Element with centroidal coordinate axes

where
$$\{X\} = \begin{Bmatrix} X_b \\ Y_b \end{Bmatrix} \tag{6.3.2}$$

and X_b and Y_b are the weight densities in the x and y directions in units of force/unit volume, respectively. These forces may arise, for instance, because of actual body weight (gravitational forces), angular velocity (called *centrifugal body forces*, as described in Chapter 9), or inertial forces in dynamics.

In Eq. (6.3.1), $[N]$ is a linear function of x and y; therefore, the integration must be carried out. Without lack of generality, the integration is simplified if the origin of the coordinates is chosen at the centroid of the element. For example, consider the element with coordinates shown in Figure 6–12. With the origin of the coordinate placed at the centroid of the element, we have, from the definition of the centroid, $\iint x\,dA = \iint y\,dA = 0$ and therefore,

$$\iint \beta_i x\,dA = \iint \gamma_i y\,dA = 0 \tag{6.3.3}$$

and
$$\alpha_i = \alpha_j = \alpha_m = \frac{2A}{3} \tag{6.3.4}$$

Using Eqs. (6.3.2)–(6.3.4) in Eq. (6.3.1), the body force at node i is then represented by

$$\{f_{bi}\} = \begin{Bmatrix} X_b \\ Y_b \end{Bmatrix} \frac{tA}{3} \tag{6.3.5}$$

Similarly, considering the j and m node body forces, we obtain the same results as in Eq. (6.3.5). In matrix form, the element body forces are

$$\{f_b\} = \begin{Bmatrix} f_{bix} \\ f_{biy} \\ f_{bjx} \\ f_{bjy} \\ f_{bmx} \\ f_{bmy} \end{Bmatrix} = \begin{Bmatrix} X_b \\ Y_b \\ X_b \\ Y_b \\ X_b \\ Y_b \end{Bmatrix} \frac{At}{3} \tag{6.3.6}$$

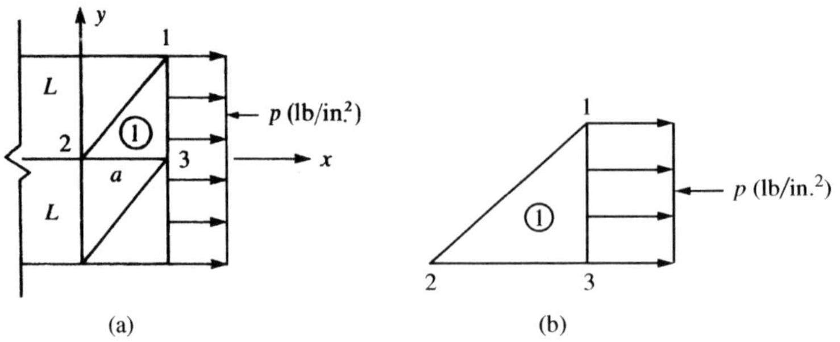

Figure 6–13 (a) Elements with uniform surface traction acting on one edge and (b) element one with uniform surface traction along edge 1–3

From the results of Eq. (6.3.6), we can conclude that the body forces are distributed to the nodes in three equal parts. The signs depend on the directions of X_b and Y_b with respect to the positive x and y global coordinates. For the case of body weight only, because of the gravitational force associated with the y direction, we have only Y_b $(X_b = 0)$.

Surface Forces

Using the third term on the right side of Eq. (6.2.46), we can evaluate the surface forces at the nodes as

$$\{f_s\} = \iint_S [N_S]^T \{T_S\}\, dS \tag{6.3.7}$$

We emphasize that the subscript S in $[N_S]$ in Eq. (6.3.7) means the shape functions evaluated along the surface where the surface traction is applied.

We will now illustrate the use of Eq. (6.3.7) by considering the example of a uniform stress p (say, in pounds per square inch) acting between nodes 1 and 3 on the edge of element 1 in Figure 6–13(b). In Eq. (6.3.7), the surface traction now becomes

$$\{T_S\} = \begin{Bmatrix} p_x \\ p_y \end{Bmatrix} = \begin{Bmatrix} p \\ 0 \end{Bmatrix} \tag{6.3.8}$$

and

$$[N_S]^T = \begin{bmatrix} N_1 & 0 \\ 0 & N_1 \\ N_2 & 0 \\ 0 & N_2 \\ N_3 & 0 \\ 0 & N_3 \end{bmatrix} \text{evaluated at } x = a,\ y = y \tag{6.3.9}$$

As the surface traction p acts along the edge at $x = a$ and $y = y$ from $y = 0$ to $y = L$, we evaluate the shape functions at $x = a$ and $y = y$ and integrate over the surface from 0 to L in the y direction and from 0 to t in the z direction, as shown by Eq. (6.3.10).

Using Eqs. (6.3.8) and (6.3.9), we express Eq. (6.3.7) as

$$\{f_s\} = \int_0^t \int_0^L \begin{bmatrix} N_1 & 0 \\ 0 & N_1 \\ N_2 & 0 \\ 0 & N_2 \\ N_3 & 0 \\ 0 & N_3 \end{bmatrix} \begin{Bmatrix} p \\ 0 \end{Bmatrix} dz\, dy \quad \text{evaluated at } x = a,\ y = y \tag{6.3.10}$$

Simplifying Eq. (6.3.10), we obtain

$$\{f_s\} = t \int_0^L \begin{bmatrix} N_1 p \\ 0 \\ N_2 p \\ 0 \\ N_3 p \\ 0 \end{bmatrix} dy \quad \text{evaluated at } x = a,\ y = y \tag{6.3.11}$$

Now, by Eqs. (6.2.18) (with $i = 1$), we have

$$N_1 = \frac{1}{2A}(\alpha_1 + \beta_1 x + \gamma_1 y) \tag{6.3.12}$$

For convenience, we choose the coordinate system for the element as shown in Figure 6–14. Using the definition Eqs. (6.2.10), we obtain

$$\alpha_i = x_j y_m - y_j x_m$$

or, with $i = 1$, $j = 2$, and $m = 3$,

$$\alpha_1 = x_2 y_3 - y_2 x_3 \tag{6.3.13}$$

Substituting the coordinates into Eq. (6.3.13), we obtain

$$\alpha_1 = 0 \tag{6.3.14}$$

Similarly, again using Eqs. (6.2.10), we obtain

$$\beta_1 = 0 \qquad \gamma_1 = a \tag{6.3.15}$$

Therefore, substituting Eqs. (6.3.14) and (6.3.15) into Eq. (6.3.12), we obtain

$$N_1 = \frac{ay}{2A} \tag{6.3.16}$$

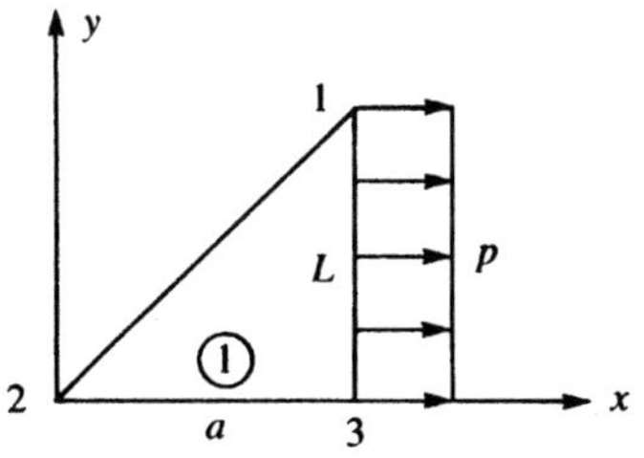

Figure 6–14 Representative element subjected to edge surface traction p

Similarly, using Eqs. (6.2.18), we can show that

$$N_2 = \frac{L(a-x)}{2A} \qquad \text{and} \qquad N_3 = \frac{Lx - ay}{2A} \tag{6.3.17}$$

On substituting Eqs. (6.3.16) and (6.3.17) for N_1, N_2, and N_3 into Eq. (6.3.11), evaluating N_1, N_2, and N_3 at $x = a$ and $y = y$ (the coordinates corresponding to the location of the surface load p), and then integrating with respect to y, we obtain

$$\{f_s\} = \frac{t}{2(aL/2)} \begin{Bmatrix} a\left(\frac{L^2}{2}\right)p \\ 0 \\ 0 \\ 0 \\ \left(L^2 - \frac{L^2}{2}\right)ap \\ 0 \end{Bmatrix} \tag{6.3.18}$$

where the shape function $N_2 = 0$ between nodes 1 and 3, as should be the case according to the definitions of the shape functions. Simplifying Eq. (6.3.18), we finally obtain

$$\{f_s\} = \begin{Bmatrix} f_{s1x} \\ f_{s1y} \\ f_{s2x} \\ f_{s2y} \\ f_{s3x} \\ f_{s3y} \end{Bmatrix} = \begin{Bmatrix} pLt/2 \\ 0 \\ 0 \\ 0 \\ pLt/2 \\ 0 \end{Bmatrix} \tag{6.3.19}$$

Figure 6–15 illustrates the results for the surface load equivalent nodal forces for both elements 1 and 2.

We can conclude that for a constant-strain triangle, a distributed load on an element edge can be treated as concentrated loads acting at the nodes associated with the loaded edge by making the two kinds of load statically equivalent [which is equivalent to applying Eq. (6.3.7)]. However, for higher-order elements such as the

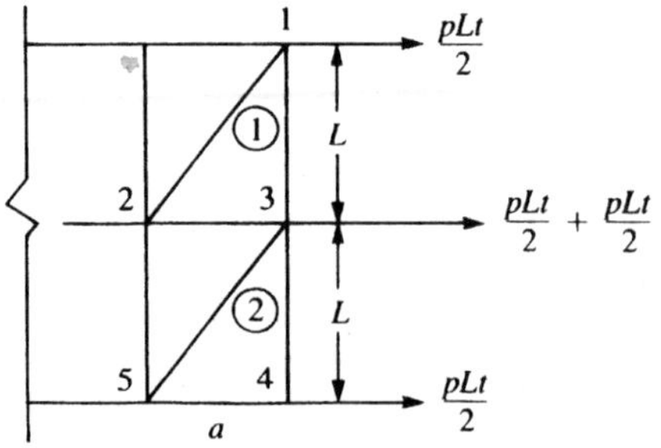

Figure 6–15 Surface traction equivalent nodal forces

linear-strain triangle (discussed in Chapter 8), the load replacement should be made by using Eq. (6.3.7), which was derived by the principle of minimum potential energy. For higher-order elements, this load replacement by use of Eq. (6.3.7) is generally not equal to the apparent statically equivalent one; however, it is consistent in that this replacement results directly from the energy approach.

We now recognize the force matrix $\{f_s\}$ defined by Eq. (6.3.7), and based on the principle of minimum potential energy, to be equivalent to that based on work equivalence, which we previously used in Chapter 4 when discussing distributed loads acting on beams.

6.4 Explicit Expression for the Constant-Strain Triangle Stiffness Matrix

Although the stiffness matrix is generally formulated internally in most computer programs by performing the matrix triple product indicated by Eq. (6.4.1), it is still a valuable learning experience to evaluate the stiffness matrix explicitly for the constant-strain triangular element. Hence, we will consider the plane strain case specifically in this development.

First, recall that the stiffness matrix is given by

$$[k] = tA[B]^T[D][B] \tag{6.4.1}$$

where, for the plane strain case, $[D]$ is given by Eq. (6.1.10) and $[B]$ is given by Eq. (6.2.34). On substituting the matrices $[D]$ and $[B]$ into Eq. (6.4.1), we obtain

$$[k] = \frac{tE}{4A(1+\nu)(1-2\nu)} \begin{bmatrix} \beta_i & 0 & \gamma_i \\ 0 & \gamma_i & \beta_i \\ \beta_j & 0 & \gamma_j \\ 0 & \gamma_j & \beta_j \\ \beta_m & 0 & \gamma_m \\ 0 & \gamma_m & \beta_m \end{bmatrix}$$

$$\times \begin{bmatrix} 1-\nu & \nu & 0 \\ \nu & 1-\nu & 0 \\ 0 & 0 & \dfrac{1-2\nu}{2} \end{bmatrix} \begin{bmatrix} \beta_i & 0 & \beta_j & 0 & \beta_m & 0 \\ 0 & \gamma_i & 0 & \gamma_j & 0 & \gamma_m \\ \gamma_i & \beta_i & \gamma_j & \beta_j & \gamma_m & \beta_m \end{bmatrix} \tag{6.4.2}$$

On multiplying the matrices in Eq. (6.4.2), we obtain Eq. (6.4.3), the explicit constant-strain triangle stiffness matrix for the plane strain case. Note that $[k]$ is a function of the difference in the x and y nodal coordinates, as indicated by the γ's and β's, of the material properties E and ν, and of the thickness t and surface area A of the element.

$$
\underline{k} = \frac{tE}{4A(1+\nu)(1-2\nu)}
$$

$$
\times \begin{bmatrix}
\beta_i^2(1-\nu)+\gamma_i^2\left(\frac{1-2\nu}{2}\right) & \beta_i\gamma_i\nu+\beta_i\gamma_i\left(\frac{1-2\nu}{2}\right) & \beta_i\beta_j(1-\nu)+\gamma_i\gamma_j\left(\frac{1-2\nu}{2}\right) & \beta_i\gamma_j\nu+\beta_j\gamma_i\left(\frac{1-2\nu}{2}\right) & \beta_i\beta_m(1-\nu)+\gamma_i\gamma_m\left(\frac{1-2\nu}{2}\right) & \beta_i\gamma_m\nu+\beta_m\gamma_i\left(\frac{1-2\nu}{2}\right) \\
 & \gamma_i^2(1-\nu)+\beta_i^2\left(\frac{1-2\nu}{2}\right) & \beta_j\gamma_i\nu+\beta_i\gamma_j\left(\frac{1-2\nu}{2}\right) & \gamma_i\gamma_j(1-\nu)+\beta_i\beta_j\left(\frac{1-2\nu}{2}\right) & \beta_m\gamma_i\nu+\beta_i\gamma_m\left(\frac{1-2\nu}{2}\right) & \gamma_i\gamma_m(1-\nu)+\beta_i\beta_m\left(\frac{1-2\nu}{2}\right) \\
 & & \beta_j^2(1-\nu)+\gamma_j^2\left(\frac{1-2\nu}{2}\right) & \beta_j\gamma_j\nu+\beta_j\gamma_j\left(\frac{1-2\nu}{2}\right) & \beta_j\beta_m(1-\nu)+\gamma_j\gamma_m\left(\frac{1-2\nu}{2}\right) & \beta_j\gamma_m\nu+\gamma_j\beta_m\left(\frac{1-2\nu}{2}\right) \\
 & & & \gamma_j^2(1-\nu)+\beta_j^2\left(\frac{1-2\nu}{2}\right) & \beta_m\gamma_j\nu+\beta_j\gamma_m\left(\frac{1-2\nu}{2}\right) & \gamma_j\gamma_m(1-\nu)+\beta_j\beta_m\left(\frac{1-2\nu}{2}\right) \\
 & & & & \beta_m^2(1-\nu)+\gamma_m^2\left(\frac{1-2\nu}{2}\right) & \gamma_m\beta_m\nu+\beta_m\gamma_m\left(\frac{1-2\nu}{2}\right) \\
\text{Symmetry} & & & & & \gamma_m^2(1-\nu)+\beta_m^2\left(\frac{1-2\nu}{2}\right)
\end{bmatrix}
\tag{6.4.3}
$$

For the plane stress case, we need only replace $1-\nu$ by 1, $(1-2\nu)/2$ by $(1-\nu)/2$, and $(1+\nu)(1-2\nu)$ outside the brackets by $1-\nu^2$ in Eq. (6.4.3).

Finally, it should be noted that for Poisson's ratio ν approaching 0.5, as in rubberlike materials and plastic solids, for instance, a material becomes incompressible [2]. For plane strain, as ν approaches 0.5, the denominator becomes zero in the material property matrix [see Eq. (6.1.10)] and hence in the stiffness matrix, Eq. (6.4.3).

A value of ν near 0.5 can cause ill-conditioned structural equations. A special formulation (called a *penalty formulation* [3]) has been used in this case.

6.5 Finite Element Solution of a Plane Stress Problem

To illustrate the finite element method for a plane stress problem, we now present a detailed solution.

Example 6.2

For a thin plate subjected to the surface traction shown in Figure 6–16, determine the nodal displacements and the element stresses. The plate thickness $t = 1$ in., $E = 30 \times 10^6$ psi, and $\nu = 0.30$.

Discretization

To illustrate the finite element method solution for the plate, we first discretize the plate into two elements, as shown in Figure 6–17. It should be understood that the coarseness of the mesh will not yield as true a predicted behavior of the plate as would a finer mesh, particularly near the fixed edge. However, since we are performing a longhand solution, we will use a coarse discretization for simplicity (but without loss of generality of the method).

In Figure 6–17, the original tensile surface traction in Figure 6–16 has been converted to nodal forces as follows:

$$F = \tfrac{1}{2}TA$$

$$F = \tfrac{1}{2}(1000 \text{ psi})(1 \text{ in.} \times 10 \text{ in.})$$

$$F = 5000 \text{ lb}$$

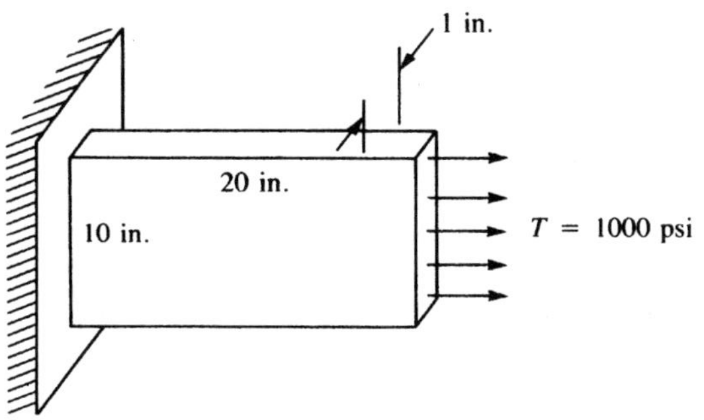

Figure 6–16 Thin plate subjected to tensile stress

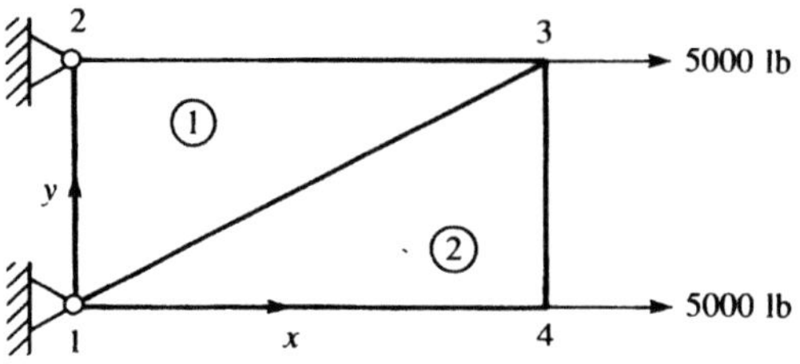

Figure 6–17 Discretized plate

In general, for higher-order elements, Eq. (6.3.7) should be used to convert distributed surface tractions to nodal forces. However, for the CST element, we have shown in Section 6.3 that a statically equivalent force replacement can be used directly, as has been done here.

The governing global matrix equation is

$$\{F\} = [K]\{d\} \tag{6.5.1}$$

Expanding matrices in Eq. (6.5.1), we obtain

$$\begin{Bmatrix} F_{1x} \\ F_{1y} \\ F_{2x} \\ F_{2y} \\ F_{3x} \\ F_{3y} \\ F_{4x} \\ F_{4y} \end{Bmatrix} = \begin{Bmatrix} R_{1x} \\ R_{1y} \\ R_{2x} \\ R_{2y} \\ 5000 \\ 0 \\ 5000 \\ 0 \end{Bmatrix} = [K]\begin{Bmatrix} d_{1x} \\ d_{1y} \\ d_{2x} \\ d_{2y} \\ d_{3x} \\ d_{3y} \\ d_{4x} \\ d_{4y} \end{Bmatrix} = [K]\begin{Bmatrix} 0 \\ 0 \\ 0 \\ 0 \\ d_{3x} \\ d_{3y} \\ d_{4x} \\ d_{4y} \end{Bmatrix} \tag{6.5.2}$$

where $[K]$ is an 8×8 matrix (two degrees of freedom per node with four nodes) before deleting rows and columns to account for the fixed boundary support conditions at nodes 1 and 2.

Assemblage of the Stiffness Matrix

We assemble the global stiffness matrix by superposition of the individual element stiffness matrices. By Eq. (6.2.52), the stiffness matrix for an element is

$$[k] = tA[B]^T[D][B] \tag{6.5.3}$$

In Figure 6–18 for element 1, we have coordinates $x_i = 0$, $y_i = 0$, $x_j = 20$, $y_j = 10$, $x_m = 0$, and $y_m = 10$, since the global coordinate axes are set up at node 1, and

$$A = \tfrac{1}{2}bh$$

$$A = \left(\tfrac{1}{2}\right)(20)(10) = 100 \text{ in}^2$$

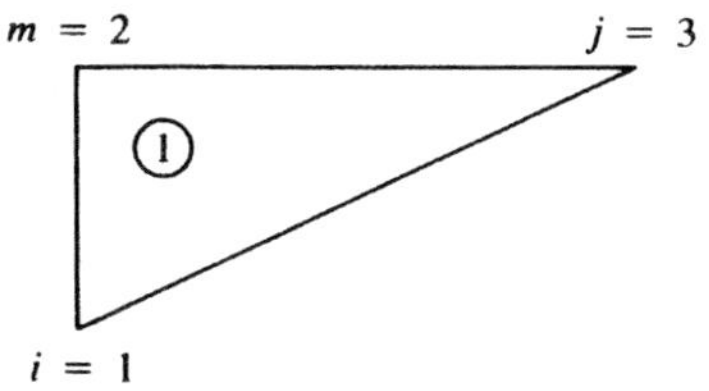

Figure 6–18 Element 1 of the discretized plate

or, in general, A can be obtained equivalently by the nodal coordinate formula of Eq. (6.2.9).

We will now evaluate $[B]$, where $[B]$ is given by Eq. (6.2.34), expanded here as

$$[B] = \frac{1}{2A}\begin{bmatrix} \beta_i & 0 & \beta_j & 0 & \beta_m & 0 \\ 0 & \gamma_i & 0 & \gamma_j & 0 & \gamma_m \\ \gamma_i & \beta_i & \gamma_j & \beta_j & \gamma_m & \beta_m \end{bmatrix} \tag{6.5.4}$$

and, from Eqs. (6.2.10),

$$\begin{aligned} \beta_i &= y_j - y_m = 10 - 10 = 0 \\ \beta_j &= y_m - y_i = 10 - 0 = 10 \\ \beta_m &= y_i - y_j = 0 - 10 = -10 \\ \gamma_i &= x_m - x_j = 0 - 20 = -20 \\ \gamma_j &= x_i - x_m = 0 - 0 = 0 \\ \gamma_m &= x_j - x_i = 20 - 0 = 20 \end{aligned} \tag{6.5.5}$$

Therefore, substituting Eqs. (6.5.5) into Eq. (6.5.4), we obtain

$$[B] = \frac{1}{200}\begin{bmatrix} 0 & 0 & 10 & 0 & -10 & 0 \\ 0 & -20 & 0 & 0 & 0 & 20 \\ -20 & 0 & 0 & 10 & 20 & -10 \end{bmatrix}\frac{1}{\text{in.}} \tag{6.5.6}$$

For plane stress, the $[D]$ matrix is conveniently expressed here as

$$[D] = \frac{E}{(1-\nu^2)}\begin{bmatrix} 1 & \nu & 0 \\ \nu & 1 & 0 \\ 0 & 0 & \dfrac{1-\nu}{2} \end{bmatrix} \tag{6.5.7}$$

With $\nu = 0.3$ and $E = 30 \times 10^6$ psi, we obtain

$$[D] = \frac{30(10^6)}{0.91}\begin{bmatrix} 1 & 0.3 & 0 \\ 0.3 & 1 & 0 \\ 0 & 0 & 0.35 \end{bmatrix}\text{psi} \tag{6.5.8}$$

Then $$[B]^T[D] = \frac{30(10^6)}{200(0.91)} \begin{bmatrix} 0 & 0 & -20 \\ 0 & -20 & 0 \\ 10 & 0 & 0 \\ 0 & 0 & 10 \\ -10 & 0 & 20 \\ 0 & 20 & -10 \end{bmatrix} \begin{bmatrix} 1 & 0.3 & 0 \\ 0.3 & 1 & 0 \\ 0 & 0 & 0.35 \end{bmatrix} \tag{6.5.9}$$

Simplifying Eq. (6.5.9) yields

$$[B]^T[D] = \frac{(0.15)(10^6)}{0.91} \begin{bmatrix} 0 & 0 & -7 \\ -6 & -20 & 0 \\ 10 & 3 & 0 \\ 0 & 0 & 3.5 \\ -10 & -3 & 7 \\ 6 & 20 & -3.5 \end{bmatrix} \tag{6.5.10}$$

Using Eqs. (6.5.10) and (6.5.6) in Eq. (6.5.3), we have the stiffness matrix for element 1 as

$$[k] = (1)(100)\frac{(0.15)(10^6)}{0.91} \begin{bmatrix} 0 & 0 & -7 \\ -6 & -20 & 0 \\ 10 & 3 & 0 \\ 0 & 0 & 3.5 \\ -10 & -3 & 7 \\ 6 & 20 & -3.5 \end{bmatrix} \times \frac{1}{2(100)} \begin{bmatrix} 0 & 0 & 10 & 0 & -10 & 0 \\ 0 & -20 & 0 & 0 & 0 & 20 \\ -20 & 0 & 0 & 10 & 20 & -10 \end{bmatrix} \tag{6.5.11}$$

Finally, simplifying Eq. (6.5.11) yields

$$[k] = \frac{75{,}000}{0.91} \begin{array}{c} \begin{array}{cccccc} u_1 & v_1 & u_3 & v_3 & u_2 & v_2 \end{array} \\ \begin{bmatrix} 140 & 0 & 0 & -70 & -140 & 70 \\ 0 & 400 & -60 & 0 & 60 & -400 \\ 0 & -60 & 100 & 0 & -100 & 60 \\ -70 & 0 & 0 & 35 & 70 & -35 \\ -140 & 60 & -100 & 70 & 240 & -130 \\ 70 & -400 & 60 & -35 & -130 & 435 \end{bmatrix} \end{array} \frac{\text{lb}}{\text{in.}} \tag{6.5.12}$$

where the labels above the columns indicate the nodal order of the degrees of freedom in the element 1 stiffness matrix.

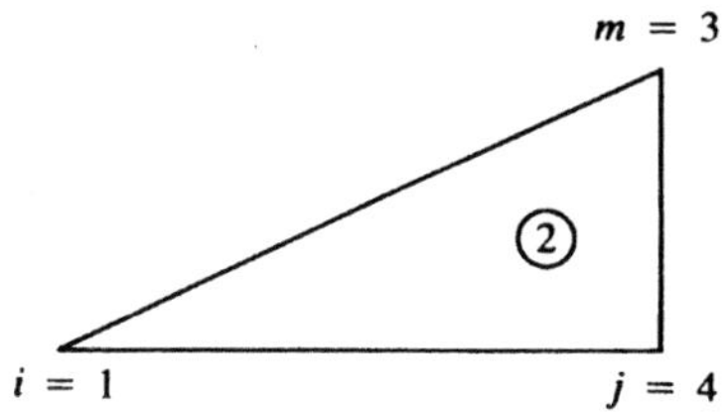

Figure 6–19 Element 2 of the discretized plate

In Figure 6–19 for element 2, we have $x_i = 0$, $y_i = 0$, $x_j = 20$, $y_j = 0$, $x_m = 20$, and $y_m = 10$. Then, from Eqs. (6.2.10), we have

$$\begin{aligned}
\beta_i &= y_j - y_m = 0 - 10 = -10 \\
\beta_j &= y_m - y_i = 10 - 0 = 10 \\
\beta_m &= y_i - y_j = 0 - 0 = 0 \\
\gamma_i &= x_m - x_j = 20 - 20 = 0 \\
\gamma_j &= x_i - x_m = 0 - 20 = -20 \\
\gamma_m &= x_j - x_i = 20 - 0 = 20
\end{aligned} \tag{6.5.13}$$

Therefore, using Eqs. (6.5.13) in Eq. (6.5.4) yields

$$[B] = \frac{1}{200}\begin{bmatrix} -10 & 0 & 10 & 0 & 0 & 0 \\ 0 & 0 & 0 & -20 & 0 & 20 \\ 0 & -10 & -20 & 10 & 20 & 0 \end{bmatrix} \frac{1}{\text{in.}} \tag{6.5.14}$$

The $[D]$ matrix is again given by

$$[D] = \frac{30(10^6)}{0.91}\begin{bmatrix} 1 & 0.3 & 0 \\ 0.3 & 1 & 0 \\ 0 & 0 & 0.35 \end{bmatrix} \text{psi} \tag{6.5.15}$$

Then, using Eqs. (6.5.14) and (6.5.15), we obtain

$$[B]^T[D] = \frac{30(10^6)}{200(0.91)}\begin{bmatrix} -10 & 0 & 0 \\ 0 & 0 & -10 \\ 10 & 0 & -20 \\ 0 & -20 & 10 \\ 0 & 0 & 20 \\ 0 & 20 & 0 \end{bmatrix}\begin{bmatrix} 1 & 0.3 & 0 \\ 0.3 & 1 & 0 \\ 0 & 0 & 0.35 \end{bmatrix} \tag{6.5.16}$$

Simplifying Eq. (6.5.16) yields

$$[B]^T[D] = \frac{(0.15)(10^6)}{0.91}\begin{bmatrix} -10 & -3 & 0 \\ 0 & 0 & -3.5 \\ 10 & 3 & -7 \\ -6 & -20 & 3.5 \\ 0 & 0 & 7 \\ 6 & 20 & 0 \end{bmatrix} \tag{6.5.17}$$

Finally, substituting Eqs. (6.5.17) and (6.5.14) into Eq. (6.5.3), we obtain the stiffness matrix for element 2 as

$$[k] = (1)(100)\frac{(0.15)(10^6)}{0.91}\begin{bmatrix} -10 & -3 & 0 \\ 0 & 0 & -3.5 \\ 10 & 3 & -7 \\ -6 & -20 & 3.5 \\ 0 & 0 & 7 \\ 6 & 20 & 0 \end{bmatrix}$$

$$\times \frac{1}{2(100)}\begin{bmatrix} -10 & 0 & 10 & 0 & 0 & 0 \\ 0 & 0 & 0 & -20 & 0 & 20 \\ 0 & -10 & -20 & 10 & 20 & 0 \end{bmatrix} \tag{6.5.18}$$

Equation (6.5.18) simplifies to

$$[k] = \frac{75{,}000}{0.91}\begin{bmatrix} \overset{u_1}{100} & \overset{v_1}{0} & \overset{u_4}{-100} & \overset{v_4}{60} & \overset{u_3}{0} & \overset{v_3}{-60} \\ 0 & 35 & 70 & -35 & -70 & 0 \\ -100 & 70 & 240 & -130 & -140 & 60 \\ 60 & -35 & -130 & 435 & 70 & -400 \\ 0 & -70 & -140 & 70 & 140 & 0 \\ -60 & 0 & 60 & -400 & 0 & 400 \end{bmatrix}\frac{\text{lb}}{\text{in.}} \tag{6.5.19}$$

where the degrees of freedom in the element 2 stiffness matrix are shown above the columns in Eq. (6.5.19). Rewriting the element stiffness matrices, Eqs. (6.5.12) and (6.5.19), expanded to the order of, and rearranged according to, increasing nodal degrees of freedom of the total $\underline{K}$ matrix (where we have factored out a constant 5), we obtain

Element 1

$$[k] = \frac{375{,}000}{0.91}\begin{bmatrix} 28 & 0 & -28 & 14 & 0 & -14 & 0 & 0 \\ 0 & 80 & 12 & -80 & -12 & 0 & 0 & 0 \\ -28 & 12 & 48 & -26 & -20 & 14 & 0 & 0 \\ 14 & -80 & -26 & 87 & 12 & -7 & 0 & 0 \\ 0 & -12 & -20 & 12 & 20 & 0 & 0 & 0 \\ -14 & 0 & 14 & -7 & 0 & 7 & 0 & 0 \\ 0 & 0 & 0 & 0 & 0 & 0 & 0 & 0 \\ 0 & 0 & 0 & 0 & 0 & 0 & 0 & 0 \end{bmatrix} \frac{\text{lb}}{\text{in.}} \quad (6.5.20)$$

(columns: u_1, v_1, u_2, v_2, u_3, v_3, u_4, v_4)

Element 2

$$[k] = \frac{375{,}000}{0.91}\begin{bmatrix} 20 & 0 & 0 & 0 & 0 & -12 & -20 & 12 \\ 0 & 7 & 0 & 0 & -14 & 0 & 14 & -7 \\ 0 & 0 & 0 & 0 & 0 & 0 & 0 & 0 \\ 0 & 0 & 0 & 0 & 0 & 0 & 0 & 0 \\ 0 & -14 & 0 & 0 & 28 & 0 & -28 & 14 \\ -12 & 0 & 0 & 0 & 0 & 80 & 12 & -80 \\ -20 & 14 & 0 & 0 & -28 & 12 & 48 & -26 \\ 12 & -7 & 0 & 0 & 14 & -80 & -26 & 87 \end{bmatrix} \frac{\text{lb}}{\text{in.}} \quad (6.5.21)$$

(columns: u_1, v_1, u_2, v_2, u_3, v_3, u_4, v_4)

Using superposition of the element stiffness matrices, Eqs. (6.5.20) and (6.5.21), now that the orders of the degrees of freedom are the same, we obtain the total global stiffness matrix as

$$[K] = \frac{375{,}000}{0.91}\begin{bmatrix} 48 & 0 & -28 & 14 & 0 & -26 & -20 & 12 \\ 0 & 87 & 12 & -80 & -26 & 0 & 14 & -7 \\ -28 & 12 & 48 & -26 & -20 & 14 & 0 & 0 \\ 14 & -80 & -26 & 87 & 12 & -7 & 0 & 0 \\ 0 & -26 & -20 & 12 & 48 & 0 & -28 & 14 \\ -26 & 0 & 14 & -7 & 0 & 87 & 12 & -80 \\ -20 & 14 & 0 & 0 & -28 & 12 & 48 & -26 \\ 12 & -7 & 0 & 0 & 14 & -80 & -26 & 87 \end{bmatrix} \frac{\text{lb}}{\text{in.}} \quad (6.5.22)$$

(columns: u_1, v_1, u_2, v_2, u_3, v_3, u_4, v_4)

[Alternatively, we could have applied the direct stiffness method to Eqs. (6.5.12) and (6.5.19) to obtain Eq. (6.5.22).] Substituting $[K]$ into $\{F\} = [K]\{d\}$ of Eq. (6.5.2), we have

$$\begin{Bmatrix} R_{1x} \\ R_{1y} \\ R_{2x} \\ R_{2y} \\ 5000 \\ 0 \\ 5000 \\ 0 \end{Bmatrix} = \frac{375{,}000}{0.91} \begin{bmatrix} 48 & 0 & -28 & 14 & 0 & -26 & -20 & 12 \\ 0 & 87 & 12 & -80 & -26 & 0 & 14 & -7 \\ -28 & 12 & 48 & -26 & -20 & 14 & 0 & 0 \\ 14 & -80 & -26 & 87 & 12 & -7 & 0 & 0 \\ 0 & -26 & -20 & 12 & 48 & 0 & -28 & 14 \\ -26 & 0 & 14 & -7 & 0 & 87 & 12 & -80 \\ -20 & 14 & 0 & 0 & -28 & 12 & 48 & -26 \\ 12 & -7 & 0 & 0 & 14 & -80 & -26 & 87 \end{bmatrix} \begin{Bmatrix} 0 \\ 0 \\ 0 \\ 0 \\ d_{3x} \\ d_{3y} \\ d_{4x} \\ d_{4y} \end{Bmatrix} \tag{6.5.23}$$

Applying the support or boundary conditions by eliminating rows and columns corresponding to displacement matrix rows and columns equal to zero [namely, rows and columns 1–4 in Eq. (6.5.23)], we obtain

$$\begin{Bmatrix} 5000 \\ 0 \\ 5000 \\ 0 \end{Bmatrix} = \frac{375{,}000}{0.91} \begin{bmatrix} 48 & 0 & -28 & 14 \\ 0 & 87 & 12 & -80 \\ -28 & 12 & 48 & -26 \\ 14 & -80 & -26 & 87 \end{bmatrix} \begin{Bmatrix} d_{3x} \\ d_{3y} \\ d_{4x} \\ d_{4y} \end{Bmatrix} \tag{6.5.24}$$

Premultiplying both sides of Eq. (6.5.24) by $\underline{K}^{-1}$, we have

$$\begin{Bmatrix} d_{3x} \\ d_{3y} \\ d_{4x} \\ d_{4y} \end{Bmatrix} = \frac{0.91}{375{,}000} \begin{bmatrix} 48 & 0 & -28 & 14 \\ 0 & 87 & 12 & -80 \\ -28 & 12 & 48 & -26 \\ 14 & -80 & -26 & 87 \end{bmatrix}^{-1} \begin{Bmatrix} 5000 \\ 0 \\ 5000 \\ 0 \end{Bmatrix} \tag{6.5.25}$$

Solving for the displacements in Eq. (6.5.25), we obtain

$$\begin{Bmatrix} d_{3x} \\ d_{3y} \\ d_{4x} \\ d_{4y} \end{Bmatrix} = \frac{0.91}{75} \begin{Bmatrix} 0.05024 \\ 0.00034 \\ 0.05470 \\ 0.00878 \end{Bmatrix} \tag{6.5.26}$$

Simplifying Eq. (6.5.26), the final displacements are given by

$$\begin{Bmatrix} d_{3x} \\ d_{3y} \\ d_{4x} \\ d_{4y} \end{Bmatrix} = \begin{Bmatrix} 609.6 \\ 4.2 \\ 663.7 \\ 104.1 \end{Bmatrix} \times 10^{-6} \text{ in.} \tag{6.5.27}$$

Comparing the finite element solution to an analytical solution, as a first approximation, we have the axial displacement given by

$$\delta = \frac{PL}{AE} = \frac{(10{,}000)20}{10(30 \times 10^6)} = 670 \times 10^{-6} \text{ in.}$$

for a one-dimensional bar subjected to tensile force. Hence, the nodal x displacement components of Eq. (6.5.27) for the two-dimensional plate appear to be reasonably correct, considering the coarseness of the mesh and the directional stiffness bias of the model. (For more on this subject see Section 7.5.) The y displacement would be expected to be downward at the top (node 3) and upward at the bottom (node 4) as a result of the Poisson effect. However, the directional stiffness bias due to the coarse mesh accounts for this unexpected poor result.

We now determine the stresses in each element by using Eq. (6.2.36):

$$\{\sigma\} = [D][B]\{d\} \tag{6.5.28}$$

In general, for element 1, we then have

$$\{\sigma\} = \frac{E}{(1-\nu^2)}\begin{bmatrix} 1 & \nu & 0 \\ \nu & 1 & 0 \\ 0 & 0 & \dfrac{1-\nu}{2} \end{bmatrix} \times \left(\frac{1}{2A}\right)\begin{bmatrix} \beta_1 & 0 & \beta_3 & 0 & \beta_2 & 0 \\ 0 & \gamma_1 & 0 & \gamma_3 & 0 & \gamma_2 \\ \gamma_1 & \beta_1 & \gamma_3 & \beta_3 & \gamma_2 & \beta_2 \end{bmatrix}\begin{Bmatrix} d_{1x} \\ d_{1y} \\ d_{3x} \\ d_{3y} \\ d_{2x} \\ d_{2y} \end{Bmatrix} \tag{6.5.29}$$

Substituting numerical values for $[B]$, given by Eq. (6.5.6); for $[D]$, given by Eq. (6.5.8); and the appropriate part of $\{d\}$, given by Eq. (6.5.27), we obtain

$$\{\sigma\} = \frac{30(10^6)(10^{-6})}{0.91(200)}\begin{bmatrix} 1 & 0.3 & 0 \\ 0.3 & 1 & 0 \\ 0 & 0 & 0.35 \end{bmatrix} \times \begin{bmatrix} 0 & 0 & 10 & 0 & -10 & 0 \\ 0 & -20 & 0 & 0 & 0 & 20 \\ -20 & 0 & 0 & 10 & 20 & -10 \end{bmatrix}\begin{Bmatrix} 0 \\ 0 \\ 609.6 \\ 4.2 \\ 0 \\ 0 \end{Bmatrix} \tag{6.5.30}$$

Simplifying Eq. (6.5.30), we obtain

$$\begin{Bmatrix} \sigma_x \\ \sigma_y \\ \tau_{xy} \end{Bmatrix} = \begin{Bmatrix} 1005 \\ 301 \\ 2.4 \end{Bmatrix} \text{psi} \tag{6.5.31}$$

In general, for element 2, we have

$$\{\sigma\} = \frac{E}{(1-\nu^2)}\left(\frac{1}{2A}\right)\begin{bmatrix} 1 & \nu & 0 \\ \nu & 1 & 0 \\ 0 & 0 & \dfrac{1-\nu}{2} \end{bmatrix} \times \begin{bmatrix} \beta_1 & 0 & \beta_4 & 0 & \beta_3 & 0 \\ 0 & \gamma_1 & 0 & \gamma_4 & 0 & \gamma_3 \\ \gamma_1 & \beta_1 & \gamma_4 & \beta_4 & \gamma_3 & \beta_3 \end{bmatrix} \begin{Bmatrix} d_{1x} \\ d_{1y} \\ d_{4x} \\ d_{4y} \\ d_{3x} \\ d_{3y} \end{Bmatrix} \tag{6.5.32}$$

Substituting numerical values into Eq. (6.5.32), we obtain

$$\{\sigma\} = \frac{30(10^6)(10^{-6})}{0.91(200)}\begin{bmatrix} 1 & 0.3 & 0 \\ 0.3 & 1 & 0 \\ 0 & 0 & 0.35 \end{bmatrix}$$

$$\times \begin{bmatrix} -10 & 0 & 10 & 0 & 0 & 0 \\ 0 & 0 & 0 & -20 & 0 & 20 \\ 0 & -10 & -20 & 10 & 20 & 0 \end{bmatrix} \begin{Bmatrix} 0 \\ 0 \\ 663.7 \\ 104.1 \\ 609.6 \\ 4.2 \end{Bmatrix} \tag{6.5.33}$$

Simplifying Eq. (6.5.33), we obtain

$$\begin{Bmatrix} \sigma_x \\ \sigma_y \\ \tau_{xy} \end{Bmatrix} = \begin{Bmatrix} 995 \\ -1.2 \\ -2.4 \end{Bmatrix} \text{psi} \tag{6.5.34}$$

The principal stresses can now be determined from Eq. (6.1.2), and the principal angle made by one of the principal stresses can be determined from Eq. (6.1.3). (The other principal stress will be directed 90° from the first.) We determine these principal stresses for element 2 (those for element 1 will be similar) as

$$\sigma_1 = \frac{\sigma_x + \sigma_y}{2} + \left[\left(\frac{\sigma_x - \sigma_y}{2}\right)^2 + \tau_{xy}^2\right]^{1/2}$$

$$\sigma_1 = \frac{995 + (-1.2)}{2} + \left[\left(\frac{995 - (-1.2)}{2}\right)^2 + (-2.4)^2\right]^{1/2} \tag{6.5.35}$$

$$\sigma_1 = 497 + 498 = 995 \text{ psi}$$

$$\sigma_2 = \frac{995 + (-1.2)}{2} - 498 = -1.1 \text{ psi}$$

The principal angle is then

$$\theta_p = \frac{1}{2}\tan^{-1}\left[\frac{2\tau_{xy}}{\sigma_x - \sigma_y}\right]$$

or

$$\theta_p = \frac{1}{2}\tan^{-1}\left[\frac{2(-2.4)}{995 - (-1.2)}\right] = 0^\circ \qquad (6.5.36)$$

Owing to the uniform stress of 1000 psi acting only in the x direction on the edge of the plate, we would expect the stress $\sigma_x(= \sigma_1)$ to be near 1000 psi in each element. Thus, the results from Eqs. (6.5.31) and (6.5.34) for σ_x are quite good. We would expect the stress σ_y to be very small (at least near the free edge). The restraint of element 1 at nodes 1 and 2 causes a relatively large element stress σ_y, whereas the restraint of element 2 at only one node causes a very small stress σ_y. The shear stresses τ_{xy} remain close to zero, as expected. Had the number of elements been increased, with smaller ones used near the support edge, even more realistic results would have been obtained. However, a finer discretization would result in a cumbersome longhand solution and thus was not used here. Use of a computer program is recommended for a detailed solution to this plate problem and certainly for solving more complex stress/strain problems. ■

The maximum distortion energy theory [4] (also called the *von Mises* or *von Mises-Hencky theory*) for ductile materials subjected to static loading predicts that a material will fail if the von Mises stress (also called *equivalent or effective stress*) reaches the yield strength, S_y, of the material. The von Mises stress as derived in [4], for instance, is given in terms of the three principal stresses by

$$\sigma_{vm} = \frac{1}{\sqrt{2}}\left[(\sigma_1 - \sigma_2)^2 + (\sigma_2 - \sigma_3)^2 + (\sigma_3 - \sigma_1)^2\right]^{1/2} \qquad (6.5.37a)$$

or equivalently in terms of the x-y-z components as

$$\sigma_{vm} = \frac{1}{\sqrt{2}}\left[(\sigma_x - \sigma_y)^2 + (\sigma_y - \sigma_z)^2 + (\sigma_z - \sigma_x)^2 + 6({\tau_{xy}}^2 + {\tau_{yz}}^2 + {\tau_{zx}}^2)\right] \qquad (6.5.37b)$$

Thus for yielding to occur, the *von Mises* stress must become equal to or greater than the yield strength of the material as given by

$$\sigma_{vm} \geq S_y \qquad (6.5.38)$$

We can see from Eqs. (6.5.37a or 6.5.37b) that the von Mises stress is a scalar that measures the intensity of the entire stress state as it includes the three principal stresses or the three normal stresses in the x, y, and z directions, along with the shear stresses on the x, y, and z planes. Other stresses, such as the maximum principal one, do not provide the most accurate way of predicting failure.

Most computer programs incorporate this failure theory and, as an optional result, the user can request a plot of the von Mises stress throughout the material model being analyzed. If the von Mises stress value is equal to or greater than the yield strength of the material being considered, then another material with greater yield strength can be selected or other design changes can be made.

For brittle materials, such as glass and cast iron, with different tension and compression properties, it is recommended to use the Coulomb-Mohr theory to predict failure. For more on this theory consult [4].

CST Element Defects

The CST element has its limitations. In bending problems, the mesh of CST elements will produce a model that is stiffer than the actual problem. As we will observe from the results shown for a beam-bending problem modeled by CST and LST (to be described in Chapter 8) elements, the CST model converges very slowly to the exact solution. This is partly due to the element predicting only constant stress within each element, when for a bending problem, the stress actually varies linearly through the depth of the beam. This problem is rectified by using the LST element as described in Chapter 8.

As shown in [3] for a beam subjected to pure bending, the CST has a spurious or false shear stress and hence a spurious shear strain in parts of the model that should not have any shear stress or shear strain. This spurious shear strain absorbs energy; therefore, some of the energy that should go into bending is lost. The CST is then too stiff in bending, and the resulting deformation is smaller than actually should be. This phenomenon of excessive stiffness developing in one or more modes of deformation is sometimes described as *shear locking* or *parasitic shear*.

Furthermore, in problems where plane strain conditions exist (recall this means when $\varepsilon_z = 0$) and the Poisson's ratio approaches 0.5, a mesh can actually lock, which means the mesh then cannot deform at all.

This brief description of some limitations in using the CST element does not stop us from using it to model plane stress and plane strain problems. It just requires us to use a fine mesh as opposed to a coarse one, particularly where bending occurs and where in general large stress gradients will result. Also, we must make sure our program can handle Poisson's ratios that approach 0.5 if that is desired, such as in rubber-like materials. For common materials, such as metals, Poisson's ratio is around 0.3 and so locking should not be of concern.

References

[1] Timoshenko, S., and Goodier, J., *Theory of Elasticity*, 3rd ed., McGraw-Hill, New York 1970.

[2] Gere, J. M., *Mechanics of Materials*, 5th ed., Brooks/Cole Publishers, Pacific Grove, CA, 2001.

[3] Cook, R. D., Malkus, D. S., Plesha, M. E., and Witt, R. J., *Concepts and Applications of Finite Element Analysis*, 4th ed., Wiley, New York, 2002.

[4] Shigley, J. E., Mischke, C. R., and Budynas, R. G., *Mechanical Engineering Design*, 7th ed., McGraw-Hill, New York, 2004.

Problems

6.1 Sketch the variations of the shape functions N_j and N_m, given by Eqs. (6.2.18), over the surface of the triangular element with nodes i, j, and m. Check that $N_i + N_j + N_m = 1$ anywhere on the element.

6.2 For a simple three-noded triangular element, show explicitly that differentiation of Eq. (6.2.47) indeed results in Eq. (6.2.48); that is, substitute the expression for $[B]$ and the plane stress condition for $[D]$ into Eq. (6.2.47), and then differentiate π_p with respect to each nodal degree of freedom in Eq. (6.2.47) to obtain Eq. (6.2.48).

6.3 Evaluate the stiffness matrix for the elements shown in Figure P6–3. The coordinates are in units of inches. Assume plane stress conditions. Let $E = 30 \times 10^6$ psi, $\nu = 0.25$, and thickness $t = 1$ in.

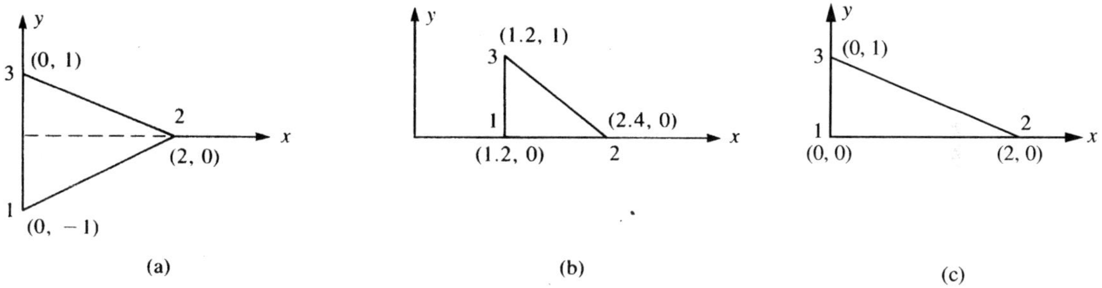

Figure P6–3

6.4 For the elements given in Problem 6.3, the nodal displacements are given as

$$u_1 = 0.0 \qquad v_1 = 0.0025 \text{ in.} \qquad u_2 = 0.0012 \text{ in.}$$

$$v_2 = 0.0 \qquad u_3 = 0.0 \qquad v_3 = 0.0025 \text{ in.}$$

Determine the element stresses $\sigma_x, \sigma_y, \tau_{xy}, \sigma_1$, and σ_2 and the principal angle θ_p. Use the values of E, ν, and t given in Problem 6.3.

6.5 Determine the von Mises stress for problem 6.4.

6.6 Evaluate the stiffness matrix for the elements shown in Figure P6–6. The coordinates are given in units of millimeters. Assume plane stress conditions. Let $E = 210$ GPa, $\nu = 0.25$, and $t = 10$ mm.

6.7 For the elements given in Problem 6.6, the nodal displacements are given as

$$u_1 = 2.0 \text{ mm} \qquad v_1 = 1.0 \text{ mm} \qquad u_2 = 0.5 \text{ mm}$$

$$v_2 = 0.0 \text{ mm} \qquad u_3 = 3.0 \text{ mm} \qquad v_3 = 1.0 \text{ mm}$$

Determine the element stresses $\sigma_x, \sigma_y, \tau_{xy}, \sigma_1$, and σ_2 and the principal angle θ_p. Use the values of E, ν, and t given in Problem 6.6.

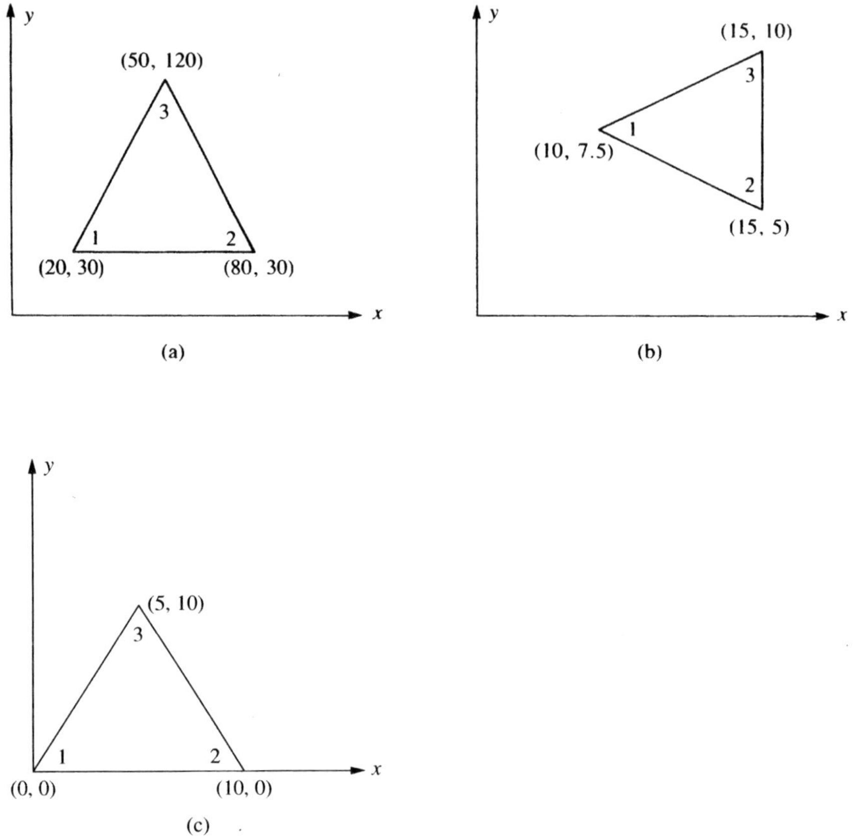

Figure P6–6

6.8 Determine the von Mises stress for problem 6.7

6.9 For the plane strain elements shown in Figure P6–9, the nodal displacements are given as

$$u_1 = 0.001 \text{ in.} \qquad v_1 = 0.005 \text{ in.} \qquad u_2 = 0.001 \text{ in.}$$
$$v_2 = 0.0025 \text{ in.} \qquad u_3 = 0.0 \text{ in.} \qquad v_3 = 0.0 \text{ in.}$$

Determine the element stresses $\sigma_x, \sigma_y, \tau_{xy}, \sigma_1$, and σ_2 and the principal angle θ_p. Let $E = 30 \times 10^6$ psi and $\nu = 0.25$, and use unit thickness for plane strain. All coordinates are in inches.

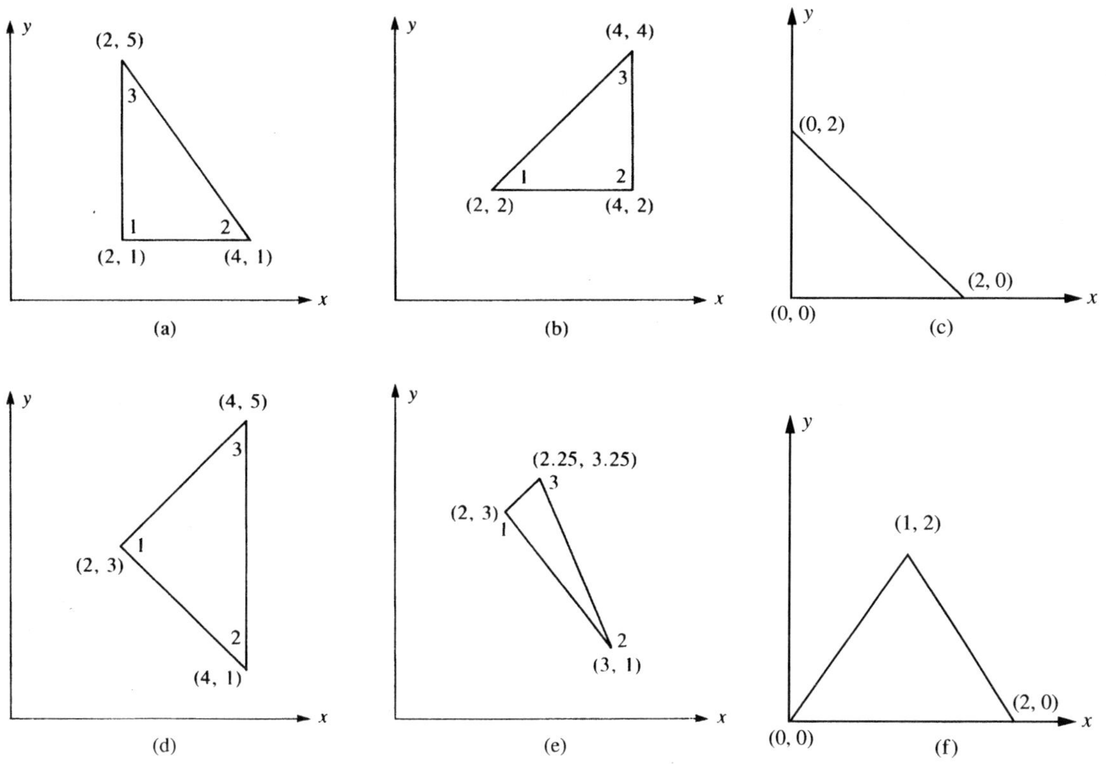

Figure P6–9

6.10 For the plane strain elements shown in Figure P6–10, the nodal displacements are given as

$$u_1 = 0.005 \text{ mm} \qquad v_1 = 0.002 \text{ mm} \qquad u_2 = 0.0 \text{ mm}$$
$$v_2 = 0.0 \text{ mm} \qquad u_3 = 0.005 \text{ mm} \qquad v_3 = 0.0 \text{ mm}$$

Determine the element stresses $\sigma_x, \sigma_y, \tau_{xy}, \sigma_1$, and σ_2 and the principal angle θ_p. Let $E = 70$ GPa and $\nu = 0.3$, and use unit thickness for plane strain. All coordinates are in millimeters.

6.11 Determine the nodal forces for (a) a linearly varying pressure p_x on the edge of the triangular element shown in Figure P6–11(a); and (b) the quadratic varying pressure shown in Figure P6–11(b) by evaluating the surface integral given by Eq. (6.3.7). Assume the element thickness is equal to t.

6.12 Determine the nodal forces for (a) the quadratic varying pressure loading shown in Figure P6–12(a) and the sinusoidal varying pressure loading shown in Figure P6–12(b)

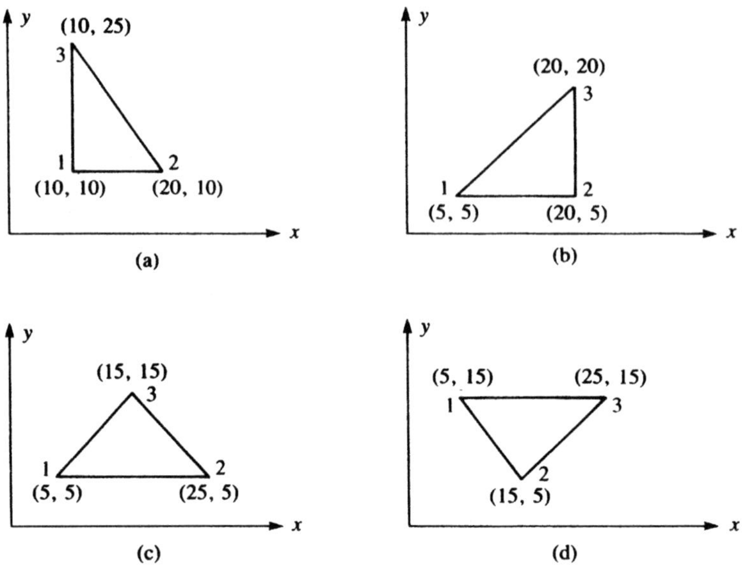

Figure P6–10

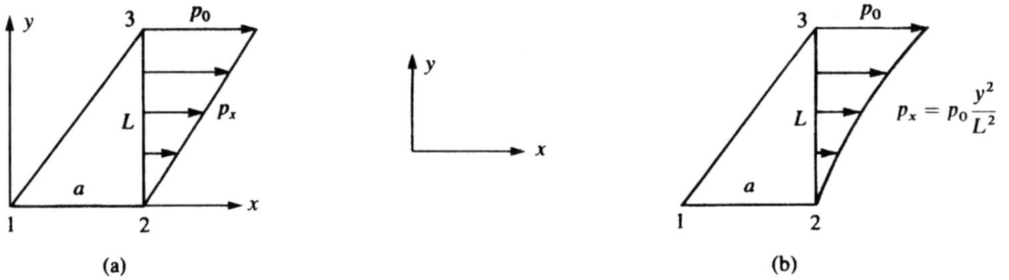

Figure P6–11

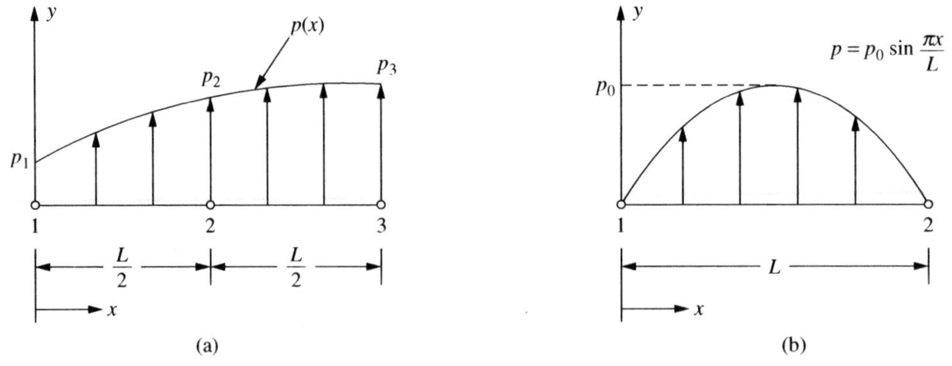

Figure P6–12

by the work equivalence method (use the surface integral expression given by Eq. (6.3.7)). Assume the element thickness to be t.

6.13 Determine the nodal displacements and the element stresses, including principal stresses, for the thin plate of Section 6.5 with a uniform shear load (instead of a tensile load) acting on the right edge, as shown in Figure P6–13. Use $E = 30 \times 10^6$ psi, $\nu = 0.30$, and $t = 1$ in. (*Hint:* The $[K]$ matrix derived in Section 6.5 and given by Eq. (6.5.22) can be used to solve the problem.)

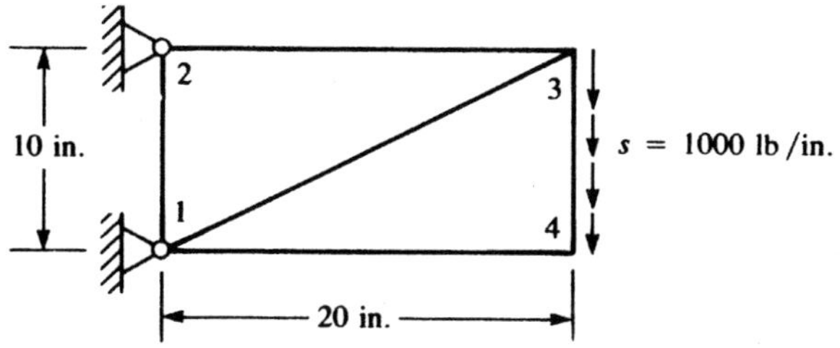

Figure P6–13

6.14 Determine the nodal displacements and the element stresses, including principal stresses, due to the loads shown for the thin plates in Figure P6–14. Use $E = 210$ GPa, $\nu = 0.30$, and $t = 5$ mm. Assume plane stress conditions apply. The recommended discretized plates are shown in the figures.

6.15 Evaluate the body force matrix for the plates shown in Figures P6–14(a) and (c). Assume the weight density to be 77.1 kN/m^3.

6.16 Why is the triangular stiffness matrix derived in Section 6.2 called a constant strain triangle?

6.17 How do the stresses vary within the constant strain triangle element?

6.18 Can you use the plane stress or plane strain element to model the following:

a. a flat slab floor of a building
b. a wall subjected to wind loading (the wall acts as a shear wall)
c. a tensile plate with a hole drilled through it
d. an eyebar
e. a soil mass subjected to a strip footing loading
f. a wrench subjected to a force in the plane of the wrench
g. a wrench subjected to twisting forces (the twisting forces act out of the plane of the wrench)
h. a triangular plate connection with loads in the plane of the triangle
i. a triangular plate connection with out-of-plane loads

6.19 The plane stress element only allows for in-plane displacements, while the frame or beam element resists displacements and rotations. How can we combine the plane stress and beam elements and still insure compatibility?

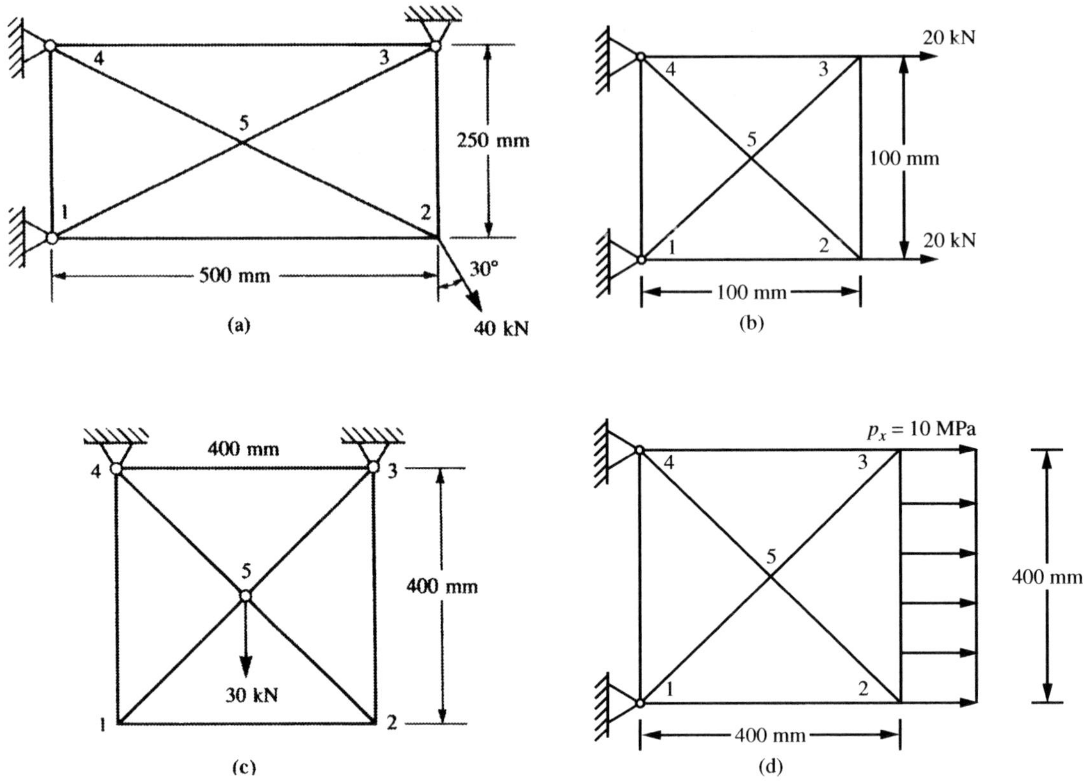

Figure P6–14

6.20 For the plane structures modeled by triangular elements shown in Figure P6–20, show that numbering in the direction that has fewer nodes, as in Figure P6–20(a) (as opposed to numbering in the direction that has more nodes), results in a reduced bandwidth. Illustrate this fact by filling in, with X's, the occupied elements in $\underline{K}$ for each mesh, as was done in Appendix B.4. Compare the bandwidths for each case.

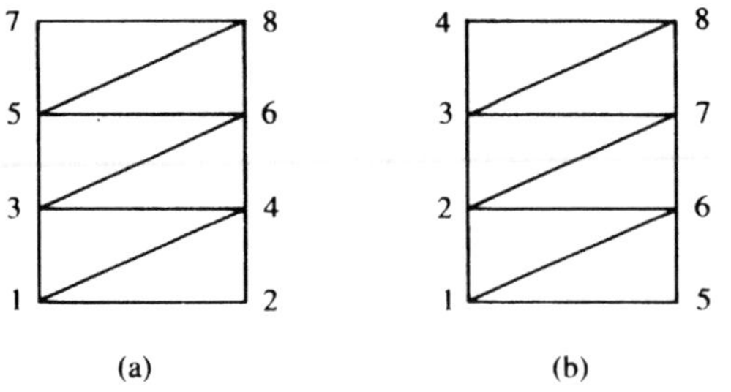

Figure P6–20

6.21 Go through the detailed steps to evaluate Eq. (6.3.6).

6.22 How would you treat a linearly varying thickness for a three-noded triangle?

6.23 Compute the stiffness matrix of element 1 of the two-triangle element model of the rectangular plate in plane stress shown in Figure P6–23. Then use it to compute the stiffness matrix of element 2.

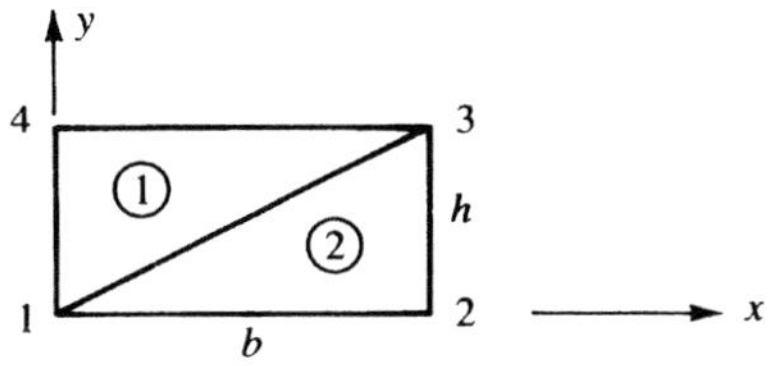

Figure P6–23

CHAPTER 9

Axisymmetric Elements

Introduction

In previous chapters, we have been concerned with line or one-dimensional elements (Chapters 2–5) and two-dimensional elements (Chapters 6–8). In this chapter, we consider a special two-dimensional element called the *axisymmetric element*. This element is quite useful when symmetry with respect to geometry and loading exists about an axis of the body being analyzed. Problems that involve soil masses subjected to circular footing loads or thick-walled pressure vessels can often be analyzed using the element developed in this chapter.

We begin with the development of the stiffness matrix for the simplest axisymmetric element, the triangular torus, whose vertical cross section is a plane triangle.

We then present the longhand solution of a thick-walled pressure vessel to illustrate the use of the axisymmetric element equations. This is followed by a description of some typical large-scale problems that have been modeled using the axisymmetric element.

9.1 Derivation of the Stiffness Matrix

In this section, we will derive the stiffness matrix and the body and surface force matrices for the axisymmetric element. However, before the development, we will first present some fundamental concepts prerequisite to the understanding of the derivation. Axisymmetric elements are triangular tori such that each element is symmetric with respect to geometry and loading about an axis such as the z axis in Figure 9–1. Hence, the z axis is called the *axis of symmetry* or the *axis of revolution*. Each vertical cross section of the element is a plane triangle. The nodal points of an axisymmetric triangular element describe circumferential lines, as indicated in Figure 9–1.

In plane stress problems, stresses exist only in the x-y plane. In axisymmetric problems, the radial displacements develop circumferential strains that induce stresses σ_r, σ_θ, σ_z, and τ_{rz}, where r, θ, and z indicate the radial, circumferential, and longitudinal

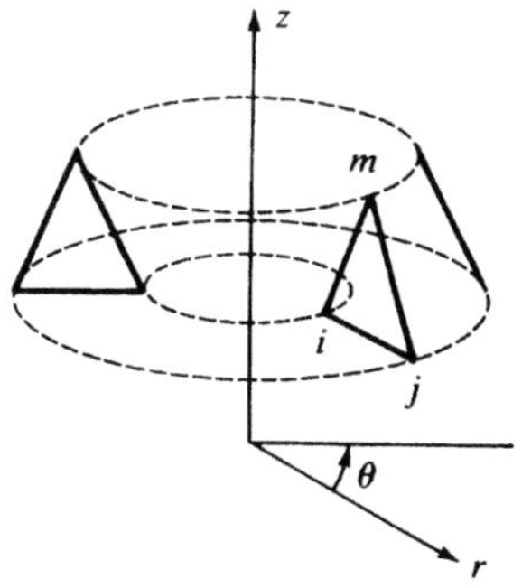

Figure 9–1 Typical axisymmetric element *ijm*

directions, respectively. Triangular torus elements are often used to idealize the axisymmetric system because they can be used to simulate complex surfaces and are simple to work with. For instance, the axisymmetric problem of a semi-infinite half-space loaded by a circular area (circular footing) shown in Figure 9–2(a), the domed pressure vessel shown in Figure 9–2(b), and the engine valve stem shown in Figure 9–2(c) can be solved using the axisymmetric element developed in this chapter.

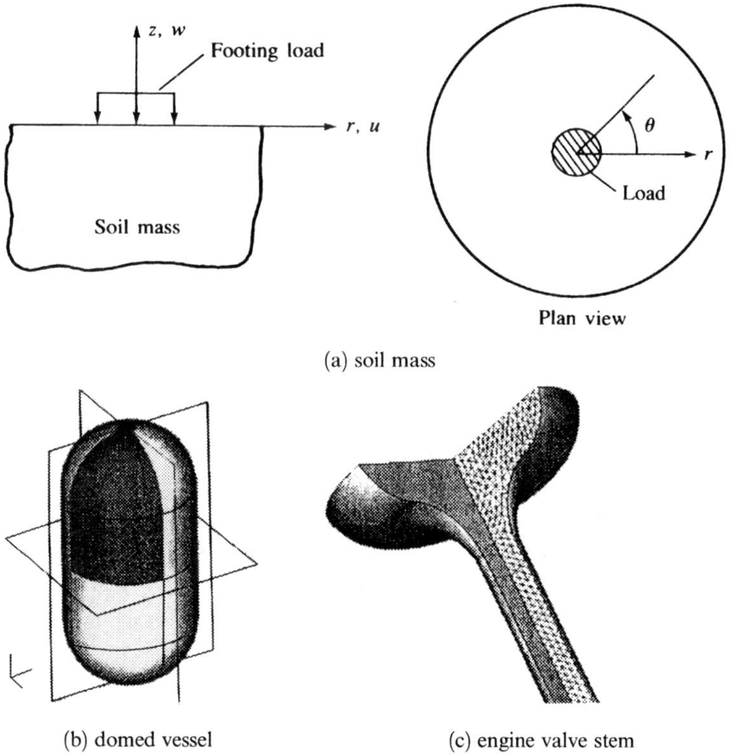

Figure 9-2 Examples of axisymmetric problems: (a) semi-infinite half-space (soil mass) modeled by axisymmetric elements, (b) a domed pressure vessel, and (c) an engine valve stem

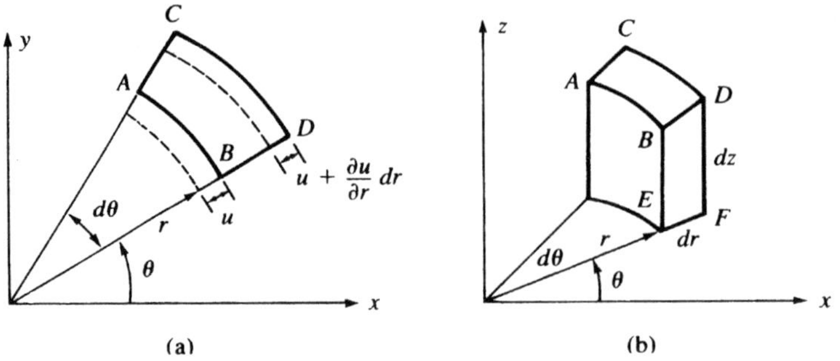

Figure 9–3 (a) Plane cross section of (b) axisymmetric element

Because of symmetry about the z axis, the stresses are independent of the θ coordinate. Therefore, all derivatives with respect to θ vanish, and the displacement component v (tangent to the θ direction), the shear strains $\gamma_{r\theta}$ and $\gamma_{\theta z}$, and the shear stresses $\tau_{r\theta}$ and $\tau_{\theta z}$ are all zero.

Figure 9–3 shows an axisymmetric ring element and its cross section to represent the general state of strain for an axisymmetric problem. It is most convenient to express the displacements of an element $ABCD$ in the plane of a cross section in cylindrical coordinates. We then let u and w denote the displacements in the radial and longitudinal directions, respectively. The side AB of the element is displaced an amount u, and side CD is then displaced an amount $u + (\partial u/\partial r)\,dr$ in the radial direction. The normal strain in the radial direction is then given by

$$\varepsilon_r = \frac{\partial u}{\partial r} \tag{9.1.1a}$$

In general, the strain in the tangential direction depends on the tangential displacement v and on the radial displacement u. However, for axisymmetric deformation behavior, recall that the tangential displacement v is equal to zero. Hence, the tangential strain is due only to the radial displacement. Having only radial displacement u, the new length of the arc $\widehat{AB}$ is $(r + u)\,d\theta$, and the tangential strain is then given by

$$\varepsilon_\theta = \frac{(r+u)\,d\theta - r\,d\theta}{r\,d\theta} = \frac{u}{r} \tag{9.1.1b}$$

Next, we consider the longitudinal element $BDEF$ to obtain the longitudinal strain and the shear strain. In Figure 9–4, the element is shown to displace by amounts u and w in the radial and longitudinal directions at point E, and to displace additional amounts $(\partial w/\partial z)\,dz$ along line BE and $(\partial u/\partial r)\,dr$ along line EF. Furthermore, observing lines EF and BE, we see that point F moves upward an amount $(\partial w/\partial r)\,dr$ with respect to point E and point B moves to the right an amount $(\partial u/\partial z)\,dz$ with respect to point E. Again, from the basic definitions of normal and shear strain, we have the longitudinal normal strain given by

$$\varepsilon_z = \frac{\partial w}{\partial z} \tag{9.1.1c}$$

and the shear strain in the r-z plane given by

$$\gamma_{rz} = \frac{\partial u}{\partial z} + \frac{\partial w}{\partial r} \tag{9.1.1d}$$

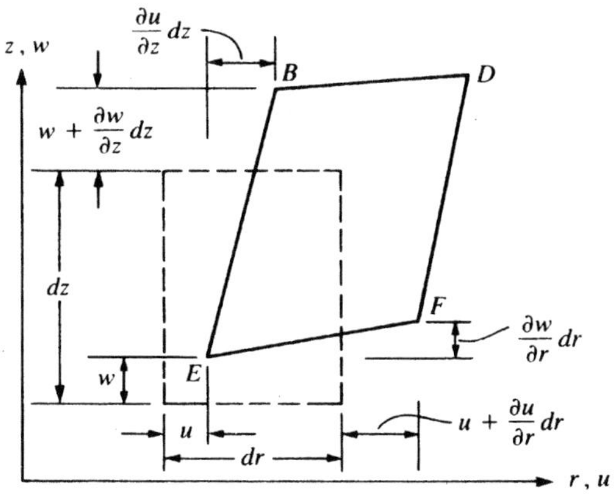

Figure 9–4 Displacement and rotations of lines of element in the *r-z* plane

Summarizing the strain/displacement relationships of Eqs. (9.1.1a–d) in one equation for easier reference, we have

$$\varepsilon_r = \frac{\partial u}{\partial r} \qquad \varepsilon_\theta = \frac{u}{r} \qquad \varepsilon_z = \frac{\partial w}{\partial z} \qquad \gamma_{rz} = \frac{\partial u}{\partial z} + \frac{\partial w}{\partial r} \tag{9.1.1e}$$

The isotropic stress/strain relationship, obtained by simplifying the general stress/strain relationships given in Appendix C, is

$$\begin{Bmatrix} \sigma_r \\ \sigma_z \\ \sigma_\theta \\ \tau_{rz} \end{Bmatrix} = \frac{E}{(1+\nu)(1-2\nu)} \begin{bmatrix} 1-\nu & \nu & \nu & 0 \\ \nu & 1-\nu & \nu & 0 \\ \nu & \nu & 1-\nu & 0 \\ 0 & 0 & 0 & \dfrac{1-2\nu}{2} \end{bmatrix} \begin{Bmatrix} \varepsilon_r \\ \varepsilon_z \\ \varepsilon_\theta \\ \gamma_{rz} \end{Bmatrix} \tag{9.1.2}$$

The theoretical development follows that of the plane stress/strain problem given in Chapter 6.

Step 1 Select Element Type

An axisymmetric solid is shown discretized in Figure 9–5(a), along with a typical triangular element. The element has three nodes with two degrees of freedom per node (that is, u_i, w_i at node i). The stresses in the axisymmetric problem are shown in Figure 9–5(b).

Step 2 Select Displacement Functions

The element displacement functions are taken to be

$$\begin{aligned} u(r,z) &= a_1 + a_2 r + a_3 z \\ w(r,z) &= a_4 + a_5 r + a_6 z \end{aligned} \tag{9.1.3}$$

so that we have the same linear displacement functions as used in the plane stress, constant-strain triangle. Again, the total number of a_i's (six) introduced in the

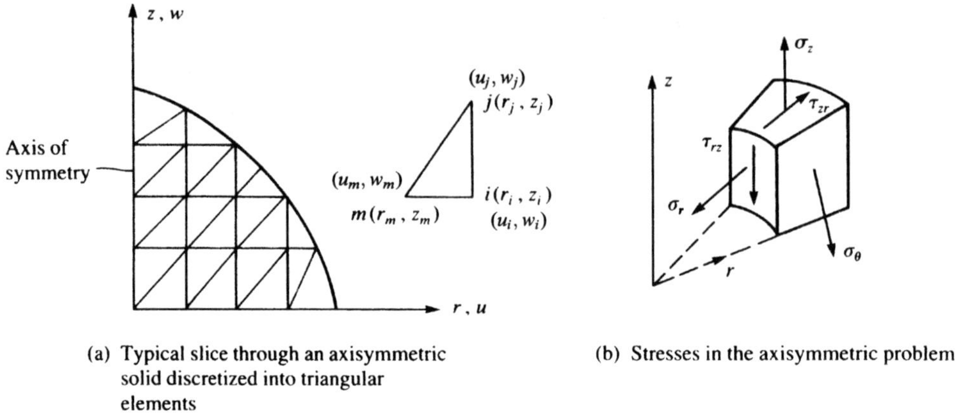

Figure 9–5 Discretized axisymmetric solid

displacement functions is the same as the total number of degrees of freedom for the element. The nodal displacements are

$$\{d\} = \begin{Bmatrix} \underline{d}_i \\ \underline{d}_j \\ \underline{d}_m \end{Bmatrix} = \begin{Bmatrix} u_i \\ w_i \\ u_j \\ w_j \\ u_m \\ w_m \end{Bmatrix} \tag{9.1.4}$$

and u evaluated at node i is

$$u(r_i, z_i) = u_i = a_1 + a_2 r_i + a_3 z_i \tag{9.1.5}$$

Using Eq. (9.1.3), the general displacement function is then expressed in matrix form as

$$\{\psi\} = \begin{Bmatrix} u \\ w \end{Bmatrix} = \begin{Bmatrix} a_1 + a_2 r + a_3 z \\ a_4 + a_5 r + a_6 z \end{Bmatrix} = \begin{bmatrix} 1 & r & z & 0 & 0 & 0 \\ 0 & 0 & 0 & 1 & r & z \end{bmatrix} \begin{Bmatrix} a_1 \\ a_2 \\ a_3 \\ a_4 \\ a_5 \\ a_6 \end{Bmatrix} \tag{9.1.6}$$

Substituting the coordinates of the nodal points shown in Figure 9–5(a) into Eq. (9.1.6), we can solve for the a_i's in a manner similar to that in Section 6.2. The resulting expressions are

$$\begin{Bmatrix} a_1 \\ a_2 \\ a_3 \end{Bmatrix} = \begin{bmatrix} 1 & r_i & z_i \\ 1 & r_j & z_j \\ 1 & r_m & z_m \end{bmatrix}^{-1} \begin{Bmatrix} u_i \\ u_j \\ u_m \end{Bmatrix} \tag{9.1.7}$$

and

$$\begin{Bmatrix} a_4 \\ a_5 \\ a_6 \end{Bmatrix} = \begin{bmatrix} 1 & r_i & z_i \\ 1 & r_j & z_j \\ 1 & r_m & z_m \end{bmatrix}^{-1} \begin{Bmatrix} w_i \\ w_j \\ w_m \end{Bmatrix} \tag{9.1.8}$$

Performing the inversion operations in Eqs. (9.1.7) and (9.1.8), we have

$$\begin{Bmatrix} a_1 \\ a_2 \\ a_3 \end{Bmatrix} = \frac{1}{2A}\begin{bmatrix} \alpha_i & \alpha_j & \alpha_m \\ \beta_i & \beta_j & \beta_m \\ \gamma_i & \gamma_j & \gamma_m \end{bmatrix}\begin{Bmatrix} u_i \\ u_j \\ u_m \end{Bmatrix} \tag{9.1.9}$$

and

$$\begin{Bmatrix} a_4 \\ a_5 \\ a_6 \end{Bmatrix} = \frac{1}{2A}\begin{bmatrix} \alpha_i & \alpha_j & \alpha_m \\ \beta_i & \beta_j & \beta_m \\ \gamma_i & \gamma_j & \gamma_m \end{bmatrix}\begin{Bmatrix} w_i \\ w_j \\ w_m \end{Bmatrix} \tag{9.1.10}$$

where

$$\begin{array}{lll} \alpha_i = r_j z_m - z_j r_m & \alpha_j = r_m z_i - z_m r_i & \alpha_m = r_i z_j - z_i r_j \\ \beta_i = z_j - z_m & \beta_j = z_m - z_i & \beta_m = z_i - z_j \\ \gamma_i = r_m - r_j & \gamma_j = r_i - r_m & \gamma_m = r_j - r_i \end{array} \tag{9.1.11}$$

We define the shape functions, similar to Eqs. (6.2.18), as

$$\begin{aligned} N_i &= \frac{1}{2A}(\alpha_i + \beta_i r + \gamma_i z) \\ N_j &= \frac{1}{2A}(\alpha_j + \beta_j r + \gamma_j z) \\ N_m &= \frac{1}{2A}(\alpha_m + \beta_m r + \gamma_m z) \end{aligned} \tag{9.1.12}$$

Substituting Eqs. (9.1.7) and (9.1.8) into Eq. (9.1.6), along with the shape function Eqs. (9.1.12), we find that the general displacement function is

$$\{\psi\} = \begin{Bmatrix} u(r,z) \\ w(r,z) \end{Bmatrix} = \begin{bmatrix} N_i & 0 & N_j & 0 & N_m & 0 \\ 0 & N_i & 0 & N_j & 0 & N_m \end{bmatrix}\begin{Bmatrix} u_i \\ w_i \\ u_j \\ w_j \\ u_m \\ w_m \end{Bmatrix} \tag{9.1.13}$$

or

$$\{\psi\} = [N]\{d\} \tag{9.1.14}$$

Step 3 Define the Strain/Displacement and Stress/Strain Relationships

When we use Eqs. (9.1.1) and (9.1.3), the strains become

$$\{\varepsilon\} = \begin{Bmatrix} a_2 \\ a_6 \\ \dfrac{a_1}{r} + a_2 + \dfrac{a_3 z}{r} \\ a_3 + a_5 \end{Bmatrix} \tag{9.1.15}$$

Rewriting Eq. (9.1.15) with the a_i's as a separate column matrix, we have

$$\begin{Bmatrix} \varepsilon_r \\ \varepsilon_z \\ \varepsilon_\theta \\ \gamma_{rz} \end{Bmatrix} = \begin{bmatrix} 0 & 1 & 0 & 0 & 0 & 0 \\ 0 & 0 & 0 & 0 & 0 & 1 \\ \dfrac{1}{r} & 1 & \dfrac{z}{r} & 0 & 0 & 0 \\ 0 & 0 & 1 & 0 & 1 & 0 \end{bmatrix} \begin{Bmatrix} a_1 \\ a_2 \\ a_3 \\ a_4 \\ a_5 \\ a_6 \end{Bmatrix} \tag{9.1.16}$$

Substituting Eqs. (9.1.7) and (9.1.8) into Eq. (9.1.16) and making use of Eq. (9.1.11), we obtain

$$\{\varepsilon\} = \frac{1}{2A} \begin{bmatrix} \beta_i & 0 & \beta_j & 0 & \beta_m & 0 \\ 0 & \gamma_i & 0 & \gamma_j & 0 & \gamma_m \\ \dfrac{\alpha_i}{r} + \beta_i + \dfrac{\gamma_i z}{r} & 0 & \dfrac{\alpha_j}{r} + \beta_j + \dfrac{\gamma_j z}{r} & 0 & \dfrac{\alpha_m}{r} + \beta_m + \dfrac{\gamma_m z}{r} & 0 \\ \gamma_i & \beta_i & \gamma_j & \beta_j & \gamma_m & \beta_m \end{bmatrix} \begin{Bmatrix} u_i \\ w_i \\ u_j \\ w_j \\ u_m \\ w_m \end{Bmatrix} \tag{9.1.17}$$

or, rewriting Eq. (9.1.17) in simplified matrix form,

$$\{\varepsilon\} = [\underline{B}_i \quad \underline{B}_j \quad \underline{B}_m] \begin{Bmatrix} u_i \\ w_i \\ u_j \\ w_j \\ u_m \\ w_m \end{Bmatrix} \tag{9.1.18}$$

where

$$[\underline{B}_i] = \frac{1}{2A} \begin{bmatrix} \beta_i & 0 \\ 0 & \gamma_i \\ \dfrac{\alpha_i}{r} + \beta_i + \dfrac{\gamma_i z}{r} & 0 \\ \gamma_i & \beta_i \end{bmatrix} \tag{9.1.19}$$

Similarly, we obtain submatrices $\underline{B}_j$ and $\underline{B}_m$ by replacing the subscript i with j and then with m in Eq. (9.1.19). Rewriting Eq. (9.1.18) in compact matrix form, we have

$$\{\varepsilon\} = [B]\{d\} \tag{9.1.20}$$

where

$$[B] = [\underline{B}_i \quad \underline{B}_j \quad \underline{B}_m] \tag{9.1.21}$$

Note that $[B]$ is a function of the r and z coordinates. Therefore, in general, the strain ε_θ will not be constant.

The stresses are given by

$$\{\sigma\} = [D][B]\{d\} \tag{9.1.22}$$

where $[D]$ is given by the first matrix on the right side of Eq. (9.1.2). (As mentioned in Chapter 6, for $\nu = 0.5$, a special formula must be used; see Reference [9].)

Step 4 Derive the Element Stiffness Matrix and Equations

The stiffness matrix is

$$[k] = \iiint_V [B]^T [D][B]\, dV \tag{9.1.23}$$

or

$$[k] = 2\pi \iint_A [B]^T [D][B] r\, dr\, dz \tag{9.1.24}$$

after integrating along the circumferential boundary. The $[B]$ matrix, Eq. (9.1.21), is a function of r and z. Therefore, $[k]$ is a function of r and z and is of order 6×6.

We can evaluate Eq. (9.1.24) for $[k]$ by one of three methods:

1. Numerical integration (Gaussian quadrature) as discussed in Chapter 10.
2. Explicit multiplication and term-by-term integration [1].
3. Evaluate $[B]$ for a centroidal point $(\bar{r}, \bar{z})$ of the element

$$r = \bar{r} = \frac{r_i + r_j + r_m}{3} \qquad z = \bar{z} = \frac{z_i + z_j + z_m}{3} \tag{9.1.25}$$

and define $[B(\bar{r}, \bar{z})] = [\bar{B}]$. Therefore, as a first approximation,

$$[k] = 2\pi \bar{r} A [\bar{B}]^T [D][\bar{B}] \tag{9.1.26}$$

If the triangular subdivisions are consistent with the final stress distribution (that is, small elements in regions of high stress gradients), then acceptable results can be obtained by method 3.

Distributed Body Forces

Loads such as gravity (in the direction of the z axis) or centrifugal forces in rotating machine parts (in the direction of the r axis) are considered to be body forces (as shown in Figure 9–6). The body forces can be found by

$$\{f_b\} = 2\pi \iint_A [N]^T \begin{Bmatrix} R_b \\ Z_b \end{Bmatrix} r\, dr\, dz \tag{9.1.27}$$

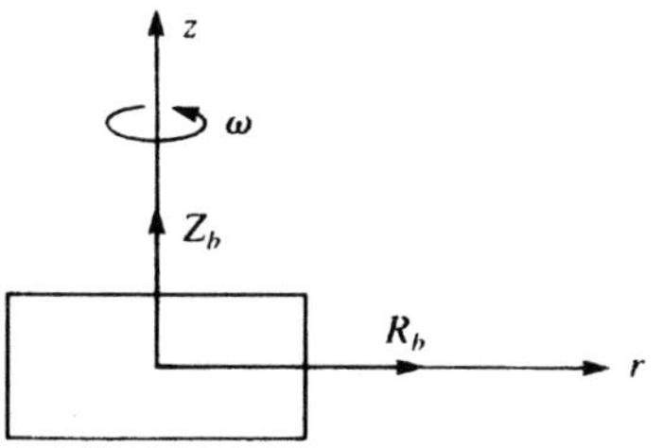

Figure 9–6 Axisymmetric element with body forces per unit volume

where $R_b = \omega^2 \rho r$ for a machine part moving with a constant angular velocity ω about the z axis, with material mass density ρ and radial coordinate r, and where Z_b is the body force per unit volume due to the force of gravity.

Considering the body force at node i, we have

$$\{f_{bi}\} = 2\pi \iint_A [N_i]^T \begin{Bmatrix} R_b \\ Z_b \end{Bmatrix} r\,dr\,dz \tag{9.1.28}$$

where

$$[N_i]^T = \begin{bmatrix} N_i & 0 \\ 0 & N_i \end{bmatrix} \tag{9.1.29}$$

Multiplying and integrating in Eq. (9.1.28), we obtain

$$\{f_{bi}\} = \frac{2\pi}{3} \begin{Bmatrix} \bar{R}_b \\ Z_b \end{Bmatrix} A\bar{r} \tag{9.1.30}$$

where the origin of the coordinates has been taken as the centroid of the element, and $\bar{R}_b$ is the radially directed body force per unit volume evaluated at the centroid of the element. The body forces at nodes j and m are identical to those given by Eq. (9.1.30) for node i. Hence, for an element, we have

$$\{f_b\} = \frac{2\pi\bar{r}A}{3} \begin{Bmatrix} \bar{R}_b \\ Z_b \\ \bar{R}_b \\ Z_b \\ \bar{R}_b \\ Z_b \end{Bmatrix} \tag{9.1.31}$$

where

$$\bar{R}_b = \omega^2 \rho \bar{r} \tag{9.1.32}$$

Equation (9.1.31) is a first approximation to the radially directed body force distribution.

Surface Forces

Surface forces can be found by

$$\{f_s\} = \iint_S [N_s]^T \{T\}\,dS \tag{9.1.33}$$

where again $[N_s]$ denotes the shape function matrix evaluated along the surface where the surface traction acts.

For radial and axial pressures p_r and p_z, respectively, we have

$$\{f_s\} = \iint_S [N_s]^T \begin{Bmatrix} p_r \\ p_z \end{Bmatrix} dS \tag{9.1.34}$$

For example, along the vertical face jm of an element, let uniform loads p_r and p_z be applied, as shown in Figure 9–7 along surface $r = r_j$. We can use Eq. (9.1.34)

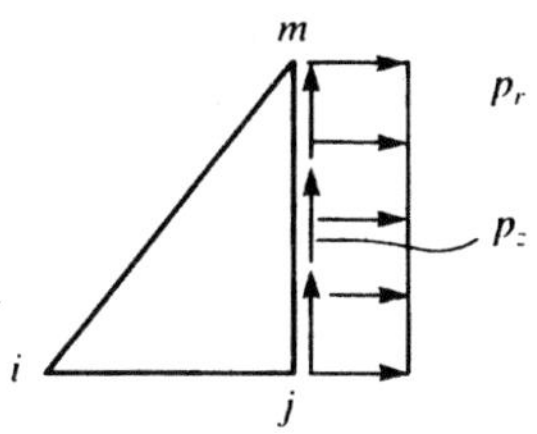

Figure 9–7 Axisymmetric element with surface forces

written for each node separately. For instance, for node j, substituting N_j from Eqs. (9.1.12) into Eq. (9.1.34), we have

$$\{f_{sj}\} = \int_{z_j}^{z_m} \frac{1}{2A} \begin{bmatrix} \alpha_j + \beta_j r + \gamma_j z & 0 \\ 0 & \alpha_j + \beta_j r + \gamma_j z \end{bmatrix} \begin{Bmatrix} p_r \\ p_z \end{Bmatrix} 2\pi r_j \, dz \qquad (9.1.35)$$

evaluated at $r = r_j, \ z = z$

Performing the integration of Eq. (9.1.35) explicitly, along with similar evaluations for f_{si} and f_{sm}, we obtain the total distribution of surface force to nodes i, j, and m as

$$\{f_s\} = \frac{2\pi r_j (z_m - z_j)}{2} \begin{Bmatrix} 0 \\ 0 \\ p_r \\ p_z \\ p_r \\ p_z \end{Bmatrix} \qquad (9.1.36)$$

Steps 5–7

Steps 5–7, which involve assembling the total stiffness matrix, total force matrix, and total set of equations; solving for the nodal degrees of freedom; and calculating the element stresses, are analogous to those of Chapter 6 for the CST element, except the stresses are not constant in each element. They are usually determined by one of two methods that we use to determine the LST element stresses. Either we determine the centroidal element stresses, or we determine the nodal stresses for the element and then average them. The latter method has been shown to be more accurate in some cases [2].

Example 9.1

For the element of an axisymmetric body rotating with a constant angular velocity $\omega = 100$ rev/min as shown in Figure 9–8, evaluate the approximate body force matrix. Include the weight of the material, where the weight density ρ_w is 0.283 lb/in^3. The coordinates of the element (in inches) are shown in the figure.

We need to evaluate Eq. (9.1.31) to obtain the approximate body force matrix. Therefore, the body forces per unit volume evaluated at the centroid of the element are

$$Z_b = 0.283 \text{ lb/in}^3$$

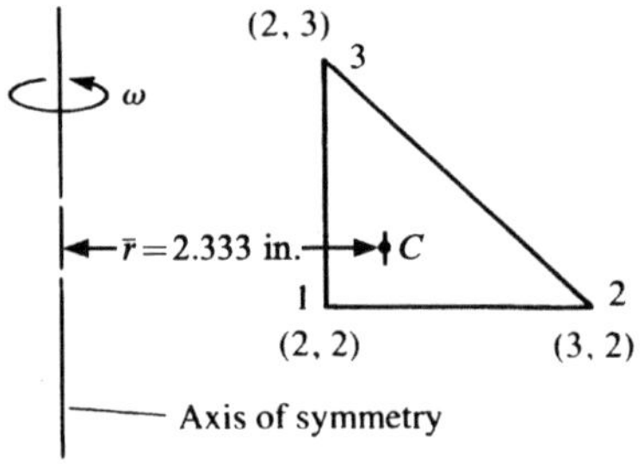

Figure 9–8 Axisymmetric element subjected to angular velocity

and by Eq. (9.1.32), we have

$$\bar{R}_b = \omega^2 \rho \bar{r} = \left[\left(100 \frac{\text{rev}}{\text{min}}\right)\left(2\pi \frac{\text{rad}}{\text{rev}}\right)\left(\frac{1 \text{ min}}{60 \text{ s}}\right)\right]^2 \frac{(0.283 \text{ lb/in}^3)}{(32.2 \times 12) \text{ in./s}^2}(2.333 \text{ in.})$$

$$\bar{R}_b = 0.187 \text{ lb/in}^3$$

$$\frac{2\pi\bar{r}A}{3} = \frac{2\pi(2.333)(0.5)}{3} = 2.44 \text{ in}^3$$

$$f_{b1r} = (2.44)(0.187) = 0.457 \text{ lb}$$

$$f_{b1z} = -(2.44)(0.283) = -0.691 \text{ lb} \qquad \text{(downward)}$$

Because we are using the first approximation Eq. (9.1.31), all r-directed nodal body forces are equal, and all z-directed body forces are equal. Therefore,

$$f_{b2r} = 0.457 \text{ lb} \qquad f_{b2z} = -0.691 \text{ lb}$$

$$f_{b3r} = 0.457 \text{ lb} \qquad f_{b3z} = -0.691 \text{ lb}$$

■

9.2 Solution of an Axisymmetric Pressure Vessel

To illustrate the use of the equations developed in Section 9.1, we will now solve an axisymmetric stress problem.

Example 9.2

For the long, thick-walled cylinder under internal pressure p equal to 1 psi shown in Figure 9–9, determine the displacements and stresses.

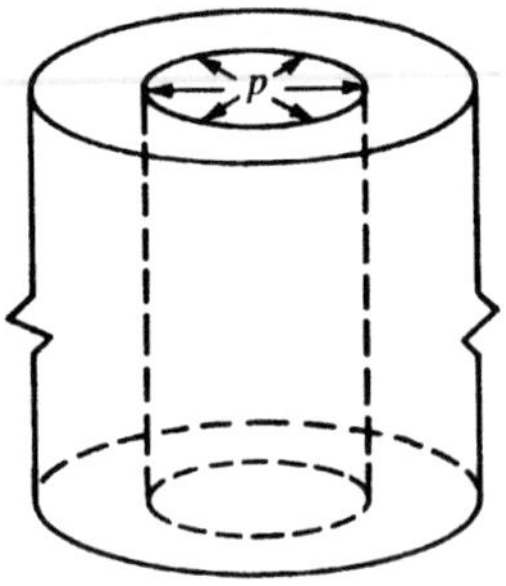

Figure 9–9 Thick-walled cylinder subjected to internal pressure

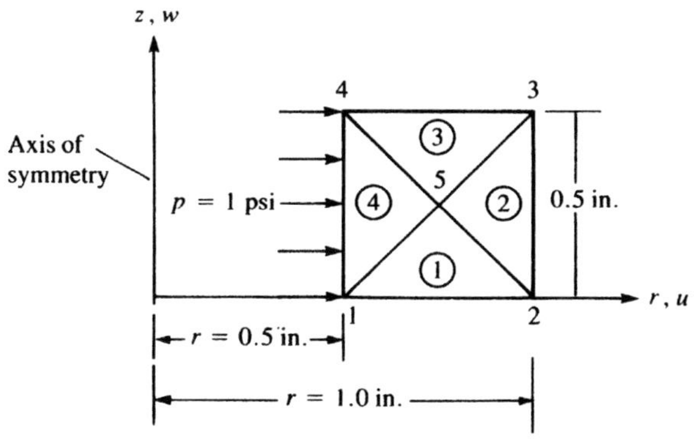

Figure 9–10 Discretized cylinder slice

Discretization

To illustrate the finite element solution for the cylinder, we first discretize the cylinder into four triangular elements, as shown in Figure 9–10. A horizontal slice of the cylinder represents the total cylinder behavior. Because we are performing a longhand solution, a coarse mesh of elements is used for simplicity's sake (but without loss of generality of the method). The governing global matrix equation is

$$\begin{Bmatrix} F_{1r} \\ F_{1z} \\ F_{2r} \\ F_{2z} \\ F_{3r} \\ F_{3z} \\ F_{4r} \\ F_{4z} \\ F_{5r} \\ F_{5z} \end{Bmatrix} = [K] \begin{Bmatrix} u_1 \\ w_1 \\ u_2 \\ w_2 \\ u_3 \\ w_3 \\ u_4 \\ w_4 \\ u_5 \\ w_5 \end{Bmatrix} \tag{9.2.1}$$

where the $[K]$ matrix is of order 10×10.

Assemblage of the Stiffness Matrix

We assemble the $[K]$ matrix in the usual manner by superposition of the individual element stiffness matrices. For simplicity's sake, we will use the first approximation method given by Eq. (9.1.26) to evaluate the element matrices. Therefore,

$$[k] = 2\pi\bar{r}A[\bar{B}]^T[D][\bar{B}] \tag{9.2.2}$$

For element 1 (Figure 9–11), the coordinates are $r_i = 0.5$, $z_i = 0$, $r_j = 1.0$, $z_j = 0$, $r_m = 0.75$, and $z_m = 0.25$ ($i = 1, j = 2$, and $m = 5$ for element 1) for the global-coordinate axes as set up in Figure 9–10.

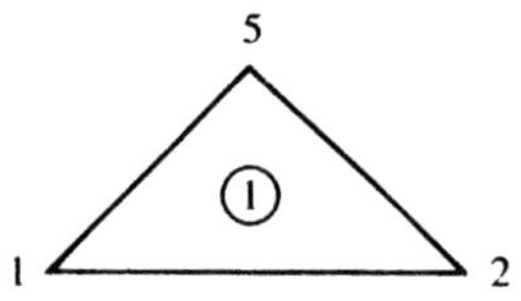

Figure 9–11 Element 1 of the discretized cylinder

We now evaluate $[\bar{B}]$, where $[\bar{B}]$ is given by Eq. (9.1.19) evaluated at the centroid of the element $r = \bar{r}$, $z = \bar{z}$, and expanded here as

$$[\bar{B}] = \frac{1}{2A}\begin{bmatrix} \beta_i & 0 & \beta_j & 0 & \beta_m & 0 \\ 0 & \gamma_i & 0 & \gamma_j & 0 & \gamma_m \\ \dfrac{\alpha_i}{\bar{r}} + \beta_i + \dfrac{\gamma_i \bar{z}}{\bar{r}} & 0 & \dfrac{\alpha_j}{\bar{r}} + \beta_j + \dfrac{\gamma_j \bar{z}}{\bar{r}} & 0 & \dfrac{\alpha_m}{\bar{r}} + \beta_m + \dfrac{\gamma_m \bar{z}}{\bar{r}} & 0 \\ \gamma_i & \beta_i & \gamma_j & \beta_j & \gamma_m & \beta_m \end{bmatrix} \tag{9.2.3}$$

where, using element coordinates in Eqs. (9.1.11), we have

$$\begin{aligned}
\alpha_i &= r_j z_m - z_j r_m = (1.0)(0.25) - (0.0)(0.75) = 0.25 \text{ in}^2 \\
\alpha_j &= r_m z_i - z_m r_i = (0.75)(0) - (0.25)(0.5) = -0.125 \text{ in}^2 \\
\alpha_m &= r_i z_j - z_i r_j = (0.5)(0.0) - (0)(1.0) = 0.0 \text{ in}^2 \\
\beta_i &= z_j - z_m = 0.0 - 0.25 = -0.25 \text{ in.} \\
\beta_j &= z_m - z_i = 0.25 - 0 = 0.25 \text{ in.} \\
\beta_m &= z_i - z_j = 0.0 - 0.0 = 0.0 \text{ in.} \\
\gamma_i &= r_m - r_j = 0.75 - 1.0 = -0.25 \text{ in.} \\
\gamma_j &= r_i - r_m = 0.5 - 0.75 = -0.25 \text{ in.} \\
\gamma_m &= r_j - r_i = 1.0 - 0.5 = 0.5 \text{ in.}
\end{aligned} \tag{9.2.4}$$

and $\bar{r} = 0.5 + \frac{1}{2}(0.5) = 0.75$ in. $\quad \bar{z} = \frac{1}{3}(0.25) = 0.0833$ in.

$$A = \tfrac{1}{2}(0.5)(0.25) = 0.0625 \text{ in}^2$$

Substituting the results from Eqs. (9.2.4) into Eq. (9.2.3), we obtain

$$[\bar{B}] = \frac{1}{0.125}\begin{bmatrix} -0.25 & 0 & 0.25 & 0 & 0 & 0 \\ 0 & -0.25 & 0 & -0.25 & 0 & 0.5 \\ 0.0556 & 0 & 0.0556 & 0 & 0.0556 & 0 \\ -0.25 & -0.25 & -0.25 & 0.25 & 0.5 & 0 \end{bmatrix}\frac{1}{\text{in.}} \tag{9.2.5}$$

For the axisymmetric stress case, the matrix $[D]$ is given in Eq. (9.1.2) as

$$[D] = \frac{E}{(1+\nu)(1-2\nu)}\begin{bmatrix} 1-\nu & \nu & \nu & 0 \\ \nu & 1-\nu & \nu & 0 \\ \nu & \nu & 1-\nu & 0 \\ 0 & 0 & 0 & \dfrac{1-2\nu}{2} \end{bmatrix} \tag{9.2.6}$$

With $\nu = 0.3$ and $E = 30 \times 10^6$ psi, we obtain

$$[D] = \frac{30(10^6)}{(1+0.3)[1-2(0.3)]}\begin{bmatrix} 1-0.3 & 0.3 & 0.3 & 0 \\ 0.3 & 1-0.3 & 0.3 & 0 \\ 0.3 & 0.3 & 1-0.3 & 0 \\ 0 & 0 & 0 & \dfrac{1-2(0.3)}{2} \end{bmatrix} \tag{9.2.7}$$

or, simplifying Eq. (9.2.7),

$$[D] = 57.7(10^6)\begin{bmatrix} 0.7 & 0.3 & 0.3 & 0 \\ 0.3 & 0.7 & 0.3 & 0 \\ 0.3 & 0.3 & 0.7 & 0 \\ 0 & 0 & 0 & 0.2 \end{bmatrix} \text{psi} \tag{9.2.8}$$

Using Eqs. (9.2.5) and (9.2.8), we obtain

$$[\bar{B}]^T[D] = \frac{57.7(10^6)}{0.125}\begin{bmatrix} -0.158 & -0.0583 & -0.0361 & -0.05 \\ -0.075 & -0.175 & -0.075 & -0.05 \\ 0.192 & 0.0917 & 0.114 & -0.05 \\ -0.075 & -0.175 & -0.075 & 0.05 \\ 0.0167 & 0.0166 & 0.0388 & 0.1 \\ 0.15 & 0.35 & 0.15 & 0 \end{bmatrix} \tag{9.2.9}$$

Substituting Eqs. (9.2.5) and (9.2.9) into Eq. (9.2.2), we obtain the stiffness matrix for element 1 as

$$[k^{(1)}] = (10^6)\begin{bmatrix} \multicolumn{2}{c}{i=1} & \multicolumn{2}{c}{j=2} & \multicolumn{2}{c}{m=5} \\ 54.46 & 29.45 & -31.63 & 2.26 & -29.37 & -31.71 \\ 29.45 & 61.17 & -11.33 & 33.98 & -31.72 & -95.15 \\ -31.63 & -11.33 & 72.59 & -38.52 & -20.31 & 49.84 \\ 2.26 & 33.98 & -38.52 & 61.17 & 22.66 & -95.15 \\ -29.37 & -31.72 & -20.31 & 22.66 & 56.72 & 9.06 \\ -31.71 & -95.15 & 49.84 & -95.15 & 9.06 & 190.31 \end{bmatrix}\frac{\text{lb}}{\text{in.}} \tag{9.2.10}$$

where the numbers above the columns indicate the nodal orders of degrees of freedom in the element 1 stiffness matrix.

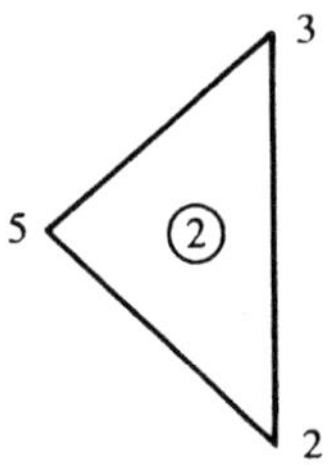

Figure 9–12 Element 2 of the discretized cylinder

For element 2 (Figure 9–12), the coordinates are $r_i = 1.0$, $z_i = 0.0$, $r_j = 1.0$, $z_j = 0.5$, $r_m = 0.75$, and $z_m = 0.25$ ($i = 2$, $j = 3$, and $m = 5$ for element 2). Therefore,

$$\begin{aligned} \alpha_i &= (1.0)(0.25) - (0.5)(0.75) = -0.125 \text{ in}^2 \\ \alpha_j &= (0.75)(0.0) - (0.25)(1.0) = -0.25 \text{ in}^2 \\ \alpha_m &= (1.0)(0.5) - (0.0)(1.0) = 0.5 \text{ in}^2 \end{aligned} \tag{9.2.11}$$

$$\begin{aligned} \beta_i &= 0.5 - 0.25 = 0.25 \text{ in.} \qquad & \beta_j &= 0.25 - 0.0 = 0.25 \text{ in.} \\ \beta_m &= 0.0 - 0.5 = -0.5 \text{ in.} \qquad & \gamma_i &= 0.75 - 1.0 = -0.25 \text{ in.} \\ \gamma_j &= 1.0 - 0.75 = 0.25 \text{ in.} \qquad & \gamma_m &= 1.0 - 1.0 = 0.0 \text{ in.} \end{aligned}$$

and $\qquad \bar{r} = 0.9167 \text{ in.} \qquad \bar{z} = 0.25 \text{ in.} \qquad A = 0.0625 \text{ in}^2$

Using Eqs. (9.2.11) in Eq. (9.2.2) and proceeding as for element 1, we obtain the stiffness matrix for element 2 as

$$[k^{(2)}] = (10^6) \begin{array}{c} \begin{array}{cccccc} \quad i = 2 & & \quad j = 3 & & \quad m = 5 & \end{array} \\ \begin{bmatrix} 85.75 & -46.07 & 52.52 & 12.84 & -118.92 & 33.23 \\ -46.07 & 74.77 & -12.84 & -41.54 & 45.32 & -33.23 \\ 52.52 & -12.84 & 85.74 & 46.07 & -118.92 & -33.23 \\ 12.84 & -41.54 & 46.07 & 74.77 & -45.32 & -33.23 \\ -118.92 & 45.32 & -118.92 & -45.32 & 216.41 & 0 \\ 33.23 & -33.23 & -33.23 & -33.23 & 0 & 66.46 \end{bmatrix} \end{array} \frac{\text{lb}}{\text{in.}} \tag{9.2.12}$$

We obtain the stiffness matrices for elements 3 and 4 in a manner similar to that used to obtain the stiffness matrices for elements 1 and 2. Thus,

$$[k^{(3)}] = (10^6) \begin{array}{c} \begin{array}{cccccc} \quad i = 3 & & \quad j = 4 & & \quad m = 5 & \end{array} \\ \begin{bmatrix} 72.58 & 38.52 & -31.63 & 11.33 & -20.31 & -49.84 \\ 38.52 & 61.17 & -2.26 & 33.98 & -22.66 & -95.15 \\ -31.63 & -2.26 & 54.46 & -29.45 & -29.37 & 31.72 \\ 11.33 & 33.98 & -29.45 & 61.17 & 31.72 & -95.15 \\ -20.31 & -22.66 & -29.37 & 31.72 & 56.72 & -9.06 \\ -49.84 & -95.15 & 31.72 & -95.15 & -9.06 & 190.31 \end{bmatrix} \end{array} \frac{\text{lb}}{\text{in.}} \tag{9.2.13}$$

and

$$
\begin{array}{c} \\ [k^{(4)}] = (10^6) \end{array}
\begin{array}{c} \qquad i=4 \qquad\qquad\qquad j=1 \qquad\qquad\qquad m=5 \\
\begin{bmatrix}
41.53 & -21.90 & 20.39 & 0.75 & -66.45 & 21.14 \\
-21.90 & 47.57 & -0.75 & -26.43 & 36.24 & -21.14 \\
20.39 & -0.75 & 41.53 & 21.90 & -66.45 & -21.14 \\
0.75 & -26.43 & 21.90 & 47.57 & -36.24 & -21.14 \\
-66.45 & 36.24 & -66.45 & -36.24 & 169.14 & 0 \\
21.14 & -21.14 & -21.14 & -21.14 & 0 & 42.28
\end{bmatrix} \end{array} \frac{\text{lb}}{\text{in.}} \tag{9.2.14}
$$

Using superposition of the element stiffness matrices [Eqs. (9.2.10) and (9.2.12)–(9.2.14)], where we rearrange the elements of each stiffness matrix in order of increasing nodal degrees of freedom, we obtain the global stiffness matrix as

$$
[K] = (10^6) \begin{bmatrix}
95.99 & 51.35 & -31.63 & 2.26 & 0 & 0 & 20.39 & -0.75 & -95.82 & -52.86 \\
51.35 & 108.74 & -11.33 & 33.98 & 0 & 0 & 0.75 & -26.43 & -67.96 & -116.3 \\
-31.63 & -11.33 & 158.34 & -84.59 & 52.52 & 12.84 & 0 & 0 & -139.2 & 83.07 \\
2.26 & 33.98 & -84.59 & 135.94 & -12.84 & -41.54 & 0 & 0 & 67.98 & -128.4 \\
0 & 0 & 52.52 & -12.84 & 158.33 & 84.59 & -31.63 & 11.33 & -139.2 & -83.07 \\
0 & 0 & 12.84 & -41.54 & 84.59 & 135.94 & -2.26 & 33.98 & -67.98 & -128.4 \\
20.39 & 0.75 & 0 & 0 & -31.63 & -2.26 & 95.99 & -51.35 & -95.82 & 52.86 \\
-0.75 & -26.43 & 0 & 0 & 11.33 & 33.98 & -51.35 & 108.74 & 67.96 & -116.3 \\
-95.82 & -67.96 & -139.2 & 67.98 & -139.2 & -67.98 & -95.82 & 67.96 & 498.99 & 0 \\
-52.86 & -116.3 & 83.07 & -128.4 & -83.07 & -128.4 & 52.86 & -116.3 & 0 & 489.36
\end{bmatrix} \frac{\text{lb}}{\text{in.}} \tag{9.2.15}
$$

The applied nodal forces are given by Eq. (9.1.36) as

$$
F_{1r} = F_{4r} = \frac{2\pi(0.5)(0.5)}{2}(1) = 0.785 \text{ lb} \tag{9.2.16}
$$

All other nodal forces are zero. Using Eq. (9.2.15) for $[K]$ and Eq. (9.2.16) for the nodal forces in Eq. (9.2.1), and solving for the nodal displacements, we obtain

$$
\begin{aligned}
u_1 &= 0.0322 \times 10^{-6} \text{ in.} & w_1 &= 0.00115 \times 10^{-6} \text{ in.} \\
u_2 &= 0.0219 \times 10^{-6} \text{ in.} & w_2 &= 0.00206 \times 10^{-6} \text{ in.} \\
u_3 &= 0.0219 \times 10^{-6} \text{ in.} & w_3 &= -0.00206 \times 10^{-6} \text{ in.} \\
u_4 &= 0.0322 \times 10^{-6} \text{ in.} & w_4 &= -0.00115 \times 10^{-6} \text{ in.} \\
u_5 &= 0.0244 \times 10^{-6} \text{ in.} & w_5 &= 0
\end{aligned} \tag{9.2.17}
$$

The results for nodal displacements are as expected because radial displacements at the inner edge are equal ($u_1 = u_4$) and those at the outer edge are equal ($u_2 = u_3$). In addition, the axial displacements at the outer nodes and inner nodes are equal but opposite in sign ($w_1 = -w_4$ and $w_2 = -w_3$) as a result of the Poisson effect and symmetry. Finally, the axial displacement at the center node is zero ($w_5 = 0$), as it should be because of symmetry.

By using Eq. (9.1.22), we now determine the stresses in each element as

$$\{\sigma\} = [D][\bar{B}]\{d\} \tag{9.2.18}$$

For element 1, we use Eq. (9.2.5) for $[\bar{B}]$, Eq. (9.2.8) for $[D]$, and Eq. (9.2.17) for $\{d\}$ in Eq. (9.2.18) to obtain

$$\sigma_r = -0.338 \text{ psi} \qquad \sigma_z = -0.0126 \text{ psi}$$
$$\sigma_\theta = 0.942 \text{ psi} \qquad \tau_{rz} = -0.1037 \text{ psi}$$

Similarly, for element 2, we obtain

$$\sigma_r = -0.105 \text{ psi} \qquad \sigma_z = -0.0747 \text{ psi}$$
$$\sigma_\theta = 0.690 \text{ psi} \qquad \tau_{rz} = 0.000 \text{ psi}$$

For element 3, the stresses are

$$\sigma_r = -0.337 \text{ psi} \qquad \sigma_z = -0.0125 \text{ psi}$$
$$\sigma_\theta = 0.942 \text{ psi} \qquad \tau_{rz} = 0.1037 \text{ psi}$$

For element 4, the stresses are

$$\sigma_r = -0.470 \text{ psi} \qquad \sigma_z = 0.1493 \text{ psi}$$
$$\sigma_\theta = 1.426 \text{ psi} \qquad \tau_{rz} = 0.000 \text{ psi}$$

Figure 9–13 shows the exact solution [10] along with the results determined here and the results from Reference [5]. Observe that agreement with the exact solution is quite good except for the limited results due to the very coarse mesh used in the longhand example, and in case 1 of Reference [5]. In Reference [5], stresses have been plotted at the center of the quadrilaterals and were obtained by averaging the stresses in the four connecting triangles. ■

9.3 Applications of Axisymmetric Elements

Numerous structural (and nonstructural) systems can be classified as axisymmetric. Some typical structural systems whose behavior is modeled accurately using the axisymmetric element developed in this chapter are represented in Figures 9–14, 9–15, and 9–17.

Figure 9–14 illustrates the finite element model of a steel-reinforced concrete pressure vessel. The vessel is a thick-walled cylinder with flat heads. An axis of symmetry (the z axis) exists such that only one-half of the r-z plane passing through the middle of the structure need be modeled. The concrete was modeled by using the axisymmetric triangular element developed in this chapter. The steel elements were laid out along the boundaries of the concrete elements so as to maintain continuity

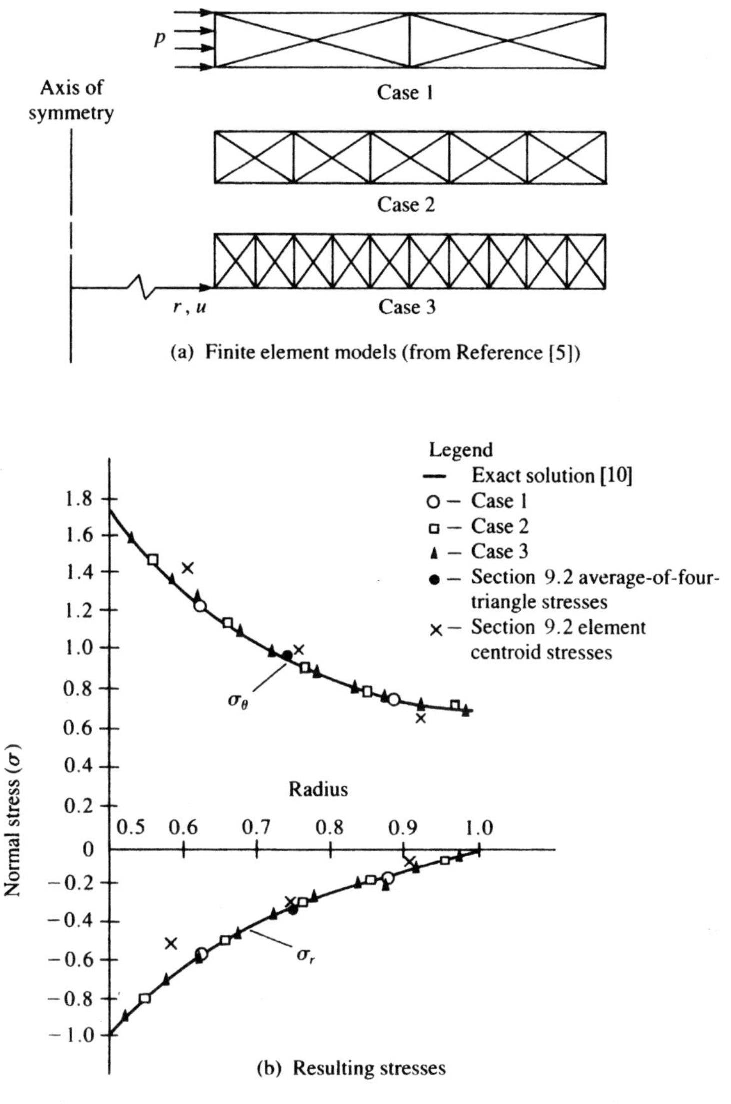

Figure 9–13 Finite element analysis of a thick-walled cylinder under internal pressure

(or perfect bond assumption) between the concrete and the steel. The vessel was then subjected to an internal pressure as shown in the figure. Note that the nodes along the axis of symmetry should be supported by rollers preventing motion perpendicular to the axis of symmetry.

Figure 9–15 shows a finite element model of a high-strength steel die used in a thin-plastic-film-making process [7]. The die is an irregularly shaped disk. An axis of symmetry with respect to geometry and loading exists as shown. The die was modeled by using simple quadrilateral axisymmetric elements. The locations of high stress were

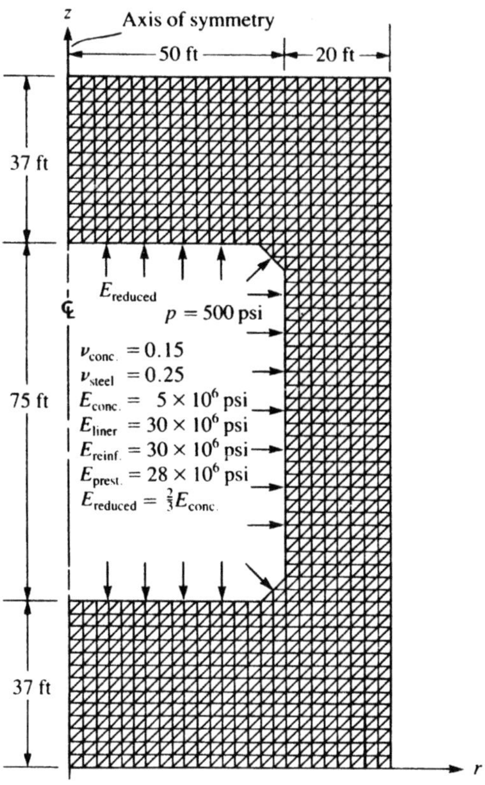

(a) Two-dimensional view of a finite element idealization for a prestressed concrete reactor vessel (PCRV)

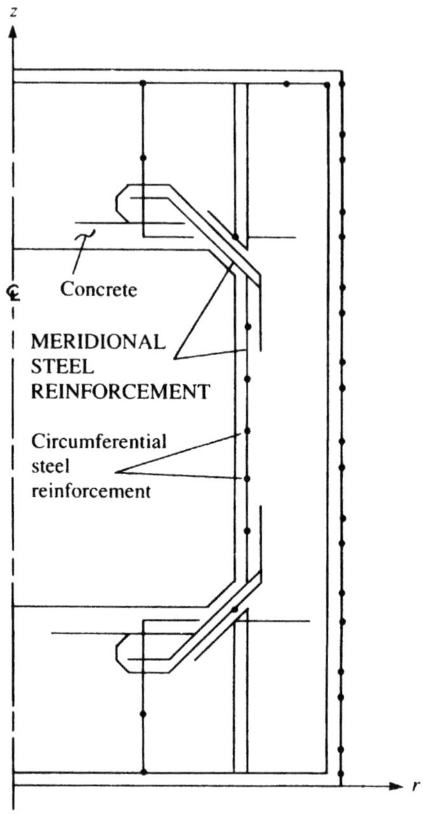

(b) Axisymmetric idealization of the steel reinforcement

Figure 9–14 Model of steel-reinforced concrete pressure vessel (from Reference [4], North Holland Physics Publishing, Amsterdam)

of primary concern. Figure 9–16 shows a plot of the von Mises stress contours for the die of Figure 9–15. The von Mises (or equivalent, or effective) stress [8] is often used as a failure criterion in design. Notice the artificially high stresses at the location of load F as explained in Section 7.1.

(Recall that the failure criterion based on the maximum distortion energy theory for ductile materials subjected to static loading predicts that a material will fail if the von Mises stress reaches the yield strength of the material.) Also recall from Eqs. (6.5.37) and (6.5.38), the von Mises stress σ_{vm} is related to the principal stresses by the expression

$$\sigma_{vm} = \frac{1}{\sqrt{2}}\sqrt{(\sigma_1 - \sigma_2)^2 + (\sigma_2 - \sigma_3)^2 + (\sigma_3 - \sigma_1)^2} \tag{9.3.1}$$

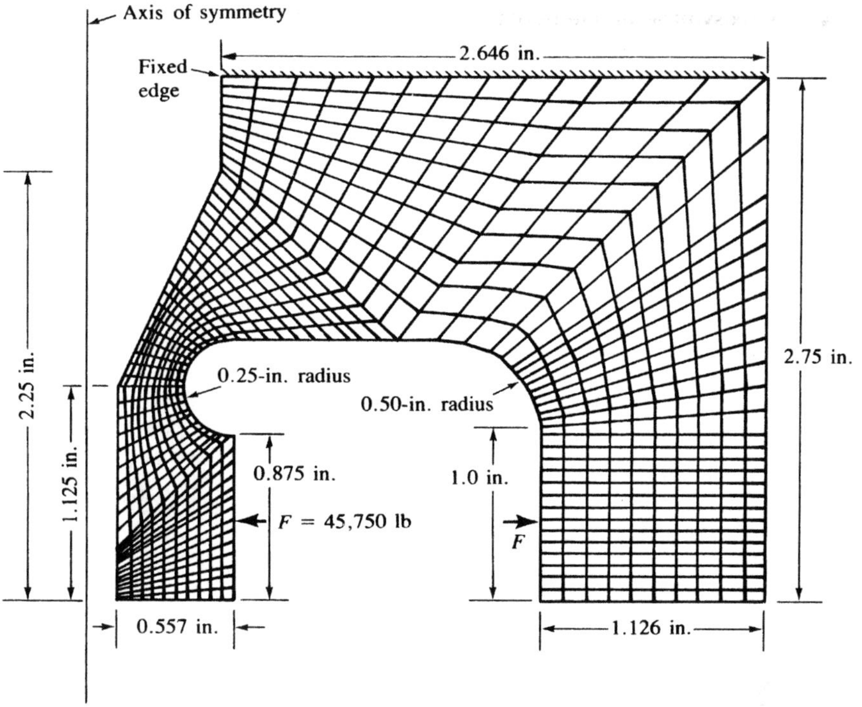

Figure 9–15 Model of a high-strength steel die (924 nodes and 830 elements)

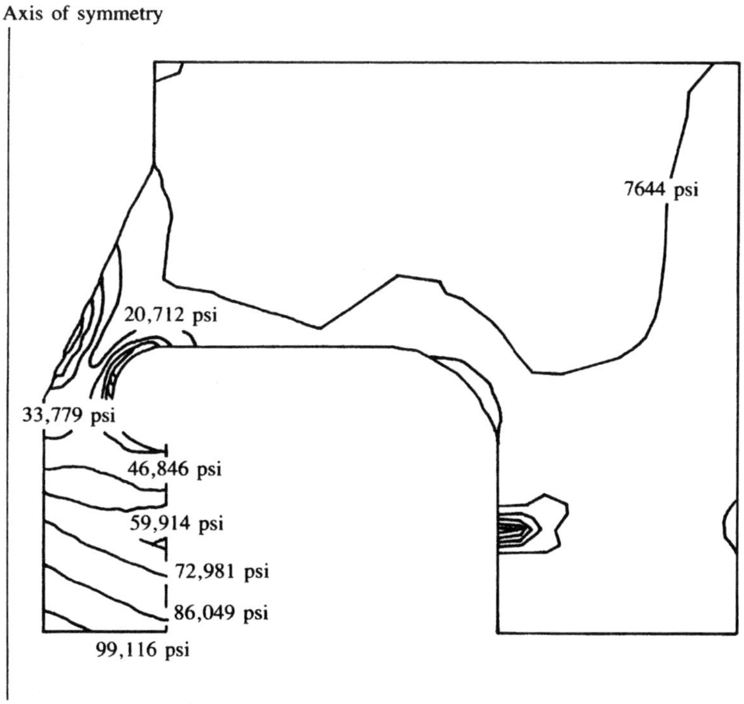

Figure 9–16 von Mises stress contour plot of axisymmetric model of Figure 9–15 (also producing a radial inward deflection of about 0.015 in.)

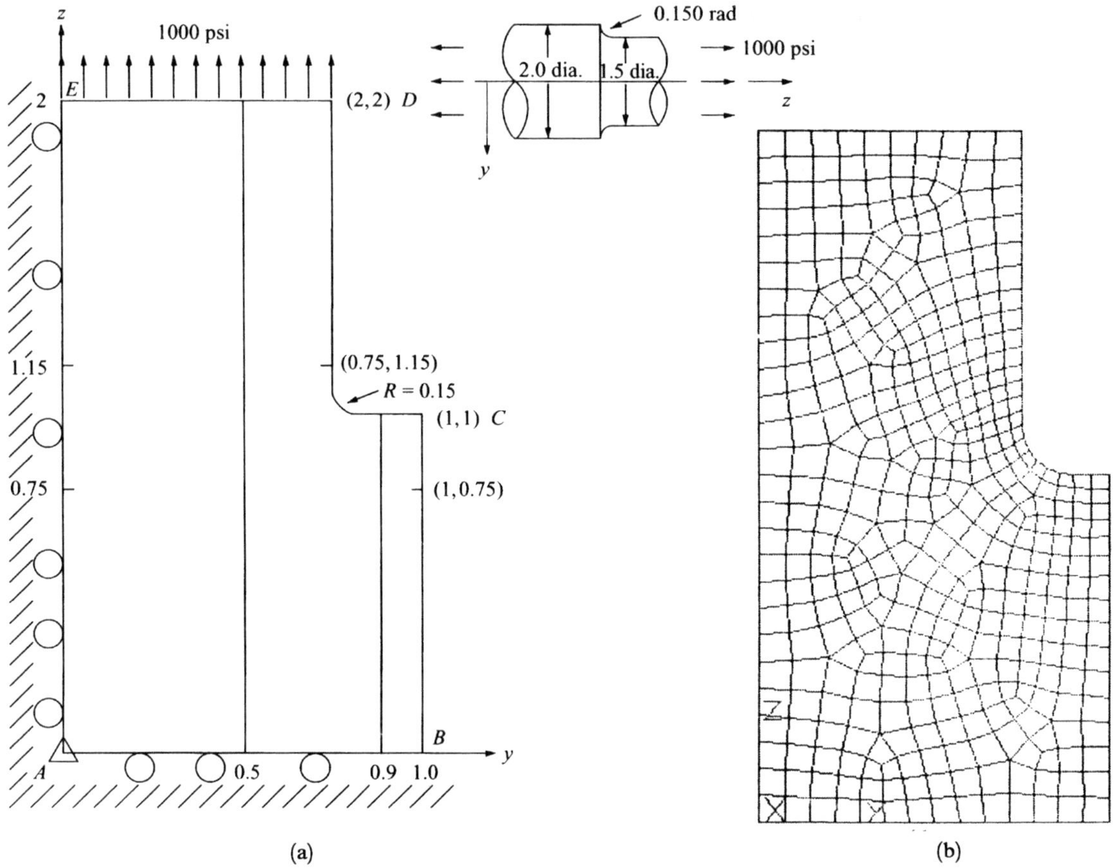

Figure 9–17 (a) Stepped shaft subjected to axial load and (b) the discretized model

where the principal stresses are given by σ_1, σ_2, and σ_3. These results were obtained from the commercial computer code ANSYS [12].

Other dies with modifications in geometry were also studied to evaluate the most suitable die before the construction of an expensive prototype. Confidence in the acceptability of the prototype was enhanced by doing these comparison studies. Finally, Figure 9–17 shows a stepped 4130 steel shaft with a fillet radius subjected to an axial pressure of 1000 psi in tension. Fatigue analysis for reversed axial loading required an accurate stress concentration factor to be applied to the average axial stress of 1000 psi. The stress concentration factor for the geometry shown was to be determined. Therefore, locations of highest stress were necessary. Figure 9–18 shows the resulting maximum principal stress plot using a computer program [11]. The largest principal stress was 1932.5 psi at the fillet. Other examples of the use of the axisymmetric element can be found in References [2]–[6].

In this chapter, we have shown the finite element analysis of axisymmetric systems using a simple three-noded triangular element to be analogous to that of the two-dimensional plane stress problem using three-noded triangular elements as

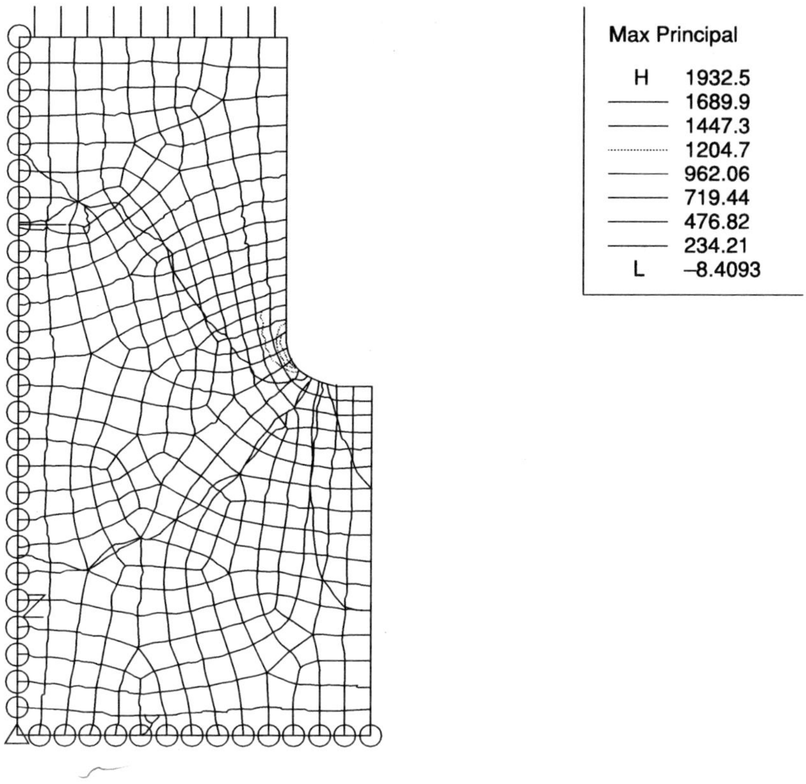

Figure 9–18 Principal stress plot for shaft of Figure 9–17

developed in Chapter 6. Therefore, the two-dimensional element in commercial computer programs with the axisymmetric element selected will allow for the analysis of axisymmetric structures.

Finally, note that other axisymmetric elements, such as a simple quadrilateral (one with four corner nodes and two degrees of freedom per node, as used in the steel die analysis of Figure 9–15) or higher-order triangular elements, such as in Reference [6], in which a cubic polynomial involving ten terms (ten a's) for both u and w, could be used for axisymmetric analysis. The three-noded triangular element was described here because of its simplicity and ability to describe geometric boundaries rather easily.

References

[1] Utku, S., "Explicit Expressions for Triangular Torus Element Stiffness Matrix," *Journal of the American Institute of Aeronautics and Astronautics*, Vol. 6, No. 6, pp. 1174–1176, June 1968.

[2] Zienkiewicz, O. C., *The Finite Element Method*, 3rd ed., McGraw-Hill, London, 1977.

[3] Clough, R., and Rashid, Y., "Finite Element Analysis of Axisymmetric Solids," *Journal of the Engineering Mechanics Division*, American Society of Civil Engineers, Vol. 91, pp. 71–85, Feb. 1965.

[4] Rashid, Y., "Analysis of Axisymmetric Composite Structures by the Finite Element Method," *Nuclear Engineering and Design*, Vol. 3, pp. 163–182, 1966.

[5] Wilson, E., "Structural Analysis of Axisymmetric Solids," *Journal of the American Institute of Aeronautics and Astronautics*, Vol. 3, No. 12, pp. 2269–2274, Dec. 1965.

[6] Chacour, S., "A High Precision Axisymmetric Triangular Element Used in the Analysis of Hydraulic Turbine Components," Transactions of the American Society of Mechanical Engineers, *Journal of Basic Engineering*, Vol. 92, pp. 819–826, 1973.

[7] Greer, R. D., *The Analysis of a Film Tower Die Utilizing the ANSYS Finite Element Package*, M.S. Thesis, Rose-Hulman Institute of Technology, Terre Haute, IN, May 1989.

[8] Gere, J. M., *Mechanics of Materials*, 5th ed., Brooks/Cole Publishers, Pacific Grove, CA, 2001.

[9] Cook, R. D., Malkus, D. S., Plesha, M. E., and Witt, R. J., *Concepts and Applications of Finite Element Analysis*, 4th ed., Wiley, New York, 2002.

[10] Cook, R. D., and Young, W. C., *Advanced Mechanics of Materials*, Macmillan, New York, 1985.

[11] Algor Interactive Systems, 150 Beta Drive, Pittsburgh, PA 15238.

[12] Swanson, J. A. ANSYS-Engineering Analysis System's User's Manual, Swanson Analysis Systems, Inc., Johnson Rd., P.O. Box 65, Houston, PA 15342.

Problems

9.1 For the elements shown in Figure P9–1, evaluate the stiffness matrices using Eq. (9.2.2). The coordinates are shown in the figures. Let $E = 30 \times 10^6$ psi and $\nu = 0.25$ for each element.

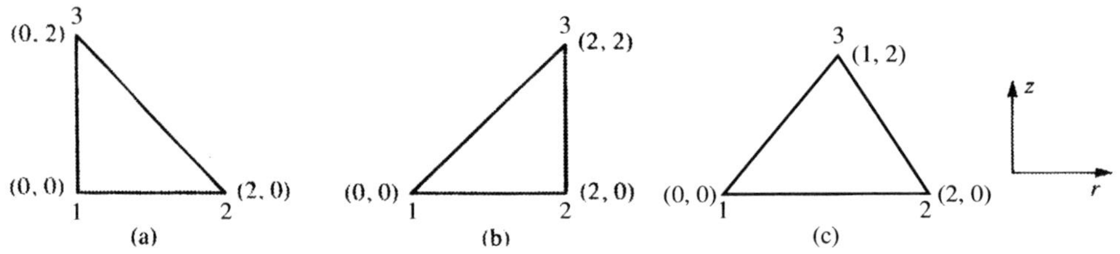

Figure P9–1

9.2 Evaluate the nodal forces used to replace the linearly varying surface traction shown in Figure P9–2. Hint: Use Eq. (9.1.34).

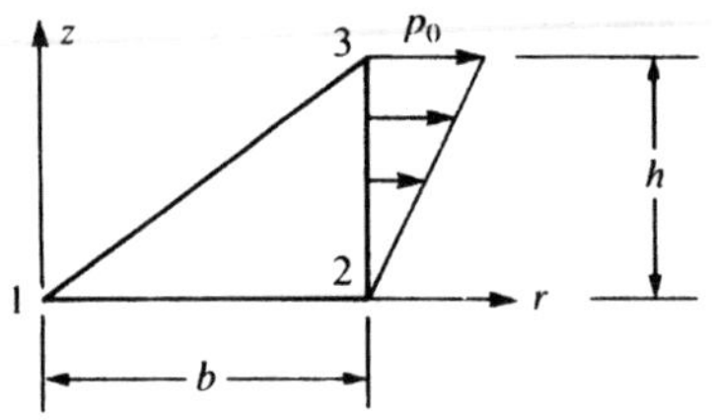

Figure P9–2

9.3 For an element of an axisymmetric body rotating with a constant angular velocity $\omega = 20$ rpm as shown in Figure P9–3, evaluate the body-force matrix. The coordinates of the element are shown in the figure. Let the weight density ρ_w be 0.283 lb/in^3.

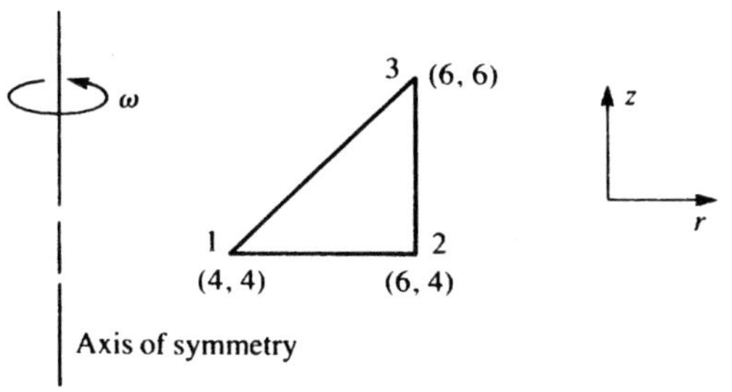

Figure P9–3

9.4 For the axisymmetric elements shown in Figure P9–4, determine the element stresses. Let $E = 30 \times 10^6$ psi and $\nu = 0.25$. The coordinates (in inches) are shown in the figures, and the nodal displacements for each element are $u_1 = 0.0001$ in., $w_1 = 0.0002$ in., $u_2 = 0.0005$ in., $w_2 = 0.0006$ in., $u_3 = 0$, and $w_3 = 0$.

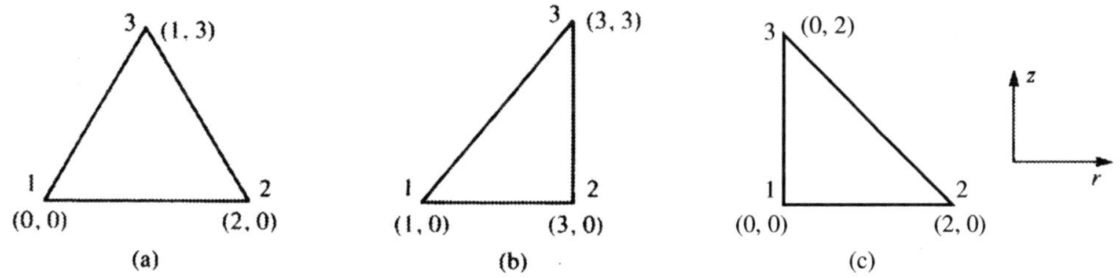

Figure P9–4

9.5 Explicitly show that the integration of Eq. (9.1.35) yields the j surface forces given by Eq. (9.1.36).

9.6 For the elements shown in Figure P9–6, evaluate the stiffness matrices using Eq. (9.2.2). The coordinates (in millimeters) are shown in the figures. Let $E = 210$ GPa and $\nu = 0.25$ for each element.

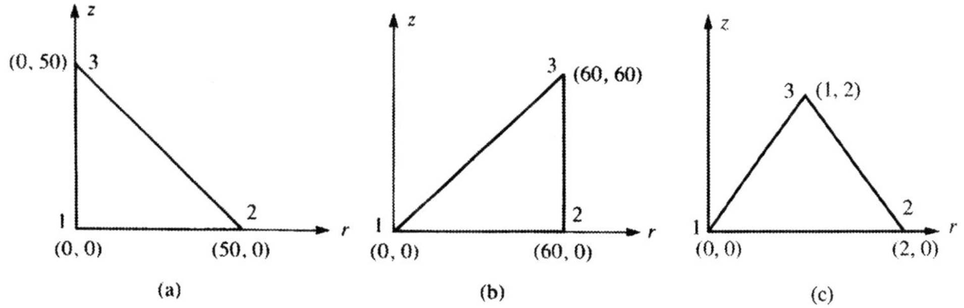

Figure P9–6

9.7 For the axisymmetric elements shown in Figure P9–7, determine the element stresses. Let $E = 210$ GPa and $\nu = 0.25$. The coordinates (in millimeters) are shown in the figures, and the nodal displacements for each element are

$$u_1 = 0.05 \text{ mm} \qquad w_1 = 0.03 \text{ mm}$$
$$u_2 = 0.02 \text{ mm} \qquad w_2 = 0.02 \text{ mm}$$
$$u_3 = 0.0 \text{ mm} \qquad w_3 = 0.0 \text{ mm}$$

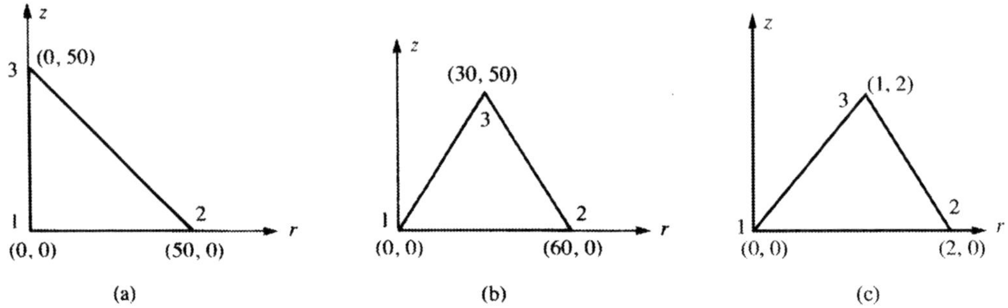

Figure P9–7

9.8 Can we connect plane stress elements with axisymmetric ones? Explain.

9.9 Is the three-noded triangular element considered in Section 9.1 a constant strain element? Why or why not?

9.10 How should one model the boundary conditions of nodes acting on the axis of symmetry?

9.11 How would you evaluate the circumferential strain, ε_θ, at $r = 0$? What is this strain in terms of the a's given in Eq. (9.1.3). Hint: Elasticity theory tells us that the radial strain must equal the circumferential strain at $r = 0$.

9.12 What will be the stresses σ_r and σ_θ at $r = 0$? Hint: Look at Eq (9.1.2) after considering problem 9.11.

Solve the following axisymmetric problems using a computer program.

9.13 The soil mass in Figure P9–13 is loaded by a force transmitted through a circular footing as shown. Determine the stresses in the soil. Compare the values of σ_r using an

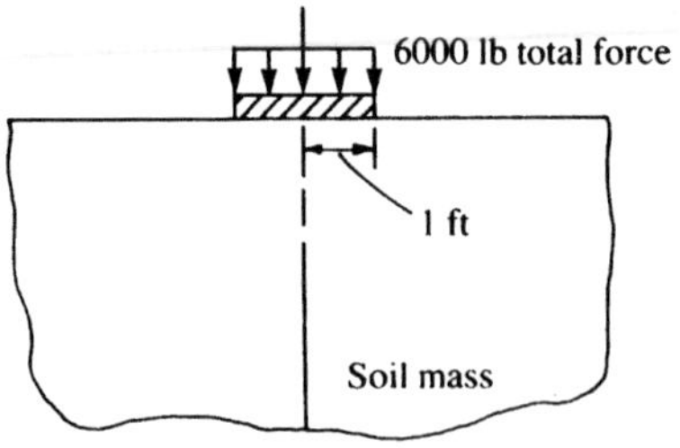

Figure P9–13

axisymmetric model with the σ_y values using a plane stress model. Let $E = 3000$ psi and $\nu = 0.45$ for the soil mass.

9.14 Perform a stress analysis of the pressure vessel shown in Figure P9–14. Let $E = 5 \times 10^6$ psi and $\nu = 0.15$ for the concrete, and let $E = 29 \times 10^6$ psi and $\nu = 0.25$ for the steel liner. The steel liner is 2 in. thick. Let the pressure p equal 500 psi.

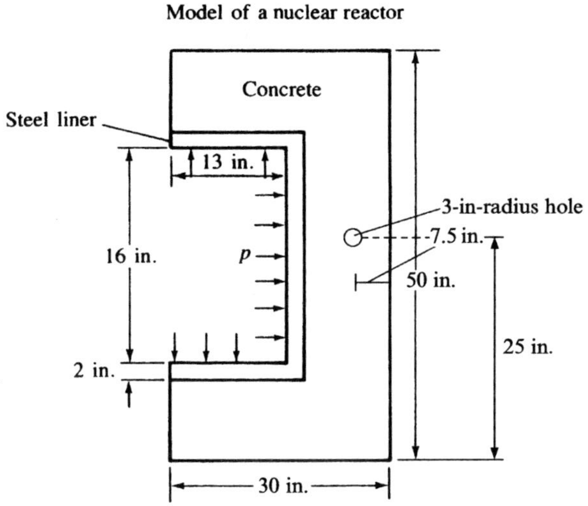

Figure P9–14

9.15 Perform a stress analysis of the concrete pressure vessel with the steel liner shown in Figure P9–15. Let $E = 30$ GPa and $\nu = 0.15$ for the concrete, and let $E = 205$ GPa and $\nu = 0.25$ for the steel liner. The steel liner is 50 mm thick. Let the pressure p equal 700 kPa.

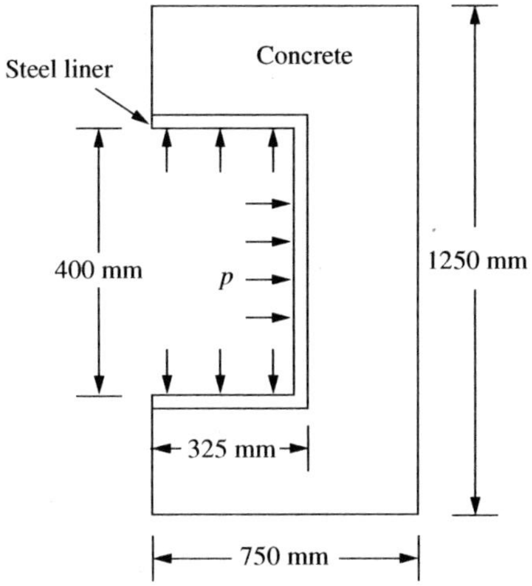

Figure P9–15

9.16 Perform a stress analysis of the disk shown in Figure P9–16 if it rotates with constant angular velocity of $\omega = 50$ rpm. Let $E = 30 \times 10^6$ psi, $\nu = 0.25$, and the weight density $\rho_w = 0.283$ lb/in^3 (mass density, $\rho = \rho_w/(g = 386$ in./s^2). (Use 8 and then 16 elements symmetrically modeled similar to Example 9.4. Compare the finite element solution to the theoretical circumferential and radial stresses given by

$$\sigma_\theta = \frac{3+\nu}{8}\rho\omega^2 a^2\left(1 - \frac{1+3\nu}{3+\nu}\frac{r^2}{a^2}\right), \qquad \sigma_r = \frac{3+\nu}{8}\rho\omega^2 a^2\left(1 - \frac{r^2}{a^2}\right)$$

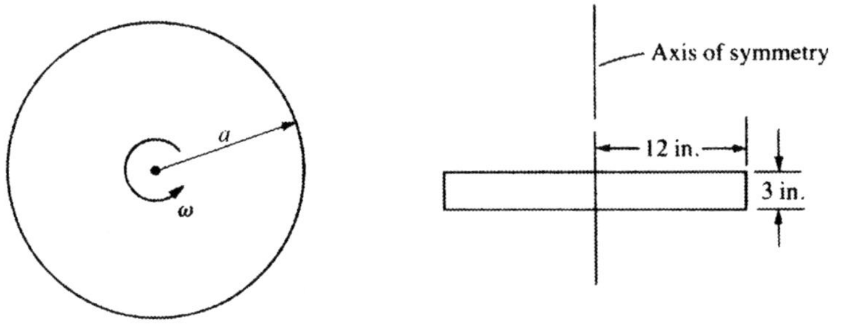

Figure P9–16

9.17 For the die casting shown in Figure P9–17, determine the maximum stresses and their locations. Let $E = 30 \times 10^6$ psi and $\nu = 0.25$. The dimensions are shown in the figure.

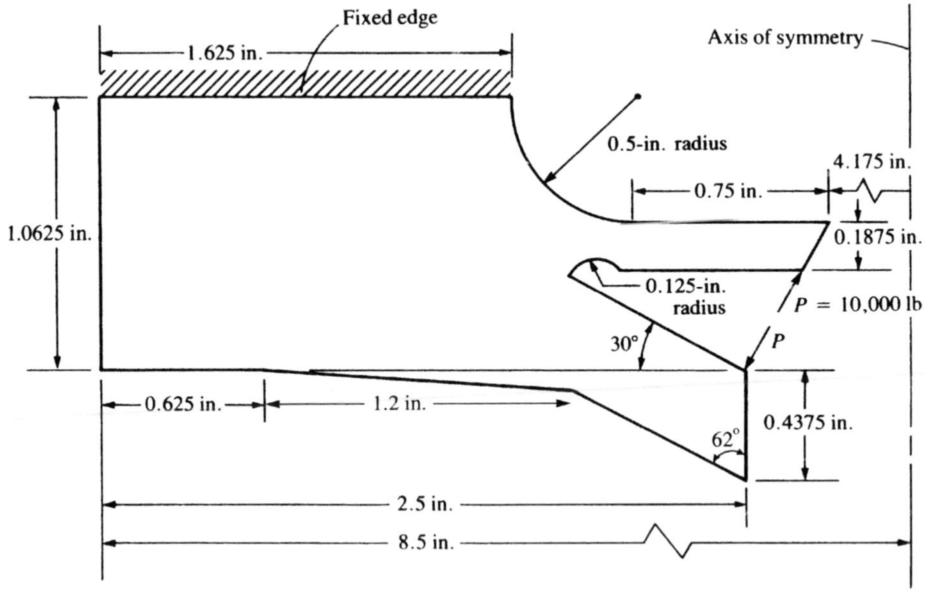

Figure P9–17

9.18 For the axisymmetric connecting rod shown in Figure P9–18, determine the stresses $\sigma_z, \sigma_r, \sigma_\theta$, and τ_{rz}. Plot stress contours (lines of constant stress) for each of the normal stresses. Let $E = 30 \times 10^6$ psi and $\nu = 0.25$. The applied loading and boundary conditions are shown in the figure. A typical discretized rod is shown in the figure for illustrative purposes only.

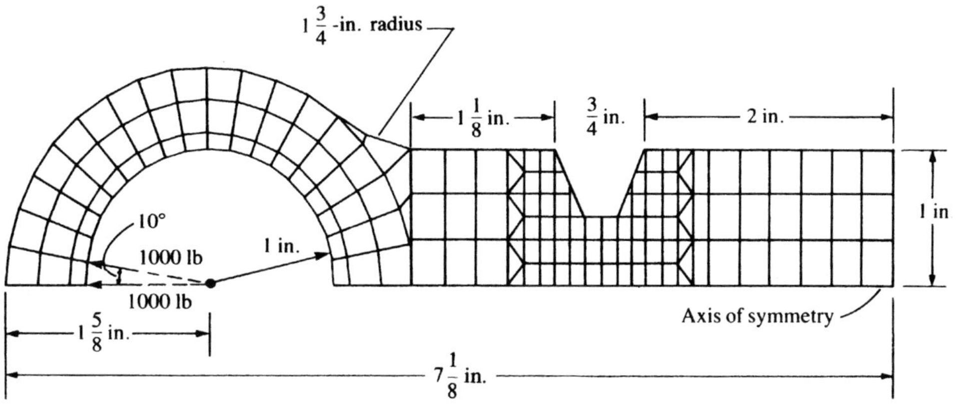

Figure P9–18

9.19 For the thick-walled open-ended cylindrical pipe subjected to internal pressure shown in Figure P9–19, use five layers of elements to obtain the circumferential stress, σ_θ,

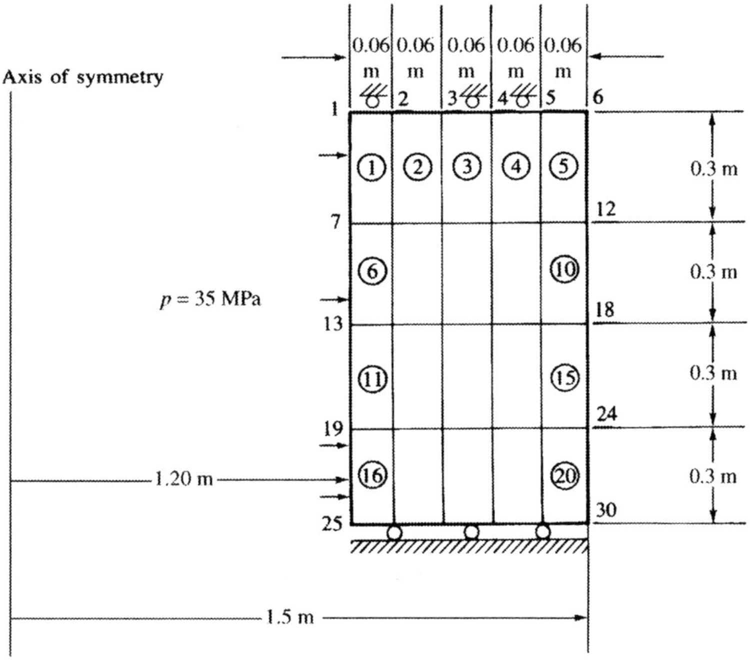

Figure P9–19

and the principal stresses and maximum radial displacement. Compare these results to the exact solution. Let $E = 205$ GPa and $\nu = 0.3$.

9.20 A steel cylindrical pressure vessel with flat plate end caps is shown in Figure P9–20 with vertical axis of symmetry. Addition of thickened sections helps to reduce stress concentrations in the corners. Analyze the design and identify the most critically stressed regions. Note that inside sharp re-entrant corners produce infinite stress concentration zones, so refining the mesh in these regions will not help you get a better answer unless you use an inelastic theory or place small fillet radii there. Recommend any design changes in your report. Let the pressure inside be 1000 kPa.

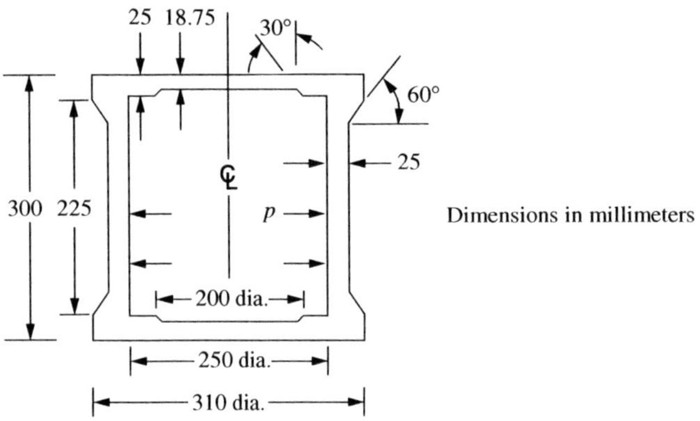

Figure P9–20

9.21 For the cylindrical vessel with hemispherical ends (heads) under uniform internal pressure of intensity $p = 500$ psi shown in Figure P9–21, determine the maximum von Mises stress and where it is located. The material is ASTM—A242 quenched and tempered alloy steel. Use a factor of safety of 3 against yielding. The inner radius is $a = 100$ inches and the thickness $t = 2$ in.

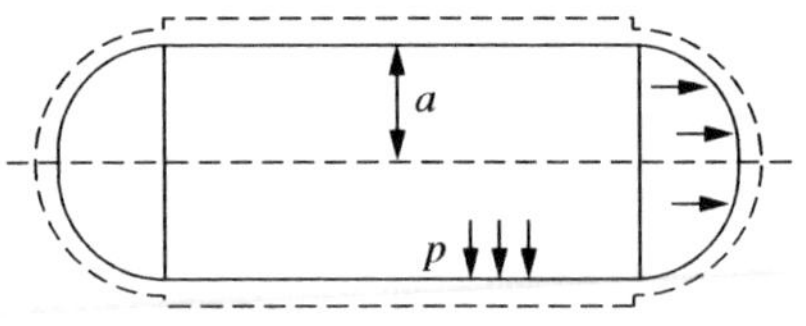

Figure P9–21

9.22 For the cylindrical vessel with *ellipsoidal heads* shown in Figure P9–22a under loading $p = 500$ psi, determine if the vessel is safe against yielding. Use the same material and factor of safety as in previous problem, 9.21. Now let $a = 100$ in. and $b = 50$ in. Which vessel has the lowest hoop stress? Recommend the preferred head shape of the two based on your answers.

For modeling purposes, the equation of an ellipse is given by $b^2x^2 + a^2y^2 = a^2b^2$, where a is the major axis and b is the minor axis of the ellipse shown in Figure P9–22(b).

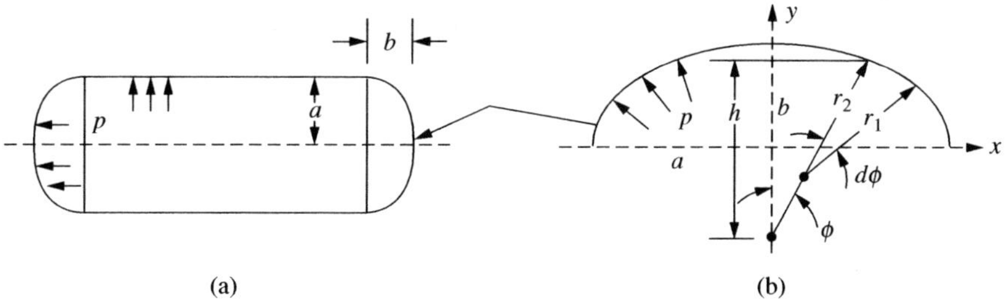

Figure P9–22

9.23 The syringe with plunger is shown in Figure P9–23. The material of the syringe is glass with $E = 69$ GPa, $\nu = 0.15$, and tensile strength of 5 MPa. The bottom hole is assumed to be closed under test conditions. Determine the deformation and stresses in the glass. Compare the maximum principal stress in the glass to the ultimate tensile strength. Do you think the syringe is safe? Why?

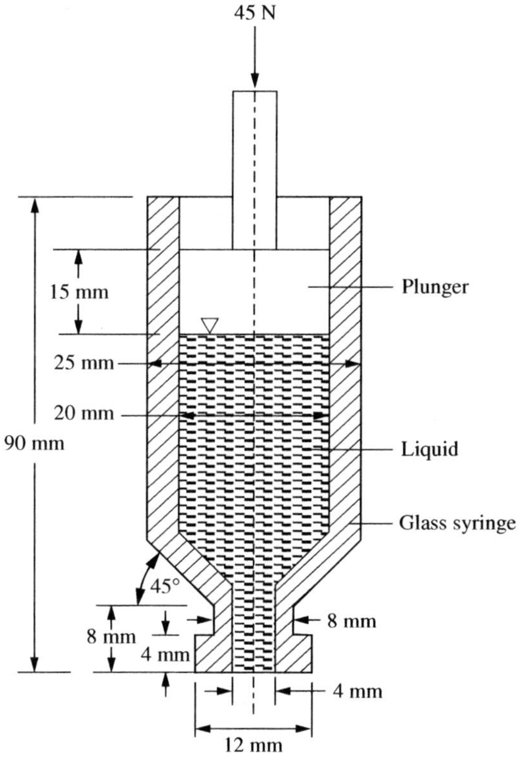

Figure P9–23

9.24 For the tapered solid circular shaft shown, a semicircular groove has been machined into the side. The shaft is made of a hot rolled 1040 steel alloy with yield strength of 71,000 psi. The shaft is subjected to a uniform axial pressure of 4000 psi. Determine the maximum principal stresses and von Mises stresses at the fillet and at the semicircular groove. Is the shaft safe from failure based on the maximum distortion energy theory?

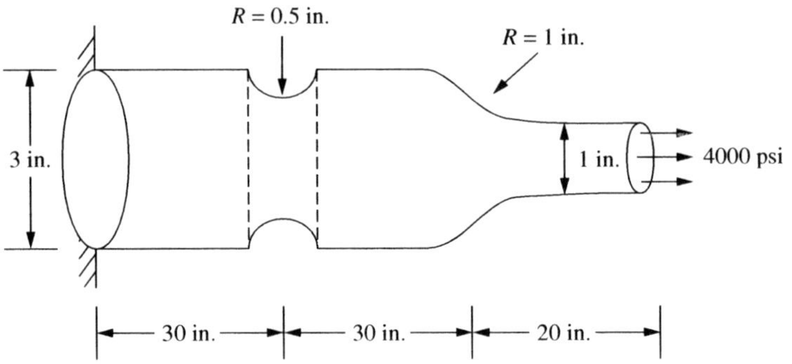

Figure P9–24

CHAPTER 10

Isoparametric Formulation

Introduction

In this chapter, we introduce the isoparametric formulation of the element stiffness matrices. After considering the linear-strain triangular element in Chapter 8, we can see that the development of element matrices and equations expressed in terms of a global coordinate system becomes an enormously difficult task (if even possible) except for the simplest of elements such as the constant-strain triangle of Chapter 6. Hence, the isoparametric formulation was developed [1]. The isoparametric method may appear somewhat tedious (and confusing initially), but it will lead to a simple computer program formulation, and it is generally applicable for two- and three-dimensional stress analysis and for nonstructural problems. The isoparametric formulation allows elements to be created that are nonrectangular and have curved sides. Furthermore, numerous commercial computer programs (as described in Chapter 1) have adapted this formulation for their various libraries of elements.

We first illustrate the isoparametric formulation to develop the simple bar element stiffness matrix. Use of the bar element makes it relatively easy to understand the method because simple expressions result.

We then consider the development of the rectangular plane stress element stiffness matrix in terms of a global-coordinate system that will be convenient for use with the element. These concepts will be useful in understanding some of the procedures used with the isoparametric formulation of the simple quadrilateral element stiffness matrix, which we will develop subsequently.

Next, we will introduce numerical integration methods for evaluating the quadrilateral element stiffness matrix and illustrate the adaptability of the isoparametric formulation to common numerical integration methods.

Finally, we will consider some higher-order elements and their associated shape functions.

10.1 Isoparametric Formulation of the Bar Element Stiffness Matrix

The term *isoparametric* is derived from the use of the same shape functions (or interpolation functions) $[N]$ to define the element's geometric shape as are used to define the displacements within the element. Thus, when the shape function is $u = a_1 + a_2 s$ for the displacement, we use $x = a_1 + a_2 s$ for the description of the nodal coordinate of a point on the bar element and, hence, the physical shape of the element.

Isoparametric element equations are formulated using a **natural** (or **intrinsic**) **coordinate system** s that is defined by element geometry and not by the element orientation in the global-coordinate system. In other words, axial coordinate s is attached to the bar and remains directed along the axial length of the bar, regardless of how the bar is oriented in space. There is a relationship (called a *transformation mapping*) between the natural coordinate system s and the global coordinate system x for each element of a specific structure, and this relationship must be used in the element equation formulations.

We will now develop the isoparametric formulation of the stiffness matrix of a simple linear bar element [with two nodes as shown in Figure 10–1(a)].

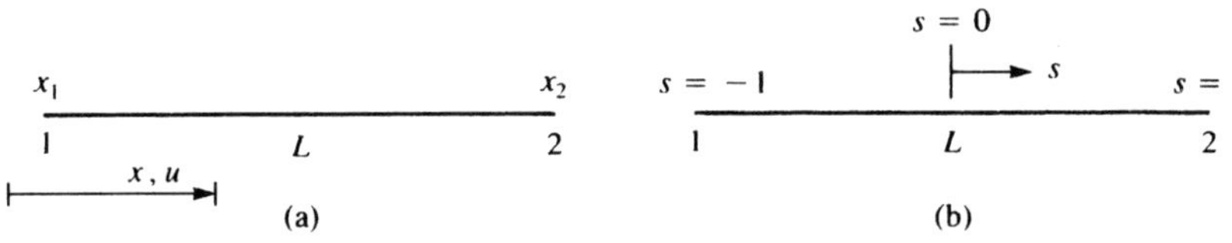

Figure 10–1 Linear bar element in (a) a global coordinate system x and (b) a natural coordinate system s

Step 1 Select Element Type

First, the natural coordinate s is attached to the element, with the origin located at the center of the element, as shown in Figure 10–1(b). The s axis need not be parallel to the x axis—this is only for convenience.

We consider the bar element to have two degrees of freedom—axial displacements u_1 and u_2 at each node associated with the global x axis.

For the special case when the s and x axes are parallel to each other, the s and x coordinates can be related by

$$x = x_c + \frac{L}{2}s \qquad (10.1.1a)$$

where x_c is the global coordinate of the element centroid.

Using the global coordinates x_1 and x_2 in Eq. (10.1.1a) with $x_c = (x_1 + x_2)/2$, we can express the natural coordinate s in terms of the global coordinates as

$$s = [x - (x_1 + x_2)/2](2/(x_2 - x_1) \qquad (10.1.1b)$$

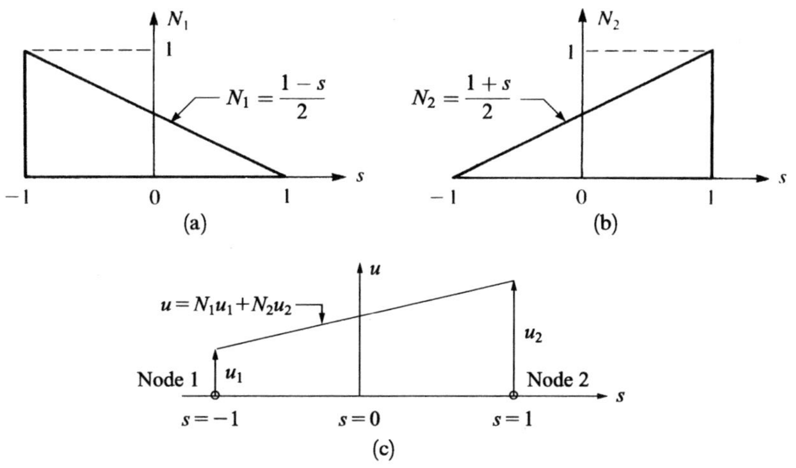

Figure 10–2 Shape function variations with natural coordinates: (a) shape function N_1, (b) shape function N_2, and (c) linear displacement field u plotted over element length

The shape functions used to define a position within the bar are found in a manner similar to that used in Chapter 3 to define displacement within a bar (Section 3.1). We begin by relating the natural coordinate to the global coordinate by

$$x = a_1 + a_2 s \tag{10.1.2}$$

where we note that s is such that $-1 \leqslant s \leqslant 1$. Solving for the a_i's in terms of x_1 and x_2, we obtain

$$x = \tfrac{1}{2}[(1-s)x_1 + (1+s)x_2] \tag{10.1.3}$$

or, in matrix form, we can express Eq. (10.1.3) as

$$\{x\} = [N_1 \quad N_2]\begin{Bmatrix} x_1 \\ x_2 \end{Bmatrix} \tag{10.1.4}$$

where the shape functions in Eq. (10.1.4) are

$$N_1 = \frac{1-s}{2} \qquad N_2 = \frac{1+s}{2} \tag{10.1.5}$$

The linear shape functions in Eqs. (10.1.5) map the s coordinate of any point in the element to the x coordinate when used in Eq. (10.1.3). For instance, when we substitute $s = -1$ into Eq. (10.1.3), we obtain $x = x_1$. These shape functions are shown in Figure 10–2, where we can see that they have the same properties as defined for the interpolation functions of Section 3.1. Hence, N_1 represents the physical shape of the coordinate x when plotted over the length of the element for $x_1 = 1$ and $x_2 = 0$, and N_2 represents the coordinate x when plotted over the length of the element for $x_2 = 1$ and $x_1 = 0$. Again, we must have $N_1 + N_2 = 1$.

These shape functions must also be continuous throughout the element domain and have finite first derivatives within the element.

Step 2 Select a Displacement Function

The displacement function within the bar is now defined by the same shape functions, Eqs. (10.1.5), as are used to define the element shape; that is,

$$\{u\} = [N_1 \quad N_2]\begin{Bmatrix} u_1 \\ u_2 \end{Bmatrix} \tag{10.1.6}$$

When a particular coordinate s of the point of interest is substituted into $[N]$, Eq. (10.1.6) yields the displacement of a point on the bar in terms of the nodal degrees of freedom u_1 and u_2 as shown in Figure 10–2(c). Since u and x are defined by the same shape functions at the same nodes, comparing Eqs. (10.1.4) and (10.1.6), the element is called *isoparametric*.

Step 3 Define the Strain/Displacement and Stress/Strain Relationships

We now want to formulate element matrix $[B]$ to evaluate $[k]$. We use the isoparametric formulation to illustrate its manipulations. For a simple bar element, no real advantage may appear evident. However, for higher-order elements, the advantage will become clear because relatively simple computer program formulations will result.

To construct the element stiffness matrix, we must determine the strain, which is defined in terms of the derivative of the displacement with respect to x. The displacement u, however, is now a function of s as given by Eq. (10.1.6). Therefore, we must apply the chain rule of differentiation to the function u as follows:

$$\frac{du}{ds} = \frac{du}{dx}\frac{dx}{ds} \tag{10.1.7}$$

We can evaluate (du/ds) and (dx/ds) using Eqs. (10.1.6) and (10.1.3). We seek $(du/dx) = \varepsilon_x$. Therefore, we solve Eq. (10.1.7) for (du/dx) as

$$\frac{du}{dx} = \frac{\left(\dfrac{du}{ds}\right)}{\left(\dfrac{dx}{ds}\right)} \tag{10.1.8}$$

Using Eq. (10.1.6) for u, we obtain

$$\frac{du}{ds} = \frac{u_2 - u_1}{2} \tag{10.1.9a}$$

and using Eq. (10.1.3) for x, we have

$$\frac{dx}{ds} = \frac{x_2 - x_1}{2} = \frac{L}{2} \tag{10.1.9b}$$

because $x_2 - x_1 = L$.

Using Eqs. (10.1.9a) and (10.1.9b) in Eq. (10.1.8), we obtain

$$\{\varepsilon_x\} = \begin{bmatrix} -\frac{1}{L} & \frac{1}{L} \end{bmatrix} \begin{Bmatrix} u_1 \\ u_2 \end{Bmatrix} \tag{10.1.10}$$

Since $\{\varepsilon\} = [B]\{d\}$, the strain/displacement matrix $[B]$ is then given in Eq. (10.1.10) as

$$[B] = \begin{bmatrix} -\frac{1}{L} & \frac{1}{L} \end{bmatrix} \tag{10.1.11}$$

We recall that use of linear shape functions results in a constant $\underline{B}$ matrix, and hence, in a constant strain within the element. For higher-order elements, such as the quadratic bar with three nodes, $[B]$ becomes a function of natural coordinate s (see Eq. (10.6.16).

The stress matrix is again given by Hooke's law as

$$\underline{\sigma} = E\underline{\varepsilon} = E\underline{B}\underline{d}$$

Step 4 Derive the Element Stiffness Matrix and Equations

The stiffness matrix is

$$[k] = \int_0^L [B]^T [D][B] A \, dx \tag{10.1.12}$$

However, in general, we must transform the coordinate x to s because $[B]$ is, in general, a function of s. This general type of transformation is given by References [4] and [5]

$$\int_0^L f(x)\, dx = \int_{-1}^1 f(s) |\underline{J}| \, ds \tag{10.1.13}$$

where $\underline{J}$ is called the *Jacobian*. In the one-dimensional case, we have $|\underline{J}| = \underline{J}$. For the simple bar element, from Eq. (10.1.9b), we have

$$|\underline{J}| = \frac{dx}{ds} = \frac{L}{2} \tag{10.1.14}$$

Observe that in Eq. (10.1.14), the Jacobian relates an element length in the global-coordinate system to an element length in the natural-coordinate system. In general, $|\underline{J}|$ is a function of s and depends on the numerical values of the nodal coordinates. This can be seen by looking at Eq. (10.3.22) for the quadrilateral element. (Section 10.3 further discusses the Jacobian.) Using Eqs. (10.1.13) and (10.1.14) in Eq. (10.1.12), we obtain the stiffness matrix in natural coordinates as

$$[k] = \frac{L}{2} \int_{-1}^1 [B]^T E [B] A \, ds \tag{10.1.15}$$

where, for the one-dimensional case, we have used the modulus of elasticity $E = [D]$ in Eq. (10.1.15). Substituting Eq. (10.1.11) in Eq. (10.1.15) and performing the simple integration, we obtain

$$[k] = \frac{AE}{L} \begin{bmatrix} 1 & -1 \\ -1 & 1 \end{bmatrix} \tag{10.1.16}$$

which is the same as Eq. (3.1.14). For higher-order one-dimensional elements, the integration in closed form becomes difficult if not impossible (see Example 10.7). Even the simple rectangular element stiffness matrix is difficult to evaluate in closed form (Section 10.3). However, the use of numerical integration, as described in Section 10.4, illustrates the distinct advantage of the isoparametric formulation of the equations.

Body Forces

We will now determine the body-force matrix using the natural coordinate system s. Using Eq. (3.10.20b), the body-force matrix is

$$\{\hat{f}_b\} = \iiint_V [N]^T \{\hat{X}_b\}\, dV \tag{10.1.17}$$

Letting $dV = A\,dx$, we have

$$\{\hat{f}_b\} = A \int_0^L [N]^T \{\hat{X}_b\}\, dx \tag{10.1.18}$$

Substituting Eqs. (10.1.5) for N_1 and N_2 into $[N]$ and noting that by Eq. (10.1.9b), $dx = (L/2)\, ds$, we obtain

$$\{\hat{f}_b\} = A \int_{-1}^{1} \begin{Bmatrix} \dfrac{1-s}{2} \\ \dfrac{1+s}{2} \end{Bmatrix} \{\hat{X}_b\} \frac{L}{2}\, ds \tag{10.1.19}$$

On integrating Eq. (10.1.19), we obtain

$$\{\hat{f}_b\} = \frac{AL\hat{X}_b}{2} \begin{Bmatrix} 1 \\ 1 \end{Bmatrix} \tag{10.1.20}$$

The physical interpretation of the results for $\{\hat{f}_b\}$ is that since AL represents the volume of the element and $\hat{X}_b$ the body force per unit volume, then $AL\hat{X}_b$ is the total body force acting on the element. The factor $\frac{1}{2}$ indicates that this body force is equally distributed to the two nodes of the element.

Surface Forces

Surface forces can be found using Eq. (3.10.20a) as

$$\{\hat{f}_s\} = \iint_S [N_s]^T \{\hat{T}_x\}\, dS \tag{10.1.21}$$

Assuming the cross section is constant and the traction is uniform over the perimeter and along the length of the element, we obtain

$$\{\hat{f}_s\} = \int_0^L [N_s]^T \{\hat{T}_x\}\, dx \tag{10.1.22}$$

where we now assume $\hat{T}_x$ is in units of force per unit length. Using the shape functions N_1 and N_2 from Eq. (10.1.5) in Eq. (10.1.22), we obtain

$$\{\hat{f}_s\} = \int_{-1}^{1} \begin{Bmatrix} \dfrac{1-s}{2} \\ \dfrac{1+s}{2} \end{Bmatrix} \{\hat{T}_x\} \frac{L}{2} ds \tag{10.1.23}$$

On integrating Eq. (10.1.23), we obtain

$$\{\hat{f}_s\} = \hat{T}_x \frac{L}{2} \begin{Bmatrix} 1 \\ 1 \end{Bmatrix} \tag{10.1.24}$$

The physical interpretation of Eq. (10.1.24) is that since $\hat{T}_x$ is in force-per-unit-length units, $\hat{T}_x L$ is now the total force. The $\frac{1}{2}$ indicates that the uniform surface traction is equally distributed to the two nodes of the element. Note that if $\hat{T}_x$ were a function of x (or s), then the amounts of force allocated to each node would generally not be equal and would be found through integration as in Example 3.12.

10.2 Rectangular Plane Stress Element

We will now develop the rectangular plane stress element stiffness matrix. We will later refer to this element in the isoparametric formulation of a general quadrilateral element.

Two advantages of the rectangular element over the triangular element are ease of data input and simpler interpretation of output stresses. A disadvantage of the rectangular element is that the simple linear-displacement rectangle with its associated straight sides poorly approximates the real boundary condition edges.

The usual steps outlined in Chapter 1 will be followed to obtain the element stiffness matrix and related equations.

Step 1 Select Element Type

Consider the rectangular element shown in Figure 10–3 (all interior angles are 90°) with corner nodes 1–4 (again labeled counterclockwise) and base and height dimensions $2b$ and $2h$, respectively.

The unknown nodal displacements are now given by

$$\{d\} = \begin{Bmatrix} u_1 \\ v_1 \\ u_2 \\ v_2 \\ u_3 \\ v_3 \\ u_4 \\ v_4 \end{Bmatrix} \tag{10.2.1}$$

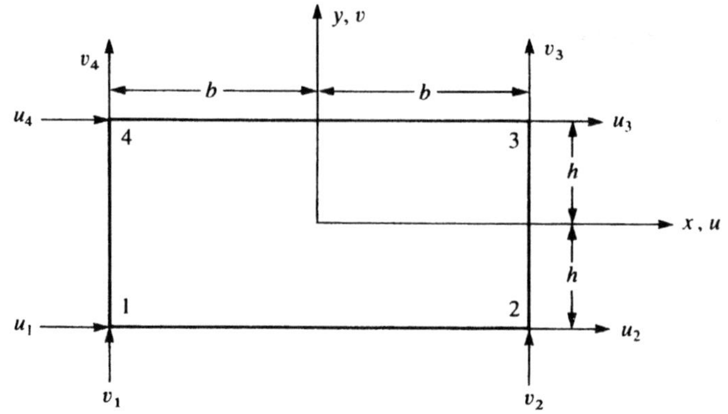

Figure 10–3 Basic four-node rectangular element with nodal degrees of freedom

Step 2 Select Displacement Functions

For a compatible displacement field, the element displacement functions u and v must be linear along each edge because only two points (the corner nodes) exist along each edge. We then select the linear displacement functions as

$$\begin{aligned} u(x,y) &= a_1 + a_2x + a_3y + a_4xy \\ v(x,y) &= a_5 + a_6x + a_7y + a_8xy \end{aligned} \tag{10.2.2}$$

We can proceed in the usual manner to eliminate the a_i's from Eqs. (10.2.2) to obtain

$$\begin{aligned} u(x,y) &= \frac{1}{4bh}[(b-x)(h-y)u_1 + (b+x)(h-y)u_2 \\ &\quad + (b+x)(h+y)u_3 + (b-x)(h+y)u_4] \\ v(x,y) &= \frac{1}{4bh}[(b-x)(h-y)v_1 + (b+x)(h-y)v_2 \\ &\quad + (b+x)(h+y)v_3 + (b-x)(h+y)v_4] \end{aligned} \tag{10.2.3}$$

These displacement expressions, Eqs. (10.2.3), can be expressed equivalently in terms of the shape functions and unknown nodal displacements as

$$\{\psi\} = [N]\{d\} \tag{10.2.4}$$

where the shape functions are given by

$$\begin{aligned} N_1 &= \frac{(b-x)(h-y)}{4bh} \qquad N_2 = \frac{(b+x)(h-y)}{4bh} \\ N_3 &= \frac{(b+x)(h+y)}{4bh} \qquad N_4 = \frac{(b-x)(h+y)}{4bh} \end{aligned} \tag{10.2.5}$$

and the N_i's are again such that $N_1 = 1$ at node 1 and $N_1 = 0$ at all the other nodes, with similar requirements for the other shape functions. In expanded form,

Eq. (10.2.4) becomes

$$\begin{Bmatrix} u \\ v \end{Bmatrix} = \begin{bmatrix} N_1 & 0 & N_2 & 0 & N_3 & 0 & N_4 & 0 \\ 0 & N_1 & 0 & N_2 & 0 & N_3 & 0 & N_4 \end{bmatrix} \begin{Bmatrix} u_1 \\ v_1 \\ u_2 \\ v_2 \\ u_3 \\ v_3 \\ u_4 \\ v_4 \end{Bmatrix} \tag{10.2.6}$$

Step 3 Define the Strain/ Displacement and Stress/ Strain Relationships

Again the element strains for the two-dimensional stress state are given by

$$\begin{Bmatrix} \varepsilon_x \\ \varepsilon_y \\ \gamma_{xy} \end{Bmatrix} = \begin{Bmatrix} \dfrac{\partial u}{\partial x} \\ \dfrac{\partial v}{\partial y} \\ \dfrac{\partial u}{\partial y} + \dfrac{\partial v}{\partial x} \end{Bmatrix} \tag{10.2.7}$$

Using Eq. (10.2.6) in Eq. (10.2.7) and taking the derivatives of u and v as indicated, we can express the strains in terms of the unknown nodal displacements as

$$\{\varepsilon\} = [B]\{d\} \tag{10.2.8}$$

where

$$[B] = \frac{1}{4bh} \begin{bmatrix} -(h-y) & 0 & (h-y) & 0 & (h+y) & 0 & -(h+y) & 0 \\ 0 & -(b-x) & 0 & -(b+x) & 0 & (b+x) & 0 & (b-x) \\ -(b-x) & -(h-y) & -(b+x) & (h-y) & (b+x) & (h+y) & (b-x) & -(h+y) \end{bmatrix} \tag{10.2.9}$$

From Eqs. (10.2.8) and (10.2.9), we observe that ε_x is a function of y, ε_y is a function of x, and γ_{xy} is a function of both x and y. The stresses are again given by the formulas in Eq. (6.2.36), where $[B]$ is now that of Eq. (10.2.9) and $\{d\}$ is that of Eq. (10.2.1).

Step 4 Derive the Element Stiffness Matrix and Equations

The stiffness matrix is determined by

$$[k] = \int_{-h}^{h}\int_{-b}^{b} [B]^T[D][B]t\,dx\,dy \tag{10.2.10}$$

with $[D]$ again given by the usual plane stress or plane strain conditions, Eq. (6.1.8) or (6.1.10). Because the $[B]$ matrix is a function of x and y, integration of Eq. (10.2.10) must be performed. The $[k]$ matrix for the rectangular element is now of order 8×8.

The element force matrix is determined by Eq. (6.2.46) as

$$\{f\} = \iiint_V [N]^T \{X\}\, dV + \{P\} + \iint_S [N_s]^T \{T\}\, dS \tag{10.2.11}$$

where $[N]$ is the rectangular matrix in Eq. (10.2.6), and N_1 through N_4 are given by Eqs. (10.2.5). The element equations are then given by

$$\{f\} = [k]\{d\} \tag{10.2.12}$$

Steps 5–7

Steps 5–7, which involve assembling the global stiffness matrix and equations, determining the unknown nodal displacements, and calculating the stress, are identical to those in Section 6.2 for the CST. However, the stresses within each element now vary in both the x and y directions.

As previously pointed out when describing defects for the CST in Chapter 6, the bilinear rectangle element described in this section also cannot provide pure bending. When this element is subjected to pure bending, it also develops false shear strain. This means that in a pure bending deformation, the bending moment needed to produce the deformation is predicted to be larger than the actual value when modeling with the rectangular element. Details of this *shear locking* are presented in [8].

To avoid the possibility of *shear locking* that occurs when the rectangular bilinear (four corner noded) element is subjected to bending, the higher order eight-noded quadratic rectangle has been developed and is described briefly in Section 10.6. This eight-noded element is created by adding mid-side nodes to the bilinear element.

10.3 Isoparametric Formulation of the Plane Element Stiffness Matrix

Recall that the term *isoparametric* is derived from the use of the same shape functions to define the element shape as are used to define the displacements within the element. Thus, when the shape function is $u = a_1 + a_2 s + a_3 t + a_4 st$ for the displacement, we use $x = a_1 + a_2 s + a_3 t + a_4 st$ for the description of a coordinate point in the plane element.

The natural-coordinate system s-t is defined by element geometry and not by the element orientation in the global-coordinate system x-y. Much as in the bar element example, there is a transformation mapping between the two coordinate systems for each element of a specific structure, and this relationship must be used in the element formulation.

We will now discuss the isoparametric formulation of the simple linear plane element stiffness matrix. This formulation is general enough to be applied to more

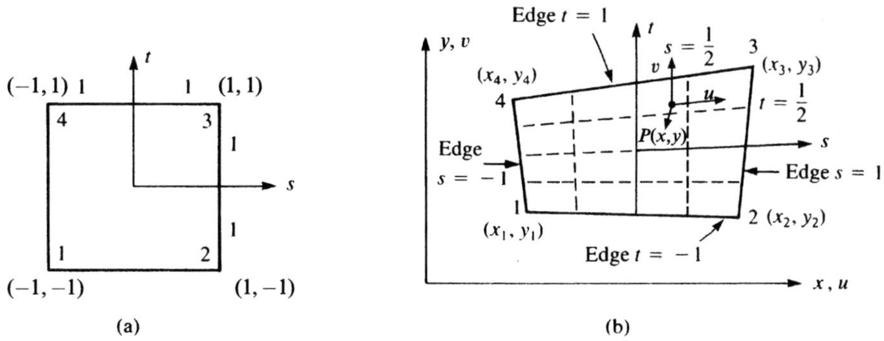

Figure 10–4 (a) Linear square element in *s*-*t* coordinates and (b) square element mapped into quadrilateral in *x*-*y* coordinates whose size and shape are determined by the eight nodal coordinates $x_1, y_1, \ldots, y_4$

complicated (higher-order) elements such as a quadratic plane element with three nodes along an edge, which can have straight or quadratic curved sides. Higher-order elements have additional nodes and use different shape functions as compared to the linear element, but the steps in the development of the stiffness matrices are the same. We will briefly discuss these elements after examining the linear plane element formulation.

Step 1 Select Element Type

First, the natural s-t coordinates are attached to the element, with the origin at the center of the element, as shown in Figure 10–4(a). The s and t axes need not be orthogonal, and neither has to be parallel to the x or y axis. The orientation of s-t coordinates is such that the four corner nodes and the edges of the quadrilateral are bounded by $+1$ or -1. This orientation will later allow us to take advantage more fully of common numerical integration schemes.

We consider the quadrilateral to have eight degrees of freedom, $u_1, v_1, \ldots, u_4$, and v_4 associated with the global x and y directions. The element then has straight sides but is otherwise of arbitrary shape, as shown in Figure 10–4(b).

For the special case when the distorted element becomes a rectangular element with sides parallel to the global x-y coordinates (see Figure 10–3), the s-t coordinates can be related to the global element coordinates x and y by

$$x = x_c + bs \qquad y = y_c + ht \tag{10.3.1}$$

where x_c and y_c are the global coordinates of the element centroid.

As we have shown for a rectangular element, the shape functions that define the displacements within the element are given by Eqs. (10.2.5). These same shape functions will now be used to map the square of Figure 10–4(a) in isoparametric coordinates s and t to the quadrilateral of Figure 10–4(b) in x and y coordinates whose size and shape are determined by the eight nodal coordinates $x_1, y_1, \ldots, x_4$, and y_4. That is, letting

$$x = a_1 + a_2 s + a_3 t + a_4 st$$
$$y = a_5 + a_6 s + a_7 t + a_8 st \tag{10.3.2}$$

and solving for the a_i's in terms of $x_1, x_2, x_3, x_4, y_1, y_2, y_3$, and y_4, we establish a form similar to Eqs. (10.2.3) such that

$$\begin{aligned} x &= \tfrac{1}{4}[(1-s)(1-t)x_1 + (1+s)(1-t)x_2 \\ &\quad + (1+s)(1+t)x_3 + (1-s)(1+t)x_4] \\ y &= \tfrac{1}{4}[(1-s)(1-t)y_1 + (1+s)(1-t)y_2 \\ &\quad + (1+s)(1+t)y_3 + (1-s)(1+t)y_4] \end{aligned} \tag{10.3.3}$$

Or, in matrix form, we can express Eqs. (10.3.3) as

$$\begin{Bmatrix} x \\ y \end{Bmatrix} = \begin{bmatrix} N_1 & 0 & N_2 & 0 & N_3 & 0 & N_4 & 0 \\ 0 & N_1 & 0 & N_2 & 0 & N_3 & 0 & N_4 \end{bmatrix} \begin{Bmatrix} x_1 \\ y_1 \\ x_2 \\ y_2 \\ x_3 \\ y_3 \\ x_4 \\ y_4 \end{Bmatrix} \tag{10.3.4}$$

where the shape functions of Eq. (10.3.4) are now

$$N_1 = \frac{(1-s)(1-t)}{4} \qquad N_2 = \frac{(1+s)(1-t)}{4}$$
$$N_3 = \frac{(1+s)(1+t)}{4} \qquad N_4 = \frac{(1-s)(1+t)}{4} \tag{10.3.5}$$

The shape functions of Eqs. (10.3.5) are linear. These shape functions are seen to map the s and t coordinates of any point in the square element of Figure 10–4(a) to those x and y coordinates in the quadrilateral element of Figure 10–4(b). For instance, consider square element node 1 coordinates, where $s = -1$ and $t = -1$. Using Eqs. (10.3.4) and (10.3.5), the left side of Eq. (10.3.4) becomes

$$x = x_1 \qquad y = y_1 \tag{10.3.6}$$

Similarly, we can map the other local nodal coordinates at nodes 2, 3, and 4 such that the square element in s-t isoparametric coordinates is mapped into a quadrilateral element in global coordinates. Also observe the property that $N_1 + N_2 + N_3 + N_4 = 1$ for all values of s and t.

We further observe that the shape functions in Eq. (10.3.5) are again such that N_1 through N_4 have the properties that N_i $(i = 1, 2, 3, 4)$ is equal to one at node i and equal to zero at all other nodes. The physical shapes of N_i as they vary over the element with natural coordinates are shown in Figure 10–5. For instance, N_1 represents the geometric shape for $x_1 = 1$, $y_1 = 1$, and x_2, y_2, x_3, y_3, x_4, and y_4 all equal to zero.

Until this point in the discussion, we have always developed the element shape functions either by assuming some relationship between the natural and global

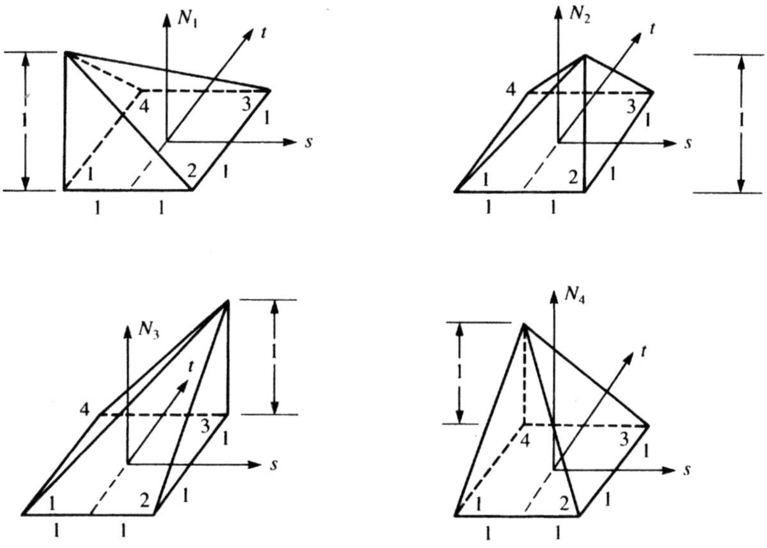

Figure 10–5 Variations of the shape functions over a linear square element

coordinates in terms of the generalized coordinates (a_i's) as in Eqs. (10.3.2) or, similarly, by assuming a displacement function in terms of the a_i's. However, physical intuition can often guide us in directly expressing shape functions based on the following two criteria set forth in Section 3.2 and used on numerous occasions:

$$\sum_{i=1}^{n} N_i = 1 \qquad (i = 1, 2, \ldots, n)$$

where $n =$ the number of shape functions corresponding to displacement shape functions N_i, and $N_i = 1$ at node i and $N_i = 0$ at all nodes other than i. In addition, a third criterion is based on Lagrangian interpolation when displacement continuity is to be satisfied, or on Hermitian interpolation when additional slope continuity needs to be satisfied, as in the beam element of Chapter 4. (For a description of the use of Lagrangian and Hermitian interpolation to develop shape functions, consult References [4] and [6].)

Step 2 Select Displacement Functions

The displacement functions within an element are now similarly defined by the same shape functions as are used to define the element shape; that is,

$$\begin{Bmatrix} u \\ v \end{Bmatrix} = \begin{bmatrix} N_1 & 0 & N_2 & 0 & N_3 & 0 & N_4 & 0 \\ 0 & N_1 & 0 & N_2 & 0 & N_3 & 0 & N_4 \end{bmatrix} \begin{Bmatrix} u_1 \\ v_1 \\ u_2 \\ v_2 \\ u_3 \\ v_3 \\ u_4 \\ v_4 \end{Bmatrix} \tag{10.3.7}$$

where u and v are displacements parallel to the global x and y coordinates, and the shape functions are given by Eqs. (10.3.5). The displacement of an interior point P located at (x, y) in the element of Figure 10–4(b) is described by u and v in Eq. (10.3.7).

Comparing Eqs. (10.2.6) and (10.3.7), we see similarities between the rectangular element with sides of lengths $2b$ and $2h$ (Figure 10–3) and the square element with sides of length 2. If we let $b = 1$ and $h = 1$, the two sets of shape functions, Eqs. (10.2.5) and (10.3.5), are identical.

Step 3 Define the Strain/Displacement and Stress/Strain Relationships

We now want to formulate element matrix $\underline{B}$ to evaluate $\underline{k}$. However, because it becomes tedious and difficult (if not impossible) to write the shape functions in terms of the x and y coordinates, as seen in Chapter 8, we will carry out the formulation in terms of the isoparametric coordinates s and t. This may appear tedious, but it is easier to use the s- and t-coordinate expressions than to attempt to use the x- and y-coordinate expressions. This approach also leads to a simple computer program formulation.

To construct an element stiffness matrix, we must determine the strains, which are defined in terms of the derivatives of the displacements with respect to the x and y coordinates. The displacements, however, are now functions of the s and t coordinates, as given by Eq. (10.3.7), with the shape functions given by Eqs. (10.3.5). Before, we could determine $(\partial f/\partial x)$ and $(\partial f/\partial y)$, where, in general, f is a function representing the displacement functions u or v. However, u and v are now expressed in terms of s and t. Therefore, we need to apply the chain rule of differentiation because it will not be possible to express s and t as functions of x and y directly. For f as a function of x and y, the chain rule yields

$$\begin{aligned} \frac{\partial f}{\partial s} &= \frac{\partial f}{\partial x}\frac{\partial x}{\partial s} + \frac{\partial f}{\partial y}\frac{\partial y}{\partial s} \\ \frac{\partial f}{\partial t} &= \frac{\partial f}{\partial x}\frac{\partial x}{\partial t} + \frac{\partial f}{\partial y}\frac{\partial y}{\partial t} \end{aligned} \qquad (10.3.8)$$

In Eq. (10.3.8), $(\partial f/\partial s)$, $(\partial f/\partial t)$, $(\partial x/\partial s)$, $(\partial y/\partial s)$, $(\partial x/\partial t)$, and $(\partial y/\partial t)$ are all known using Eqs. (10.3.7) and (10.3.4). We still seek $(\partial f/\partial x)$ and $(\partial f/\partial y)$. The strains can then be found; for example, $\varepsilon_x = (\partial u/\partial x)$. Therefore, we solve Eqs. (10.3.8) for $(\partial f/\partial x)$ and $(\partial f/\partial y)$ using Cramer's rule, which involves evaluation of determinants (Appendix B), as

$$\frac{\partial f}{\partial x} = \frac{\begin{vmatrix} \dfrac{\partial f}{\partial s} & \dfrac{\partial y}{\partial s} \\[2ex] \dfrac{\partial f}{\partial t} & \dfrac{\partial y}{\partial t} \end{vmatrix}}{\begin{vmatrix} \dfrac{\partial x}{\partial s} & \dfrac{\partial y}{\partial s} \\[2ex] \dfrac{\partial x}{\partial t} & \dfrac{\partial y}{\partial t} \end{vmatrix}} \qquad \frac{\partial f}{\partial y} = \frac{\begin{vmatrix} \dfrac{\partial x}{\partial s} & \dfrac{\partial f}{\partial s} \\[2ex] \dfrac{\partial x}{\partial t} & \dfrac{\partial f}{\partial t} \end{vmatrix}}{\begin{vmatrix} \dfrac{\partial x}{\partial s} & \dfrac{\partial y}{\partial s} \\[2ex] \dfrac{\partial x}{\partial t} & \dfrac{\partial y}{\partial t} \end{vmatrix}} \qquad (10.3.9)$$

where the determinant in the denominator is the determinant of the *Jacobian* matrix $\underline{J}$. Hence, the Jacobian matrix is given by

$$[J] = \begin{bmatrix} \dfrac{\partial x}{\partial s} & \dfrac{\partial y}{\partial s} \\[2ex] \dfrac{\partial x}{\partial t} & \dfrac{\partial y}{\partial t} \end{bmatrix} \tag{10.3.10}$$

We now want to express the element strains as

$$\underline{\varepsilon} = \underline{B}\underline{d} \tag{10.3.11}$$

where $\underline{B}$ must now be expressed as a function of s and t. We start with the usual relationship between strains and displacements given in matrix form as

$$\begin{Bmatrix} \varepsilon_x \\ \varepsilon_y \\ \gamma_{xy} \end{Bmatrix} = \begin{bmatrix} \dfrac{\partial(\ )}{\partial x} & 0 \\[2ex] 0 & \dfrac{\partial(\ )}{\partial y} \\[2ex] \dfrac{\partial(\ )}{\partial y} & \dfrac{\partial(\ )}{\partial x} \end{bmatrix} \begin{Bmatrix} u \\ v \end{Bmatrix} \tag{10.3.12}$$

where the rectangular matrix on the right side of Eq. (10.3.12) is an *operator matrix*; that is, $\partial(\ )/\partial x$ and $\partial(\ )/\partial y$ represent the partial derivatives of any variable we put inside the parentheses.

Using Eqs. (10.3.9) and evaluating the determinant in the numerators, we have

$$\begin{aligned} \frac{\partial(\ )}{\partial x} &= \frac{1}{|\underline{J}|}\left[\frac{\partial y}{\partial t}\frac{\partial(\ )}{\partial s} - \frac{\partial y}{\partial s}\frac{\partial(\ )}{\partial t}\right] \\ \frac{\partial(\ )}{\partial y} &= \frac{1}{|\underline{J}|}\left[\frac{\partial x}{\partial s}\frac{\partial(\ )}{\partial t} - \frac{\partial x}{\partial t}\frac{\partial(\ )}{\partial s}\right] \end{aligned} \tag{10.3.13}$$

where $|\underline{J}|$ is the determinant of $\underline{J}$ given by Eq. (10.3.10). Using Eq. (10.3.13) in Eq. (10.3.12) we obtain the strains expressed in terms of the natural coordinates (s-t) as

$$\begin{Bmatrix} \varepsilon_x \\ \varepsilon_y \\ \gamma_{xy} \end{Bmatrix} = \frac{1}{|\underline{J}|} \begin{bmatrix} \dfrac{\partial y}{\partial t}\dfrac{\partial(\ )}{\partial s} - \dfrac{\partial y}{\partial s}\dfrac{\partial(\ )}{\partial t} & 0 \\[2ex] 0 & \dfrac{\partial x}{\partial s}\dfrac{\partial(\ )}{\partial t} - \dfrac{\partial x}{\partial t}\dfrac{\partial(\ )}{\partial s} \\[2ex] \dfrac{\partial x}{\partial s}\dfrac{\partial(\ )}{\partial t} - \dfrac{\partial x}{\partial t}\dfrac{\partial(\ )}{\partial s} & \dfrac{\partial y}{\partial t}\dfrac{\partial(\ )}{\partial s} - \dfrac{\partial y}{\partial s}\dfrac{\partial(\ )}{\partial t} \end{bmatrix} \begin{Bmatrix} u \\ v \end{Bmatrix} \tag{10.3.14}$$

Using Eq. (10.3.7), we can express Eq. (10.3.14) in terms of the shape functions and global coordinates in compact matrix form as

$$\underline{\varepsilon} = \underline{D}'\underline{N}\underline{d} \tag{10.3.15}$$

where $\underline{D}'$ is an operator matrix given by

$$\underline{D}' = \frac{1}{|\underline{J}|}\begin{bmatrix} \dfrac{\partial y}{\partial t}\dfrac{\partial(\;)}{\partial s} - \dfrac{\partial y}{\partial s}\dfrac{\partial(\;)}{\partial t} & 0 \\ 0 & \dfrac{\partial x}{\partial s}\dfrac{\partial(\;)}{\partial t} - \dfrac{\partial x}{\partial t}\dfrac{\partial(\;)}{\partial s} \\ \dfrac{\partial x}{\partial s}\dfrac{\partial(\;)}{\partial t} - \dfrac{\partial x}{\partial t}\dfrac{\partial(\;)}{\partial s} & \dfrac{\partial y}{\partial t}\dfrac{\partial(\;)}{\partial s} - \dfrac{\partial y}{\partial s}\dfrac{\partial(\;)}{\partial t} \end{bmatrix} \tag{10.3.16}$$

and $\underline{N}$ is the 2×8 shape function matrix given as the first matrix on the right side of Eq. (10.3.7) and $\underline{d}$ is the column matrix on the right side of Eq. (10.3.7).

Defining $\underline{B}$ as

$$\underset{(3\times 8)}{\underline{B}} = \underset{(3\times 2)}{\underline{D}'}\;\underset{(2\times 8)}{\underline{N}} \tag{10.3.17}$$

we have $\underline{B}$ expressed as a function of s and t and thus have the strains in terms of s and t. Here $\underline{B}$ is of order 3×8, as indicated in Eq. (10.3.17).

The explicit form of $\underline{B}$ can be obtained by substituting Eq. (10.3.16) for $\underline{D}'$ and Eqs. (10.3.5) for the shape functions into Eq. (10.3.17). The matrix multiplications yield

$$\underline{B}(s,t) = \frac{1}{|\underline{J}|}[\underline{B}_1 \quad \underline{B}_2 \quad \underline{B}_3 \quad \underline{B}_4] \tag{10.3.18}$$

where the submatrices of $\underline{B}$ are given by

$$\underline{B}_i = \begin{bmatrix} a(N_{i,s}) - b(N_{i,t}) & 0 \\ 0 & c(N_{i,t}) - d(N_{i,s}) \\ c(N_{i,t}) - d(N_{i,s}) & a(N_{i,s}) - b(N_{i,t}) \end{bmatrix} \tag{10.3.19}$$

Here i is a dummy variable equal to 1, 2, 3, and 4, and

$$\begin{aligned} a &= \tfrac{1}{4}[y_1(s-1) + y_2(-1-s) + y_3(1+s) + y_4(1-s)] \\ b &= \tfrac{1}{4}[y_1(t-1) + y_2(1-t) + y_3(1+t) + y_4(-1-t)] \\ c &= \tfrac{1}{4}[x_1(t-1) + x_2(1-t) + x_3(1+t) + x_4(-1-t)] \\ d &= \tfrac{1}{4}[x_1(s-1) + x_2(-1-s) + x_3(1+s) + x_4(1-s)] \end{aligned} \tag{10.3.20}$$

Using the shape functions defined by Eqs. (10.3.5), we have

$$N_{1,s} = \tfrac{1}{4}(t-1) \qquad N_{1,t} = \tfrac{1}{4}(s-1) \qquad \text{(and so on)} \tag{10.3.21}$$

where the comma followed by the variable s or t indicates differentiation with respect to that variable; that is, $N_{1,s} \equiv \partial N_1/\partial s$, and so on. The determinant $|\underline{J}|$ is a polynomial in s and t and is tedious to evaluate even for the simplest case of the linear plane element. However, using Eq. (10.3.10) for $[J]$ and Eqs. (10.3.3) for x and y, we can evaluate $|\underline{J}|$ as

$$|\underline{J}| = \frac{1}{8}\{X_c\}^T \begin{bmatrix} 0 & 1-t & t-s & s-1 \\ t-1 & 0 & s+1 & -s-t \\ s-t & -s-1 & 0 & t+1 \\ 1-s & s+t & -t-1 & 0 \end{bmatrix} \{Y_c\} \tag{10.3.22}$$

where

$$\{X_c\}^T = [x_1 \quad x_2 \quad x_3 \quad x_4] \tag{10.3.23}$$

and

$$\{Y_c\} = \begin{Bmatrix} y_1 \\ y_2 \\ y_3 \\ y_4 \end{Bmatrix} \tag{10.3.24}$$

We observe that $|\underline{J}|$ is a function of s and t and the known global coordinates $x_1, x_2, \ldots, y_4$. Hence, $\underline{B}$ is a function of s and t in both the numerator and the denominator [because of $|\underline{J}|$ given by Eq. (10.3.22)] and of the known global coordinates x_1 through y_4.

The stress/strain relationship is again $\underline{\sigma} = \underline{D}\underline{B}\underline{d}$, where because the $\underline{B}$ matrix is a function of s and t, so also is the stress matrix $\underline{\sigma}$.

Step 4 Derive the Element Stiffness Matrix and Equations

We now want to express the stiffness matrix in terms of s-t coordinates. For an element with a constant thickness h, we have

$$[k] = \iint_A [B]^T[D][B]h\,dx\,dy \tag{10.3.25}$$

However, $\underline{B}$ is now a function of s and t, as seen by Eqs. (10.3.18)–(10.3.20), and so we must integrate with respect to s and t. Once again, to transform the variables and the region from x and y to s and t, we must have a standard procedure that involves the determinant of $\underline{J}$. This general type of transformation [4, 5] is given by

$$\iint_A f(x,y)\,dx\,dy = \iint_A f(s,t)|\underline{J}|\,ds\,dt \tag{10.3.26}$$

where the inclusion of $|\underline{J}|$ in the integrand on the right side of Eq. (10.3.26) results from a theorem of integral calculus (see Reference [5] for the complete proof of this theorem). Using Eq. (10.3.26) in Eq. (10.3.25), we obtain

$$[k] = \int_{-1}^{1}\int_{-1}^{1} [B]^T[D][B]h|\underline{J}|\,ds\,dt \tag{10.3.27}$$

The $|\underline{J}|$ and $\underline{B}$ are such as to result in complicated expressions within the integral of Eq. (10.3.27), and so the integration to determine the element stiffness matrix is usually done numerically. A method for numerically integrating Eq. (10.3.27) is given in Section 10.4. The stiffness matrix in Eq. (10.3.27) is of the order 8×8.

Body Forces

The element body-force matrix will now be determined from

$$\underset{(8\times 1)}{\{f_b\}} = \int_{-1}^{1}\int_{-1}^{1} \underset{(8\times 2)}{[N]^T} \underset{(2\times 1)}{\{X\}}\; h|\underline{J}|\, ds\, dt \tag{10.3.28}$$

Like the stiffness matrix, the body-force matrix in Eq. (10.3.28) has to be evaluated by numerical integration.

Surface Forces

The surface-force matrix, say, along edge $t = 1$ (Figure 10–6) with overall length L, is

$$\underset{(4\times 1)}{\{f_s\}} = \int_{-1}^{1} \underset{(4\times 2)}{[N_s]^T} \underset{(2\times 1)}{\{T\}}\; h\frac{L}{2} ds \tag{10.3.29}$$

or

$$\begin{Bmatrix} f_{s3s} \\ f_{s3t} \\ f_{s4s} \\ f_{s4t} \end{Bmatrix} = \int_{-1}^{1} \begin{bmatrix} N_3 & 0 & N_4 & 0 \\ 0 & N_3 & 0 & N_4 \end{bmatrix}^T \Bigg|_{\substack{\text{evaluated} \\ \text{along } t=1}} \begin{Bmatrix} p_s \\ p_t \end{Bmatrix} h\frac{L}{2} ds \tag{10.3.30}$$

because $N_1 = 0$ and $N_2 = 0$ along edge $t = 1$, and hence, no nodal forces exist at nodes 1 and 2. For the case of uniform (constant) p_s and p_t along edge $t = 1$, the total surface-force matrix is

$$\{f_s\} = h\frac{L}{2}[0 \quad 0 \quad 0 \quad 0 \quad p_s \quad p_t \quad p_s \quad p_t]^T \tag{10.3.31}$$

Surface forces along other edges can be obtained similar to Eq. (10.3.30) by merely using the proper shape functions associated with the edge where the tractions are applied.

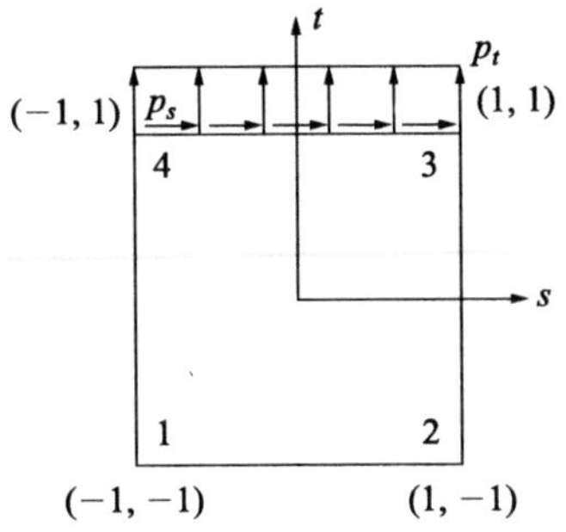

Figure 10–6 Surface traction: p_s and p_t acting at edge $t = 1$

Example 10.1

For the four-noded linear plane element shown in Figure 10–7 with a uniform surface traction along side 2–3, evaluate the force matrix by using the energy equivalent nodal forces obtained from the integral similar to Eq. (10.3.29). Let the thickness of the element be $h = 0.1$ in.

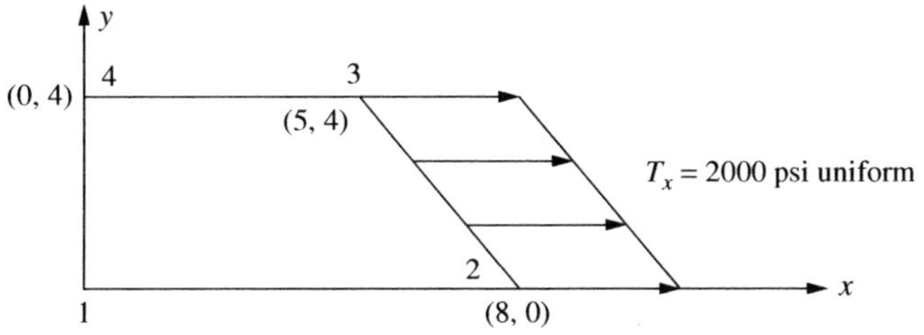

Figure 10–7 Element subjected to uniform surface traction

Using Eq. (10.3.29), we have

$$\{f_s\} = \int_{-1}^{1} [N_s]^T \{T\} h \frac{L}{2} dt \tag{10.3.32}$$

With length of side 2–3 given by

$$L = \sqrt{(5-8)^2 + (4-0)^2} = \sqrt{9+16} = 5 \tag{10.3.33}$$

Shape functions N_2 and N_3 must be used, as we are evaluating the surface traction along side 2–3 (at $s = 1$). Therefore, Eq. (10.3.33) becomes

$$\{f_s\} = \int_{-1}^{1} [N_s]^T \{T\} h \frac{L}{2} dt = \int_{-1}^{1} \begin{bmatrix} N_2 & 0 & N_3 & 0 \\ 0 & N_2 & 0 & N_3 \end{bmatrix}^T \begin{Bmatrix} p_s \\ p_t \end{Bmatrix} h \frac{L}{2} dt \tag{10.3.34}$$

evaluated along $s = 1$

The shape functions for the four-noded linear plane element are taken from Eq. (10.3.5) as

$$N_2 = \frac{(1+s)(1-t)}{4} = \frac{s-t-st+1}{4} \qquad N_3 = \frac{(1+s)(1+t)}{4} = \frac{s+t+st+1}{4} \tag{10.3.35}$$

The surface traction matrix is given by

$$\{T\} = \begin{Bmatrix} p_s \\ p_t \end{Bmatrix} = \begin{Bmatrix} 2000 \\ 0 \end{Bmatrix} \tag{10.3.36}$$

Substituting Eq. (10.3.33) for L and Eq. (10.3.36) for the surface traction matrix and the thickness $h = 0.1$ inch into Eq. (10.3.32), we obtain

$$\{f_s\} = \int_{-1}^{1} [N_s]^T \{T\} h \frac{L}{2}\, dt = \int_{-1}^{1} \begin{bmatrix} N_2 & 0 \\ 0 & N_2 \\ N_3 & 0 \\ 0 & N_3 \end{bmatrix} \begin{Bmatrix} 2000 \\ 0 \end{Bmatrix} 0.1 \frac{5}{2}\, dt \tag{10.3.37}$$

evaluated along $s = 1$

Simplifying Eq. (10.3.37), we obtain

$$\{f_s\} = 0.25 \int_{-1}^{1} \begin{bmatrix} 2000N_2 \\ 0 \\ 2000N_3 \\ 0 \end{bmatrix} dt = 500 \int_{-1}^{1} \begin{bmatrix} N_2 \\ 0 \\ N_3 \\ 0 \end{bmatrix} dt \tag{10.3.38}$$

evaluated along $s = 1$

Substituting the shape functions from Eq. (10.3.35) into Eq. (10.3.38), we have

$$\{f_s\} = 500 \int_{-1}^{1} \begin{bmatrix} \dfrac{s - t - st + 1}{4} \\ 0 \\ \dfrac{s + t + st + 1}{4} \\ 0 \end{bmatrix} dt \tag{10.3.39}$$

evaluated along $s = 1$

Upon substituting $s = 1$ into the integrand in Eq. (10.3.39) and performing the explicit integration in Eq. (10.3.40), we obtain

$$\{f_s\} = 500 \int_{-1}^{1} \begin{bmatrix} \dfrac{2 - 2t}{4} \\ 0 \\ \dfrac{2t + 2}{4} \\ 0 \end{bmatrix} dt = 500 \begin{bmatrix} 0.50t - \dfrac{t^2}{4} \\ 0 \\ 0.50t + \dfrac{t^2}{4} \\ 0 \end{bmatrix}_{-1}^{1} \tag{10.3.40}$$

Evaluating the resulting integration expression for each limit, we obtain the final expression for the surface traction matrix as

$$\{f_s\} = 500 \begin{bmatrix} 0.50 - 0.25 \\ 0 \\ 0.50 + 0.25 \\ 0 \end{bmatrix} - 500 \begin{bmatrix} -0.50 - 0.25 \\ 0 \\ -0.50 + 0.25 \\ 0 \end{bmatrix} = 500 \begin{bmatrix} 1 \\ 0 \\ 1 \\ 0 \end{bmatrix} lb \tag{10.3.41}$$

Or in explicit form the surface tractions at nodes 2 and 3 are

$$\begin{Bmatrix} f_{s2s} \\ f_{s2t} \\ f_{s3s} \\ f_{s3t} \end{Bmatrix} = \begin{bmatrix} 500 \\ 0 \\ 500 \\ 0 \end{bmatrix} lb \tag{10.3.42}$$

■

10.4 Gaussian and Newton-Cotes Quadrature (Numerical Integration)

In this section, we will describe Gauss's method, one of the many schemes for numerical evaluation of definite integrals, because it has proved most useful for finite element work.

For completion sake, we will also describe the more common numerical integration method of Newton-Cotes. The Newton-Cotes methods for one and two intervals of integration are the well-known trapezoid and Simpson's one-third rule, respectively. After describing both methods, we will then understand why the Gaussian quadrature method is used in finite element work.

Gaussian Quadrature:

To evaluate the integral

$$I = \int_{-1}^{1} y\,dx \tag{10.4.1a}$$

where $y = y(x)$, we might choose (sample or evaluate) y at the midpoint $y(0) = y_1$ and multiply by the length of the interval, as shown in Figure 10–8, to arrive at $I = 2y_1$, a result that is exact if the curve happens to be a straight line. This is an example of what is called **one-point Gaussian quadrature** because only one sampling point was used. Therefore,

$$I = \int_{-1}^{1} y(x)\,dx \cong 2y(0) \tag{10.4.1b}$$

which is the familiar midpoint rule. Generalization of the formula [Eq. (10.4.1b)] leads to

$$I = \int_{-1}^{1} y\,dx = \sum_{i=1}^{n} W_i y_i \tag{10.4.2}$$

That is, to approximate the integral, we evaluate the function at several sampling points n, multiply each value y_i by the appropriate weight W_i, and add the terms. Gauss's method chooses the sampling points so that for a given number of points, the best possible accuracy is obtained. Sampling points are located symmetrically with respect to the center of the interval. Symmetrically paired points are given the

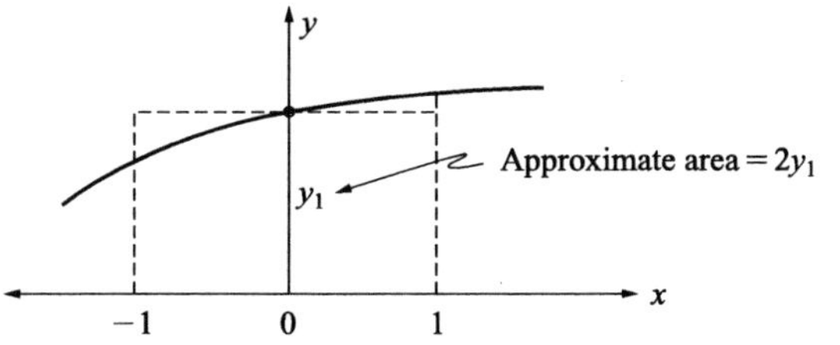

Figure 10–8 Gaussian quadrature using one sampling point

same weight W_i. Table 10–1 gives appropriate sampling points and weighting coefficients for the first three orders—that is, one, two, or three sampling points (see Reference [2] for more complete tables). For example, using two points (Figure 10–9), we simply have $I = y_1 + y_2$ because $W_1 = W_2 = 1.000$. This is the exact result if $y = f(x)$ is a polynomial containing terms up to and including x^3. In general, Gaussian quadrature using n points (Gauss points) is exact if the integrand is a polynomial of degree $2n - 1$ or less. In using n points, we effectively replace the given function $y = f(x)$ by a polynomial of degree $2n - 1$. The accuracy of the numerical integration depends on how well the polynomial fits the given curve.

If the function $f(x)$ is not a polynomial, Gaussian quadrature is inexact, but it becomes more accurate as more Gauss points are used. Also, it is important to understand that the ratio of two polynomials is, in general, not a polynomial; therefore, Gaussian quadrature will not yield exact integration of the ratio.

Table 10–1 Table for Gauss points for integration from minus one to one, $\int_{-1}^{1} y(x)\,dx = \sum_{i=1}^{n} W_i y_i$

Number of Points	Locations, x_i	Associated Weights, W_i
1	$x_1 = 0.000\ldots$	2.000
2	$x_1, x_2 = \pm 0.57735026918962$	1.000
3	$x_1, x_3 = \pm 0.77459666924148$	$\frac{5}{9} = 0.555\ldots$
	$x_2 = 0.000\ldots$	$\frac{8}{9} = 0.888\ldots$
4	$x_1, x_4 = \pm 0.8611363116$	0.3478548451
	$x_2, x_3 = \pm 0.3399810436$	0.6521451549

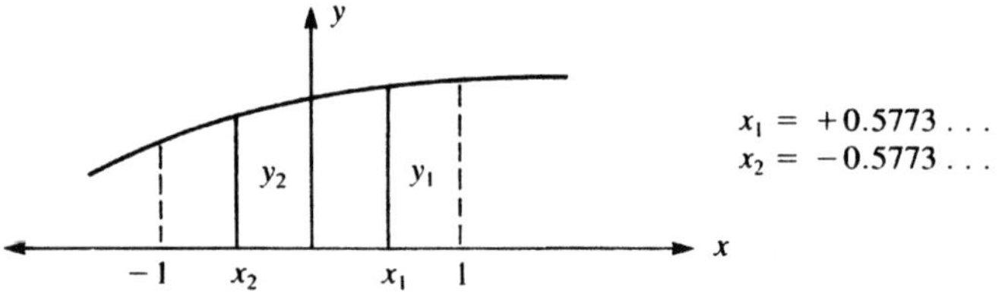

Figure 10–9 Gaussian quadrature using two sampling points

Two-Point Formula

To illustrate the derivation of a two-point ($n = 2$) Gauss formula based on Eq. (10.4.2), we have

$$I = \int_{-1}^{1} y\,dx = W_1 y_1 + W_2 y_2 = W_1 y(x_1) + W_2 y(x_2) \tag{10.4.3}$$

There are four unknown parameters to determine: W_1, W_2, x_1, and x_2. Therefore, we assume a cubic function for y as follows:

$$y = C_0 + C_1 x + C_2 x^2 + C_3 x^3 \tag{10.4.4}$$

In general, with four parameters in the two-point formula, we would expect the Gauss formula to exactly predict the area under the curve. That is,

$$A = \int_{-1}^{1} (C_0 + C_1 x + C_2 x^2 + C_3 x^3)\, dx = 2C_0 + \frac{2C_2}{3} \tag{10.4.5}$$

However, we will assume, based on Gauss's method, that $W_1 = W_2$ and $x_1 = x_2$ as we use two symmetrically located Gauss points at $x = \pm a$ with equal weights. The area predicted by Gauss's formula is

$$A_G = Wy(-a) + Wy(a) = 2W(C_0 + C_2 a^2) \tag{10.4.6}$$

where $y(-a)$ and $y(a)$ are evaluated using Eq. (10.4.4). If the error, $e = A - A_G$, is to vanish for any C_0 and C_2, we must have, using Eqs. (10.4.5) and (10.4.6) in the error expression,

$$\frac{\partial e}{\partial C_0} = 0 = 2 - 2W \qquad \text{or} \qquad W = 1 \tag{10.4.7a}$$

and

$$\frac{\partial e}{\partial C_2} = 0 = \frac{2}{3} - 2a^2 W \qquad \text{or} \qquad a = \sqrt{\frac{1}{3}} = 0.5773\ldots \tag{10.4.7b}$$

Now $W = 1$ and $a = 0.5773\ldots$ are the W_i's and a_i's (x_i's) for the two-point Gaussian quadrature given in Table 10–1.

Example 10.2

Evaluate the integrals (a) $I = \int_{-1}^{1} [x^2 + \cos(x/2)]\, dx$ and (b) $I = \int_{-1}^{1} (3^x - x) dx$ using three-point Gaussian quadrature.

(a) Using Table 10–1 for the three Gauss points and weights, we have $x_1 = x_3 = \pm 0.77459\ldots$, $x_2 = 0.000\ldots$, $W_1 = W_3 = \frac{5}{9}$, and $W_2 = \frac{8}{9}$. The integral then becomes

$$I = \left[(-0.77459)^2 + \cos\left(-\frac{0.77459}{2}\ \text{rad}\right)\right]\frac{5}{9} + \left[0^2 + \cos\frac{0}{2}\right]\frac{8}{9}$$

$$+ \left[(0.77459)^2 + \cos\left(\frac{0.77459}{2}\ \text{rad}\right)\right]\frac{5}{9}$$

$$= 1.918 + 0.667 = 2.585$$

Compared to the exact solution, we have $I_{\text{exact}} = 2.585$.

In this example, three-point Gaussian quadrature yields the exact answer to four significant figures.

(b) Using Table 10–1 for the three Gauss points and weights as in part (a), the integral then becomes

$$I = [3^{(-0.77459)} - (-0.77459)]\frac{5}{9} + [3^0 - 0]\frac{8}{9} + [3^{(0.77459)} - (0.77459)]\frac{5}{9}$$

$$= 0.66755 + 0.88889 + 0.86065 = 2.4229 (2.423 \text{ to four significant figures})$$

Compared to the exact solution, we have $I_{\text{exact}} = 2.427$. The error is $2.427 - 2.423 = 0.004$. ■

In two dimensions, we obtain the quadrature formula by integrating first with respect to one coordinate and then with respect to the other as

$$I = \int_{-1}^{1}\int_{-1}^{1} f(s,t)\,ds\,dt = \int_{-1}^{1}\left[\sum_i W_i f(s_i, t)\right] dt$$

$$= \sum_j W_j\left[\sum_i W_i f(s_i, t_j)\right] = \sum_i\sum_j W_i W_j f(s_i, t_j) \qquad (10.4.8)$$

In Eq. (10.4.8), we need not use the same number of Gauss points in each direction (that is, i does not have to equal j), but this is usually done. Thus, for example, a four-point Gauss rule (often described as a 2×2 rule) is shown in Figure 10–10. Equation (10.4.8) with $i = 1, 2$ and $j = 1, 2$ yields

$$I = W_1 W_1 f(s_1, t_1) + W_1 W_2 f(s_1, t_2) + W_2 W_1 f(s_2, t_1) + W_2 W_2 f(s_2, t_2) \qquad (10.4.9)$$

where the four sampling points are at $s_i, t_i = \pm 0.5773\ldots = \pm 1/\sqrt{3}$, and the weights are all 1.000. Hence, the double summation in Eq. (10.4.8) can really be interpreted as a single summation over the four points for the rectangle.

In general, in three dimensions, we have

$$I = \int_{-1}^{1}\int_{-1}^{1}\int_{-1}^{1} f(s,t,z)\,ds\,dt\,dz = \sum_i\sum_j\sum_k W_i W_j W_k f(s_i, t_j, z_k) \qquad (10.4.10)$$

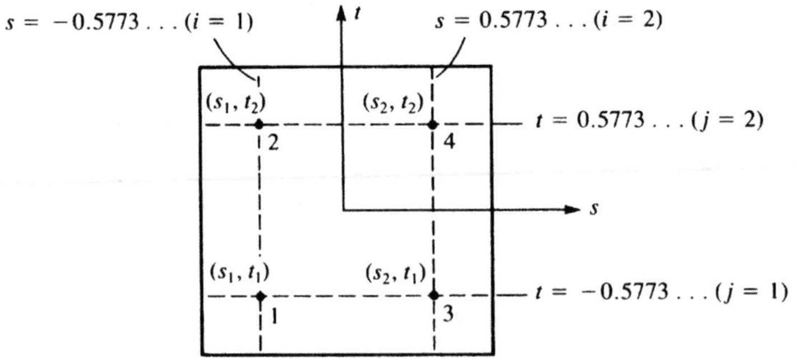

Figure 10–10 Four-point Gaussian quadrature in two dimensions

Newton-Cotes Numerical Integration:

We now describe the common numerical integration method called the Newton-Cotes method for evaluation of definite integrals. However, the method does not yield as accurate of results as the Gaussian quadrature method and so is not normally used in finite element method evaluations, such as to evaluate the stiffness matrix.

To evaluate the integral

$$I = \int_{-1}^{1} y\,dx$$

we assume the sampling points of $y(x)$ are spaced at equal intervals. Since the limits of integration are from -1 to 1 using the isoparametric formulation, the Newton-Cotes formula is given by

$$I = \int_{-1}^{1} y\,dx = h\sum_{i=0}^{n} C_i y_i = h[C_0 y_0 + C_1 y_1 + C_2 y_2 + C_3 y_3 + \ldots + C_n y_n] \quad (10.4.11)$$

where the C_i are the Newton-Cotes constants for numerical integration with i intervals (the number of intervals will be one less than the number of sampling points, n) and h is the interval between the limits of integration (for limits of integration between -1 and 1 this makes $h = 2$). The Newton-Cotes constants have been published and are summarized in Table 10–2 for $i = 1$ to 6. The case $i = 1$ corresponds to the well-known trapezoid rule illustrated by Figure 10–11. The case $i = 2$ corresponds to the well-known Simpson one-third rule. It is shown [9] that the formulas for $i = 3$ and $i = 5$ have the same accuracy as the formulas for $i = 2$ and $i = 4$, respectively. Therefore, it is recommended that the even formulas with $i = 2$ and $i = 4$ be used in practice. To obtain greater accuracy one can then use a smaller interval (include more evaluations of the function to be integrated). This can be accomplished by using a higher-order Newton-Cotes formula, thus increasing the number of intervals i.

It is shown [9] that we need to use n equally spaced sampling points to integrate exactly a polynomial of order at most $n - 1$. On the other hand, using Gaussian quadrature

Table 10–2 Table for Newton-Cotes intervals and points for integration, $\int_{-1}^{1} y(x)\,dx = h\sum_{i=0}^{n} C_i y_i$

Intervals, i	No. of Points, n	C_0	C_1	C_2	C_3	C_4	C_5	C_6
1	2	1/2	1/2			(trapezoid rule)		
2	3	1/6	4/6	1/6		(Simpson's 1/3 rule)		
3	4	1/8	3/8	3/8	1/8	(Simpson's 3/8 rule)		
4	5	7/90	32/90	12/90	32/90	7/90		
5	6	19/288	75/288	50/288	50/288	75/288	19/288	
6	7	41/840	216/840	27/840	272/840	27/840	216/840	41/840

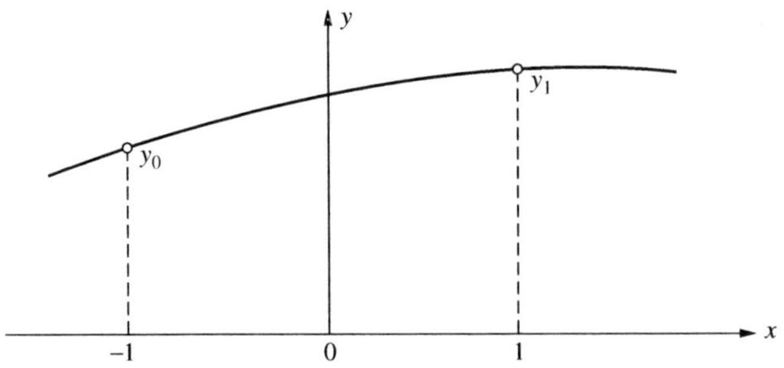

Figure 10–11 Approximation of numerical integration (approximate area under curve) using $i = 1$ interval, $n = 2$ sampling points (trapezoid rule), for

$$I = \int_{-1}^{1} y(x)\,dx = h\sum_{i=0}^{2} C_i y_i$$

we have previously stated that we use unequally spaced sampling points n and integrate exactly a polynomial of order at most $2n - 1$. For instance, using the Newton-Cotes formula with $n = 2$ sampling points, the highest order polynomial we can integrate exactly is a linear one. However, using Gaussian quadrature, we can integrate a cubic polynomial exactly. Gaussian quadrature is then more accurate with fewer sampling points than Newton-Cotes quadrature. This is because Gaussian quadrature is based on optimizing the position of the sampling points (not making them equally spaced as in the Newton-Cotes method) and also optimizing the weights W_i given in Table 10–1. After the function is evaluated at the sampling points, the corresponding weights are multiplied by these evaluated functions as was illustrated in Example 10.2.

Example 10.3 is used to illustrate the Newton-Cotes method and compare its accuracy to that of the Gaussian quadrature method previously described.

Example 10.3

Solve Example 10.2 using the Newton-Cotes method with $i = 2$ intervals ($n = 3$ sampling points). That is, evaluate the integrals (a) $I = \int_{-1}^{1}[x^2 + \cos(x/2)]dx$ and (b) $I = \int_{-1}^{1}(3^x - x)dx$ using the Newton-Cotes method.

Using Table 10–2 with three sampling points means we evaluate the function inside the integrand at $x = -1$, $x = 0$, and $x = 1$, and multiply each evaluated function by the respective Newton-Cotes numbers, 1/6, 4/6, and 1/6. We then add these three products together and finally multiply this sum by the interval of integration ($h = 2$) as follows:

$$I = 2\left[\frac{1}{6}y_0 + \frac{4}{6}y_1 + \frac{1}{6}y_2\right] \tag{10.4.12}$$

(a): Using Eq. (10.4.12), we obtain

$$\begin{aligned} y_0 &= x^2 + \cos(x/2) \text{ evaluated at } x = -1, \text{ etc. as follows:} \\ y_0 &= (-1)^2 + \cos(-1/2 \text{ rad}) = 1.8775826 \\ y_1 &= (0)^2 + \cos(0/2) = 1 \\ y_2 &= (1)^2 + \cos(1/2 \text{ rad}) = 1.8775826 \end{aligned} \qquad (10.4.13)$$

Substituting $y_0 - y_2$ from Eq. (10.4.13) into Eq. (10.4.12), we obtain the evaluation of the integral as

$$I = 2\left[\frac{1}{6}(1.8775826) + \frac{4}{6}(1) + \frac{1}{6}(1.8775826)\right] = 2.585$$

This solution compares exactly to the evaluation performed using Gaussian quadrature and to the exact solution. However, for higher-order functions the Gaussian quadrature method yields more accurate results than the Newton-Cotes method as illustrated by part (b) as follows:

(b): Using Eq. (10.4.12), we obtain

$$\begin{aligned} y_0 &= 3^{(-1)} - (-1) = \frac{4}{3} \\ y_1 &= 3^0 - 0 = 1 \\ y_2 &= 3^1 - (1) = 2 \end{aligned}$$

Substituting $y_0 - y_2$ into Eq. (10.4.12) we obtain I as

$$I = 2\left[\frac{1}{6}\left(\frac{4}{3}\right) + \frac{4}{6}(1) + \frac{1}{6}(2)\right] = 2.444$$

The error is $2.444 - 2.427 = 0.017$. This error is larger than that found using Gaussian quadrature (see Example 10.2 (b)). ■

10.5 Evaluation of the Stiffness Matrix and Stress Matrix by Gaussian Quadrature

Evaluation of the Stiffness Matrix

For the two-dimensional element, we have shown in previous chapters that

$$\underline{k} = \iint_A \underline{B}^T(x, y)\underline{D}\underline{B}(x, y)h\,dx\,dy \qquad (10.5.1)$$

where, in general, the integrand is a function of x and y and nodal coordinate values.

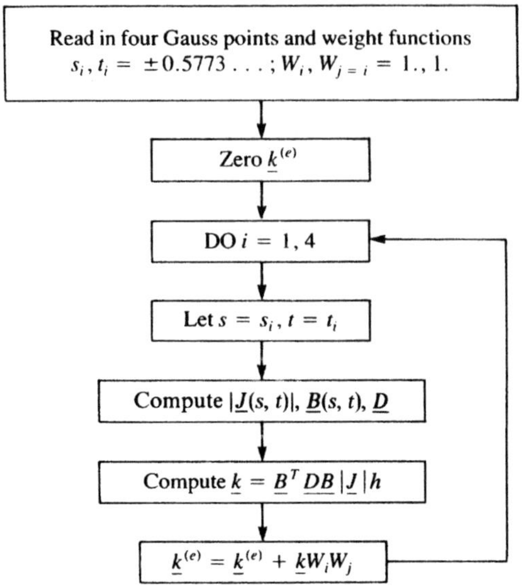

Figure 10–12 Flowchart to evaluate $\underline{k}^{(e)}$ by four-point Gaussian quadrature

We have shown in Section 10.3 that $\underline{k}$ for a quadrilateral element can be evaluated in terms of a local set of coordinates s-t, with limits from minus one to one within the element, and in terms of global nodal coordinates as given by Eq. (10.3.27). We repeat Eq. (10.3.27) here for convenience as

$$\underline{k} = \int_{-1}^{1}\int_{-1}^{1} \underline{B}^T(s,t)\underline{D}\underline{B}(s,t)|\underline{J}|h\,ds\,dt \tag{10.5.2}$$

where $|\underline{J}|$ is defined by Eq. (10.3.22) and $\underline{B}$ is defined by Eq. (10.3.18). In Eq. (10.5.2), each coefficient of the integrand $\underline{B}^T\underline{D}\underline{B}|\underline{J}|$ must be evaluated by numerical integration in the same manner as $f(s,t)$ was integrated in Eq. (10.4.9).

A flowchart to evaluate $\underline{k}$ of Eq. (10.5.2) for an element using four-point Gaussian quadrature is given in Figure 10–12. The four-point Gaussian quadrature rule is relatively easy to use. Also, it has been shown to yield good results [7]. In Figure 10–12, in explicit form for four-point Gaussian quadrature (now using the single summation notation with $i = 1, 2, 3, 4$), we have

$$\begin{aligned}\underline{k} = {} & \underline{B}^T(s_1,t_1)\underline{D}\underline{B}(s_1,t_1)|\underline{J}(s_1,t_1)|hW_1W_1 \\ & + \underline{B}^T(s_2,t_2)\underline{D}\underline{B}(s_2,t_2)|\underline{J}(s_2,t_2)|hW_2W_2 \\ & + \underline{B}^T(s_3,t_3)\underline{D}\underline{B}(s_3,t_3)|\underline{J}(s_3,t_3)|hW_3W_3 \\ & + \underline{B}^T(s_4,t_4)\underline{D}\underline{B}(s_4,t_4)|\underline{J}(s_4,t_4)|hW_4W_4\end{aligned} \tag{10.5.3}$$

where $s_1 = t_1 = -0.5773$, $s_2 = -0.5773$, $t_2 = 0.5773$, $s_3 = 0.5773$, $t_3 = -0.5773$, and $s_4 = t_4 = 0.5773$ as shown in Figure 10–9, and $W_1 = W_2 = W_3 = W_4 = 1.000$.

Example 10.4

Evaluate the stiffness matrix for the quadrilateral element shown in Figure 10–13 using the four-point Gaussian quadrature rule. Let $E = 30 \times 10^6$ psi and $\nu = 0.25$. The global coordinates are shown in inches. Assume $h = 1$ in.

Using Eq. (10.5.3), we evaluate the $\underline{k}$ matrix. Using the four-point rule, the four points are

$$\begin{aligned}(s_1, t_1) &= (-0.5773, -0.5773)\\ (s_2, t_2) &= (-0.5773, 0.5773)\\ (s_3, t_3) &= (0.5773, -0.5773)\\ (s_4, t_4) &= (0.5773, 0.5773)\end{aligned} \tag{10.5.4a}$$

with weights $W_1 = W_2 = W_3 = W_4 = 1.000$.

Therefore, by Eq. (10.5.3), we have

$$\begin{aligned}\underline{k} = {} & \underline{B}^T(-0.5773, -0.5773)\underline{D}\underline{B}(-0.5773, -0.5773)\\ & \times |\underline{J}(-0.5773, -0.5773)|(1)(1.000)(1.000)\\ & + \underline{B}^T(-0.5773, 0.5773)\underline{D}\underline{B}(-0.5773, 0.5773)\\ & \times |\underline{J}(-0.5773, 0.5773)|(1)(1.000)(1.000)\\ & + \underline{B}^T(0.5773, -0.5773)\underline{D}\underline{B}(0.5773, -0.5773)\\ & \times |\underline{J}(0.5773, -0.5773)|(1)(1.000)(1.000)\\ & + \underline{B}^T(0.5773, 0.5773)\underline{D}\underline{B}(0.5773, 0.5773)\\ & \times |\underline{J}(0.5773, 0.5773)|(1)(1.000)(1.000)\end{aligned} \tag{10.5.4b}$$

To evaluate $\underline{k}$, we first evaluate $|\underline{J}|$ at each Gauss point by using Eq. (10.3.22). For instance, one part of $|\underline{J}|$ is given by

$$\begin{aligned}&|\underline{J}(-0.5773, -0.5773)| = \tfrac{1}{8}[3 \quad 5 \quad 5 \quad 3]\\ &\times \begin{bmatrix} 0 & 1-(-0.5773) & -0.5773-(-0.5773) & -0.5773-1 \\ -0.5773-1 & 0 & -0.5773+1 & -0.5773-(-0.5773) \\ -0.5773-(-0.5773) & -0.5773-1 & 0 & -0.5773+1 \\ 1-(-0.5773) & -0.5773+(-0.5773) & -0.5773-1 & 0 \end{bmatrix}\\ &\times \begin{Bmatrix} 2 \\ 2 \\ 4 \\ 4 \end{Bmatrix} = 1.000\end{aligned} \tag{10.5.4c}$$

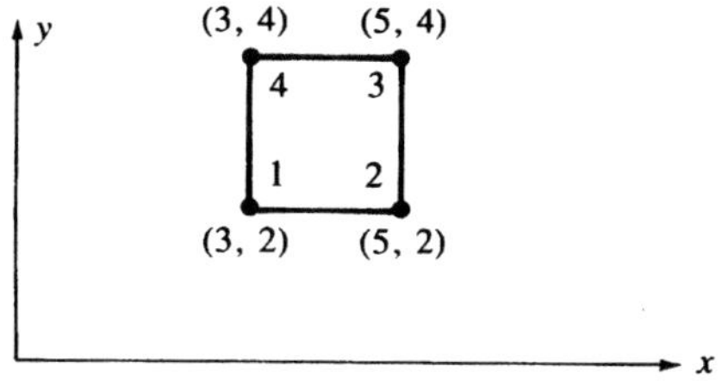

Figure 10–13 Quadrilateral element for stiffness evaluation

Similarly,

$$|\underline{J}(-0.5773, 0.5773)| = 1.000$$

$$|\underline{J}(0.5773, -0.5773)| = 1.000 \qquad (10.5.4d)$$

$$|\underline{J}(0.5773, 0.5773)| = 1.000$$

Even though $|\underline{J}| = 1$ in this example, in general, $|\underline{J}| \neq 1$ and varies in space.

Then, using Eqs. (10.3.18) and (10.3.19), we evaluate $\underline{B}$. For instance, one part of $\underline{B}$ is

$$\underline{B}(-0.5773, -0.5773) = \frac{1}{|\underline{J}(-0.5773, -0.5773)|}[\underline{B}_1 \quad \underline{B}_2 \quad \underline{B}_3 \quad \underline{B}_4]$$

where, by Eq. (10.3.19),

$$\underline{B}_1 = \begin{bmatrix} aN_{1,s} - bN_{1,t} & 0 \\ 0 & cN_{1,t} - dN_{1,s} \\ cN_{1,t} - dN_{1,s} & aN_{1,s} - bN_{1,t} \end{bmatrix} \qquad (10.5.4e)$$

and by Eqs. (10.3.20) and (10.3.21), $a, b, c, d, N_{1,s}$, and $N_{1,t}$ are evaluated. For instance,

$$\begin{aligned} a &= \tfrac{1}{4}[y_1(s-1) + y_2(-1-s) + y_3(1+s) + y_4(1-s)] \\ &= \tfrac{1}{4}\{2(-0.5773 - 1) + 2[-1 - (0.5773)]\} + 4[1 + (-0.5773)] + 4[1 - (-0.5773)] \\ &= 1.00 \end{aligned} \qquad (10.5.4f)$$

with similar computations used to obtain b, c, and d. Also,

$$\begin{aligned} N_{1,s} &= \tfrac{1}{4}(t-1) = \tfrac{1}{4}(-0.5773 - 1) = -0.3943 \\ N_{1,t} &= \tfrac{1}{4}(s-1) = \tfrac{1}{4}(-0.5773 - 1) = -0.3943 \end{aligned} \qquad (10.5.4g)$$

Similarly, $\underline{B}_2$, $\underline{B}_3$, and $\underline{B}_4$ must be evaluated like $\underline{B}_1$, at $(-0.5773, -0.5773)$. We then repeat the calculations to evaluate $\underline{B}$ at the other Gauss points [Eq. (10.5.4a)].

Using a computer program written specifically to evaluate $\underline{B}$ at each Gauss point and then $\underline{k}$, we obtain the final form of $\underline{B}(-0.5773, -0.5773)$ as

$$\underline{B}(-0.5773, -0.5773) =$$

$$\begin{bmatrix} -0.1057 & 0 & 0.1057 & 0 & 0 & -0.1057 & 0 & -0.3943 \\ -0.1057 & -0.1057 & -0.3943 & 0.1057 & 0.3943 & 0 & -0.3943 & 0 \\ 0 & 0.3943 & 0 & 0.1057 & 0.3943 & 0.3943 & 0.1057 & -0.3943 \end{bmatrix}$$

(10.5.4h)

with similar expressions for $\underline{B}(-0.5773, 0.5773)$, and so on.

From Eq. (6.1.8), the matrix $\underline{D}$ is

$$\underline{D} = \frac{E}{1-\nu^2}\begin{bmatrix} 1 & \nu & 0 \\ \nu & 1 & 0 \\ 0 & 0 & \dfrac{1-\nu}{2} \end{bmatrix} = \begin{bmatrix} 32 & 8 & 0 \\ 8 & 32 & 0 \\ 0 & 0 & 12 \end{bmatrix} \times 10^6 \text{ psi} \quad (10.5.4\text{i})$$

Finally, using Eq. (10.5.4b), the matrix $\underline{k}$ becomes

$$\underline{k} = 10^4 \begin{bmatrix} 1466 & 500 & -866 & -99 & -733 & -500 & 133 & 99 \\ 500 & 1466 & 99 & 133 & -500 & -733 & -99 & -866 \\ -866 & 99 & 1466 & -500 & 133 & -99 & -733 & 500 \\ -99 & 133 & -500 & 1466 & 99 & -866 & 500 & -733 \\ -733 & -500 & 133 & 99 & 1466 & 500 & -866 & -99 \\ -500 & -733 & -99 & -866 & 500 & 1466 & 99 & 133 \\ 133 & -99 & -733 & 500 & -866 & 99 & 1466 & -500 \\ 99 & -866 & 500 & -733 & -99 & 133 & -500 & 1466 \end{bmatrix}$$

(10.5.4j) ■

Evaluation of Element Stresses

The stresses $\underline{\sigma} = \underline{D}\underline{B}\underline{d}$ are not constant within the quadrilateral element. Because $\underline{B}$ is a function of s and t coordinates, $\underline{\sigma}$ is also a function of s and t. In practice, the stresses are evaluated at the same Gauss points used to evaluate the stiffness matrix $\underline{k}$. For a quadrilateral using 2×2 integration, we get four sets of stress data. To reduce the data, it is often practical to evaluate $\underline{\sigma}$ at $s = 0$, $t = 0$ instead. Another method mentioned in Section 7.4 is to evaluate the stresses in all elements at a shared (common) node and then use an average of these element nodal stresses to represent the stress at the node. Most computer programs use this method. Stress plots obtained in these programs are based on this average nodal stress method. Example 10.5 illustrates the use of Gaussian quadrature to evaluate the stress matrix at the $s = 0$, $t = 0$ location of the element.

Example 10.5

For the rectangular element shown in Figure 10–13, assume plane stress conditions with $E = 30 \times 10^6$ psi, $\nu = 0.3$, and displacements $u_1 = 0$, $v_1 = 0$, $u_2 = 0.001$ in., $v_2 = 0.0015$ in., $u_3 = 0.003$ in., $v_3 = 0.0016$ in., $u_4 = 0$, and $v_4 = 0$. Evaluate the stresses, σ_x, σ_y, and τ_{xy} at $s = 0$, $t = 0$.

Using Eqs. (10.3.18)–(10.3.20), we evaluate $\underline{B}$ at $s = 0$, $t = 0$.

$$\underline{B} = \frac{1}{|\underline{J}|}[\underline{B}_1 \quad \underline{B}_2 \quad \underline{B}_3 \quad \underline{B}_4] \quad (10.3.18) \text{ (repeated)}$$

$$\underline{B}(0,0) = \frac{1}{|\underline{J}(0,0)|}[\underline{B}_1(0,0) \quad \underline{B}_2(0,0) \quad \underline{B}_3(0,0) \quad \underline{B}_4(0,0)]$$

By Eq. (10.3.22), $|\underline{J}|$ is

$$|\underline{J}(0,0)| = \tfrac{1}{8}[3 \quad 5 \quad 5 \quad 3]\begin{bmatrix} 0 & 1 & 0 & -1 \\ -1 & 0 & 1 & 0 \\ 0 & -1 & 0 & 1 \\ 1 & 0 & -1 & 0 \end{bmatrix}\begin{Bmatrix} 2 \\ 2 \\ 4 \\ 4 \end{Bmatrix}$$

$$= \tfrac{1}{8}[-2 \quad -2 \quad 2 \quad 2]\begin{Bmatrix} 2 \\ 2 \\ 4 \\ 4 \end{Bmatrix}$$

$$|\underline{J}(0,0)| = 1 \tag{10.5.5a}$$

By Eq. (10.3.19), we have

$$\underline{B}_i = \begin{bmatrix} aN_{i,s} - bN_{i,t} & 0 \\ 0 & cN_{i,t} - dN_{i,s} \\ cN_{i,t} - dN_{i,s} & aN_{i,s} - bN_{i,t} \end{bmatrix} \tag{10.5.5b}$$

By Eq. (10.3.20), we obtain

$$a = 1 \qquad b = 0 \qquad c = 1 \qquad d = 0$$

Differentiating the shape functions in Eq. (10.3.5) with respect to s and t and then evaluating at $s = 0$, $t = 0$, we obtain

$$\begin{array}{llll} N_{1,s} = -\tfrac{1}{4} & N_{1,t} = -\tfrac{1}{4} & N_{2,s} = \tfrac{1}{4} & N_{2,t} = -\tfrac{1}{4} \\ N_{3,s} = \tfrac{1}{4} & N_{3,t} = \tfrac{1}{4} & N_{4,s} = -\tfrac{1}{4} & N_{4,t} = \tfrac{1}{4} \end{array} \tag{10.5.5c}$$

Therefore, substituting Eqs. (10.5.5c) into Eq. (10.5.5b), we obtain

$$\underline{B}_1 = \begin{bmatrix} -\tfrac{1}{4} & 0 \\ 0 & -\tfrac{1}{4} \\ -\tfrac{1}{4} & -\tfrac{1}{4} \end{bmatrix} \quad \underline{B}_2 = \begin{bmatrix} \tfrac{1}{4} & 0 \\ 0 & -\tfrac{1}{4} \\ -\tfrac{1}{4} & \tfrac{1}{4} \end{bmatrix} \quad \underline{B}_3 = \begin{bmatrix} \tfrac{1}{4} & 0 \\ 0 & \tfrac{1}{4} \\ \tfrac{1}{4} & \tfrac{1}{4} \end{bmatrix} \quad \underline{B}_4 = \begin{bmatrix} -\tfrac{1}{4} & 0 \\ 0 & \tfrac{1}{4} \\ \tfrac{1}{4} & -\tfrac{1}{4} \end{bmatrix} \tag{10.5.5d}$$

The element stress matrix $\underline{\sigma}$ is then obtained by substituting Eqs. (10.5.5a) for $|\underline{J}| = 1$ and (10.5.5d) into Eq. (10.3.18) for $\underline{B}$ and the plane stress $\underline{D}$ matrix from Eq. (6.1.8) into the definition for $\underline{\sigma}$ as

$$\underline{\sigma} = \underline{D}\underline{B}\underline{d} = (30)\frac{10^6\begin{bmatrix} 1 & 0.3 & 0 \\ 0.3 & 1 & 0 \\ 0 & 0 & 0.35 \end{bmatrix}}{1 - 0.09}$$

(continued)

$$\times \begin{bmatrix} -0.25 & 0 & 0.25 & 0 & 0.25 & 0 & -0.25 & 0 \\ 0 & -0.25 & 0 & -0.25 & 0 & 0.25 & 0 & 0.25 \\ -0.25 & -0.25 & -0.25 & 0.25 & 0.25 & 0.25 & 0.25 & -0.25 \end{bmatrix} \begin{Bmatrix} 0 \\ 0 \\ 0.001 \\ 0.0015 \\ 0.003 \\ 0.0016 \\ 0 \\ 0 \end{Bmatrix}$$

$$\underline{\sigma} = \begin{Bmatrix} 3.321 \cdot 10^4 \\ 1.071 \cdot 10^4 \\ 1.471 \cdot 10^4 \end{Bmatrix} \text{ psi} \qquad \blacksquare$$

10.6 Higher-Order Shape Functions

In general, higher-order element shape functions can be developed by adding additional nodes to the sides of the linear element. These elements result in higher-order strain variations within each element, and convergence to the exact solution thus occurs at a faster rate using fewer elements. (However, a trade-off exists because a more complicated element takes up so much computation time that even with few elements in the model, the computation time can become larger than for the simple linear element model.) Another advantage of the use of higher-order elements is that curved boundaries of irregularly shaped bodies can be approximated more closely than by the use of simple straight-sided linear elements.

To illustrate the concept of higher order elements, we will begin with the three-noded linear strain quadratic displacement (and quadratic shape functions) shown in Figure 10–14. Figure 10–14 shows a quadratic isoparametric bar element (also called a linear strain bar) with three coordinates of nodes, x_1, x_2, and x_3, in the global coordinates.

Example 10.6

For the three-noded linear strain bar isoparametric element shown in Figure 10–14, determine (a) the shape functions, N_1, N_2, and N_3, and (b) the strain/displacement matrix [B]. Assume the general axial displacement function to be a quadratic taken as $u = a_1 + a_2 s + a_3 s^2$.

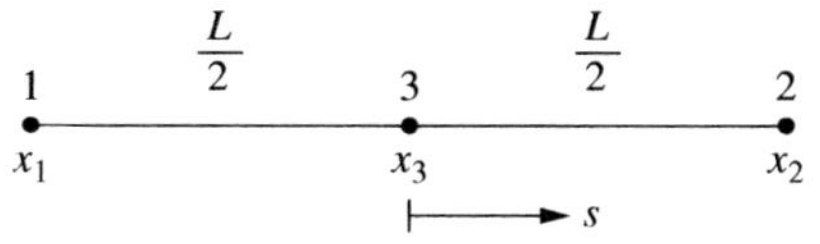

Figure 10–14 Three-noded linear strain bar element

(a) As we are formulating shape functions for an isoparametric element, we assume the following axial coordinate function for x as

$$x = a_1 + a_2 s + a_3 s^2 \tag{10.6.1}$$

Evaluating the a_i's in terms of the nodal coordinates, we obtain

$$\begin{aligned} x(-1) &= a_1 - a_2 + a_3 = x_1 \quad &\text{or} \quad & x_1 = a_1 - a_2 + a_3 \\ x(0) &= a_1 = x_3 \quad &\text{or} \quad & x_3 = a_1 \\ x(1) &= a_1 + a_2 + a_3 = x_2 \quad &\text{or} \quad & x_2 = a_1 - a_2 + a_3 \end{aligned} \tag{10.6.2}$$

Substituting $a_1 = x_3$ from the second of Eqs. (10.6.2), into the first and third of Eqs. (10.6.2), we obtain a_2 and a_3 as follows:

$$\begin{aligned} x_1 &= x_3 - a_2 + a_3 \\ x_2 &= x_3 + a_2 + a_3 \end{aligned} \tag{10.6.3}$$

Adding Eqs. (10.6.3) together and solving for a_3 gives the following:

$$a_3 = (x_1 + x_2 - 2x_3)/2 \tag{10.6.4}$$

$$x_1 = x_3 - a_2 + ((x_1 + x_2 - 2x_3)/2) \tag{10.6.5}$$

$$a_2 = x_3 - x_1 + ((x_1 + x_2 - 2x_3)/2) = (x_2 - x_1)/2$$

Substituting the values for a_1, a_2 and a_3 into the general equation for x given by Eq. (10.6.1), we obtain

$$x = a_1 + a_2 s + a_3 s^2 = x_3 + \frac{x_2 - x_1}{2} s + \frac{x_1 + x_2 - 2x_3}{2} s^2 \tag{10.6.6}$$

Combining like terms in x_1, x_2, and x_3, from Eq. (10.6.6), we obtain the final form of x as:

$$x = \left(\frac{s(s-1)}{2}\right) x_1 + \frac{s(s+1)}{2} x_2 + (1 - s^2) x_3 \tag{10.6.7}$$

Recall that the function x can be expressed in terms of the shape function matrix and the axial coordinates, we have from Eq. (10.6.7)

$$\{x\} = [N_1 \quad N_2 \quad N_3] \begin{Bmatrix} x_1 \\ x_2 \\ x_3 \end{Bmatrix} = \left[\left(\frac{s(s-1)}{2}\right) \quad \frac{s(s+1)}{2} \quad (1 - s^2)\right] \begin{Bmatrix} x_1 \\ x_2 \\ x_3 \end{Bmatrix} \tag{10.6.8}$$

Therefore the shape functions are

$$N_1 = \frac{s(s-1)}{2} \qquad N_2 = \frac{s(s+1)}{2} \qquad N_3 = (1 - s^2) \tag{10.6.9}$$

(b) We now determine the strain/displacement matrix $[B]$ as follows:
From our basic definition of axial strain we have

$$\{\varepsilon_x\} = \frac{du}{dx} = \frac{du}{ds}\frac{ds}{dx} = [B] \begin{Bmatrix} u_1 \\ u_2 \\ u_3 \end{Bmatrix} \tag{10.6.10}$$

Using an isoparametric formulation means the displacement function is of the same form as the axial coordinate function. Therefore, using Eq. (10.6.6), we have

$$u = u_3 + \frac{u_2}{2}s - \frac{u_1}{2}s + \frac{u_1}{2}s^2 + \frac{u_2}{2}s^2 - \frac{2u_3}{2}s^2 \tag{10.6.11}$$

Differentiating u with respect to s, we obtain

$$\frac{du}{ds} = \frac{u_2}{2} - \frac{u_1}{2} + u_1 s + u_2 s - 2u_3 s = \left(s - \frac{1}{2}\right)u_1 + \left(s + \frac{1}{2}\right)u_2 + (-2s)u_3 \tag{10.6.12}$$

We have previously proven that $dx/ds = L/2 = [J]$ (see Eq. (10.1.9b). This relationship holds for the higher-order one-dimensional elements as well as for the two-noded constant strain bar element. Using this relationship and Eq. (10.6.12) in Eq. (10.6.10), we obtain

$$\begin{aligned}\frac{du}{dx} = \frac{du}{ds}\frac{ds}{dx} &= \left(\frac{2}{L}\right)\left(\left(s - \frac{1}{2}\right)u_1 + \left(s + \frac{1}{2}\right)u_2 + (-2s)u_3\right) \\ &= \left(\frac{2s-1}{L}\right)u_1 + \left(\frac{2s+1}{L}\right)u_2 + \left(\frac{-4s}{L}\right)u_3\end{aligned} \tag{10.6.13}$$

In matrix form, Eq. (10.6.13) becomes

$$\frac{du}{dx} = \begin{bmatrix} \frac{2s-1}{L} & \frac{2s+1}{L} & \frac{-4s}{L} \end{bmatrix} \begin{Bmatrix} u_1 \\ u_2 \\ u_3 \end{Bmatrix} \tag{10.6.14}$$

As Eq. (10.6.14) represents the axial strain, we have

$$\{\varepsilon_x\} = \frac{du}{dx} = \begin{bmatrix} \frac{2s-1}{L} & \frac{2s+1}{L} & \frac{-4s}{L} \end{bmatrix} \begin{Bmatrix} u_1 \\ u_2 \\ u_3 \end{Bmatrix} = [B]\begin{Bmatrix} u_1 \\ u_2 \\ u_3 \end{Bmatrix} \tag{10.6.15}$$

Therefore the $[B]$ is given by

$$[B] = \begin{bmatrix} \frac{2s-1}{L} & \frac{2s+1}{L} & \frac{-4s}{L} \end{bmatrix} \tag{10.6.16}$$

■

Example 10.7

For the three-noded bar element shown previously in Figure 10–14, evaluate the stiffness matrix analytically. Use the $[B]$ from Example 10.6.

From example 10.6, Eq. (10.6.16), we have

$$[B] = \begin{bmatrix} \frac{2s-1}{L} & \frac{2s+1}{L} & \frac{-4s}{L} \end{bmatrix}, \quad [J] = \frac{L}{2} \quad \text{(see Eq. (10.1.9b))} \tag{10.6.17}$$

Substituting the expression for $[B]$ into Eq. (10.1.15) for the stiffness matrix, we obtain

$$[k] = \frac{L}{2}\int_{-1}^{1}[B]^T E[B]A\,ds = \frac{AEL}{2}\int_{-1}^{1}\begin{bmatrix} \frac{(2s-1)^2}{L^2} & \frac{(2s-1)(2s+1)}{L^2} & \frac{(2s-1)(-4s)}{L^2} \\ \frac{(2s+1)(2s-1)}{L^2} & \frac{(2s+1)^2}{L^2} & \frac{(2s+1)(-4s)}{L^2} \\ \frac{(-4s)(2s-1)}{L^2} & \frac{(-4s)(2s+1)}{L^2} & \frac{(-4s)^2}{L^2} \end{bmatrix} ds \tag{10.6.18}$$

Simplifying the terms in Eq. (10.6.18) for easier integration, we have

$$[k] = \frac{AE}{2L}\int_{-1}^{1}\begin{bmatrix} 4s^2-4s+1 & 4s^2-1 & -8s^2+4s \\ 4s^2-1 & 4s^2+4s+1 & -8s^2-4s \\ -8s^2+4s & -8s^2-4s & 16s^2 \end{bmatrix} ds \tag{10.6.19}$$

Upon explicit integration of Eq. (10.6.19), we obtain

$$[k] = \frac{AE}{2L}\left.\begin{bmatrix} \frac{4}{3}s^3-2s^2+s & \frac{4}{3}s^3-s & -\frac{8}{3}s^3+2s^2 \\ \frac{4}{3}s^3-s & \frac{4}{3}s^3+2s^2+s & -\frac{8}{3}s^3-2s^2 \\ -\frac{8}{3}s^3+2s^2 & -\frac{8}{3}s^3-2s^2 & \frac{16}{3}s^3 \end{bmatrix}\right|_{-1}^{1} \tag{10.6.20}$$

Evaluating Eq. (10.6.20) at the limits 1 and -1, we have

$$[k] = \frac{AE}{2L}\begin{bmatrix} \frac{4}{3}-2+1 & \frac{4}{3}-1 & -\frac{8}{3}+2 \\ \frac{4}{3}-1 & \frac{4}{3}+2+1 & -\frac{8}{3}-2 \\ -\frac{8}{3}+2 & -\frac{8}{3}-2 & \frac{16}{3} \end{bmatrix} - \begin{bmatrix} -\frac{4}{3}-2-1 & -\frac{4}{3}+1 & \frac{8}{3}+2 \\ -\frac{4}{3}+1 & -\frac{4}{3}+2-1 & \frac{8}{3}-2 \\ \frac{8}{3}+2 & \frac{8}{3}-2 & -\frac{16}{3} \end{bmatrix} \tag{10.6.21}$$

Simplifying Eq. (10.6.21), we obtain the final stiffness matrix as

$$[k] = \frac{AE}{2L}\begin{bmatrix} 4.67 & 0.667 & -5.33 \\ 0.667 & 4.67 & -5.33 \\ -5.33 & -5.33 & 10.67 \end{bmatrix} \tag{10.6.22}$$

■

Example 10.8

We now illustrate how to evaluate the stiffness matrix for the three-noded bar element shown in Figure 10–15 by using two-point Guassian quadrature. We can then compare this result to that obtained by the explicit integration performed in Example 10.7.

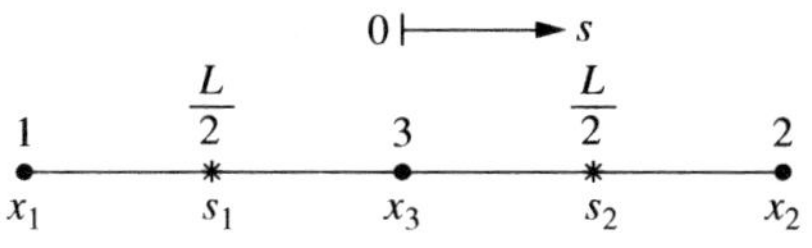

Figure 10–15 Three-noded bar with two Gauss points

Starting with Eq. (10.6.18), we have for the stiffness matrix

$$[k] = \frac{L}{2}\int_{-1}^{1}[B]^T E[B]A\,ds = \frac{AEL}{2}\int_{-1}^{1}\begin{bmatrix} \frac{(2s-1)^2}{L^2} & \frac{(2s-1)(2s+1)}{L^2} & \frac{(2s-1)(-4s)}{L^2} \\ \frac{(2s+1)(2s-1)}{L^2} & \frac{(2s+1)^2}{L^2} & \frac{(2s+1)(-4s)}{L^2} \\ \frac{(-4s)(2s-1)}{L^2} & \frac{(-4s)(2s+1)}{L^2} & \frac{(-4s)^2}{L^2} \end{bmatrix} ds \tag{10.6.23}$$

Using two-point Guassian quadrature, we evaluate the stiffness matrix at the two points shown in Figure 10–15 given by

$$s_1 = -0.57735, \quad s_2 = 0.57735 \tag{10.6.24}$$

with weights given by

$$W_1 = 1, \quad W_2 = 1 \tag{10.6.25}$$

We then evaluate each term in the integrand of Eq. (10.6.23) at each Gauss point and multiply each term by its weight (here each weight is 1). We then add those Gauss point evaluations together to obtain the final term for each element of the stiffness matrix. For two-point evaluation, there will be two terms added together to obtain each element of the stiffness matrix. We proceed to evaluate the stiffness matrix term by term as follows:

The one–one element:

$$\sum_{i=1}^{2} W_i(2s_i - 1)^2 = (1)[2(-0.57735) - 1]^2 + (1)[2(0.57735) - 1]^2 = 4.6667$$

The one–two element:

$$\sum_{i=1}^{2} W_i(2s_i - 1)(2s_i + 1) = (1)[(2)(-0.57735) - 1][(2)(-0.57735) + 1]$$
$$+ (1)[(2)(0.57735) - 1][(2)(0.57735) + 1] = 0.6667$$

The one–three element:

$$\sum_{i=1}^{2} W_i(-4s_i(2s_i - 1)) = (1)(-4)(-0.57735)[(2)(-0.57735) - 1]$$
$$+ (1)(-4)(0.57735)[(2)(0.57735) - 1] = -5.3333$$

The two–two element:

$$\sum_{i=1}^{2} W_i(2s_i+1)^2 = (1)[(2)(-0.57735)+1]^2 + (1)[(2)(0.57735)+1]^2 = 4.6667$$

The two–three element:

$$\sum_{i=1}^{2} W_i[-4s_i(2s_i+1)] = (1)(-4)(-0.57735)[(2)(-0.57735)+1] + (1)(-4)(0.57735)[(2)(0.57735)+1] = -5.3333$$

The three–three element:

$$\sum_{i=1}^{2} W_i(16s_i^2) = (1)(16)(-0.57735)^2 + (1)(16)(0.57735)^2 = 10.6667$$

By symmetry, the two–one element equals the one–two element, etc. Therefore, from the evaluations of the terms above, the final stiffness matrix is

$$\underline{k} = \frac{AE}{2L}\begin{bmatrix} 4.67 & 0.667 & -5.33 \\ 0.667 & 4.67 & -5.33 \\ -5.33 & -5.33 & 10.67 \end{bmatrix} \tag{10.6.26}$$

Equation (10.6.26) is identical to Eq. (10.6.22) obtained analytically by direct explicit integration of each term in the stiffness matrix. ■

To further illustrate the concept of higher-order elements, we will consider the quadratic and cubic element shape functions as described in Reference [3]. Figure 10–16 shows a quadratic isoparametric element with four corner nodes and four additional midside nodes. This eight-noded element is often called a "Q8" element.

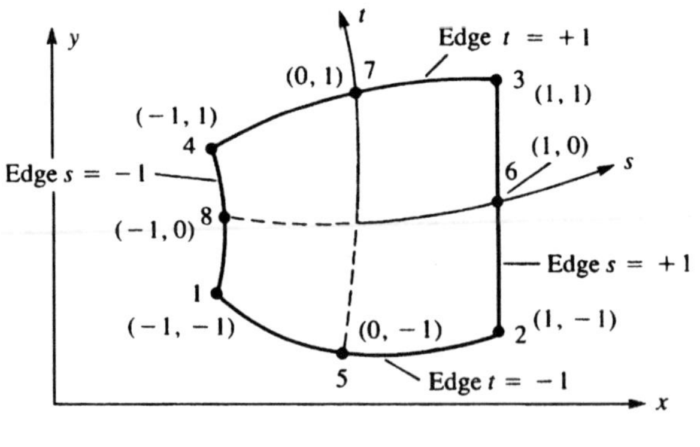

Figure 10–16 Quadratic isoparametric element

The shape functions of the quadratic element are based on the incomplete cubic polynomial such that coordinates x and y are

$$
\begin{aligned}
x &= a_1 + a_2 s + a_3 t + a_4 st + a_5 s^2 + a_6 t^2 + a_7 s^2 t + a_8 st^2 \\
y &= a_9 + a_{10} s + a_{11} t + a_{12} st + a_{13} s^2 + a_{14} t^2 + a_{15} s^2 t + a_{16} st^2
\end{aligned}
\qquad (10.6.27)
$$

These functions have been chosen so that the number of generalized degrees of freedom (2 per node times 8 nodes equals 16) are identical to the total number of a_i's. The literature also refers to this eight-noded element as a "serendipity" element as it is based on an incomplete cubic, but it yields good results in such cases as beam bending. We are also reminded that because we are considering an isoparametric formulation, displacements u and v are of identical form as x and y, respectively, in Eq. (10.6.27).

To describe the shape functions, two forms are required—one for corner nodes and one for midside nodes, as given in Reference [3]. For the corner nodes ($i = 1, 2, 3, 4$),

$$
\begin{aligned}
N_1 &= \tfrac{1}{4}(1-s)(1-t)(-s-t-1) \\
N_2 &= \tfrac{1}{4}(1+s)(1-t)(s-t-1) \\
N_3 &= \tfrac{1}{4}(1+s)(1+t)(s+t-1) \\
N_4 &= \tfrac{1}{4}(1-s)(1+t)(-s+t-1)
\end{aligned}
\qquad (10.6.28)
$$

or, in compact index notation, we express Eqs. (10.6.28) as

$$
N_i = \tfrac{1}{4}(1+ss_i)(1+tt_i)(ss_i+tt_i-1) \qquad (10.6.29)
$$

where i is the number of the shape function and

$$
\begin{aligned}
s_i &= -1, 1, 1, -1 \qquad (i = 1, 2, 3, 4) \\
t_i &= -1, -1, 1, 1 \qquad (i = 1, 2, 3, 4)
\end{aligned}
\qquad (10.6.30)
$$

For the midside nodes ($i = 5, 6, 7, 8$),

$$
\begin{aligned}
N_5 &= \tfrac{1}{2}(1-t)(1+s)(1-s) \\
N_6 &= \tfrac{1}{2}(1+s)(1+t)(1-t) \\
N_7 &= \tfrac{1}{2}(1+t)(1+s)(1-s) \\
N_8 &= \tfrac{1}{2}(1-s)(1+t)(1-t)
\end{aligned}
\qquad (10.6.31)
$$

or, in index notation,

$$
\begin{aligned}
N_i &= \tfrac{1}{2}(1-s^2)(1+tt_i) \qquad t_i = -1, 1 \qquad (i = 5, 7) \\
N_i &= \tfrac{1}{2}(1+ss_i)(1-t^2) \qquad s_i = 1, -1 \qquad (i = 6, 8)
\end{aligned}
\qquad (10.6.32)
$$

We can observe from Eqs. (10.6.28) and (10.6.31) that an edge (and displacement) can vary with s^2 (along t constant) or with t^2 (along s constant). Furthermore, $N_i = 1$ at node i and $N_i = 0$ at the other nodes, as it must be according to our usual definition of shape functions.

The displacement functions are given by

$$\begin{Bmatrix} u \\ v \end{Bmatrix} = \begin{bmatrix} N_1 & 0 & N_2 & 0 & N_3 & 0 & N_4 & 0 & N_5 & 0 & N_6 & 0 & N_7 & 0 & N_8 & 0 \\ 0 & N_1 & 0 & N_2 & 0 & N_3 & 0 & N_4 & 0 & N_5 & 0 & N_6 & 0 & N_7 & 0 & N_8 \end{bmatrix} \times \begin{Bmatrix} u_1 \\ v_1 \\ u_2 \\ v_2 \\ \vdots \\ v_8 \end{Bmatrix} \tag{10.6.33}$$

and the strain matrix is now

$$\underline{\varepsilon} = \underline{D}'\underline{N}\underline{d}$$

with

$$\underline{B} = \underline{D}'\underline{N}$$

We can develop the matrix $\underline{B}$ using Eq. (10.3.17) with $\underline{D}'$ from Eq. (10.3.16) and with $\underline{N}$ now the 2×16 matrix given in Eq. (10.6.33), where the N's are defined in explicit form by Eq. (10.6.28) and (10.6.31).

To evaluate the matrix $\underline{B}$ and the matrix $\underline{k}$ for the eight-noded quadratic isoparametric element, we now use the nine-point Gauss rule (often described as a 3×3 rule). Results using 2×2 and 3×3 rules have shown significant differences, and the 3×3 rule is recommended by Bathe and Wilson [7]. Table 10–1 indicates the locations of points and the associated weights. The 3×3 rule is shown in Figure 10–17.

By adding a ninth node at $s = 0$, $t = 0$ in Figure 10–16, we can create an element called a "Q9." This is an internal node that is not connected to any other nodes. We then add the $a_{17}s^2t^2$ and $a_{18}s^2t^2$ terms to x and y, respectively in Eq. (10.6.27) and to u and y. The element is then called a Lagrange element as the shape functions can be derived using Lagrange interpolation formulas. For more on this subject consult [8].

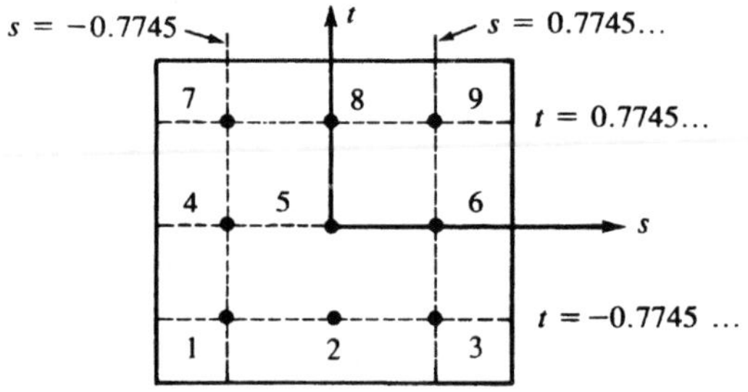

Figure 10–17 3×3 rule in two dimensions

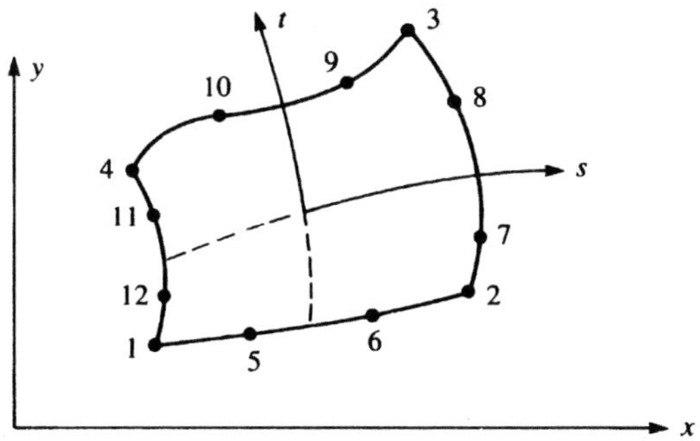

Figure 10–18 Cubic isoparametric element

The cubic element in Figure 10–18 has four corner nodes and additional nodes taken to be at one-third and two-thirds of the length along each side. The shape functions of the cubic element (as derived in Reference [3]) are based on the incomplete quartic polynomial such that

$$x = a_1 + a_2 s + a_3 t + a_4 s^2 + a_5 st + a_6 t^2 + a_7 s^2 t + a_8 st^2 + a_9 s^3 + a_{10} t^3 + a_{11} s^3 t + a_{12} st^3 \tag{10.6.34}$$

with a similar polynomial for y. For the corner nodes ($i = 1, 2, 3, 4$),

$$N_i = \tfrac{1}{32}(1 + ss_i)(1 + tt_i)[9(s^2 + t^2) - 10] \tag{10.6.35}$$

with s_i and t_i given by Eqs. (10.6.30). For the nodes on sides $s = \pm 1$ ($i = 7, 8, 11, 12$),

$$N_i = \tfrac{9}{32}(1 + ss_i)(1 + 9tt_i)(1 - t^2) \tag{10.6.36}$$

with $s_i = \pm 1$ and $t_i = \pm\frac{1}{3}$. For the nodes on sides $t = \pm 1$ ($i = 5, 6, 9, 10$),

$$N_i = \tfrac{9}{32}(1 + tt_i)(1 + 9ss_i)(1 - s^2) \tag{10.6.37}$$

with $t_i = \pm 1$ and $s_i = \pm\frac{1}{3}$.

Having the shape functions for the quadratic element given by Eqs. (10.6.28) and (10.6.31) or for the cubic element given by Eqs. (10.6.35)–(10.6.37), we can again use Eq. (10.3.17) to obtain $\underline{B}$ and then Eq. (10.3.27) to set up $\underline{k}$ for numerical integration for the plane element. The cubic element requires a 3×3 rule (nine points) to evaluate the matrix $\underline{k}$ exactly. We then conclude that what is really desired is a library of shape functions that can be used in the general equations developed for stiffness matrices, distributed load, and body force and can be applied not only to stress analysis but to nonstructural problems as well.

Since in this discussion the element shape functions N_i relating x and y to nodal coordinates x_i and y_i are of the same form as the shape functions relating u and v to nodal displacements u_i and v_i, this is said to be an *isoparametric formulation*. For instance, for the linear element $x = \sum_{i=1}^{4} N_i x_i$ and the displacement function $u = \sum_{i=1}^{4} N_i u_i$, use the same shape functions N_i given by Eq. (10.3.5). If instead the shape functions for the coordinates are of lower order (say, linear for x) than the shape functions used for displacements (say, quadratic for u), this is called a *subparametric formulation*.

Finally, referring to Figure 10–18, note that an element can have a linear shape along, say, one edge (1–2), a quadratic along, say, two edges (2–3 and 1–4), and a cubic along the other edge (3–4). Hence, the simple linear element can be mixed with different higher-order elements in regions of a model where rapid stress variation is expected. The advantage of the use of higher-order elements is further illustrated in Reference [3].

References

[1] Irons, B. M., "Engineering Applications of Numerical Integration in Stiffness Methods," *Journal of the American Institute of Aeronautics and Astronautics*, Vol. 4, No. 11, pp. 2035–2037, 1966.

[2] Stroud, A. H., and Secrest, D., *Gaussian Quadrature Formulas*, Prentice-Hall, Englewood Cliffs, NJ, 1966.

[3] Ergatoudis, I., Irons, B. M., and Zienkiewicz, O. C., "Curved Isoparametric, Quadrilateral Elements for Finite Element Analysis," *International Journal of Solids and Structures*, Vol. 4, pp. 31–42, 1968.

[4] Zienkiewicz, O. C., *The Finite Element Method*, 3rd ed., McGraw-Hill, London, 1977.

[5] Thomas, B. G., and Finney, R. L., *Calculus and Analytic Geometry*, Addison-Wesley, Reading, MA, 1984.

[6] Gallagher, R., *Finite Element Analysis Fundamentals*, Prentice-Hall, Englewood Cliffs, NJ, 1975.

[7] Bathe, K. J., and Wilson, E. L., *Numerical Methods in Finite Element Analysis*, Prentice-Hall, Englewood Cliffs, NJ, 1976.

[8] Cook, R. D., Malkus, D. S., Plesha, M. E., and Witt, R. J., *Concepts and Applications of Finite Element Analysis*, 4th ed., Wiley, New York, 2002.

[9] Bathe, Klaus-Jurgen, *Finite Element Procedures in Engineering Analysis*, Prentice-Hall, Englewood Cliffs, New Jersey, 1982.

Problems

10.1 For the three-noded linear strain bar with three coordinates of nodes x_1, x_2, and x_3, shown in Figure P10–1 in the global-coordinate system show that the Jacobian determinate is $|\underline{J}| = L/2$.

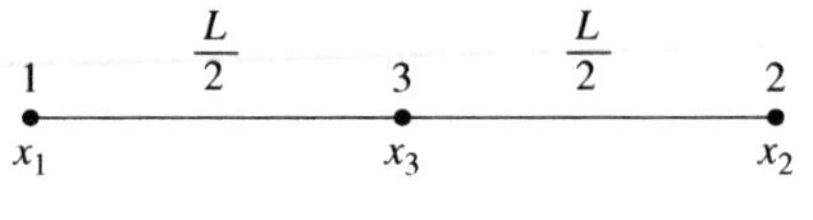

Figure P10–1

10.2 For the two-noded one-dimensional isoparametric element shown in Figure P10–2 (a) and (b), with shape functions given by Eq. (10.1.5), determine (a) intrinsic coordinate s

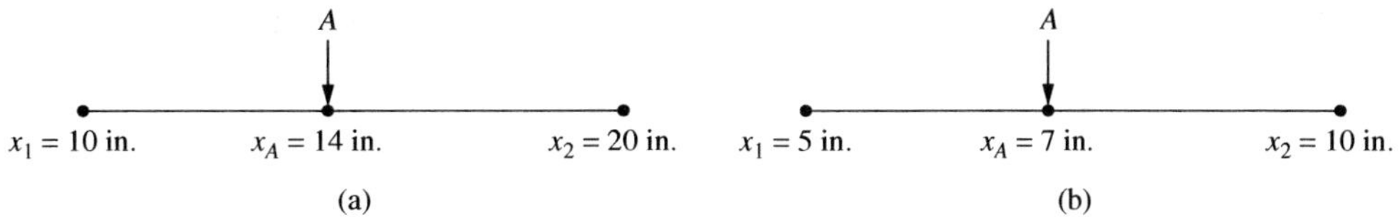

Figure P10–2

at point A and (b) shape functions N_1 and N_2 at point A. If the displacements at nodes one and two are respectively, $u_1 = 0.006$ in. and $u_2 = -0.006$ in., determine (c) the value of the displacement at point A and (d) the strain in the element.

10.3 Answer the same questions as posed in problem 10.2 with the data listed under the Figure P10–3.

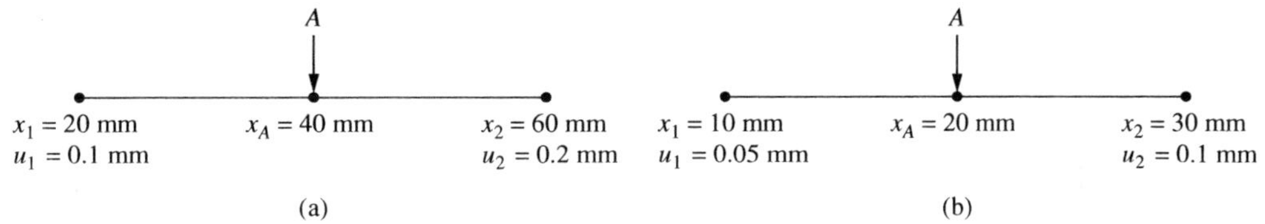

Figure P10–3

10.4 For the four-noded bar element in Figure P10–4, show that the Jacobian determinate is $|\underline{J}| = L/2$. Also determine the shape functions $N_1 - N_4$ and the strain/displacement matrix $\underline{B}$. Assume $u = a_1 + a_2 s + a_3 s^2 + a_4 s^3$.

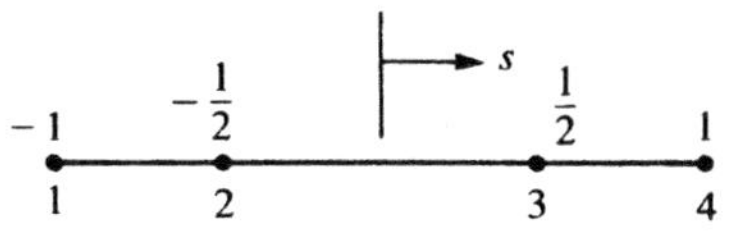

Figure P10–4

10.5 Using the three-noded bar element shown in Figure P10–5 (a) and (b), with shape functions given by Eq. (10.6.9), determine (a) the intrinsic coordinate s at point A and (b) the shape functions, N_1, N_2, and N_3 at A. For the displacements of the nodes shown in Figure P10–5, determine (c) the displacement at A and (d) the axial strain expression in the element.

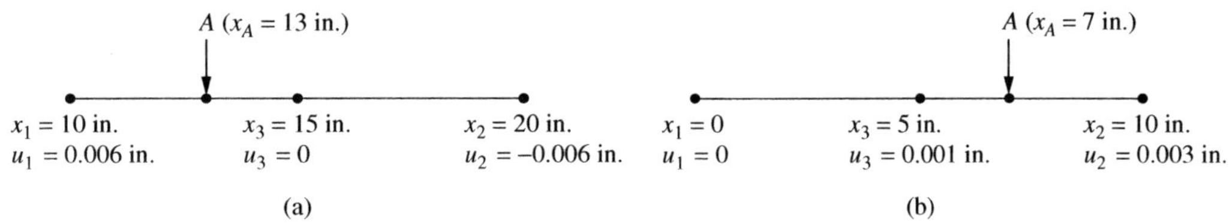

Figure P10–5

10.6 Using the three-noded bar element shown in Figure P10–5 (a) and (b), with shape functions given by Eq. (10.6.9), determine (a) the intrinsic coordinate s at point A and (b) the shape functions, N_1, N_2, and N_3 at point A. For the displacements of the nodes shown in Figure P10–6, determine (c) the displacement at A and (d) the axial strain expression in the element.

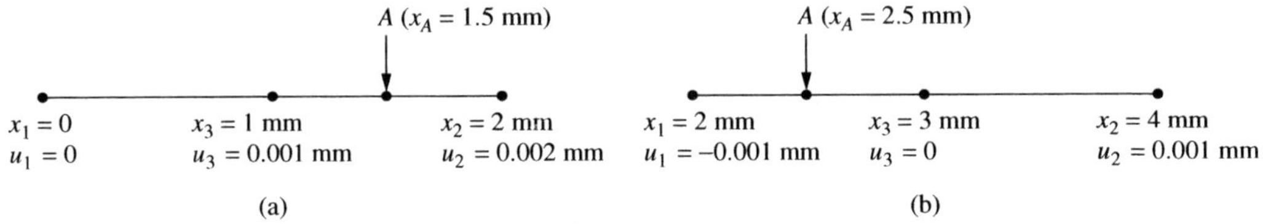

Figure P10–6

10.7 For the bar subjected to the linearly varying axial line load shown in Figure P10–7, use the linear strain (three-noded element) with two elements in the model, to determine the nodal displacements and nodal stresses. Compare your answer with that in Figure 3–31 and Eqs. (3.11.6) and (3.11.7). Let $A = 2$ in.^{2} and $E = 30 \times 10^6$ psi. Hint: Use Eq. (10.6.22) for the element stiffness matrix.

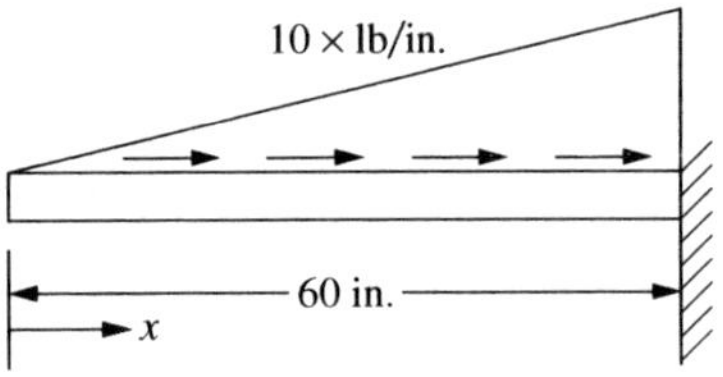

Figure P10–7

10.8 Use the three-noded bar element and find the axial displacement at the end of the rod shown in Figure P10–8. Determine the stress at $x = 0$, $x = L/2$ and $x = L$. Let $A = 2 \times 10^{-4}$ m^2, $E = 205$ GPa, and $L = 4$ m. Hint: use Eq. (10.6.22) for the element stiffness matrix.

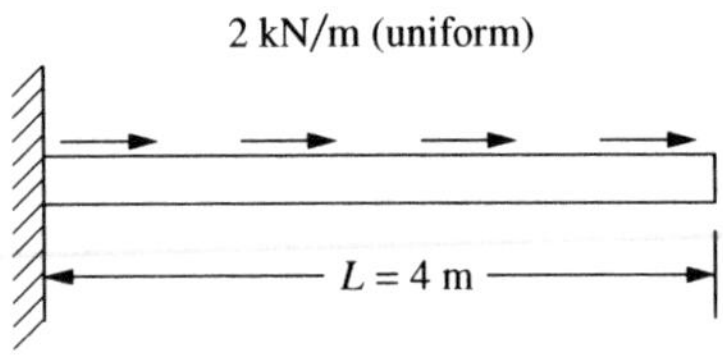

Figure P10–8

10.9 Show that the sum $N_1 + N_2 + N_3 + N_4$ is equal to 1 anywhere on a rectangular element, where N_1 through N_4 are defined by Eqs. (10.2.5).

10.10 For the rectangular element of Figure 10–3 on page 450, the nodal displacements are given by

$$u_1 = 0 \qquad v_1 = 0 \qquad u_2 = 0.005 \text{ in.}$$
$$v_2 = 0.0025 \text{ in.} \qquad u_3 = 0.0025 \text{ in.} \qquad v_3 = -0.0025 \text{ in.}$$
$$u_4 = 0 \qquad v_4 = 0$$

For $b = 2$ in., $h = 1$ in., $E = 30 \times 10^6$ psi, and $\nu = 0.3$, determine the element strains and stresses at the centroid of the element and at the corner nodes.

10.11 Derive $|\underline{J}|$ given by Eq. (10.3.22) for a four-noded isoparametric quadrilateral element.

10.12 Show that for the quadrilateral element described in Section 10.3, $[J]$ can be expressed as

$$[J] = \begin{bmatrix} N_{1,s} & N_{2,s} & N_{3,s} & N_{4,s} \\ N_{1,t} & N_{2,t} & N_{3,t} & N_{4,t} \end{bmatrix} \begin{bmatrix} x_1 & y_1 \\ x_2 & y_2 \\ x_3 & y_3 \\ x_4 & y_4 \end{bmatrix}$$

10.13 Derive Eq. (10.3.18) with B_i given by Eq. (10.3.19) by substituting Eq. (10.3.16) for $\underline{D}'$ and Eqs. (10.3.5) for the shape functions into Eq. (10.3.17).

10.14 Use Eq. (10.3.30) with $p_s = 0$ and $p_t = p$ (a constant) alongside 3-4 of the element shown in Figure 10–6 on page 460 to obtain the nodal forces.

10.15 For the element shown in Figure P10–15, replace the distributed load with the energy equivalent nodal forces by evaluating a force matrix similar to Eq. (10.3.29). Let $h = 0.1$ in thick.

10.16 Use Gaussian quadature with two and three Gauss points and Table 10–1 to evaluate the following integrals:

(a) $\int_{-1}^{1} \cos\frac{s}{2}\,ds$ (b) $\int_{-1}^{1} s^2\,ds$ (c) $\int_{-1}^{1} s^4\,ds$

(d) $\int_{-1}^{1} \frac{\cos s}{1-s^2}\,ds$ (e) $\int_{-1}^{1} s^3\,ds$ (f) $\int_{-1}^{1} s\cos s\,ds$ (g) $\int_{-1}^{1} (4^s - 2s)ds$

Then use the Newton-Cotes quadrature with two and three sampling points and Table 10–2 to evaluate the same integrals.

10.17 For the quadrilateral elements shown in Figure P10–17, write a computer program to evaluate the stiffness matrices using four-point Gaussian quadrature as outlined in Section 10.5. Let $E = 30 \times 10^6$ psi and $\nu = 0.25$. The global coordinates (in inches) are shown in the figures.

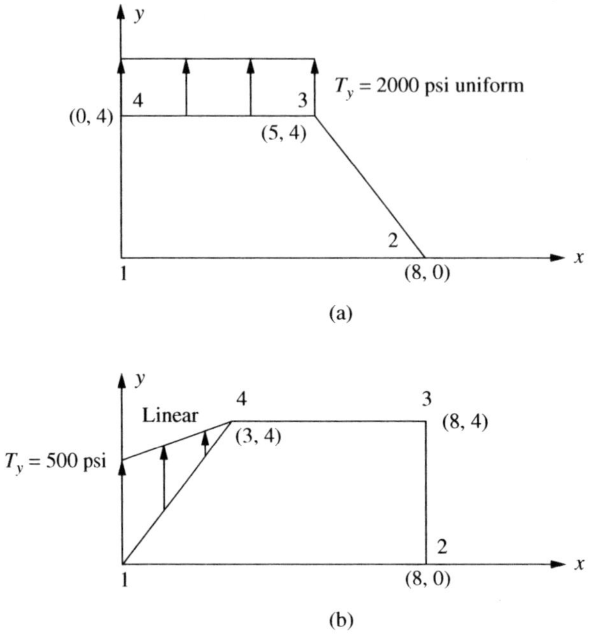

Figure P10–15

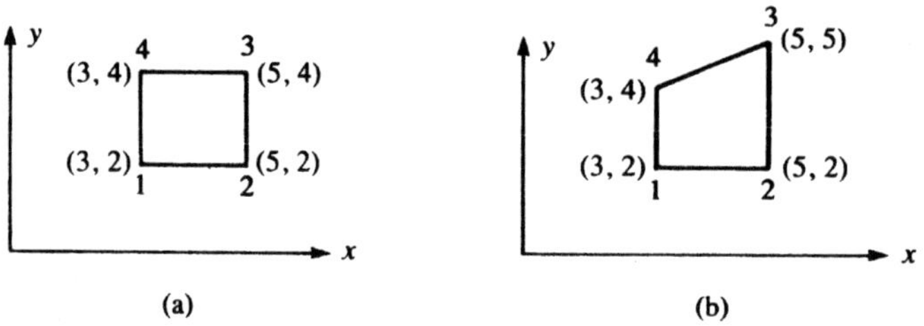

Figure P10–17

10.18 For the quadrilateral elements shown in Figure P10–18, evaluate the stiffness matrices using four-point Gaussian quadrature as outlined in Section 10.5. Let $E = 210$ GPa and $\nu = 0.25$. The global coordinates (in millimeters) are shown in the figures.

10.19 Evaluate the matrix $\underline{B}$ for the quadratic quadrilateral element shown in Figure 10–16 on page 480 (Section 10.6).

10.20 Evaluate the stiffness matrix for the four-noded bar of Problem 10.4 using three-point Gaussian quadrature.

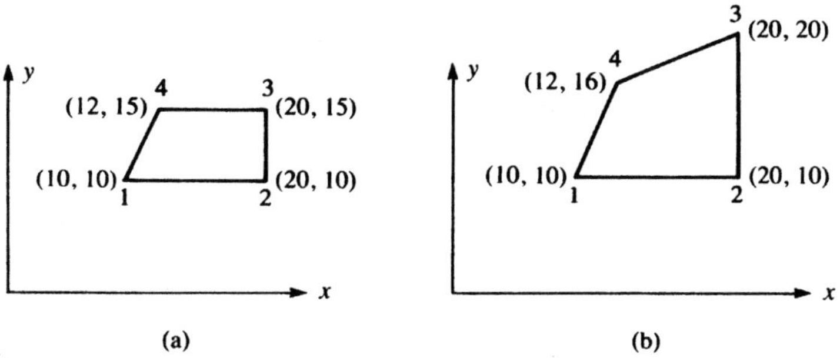

Figure P10–18

10.21 For the rectangular element in Figure P10–21, with the nodal displacements given in Problem 10.10, determine the $\underline{\sigma}$ matrix at $s = 0$, $t = 0$ using the isoparametric formulation described in Section 10.5. (Also see Example 10.5.)

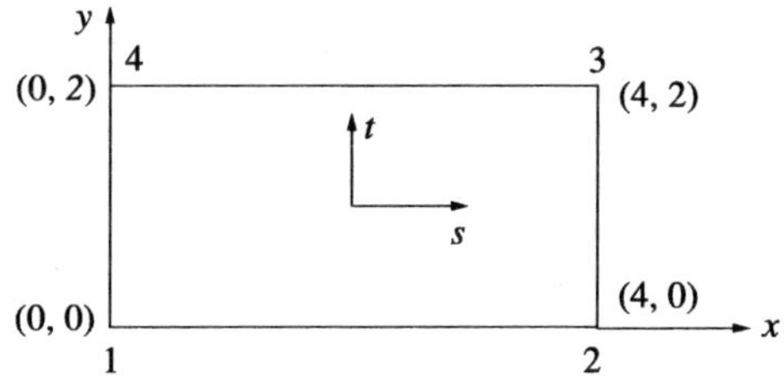

Figure P10–21

10.22 For the three-noded bar (Figure P10–1), what Gaussian quadrature rule (how many Gauss points) would you recommend to evaluate the stiffness matrix? Why?

CHAPTER 11

Three-Dimensional Stress Analysis

Introduction

In this chapter, we consider the three-dimensional, or solid, element. This element is useful for the stress analysis of general three-dimensional bodies that require more precise analysis than is possible through two-dimensional and/or axisymmetric analysis. Examples of three-dimensional problems are arch dams, thick-walled pressure vessels, and solid forging parts as used, for instance, in the heavy equipment and automotive industries. Figure 11–1 shows finite element models of some typical automobile parts. Also see Figure 1–7 for a model of a swing casting for a backhoe frame, Figure 1–9 for a model of a pelvis bone with an implant, and Figures 11–7 through 11–10 of a forging part, a foot pedal, a hollow pipe section, and an alternator bracket, respectively.

The tetrahedron is the basic three-dimensional element, and it is used in the development of the shape functions, stiffness matrix, and force matrices in terms of a global coordinate system. We follow this development with the isoparametric formulation of the stiffness matrix for the hexahedron, or brick element. Finally, we will provide some typical three-dimensional applications.

In the last section of this chapter, we show some three-dimensional problems solved using a computer program.

11.1 Three-Dimensional Stress and Strain

We begin by considering the three-dimensional infinitesimal element in Cartesian coordinates with dimensions dx, dy, and dz and normal and shear stresses as shown in Figure 11–2. This element conveniently represents the state of stress on three mutually perpendicular planes of a body in a state of three-dimensional stress. As usual, normal stresses are perpendicular to the faces of the element and are represented by

(a)

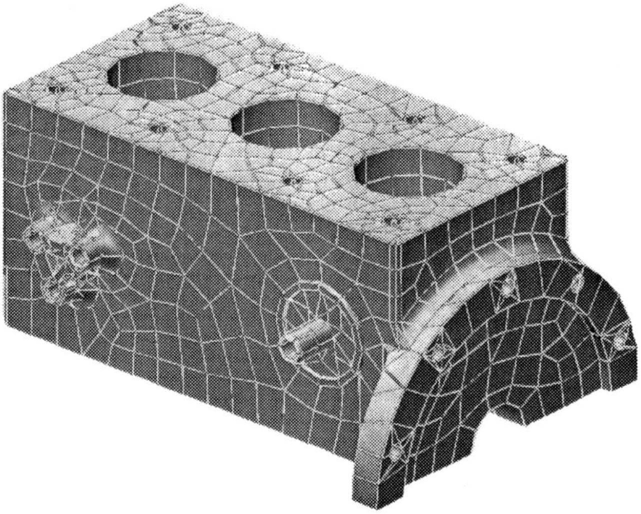

(b)

Figure 11–1 (a) wheel rim; (b) engine block. ((a) Courtesy of Mark Blair; (b) courtesy of Mark Guard.)

σ_x, σ_y, and σ_z. Shear stresses act in the faces (planes) of the element and are represented by $\tau_{xy}, \tau_{yz}, \tau_{zx}$, and so on.

From moment equilibrium of the element, we show in Appendix C that

$$\tau_{xy} = \tau_{yx} \qquad \tau_{yz} = \tau_{zy} \qquad \tau_{zx} = \tau_{xz}$$

Hence, there are only three independent shear stresses, along with the three normal stresses.

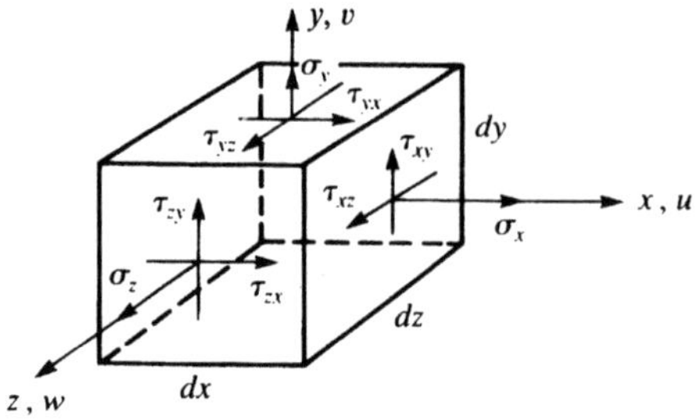

Figure 11–2 Three-dimensional stresses on an element

The element strain/displacement relationships are obtained in Appendix C. They are repeated here, for convenience, as

$$\varepsilon_x = \frac{\partial u}{\partial x} \qquad \varepsilon_y = \frac{\partial v}{\partial y} \qquad \varepsilon_z = \frac{\partial w}{\partial z} \tag{11.1.1}$$

where u, v, and w are the displacements associated with the x, y, and z directions. The shear strains γ are now given by

$$\begin{aligned} \gamma_{xy} &= \frac{\partial u}{\partial y} + \frac{\partial v}{\partial x} = \gamma_{yx} \\ \gamma_{yz} &= \frac{\partial v}{\partial z} + \frac{\partial w}{\partial y} = \gamma_{zy} \\ \gamma_{zx} &= \frac{\partial w}{\partial x} + \frac{\partial u}{\partial z} = \gamma_{xz} \end{aligned} \tag{11.1.2}$$

where, as for shear stresses, only three independent shear strains exist.

We again represent the stresses and strains by column matrices as

$$\{\sigma\} = \begin{Bmatrix} \sigma_x \\ \sigma_y \\ \sigma_z \\ \tau_{xy} \\ \tau_{yz} \\ \tau_{zx} \end{Bmatrix} \qquad \{\varepsilon\} = \begin{Bmatrix} \varepsilon_x \\ \varepsilon_y \\ \varepsilon_z \\ \gamma_{xy} \\ \gamma_{yz} \\ \gamma_{zx} \end{Bmatrix} \tag{11.1.3}$$

The stress/strain relationships for an isotropic material are again given by

$$\{\sigma\} = [D]\{\varepsilon\} \tag{11.1.4}$$

where $\{\sigma\}$ and $\{\varepsilon\}$ are defined by Eqs. (11.1.3), and the constitutive matrix $[D]$ (see also Appendix C) is now given by

$$[D] = \frac{E}{(1+\nu)(1-2\nu)} \begin{bmatrix} 1-\nu & \nu & \nu & 0 & 0 & 0 \\ & 1-\nu & \nu & 0 & 0 & 0 \\ & & 1-\nu & 0 & 0 & 0 \\ & & & \frac{1-2\nu}{2} & 0 & 0 \\ & & & & \frac{1-2\nu}{2} & 0 \\ \text{Symmetry} & & & & & \frac{1-2\nu}{2} \end{bmatrix} \tag{11.1.5}$$

11.2 Tetrahedral Element

We now develop the tetrahedral stress element stiffness matrix by again using the steps outlined in Chapter 1. The development is seen to be an extension of the plane element previously described in Chapter 6. This extension was suggested in References [1] and [2].

Step 1 Select Element Type

Consider the tetrahedral element shown in Figure 11–3 with corner nodes 1–4. This element is a four-noded solid. The nodes of the element must be numbered such that when viewed from the last node (say, node 4), the first three nodes are numbered in a counterclockwise manner, such as 1, 2, 3, 4 or 2, 3, 1, 4. This ordering of nodes avoids the calculation of negative volumes and is consistent with the counterclockwise node numbering associated with the CST element in Chapter 6. (Using an isoparametric formulation to evaluate the $\underline{k}$ matrix for the tetrahedral element enables us to use the element node numbering in any order. The isoparametric formulation of $\underline{k}$ is left

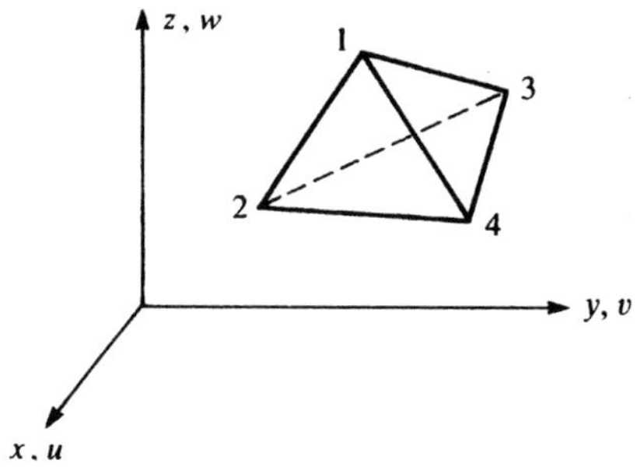

Figure 11–3 Tetrahedral solid element

to Section 11.3.) The unknown nodal displacements are now given by

$$\{d\} = \begin{Bmatrix} u_1 \\ v_1 \\ w_1 \\ \vdots \\ u_4 \\ v_4 \\ w_4 \end{Bmatrix} \tag{11.2.1}$$

Hence, there are 3 degrees of freedom per node, or 12 total degrees of freedom per element.

Step 2 Select Displacement Functions

For a compatible displacement field, the element displacement functions u, v, and w must be linear along each edge because only two points (the corner nodes) exist along each edge, and the functions must be linear in each plane side of the tetrahedron. We then select the linear displacement functions as

$$\begin{aligned} u(x,y,z) &= a_1 + a_2 x + a_3 y + a_4 z \\ v(x,y,z) &= a_5 + a_6 x + a_7 y + a_8 z \\ w(x,y,z) &= a_9 + a_{10} x + a_{11} y + a_{12} z \end{aligned} \tag{11.2.2}$$

In the same manner as in Chapter 6, we can express the a_i's in terms of the known nodal coordinates $(x_1, y_1, z_1, \ldots, z_4)$ and the unknown nodal displacements $(u_1, v_1, w_1, \ldots, w_4)$ of the element. Skipping the straightforward but tedious details, we obtain

$$\begin{aligned} u(x,y,z) = \frac{1}{6V}\{(\alpha_1 + \beta_1 x + \gamma_1 y + \delta_1 z)u_1 \\ + (\alpha_2 + \beta_2 x + \gamma_2 y + \delta_2 z)u_2 \\ + (\alpha_3 + \beta_3 x + \gamma_3 y + \delta_3 z)u_3 \\ + (\alpha_4 + \beta_4 x + \gamma_4 y + \delta_4 z)u_4\} \end{aligned} \tag{11.2.3}$$

where $6V$ is obtained by evaluating the determinant

$$6V = \begin{vmatrix} 1 & x_1 & y_1 & z_1 \\ 1 & x_2 & y_2 & z_2 \\ 1 & x_3 & y_3 & z_3 \\ 1 & x_4 & y_4 & z_4 \end{vmatrix} \tag{11.2.4}$$

and V represents the volume of the tetrahedron. The coefficients $\alpha_i, \beta_i, \gamma_i$, and δ_i $(i = 1, 2, 3, 4)$ in Eq. (11.2.3) are given by

$$\alpha_1 = \begin{vmatrix} x_2 & y_2 & z_2 \\ x_3 & y_3 & z_3 \\ x_4 & y_4 & z_4 \end{vmatrix} \qquad \beta_1 = -\begin{vmatrix} 1 & y_2 & z_2 \\ 1 & y_3 & z_3 \\ 1 & y_4 & z_4 \end{vmatrix}$$

$$\gamma_1 = \begin{vmatrix} 1 & x_2 & z_2 \\ 1 & x_3 & z_3 \\ 1 & x_4 & z_4 \end{vmatrix} \qquad \delta_1 = -\begin{vmatrix} 1 & x_2 & y_2 \\ 1 & x_3 & y_3 \\ 1 & x_4 & y_4 \end{vmatrix} \tag{11.2.5}$$

$$\alpha_2 = -\begin{vmatrix} x_1 & y_1 & z_1 \\ x_3 & y_3 & z_3 \\ x_4 & y_4 & z_4 \end{vmatrix} \qquad \beta_2 = \begin{vmatrix} 1 & y_1 & z_1 \\ 1 & y_3 & z_3 \\ 1 & y_4 & z_4 \end{vmatrix}$$

and

$$\gamma_2 = -\begin{vmatrix} 1 & x_1 & z_1 \\ 1 & x_3 & z_3 \\ 1 & x_4 & z_4 \end{vmatrix} \qquad \delta_2 = \begin{vmatrix} 1 & x_1 & y_1 \\ 1 & x_3 & y_3 \\ 1 & x_4 & y_4 \end{vmatrix} \tag{11.2.6}$$

$$\alpha_3 = \begin{vmatrix} x_1 & y_1 & z_1 \\ x_2 & y_2 & z_2 \\ x_4 & y_4 & z_4 \end{vmatrix} \qquad \beta_3 = -\begin{vmatrix} 1 & y_1 & z_1 \\ 1 & y_2 & z_2 \\ 1 & y_4 & z_4 \end{vmatrix}$$

and

$$\gamma_3 = \begin{vmatrix} 1 & x_1 & z_1 \\ 1 & x_2 & z_2 \\ 1 & x_4 & z_4 \end{vmatrix} \qquad \delta_3 = -\begin{vmatrix} 1 & x_1 & y_1 \\ 1 & x_2 & y_2 \\ 1 & x_4 & y_4 \end{vmatrix} \tag{11.2.7}$$

$$\alpha_4 = -\begin{vmatrix} x_1 & y_1 & z_1 \\ x_2 & y_2 & z_2 \\ x_3 & y_3 & z_3 \end{vmatrix} \qquad \beta_4 = \begin{vmatrix} 1 & y_1 & z_1 \\ 1 & y_2 & z_2 \\ 1 & y_3 & z_3 \end{vmatrix}$$

and

$$\gamma_4 = -\begin{vmatrix} 1 & x_1 & z_1 \\ 1 & x_2 & z_2 \\ 1 & x_3 & z_3 \end{vmatrix} \qquad \delta_4 = \begin{vmatrix} 1 & x_1 & y_1 \\ 1 & x_2 & y_2 \\ 1 & x_3 & y_3 \end{vmatrix} \tag{11.2.8}$$

Expressions for v and w are obtained by simply substituting v_i's for all u_i's and then w_i's for all u_i's in Eq. (11.2.3).

The displacement expression for u given by Eq. (11.2.3), with similar expressions for v and w, can be written equivalently in expanded form in terms of the shape functions and unknown nodal displacements as

$$\begin{Bmatrix} u \\ v \\ w \end{Bmatrix} = \begin{bmatrix} N_1 & 0 & 0 & N_2 & 0 & 0 & N_3 & 0 & 0 & N_4 & 0 & 0 \\ 0 & N_1 & 0 & 0 & N_2 & 0 & 0 & N_3 & 0 & 0 & N_4 & 0 \\ 0 & 0 & N_1 & 0 & 0 & N_2 & 0 & 0 & N_3 & 0 & 0 & N_4 \end{bmatrix} \begin{Bmatrix} u_1 \\ v_1 \\ w_1 \\ \vdots \\ u_4 \\ v_4 \\ w_4 \end{Bmatrix} \tag{11.2.9}$$

where the shape functions are given by

$$N_1 = \frac{(\alpha_1 + \beta_1 x + \gamma_1 y + \delta_1 z)}{6V} \qquad N_2 = \frac{(\alpha_2 + \beta_2 x + \gamma_2 y + \delta_2 z)}{6V}$$
$$N_3 = \frac{(\alpha_3 + \beta_3 x + \gamma_3 y + \delta_3 z)}{6V} \qquad N_4 = \frac{(\alpha_4 + \beta_4 x + \gamma_4 y + \delta_4 z)}{6V} \tag{11.2.10}$$

and the rectangular matrix on the right side of Eq. (11.2.9) is the shape function matrix $[N]$.

Step 3 Define the Strain/ Displacement and Stress/ Strain Relationships

The element strains for the three-dimensional stress state are given by

$$\{\varepsilon\} = \begin{Bmatrix} \varepsilon_x \\ \varepsilon_y \\ \varepsilon_z \\ \gamma_{xy} \\ \gamma_{yz} \\ \gamma_{zx} \end{Bmatrix} = \begin{Bmatrix} \dfrac{\partial u}{\partial x} \\ \dfrac{\partial v}{\partial y} \\ \dfrac{\partial w}{\partial z} \\ \dfrac{\partial u}{\partial y} + \dfrac{\partial v}{\partial x} \\ \dfrac{\partial v}{\partial z} + \dfrac{\partial w}{\partial y} \\ \dfrac{\partial w}{\partial x} + \dfrac{\partial u}{\partial z} \end{Bmatrix} \tag{11.2.11}$$

Using Eq. (11.2.9) in Eq. (11.2.11), we obtain

$$\{\varepsilon\} = [B]\{d\} \tag{11.2.12}$$

where

$$[B] = [\underline{B}_1 \quad \underline{B}_2 \quad \underline{B}_3 \quad \underline{B}_4] \tag{11.2.13}$$

The submatrix $\underline{B}_1$ in Eq. (11.2.13) is defined by

$$\underline{B}_1 = \begin{bmatrix} N_{1,x} & 0 & 0 \\ 0 & N_{1,y} & 0 \\ 0 & 0 & N_{1,z} \\ N_{1,y} & N_{1,x} & 0 \\ 0 & N_{1,z} & N_{1,y} \\ N_{1,z} & 0 & N_{1,x} \end{bmatrix} \tag{11.2.14}$$

where, again, the comma after the subscript indicates differentation with respect to the variable that follows. Submatrices $\underline{B}_2, \underline{B}_3$, and $\underline{B}_4$ are defined by simply indexing the subscript in Eq. (11.2.14) from 1 to 2, 3, and then 4, respectively. Substituting the shape functions from Eqs. (11.2.10) into Eq. (11.2.14), $\underline{B}_1$ is expressed as

$$\underline{B}_1 = \frac{1}{6V} \begin{bmatrix} \beta_1 & 0 & 0 \\ 0 & \gamma_1 & 0 \\ 0 & 0 & \delta_1 \\ \gamma_1 & \beta_1 & 0 \\ 0 & \delta_1 & \gamma_1 \\ \delta_1 & 0 & \beta_1 \end{bmatrix} \tag{11.2.15}$$

with similar expressions for $\underline{B}_2, \underline{B}_3$, and $\underline{B}_4$.

The element stresses are related to the element strains by

$$\{\sigma\} = [D]\{\varepsilon\} \tag{11.2.16}$$

where the constitutive matrix for an elastic material is now given by Eq. (11.1.5).

Step 4 Derive the Element Stiffness Matrix and Equations

The element stiffness matrix is given by

$$[k] = \iiint\limits_V [B]^T [D][B]\, dV \tag{11.2.17}$$

Because both matrices $[B]$ and $[D]$ are constant for the simple tetrahedral element, Eq. (11.2.17) can be simplified to

$$[k] = [B]^T [D][B]\, V \tag{11.2.18}$$

where, again, V is the volume of the element. The element stiffness matrix is now of order 12×12.

Body Forces

The element body force matrix is given by

$$\{f_b\} = \iiint\limits_V [N]^T \{X\}\, dV \tag{11.2.19}$$

where $[N]$ is given by the 3×12 matrix in Eq. (11.2.9), and

$$\{X\} = \begin{Bmatrix} X_b \\ Y_b \\ Z_b \end{Bmatrix} \tag{11.2.20}$$

For constant body forces, the nodal components of the total resultant body forces can be shown to be distributed to the nodes in four equal parts. That is,

$$\{f_b\} = \tfrac{1}{4}[X_b \; Y_b \; Z_b \; X_b \; Y_b \; Z_b \; X_b \; Y_b \; Z_b \; X_b \; Y_b \; Z_b]^T$$

The element body force is then a 12×1 matrix.

Surface Forces

Again, the surface forces are given by

$$\{f_s\} = \iint_S [N_s]^T \{T\}\, dS \tag{11.2.21}$$

where $[N_s]$ is the shape function matrix evaluated on the surface where the surface traction occurs.

For example, consider the case of uniform pressure p acting on the face with nodes 1–3 of the element shown in Figure 11–3 or 11–4. The resulting nodal forces become

$$\{f_s\} = \iint_S [N]^T \Big|_{\substack{\text{evaluated on} \\ \text{surface } 1,2,3}} \begin{Bmatrix} p_x \\ p_y \\ p_z \end{Bmatrix} dS \tag{11.2.22}$$

where p_x, p_y, and p_z are the x, y, and z components, respectively, of p. Simplifying and integrating Eq. (11.2.22), we can show that

$$\{f_s\} = \frac{S_{123}}{3} \begin{Bmatrix} p_x \\ p_y \\ p_z \\ p_x \\ p_y \\ p_z \\ p_x \\ p_y \\ p_z \\ 0 \\ 0 \\ 0 \end{Bmatrix} \tag{11.2.23}$$

where S_{123} is the area of the surface associated with nodes 1–3. The use of volume coordinates, as explained in Reference [8], facilitates the integration of Eq. (11.2.22).

Example 11.1

Evaluate the matrices necessary to determine the stiffness matrix for the tetrahedral element shown in Figure 11–4. Let $E = 30 \times 10^6$ psi and $\nu = 0.30$. The coordinates are shown in the figure in units of inches.

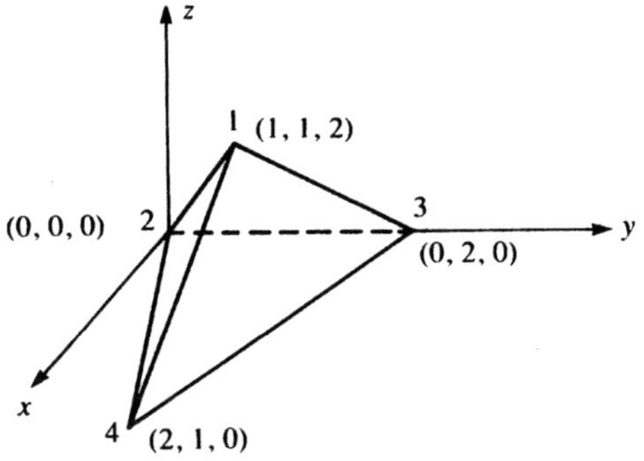

Figure 11–4 Tetrahedral element

To evaluate the element stiffness matrix, we first determine the element volume V and all α's, β's, γ's, and δ's from Eqs. (11.2.4)–(11.2.8). From Eq. (11.2.4), we have

$$6V = \begin{vmatrix} 1 & 1 & 1 & 2 \\ 1 & 0 & 0 & 0 \\ 1 & 0 & 2 & 0 \\ 1 & 2 & 1 & 0 \end{vmatrix} = 8 \text{ in}^3 \tag{11.2.24}$$

From Eqs. (11.2.5), we obtain

$$\alpha_1 = \begin{vmatrix} 0 & 0 & 0 \\ 0 & 2 & 0 \\ 2 & 1 & 0 \end{vmatrix} = 0 \qquad \beta_1 = -\begin{vmatrix} 1 & 0 & 0 \\ 1 & 2 & 0 \\ 1 & 1 & 0 \end{vmatrix} = 0 \tag{11.2.25}$$

and similarly,

$$\gamma_1 = 0 \qquad \delta_1 = 4$$

From Eqs. (11.2.6)–(11.2.8), we obtain

$$\begin{array}{llll} \alpha_2 = 8 & \beta_2 = -2 & \gamma_2 = -4 & \delta_2 = -1 \\ \alpha_3 = 0 & \beta_3 = -2 & \gamma_3 = 4 & \delta_3 = -1 \\ \alpha_4 = 0 & \beta_4 = 4 & \gamma_4 = 0 & \delta_4 = -2 \end{array} \tag{11.2.26}$$

Note that α's typically have units of cubic inches or cubic meters, where β's, γ's, and δ's have units of square inches or square meters.

Next, the shape functions are determined using Eqs. (11.2.10) and the results from Eqs. (11.2.25) and (11.2.26) as

$$N_1 = \frac{4z}{8} \qquad N_2 = \frac{8 - 2x - 4y - z}{8}$$
$$N_3 = \frac{-2x + 4y - z}{8} \qquad N_4 = \frac{4x - 2z}{8} \tag{11.2.27}$$

Note that $N_1 + N_2 + N_3 + N_4 = 1$ is again satisfied.

The 6×3 submatrices of the matrix $\underline{B}$, Eq. (11.2.13), are now evaluated using Eqs. (11.2.14) and (11.2.27) as

$$\underline{B}_1 = \begin{bmatrix} 0 & 0 & 0 \\ 0 & 0 & 0 \\ 0 & 0 & \frac{1}{2} \\ 0 & 0 & 0 \\ 0 & \frac{1}{2} & 0 \\ \frac{1}{2} & 0 & 0 \end{bmatrix} \qquad \underline{B}_2 = \begin{bmatrix} -\frac{1}{4} & 0 & 0 \\ 0 & -\frac{1}{2} & 0 \\ 0 & 0 & -\frac{1}{8} \\ -\frac{1}{2} & -\frac{1}{4} & 0 \\ 0 & -\frac{1}{8} & -\frac{1}{2} \\ -\frac{1}{8} & 0 & -\frac{1}{4} \end{bmatrix} \tag{11.2.28}$$

$$\underline{B}_3 = \begin{bmatrix} -\frac{1}{4} & 0 & 0 \\ 0 & \frac{1}{2} & 0 \\ 0 & 0 & -\frac{1}{8} \\ \frac{1}{2} & -\frac{1}{4} & 0 \\ 0 & -\frac{1}{8} & \frac{1}{2} \\ -\frac{1}{8} & 0 & -\frac{1}{4} \end{bmatrix} \qquad \underline{B}_4 = \begin{bmatrix} \frac{1}{2} & 0 & 0 \\ 0 & 0 & 0 \\ 0 & 0 & -\frac{1}{4} \\ 0 & \frac{1}{2} & 0 \\ 0 & -\frac{1}{4} & 0 \\ -\frac{1}{4} & 0 & \frac{1}{2} \end{bmatrix}$$

Next, the matrix $\underline{D}$ is evaluated using Eq. (11.1.5) as

$$[D] = \frac{30 \times 10^6}{(1 + 0.3)(1 - 0.6)} \begin{bmatrix} 0.7 & 0.3 & 0.3 & 0 & 0 & 0 \\ & 0.7 & 0.3 & 0 & 0 & 0 \\ & & 0.7 & 0 & 0 & 0 \\ & & & 0.2 & 0 & 0 \\ & & & & 0.2 & 0 \\ \text{Symmetry} & & & & & 0.2 \end{bmatrix} \tag{11.2.29}$$

Finally, substituting the results from Eqs. (11.2.24), (11.2.28), and (11.2.29) into Eq. (11.2.18), we obtain the element stiffness matrix. The resulting 12×12 matrix, being cumbersome to obtain by longhand calculations, is best left for the computer to evaluate. ■

11.3 Isoparametric Formulation

We now describe the isoparametric formulation of the stiffness matrix for some three-dimensional hexahedral elements.

Linear Hexahedral Element

The basic (linear) hexahedral element [Figure 11–5(a)] now has eight corner nodes with isoparametric natural coordinates given by s, t, and z' as shown in Figure 11–5(b). The element faces are now defined by $s, t, z' = \pm 1$. (We use s, t, and z' for the coordinate axes because they are probably simpler to use than Greek letters ξ, η, and ζ).

The formulation of the stiffness matrix follows steps analogous to the isoparametric formulation of the stiffness matrix for the plane element in Chapter 10.

The function use to describe the element geometry for x in terms of the generalized degrees of freedom a_i's is

$$x = a_1 + a_2 s + a_3 t + a_4\, z' + a_5 st + a_6 tz' + a_7 z's + a_8 stz' \tag{11.3.1}$$

The same form as Eq. (11.3.1) is used for y and z as well. Just start with a_9 through a_{16} for y and a_{17} through a_{24} for z.

First, we expand Eq. (10.3.4) to include the z coordinate as follows:

$$\begin{Bmatrix} x \\ y \\ z \end{Bmatrix} = \sum_{i=1}^{8} \left(\begin{bmatrix} N_i & 0 & 0 \\ 0 & N_i & 0 \\ 0 & 0 & N_i \end{bmatrix} \begin{Bmatrix} x_i \\ y_i \\ z_i \end{Bmatrix} \right) \tag{11.3.2}$$

where the shape functions are now given by

$$N_i = \frac{(1 + ss_i)(1 + tt_i)(1 + z'z'_i)}{8} \tag{11.3.3}$$

with $s_i, t_i, z'_i = \pm 1$ and $i = 1, 2, \ldots, 8$. For instance,

$$N_1 = \frac{(1 + ss_1)(1 + tt_1)(1 + z'z'_1)}{8} \tag{11.3.4}$$

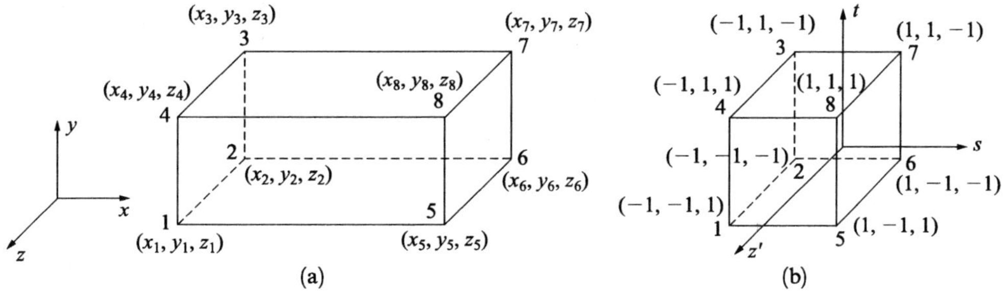

Figure 11–5 Linear hexahedral element (a) in a global-coordinate system and (b) element mapped into a cube of two unit sides placed symmetrically with natural or intrinsic coordinates s, t, and z'

and when, from Figure 11–5, $s_1 = -1$, $t_1 = -1$, and $z'_1 = +1$ are used in Eq. (11.3.4), we obtain

$$N_1 = \frac{(1-s)(1-t)(1+z')}{8} \tag{11.3.5}$$

Explicit forms of the other shape functions follow similarly. The shape functions in Eq. (11.3.3) map the natural coordinates (s, t, z') of any point in the element to any point in the global coordinates (x, y, z) when used in Eq. (11.3.2). For instance, when we let $i = 8$ and substitute $s_8 = 1$, $t_8 = 1$, $z'_8 = 1$ into Eq. (11.3.3) for N_8, we obtain

$$N_8 = \frac{(1+s)(1+t)(1+z')}{8}$$

Similar expressions are obtained for the other shape functions. Then evaluating all shape functions at node 8, we obtain $N_8 = 1$, and all other shape functions equal zero at node 8. [From Eq. (11.3.5), we see that $N_1 = 0$ when $s = 1$ or when $t = 1$.] Therefore, using Eq. (11.3.2), we obtain

$$x = x_8 \qquad y = y_8 \qquad z = z_8$$

We see that indeed Eq. (11.3.2) maps any point in the natural-coordinate system to one in the global-coordinate system.

The displacement functions in terms of the generalized degrees of freedom are of the same form as used to describe the element geometry given by Eq. (11.3.1). Therefore we use the same shape functions as used to describe the geometry (Eq. (11.3.3)). The displacement functions now include w such that

$$\begin{Bmatrix} u \\ v \\ w \end{Bmatrix} = \sum_{i=1}^{8} \left(\begin{bmatrix} N_i & 0 & 0 \\ 0 & N_i & 0 \\ 0 & 0 & N_i \end{bmatrix} \begin{Bmatrix} u_i \\ v_i \\ w_i \end{Bmatrix} \right) \tag{11.3.6}$$

with the same shape functions as defined by Eq. (11.3.3) and the size of the shape function matrix now 3×24.

The Jacobian matrix [Eq. (10.3.10)] is now expanded to

$$[J] = \begin{bmatrix} \dfrac{\partial x}{\partial s} & \dfrac{\partial y}{\partial s} & \dfrac{\partial z}{\partial s} \\ \dfrac{\partial x}{\partial t} & \dfrac{\partial y}{\partial t} & \dfrac{\partial z}{\partial t} \\ \dfrac{\partial x}{\partial z'} & \dfrac{\partial y}{\partial z'} & \dfrac{\partial z}{\partial z'} \end{bmatrix} \tag{11.3.7}$$

Because the strain/displacement relationships, given by Eq. (11.2.11) in terms of global coordinates, include differentiation with respect to z, we expand Eq. (10.3.9) as follows:

$$\frac{\partial f}{\partial x} = \frac{\begin{vmatrix} \dfrac{\partial f}{\partial s} & \dfrac{\partial y}{\partial s} & \dfrac{\partial z}{\partial s} \\[2ex] \dfrac{\partial f}{\partial t} & \dfrac{\partial y}{\partial t} & \dfrac{\partial z}{\partial t} \\[2ex] \dfrac{\partial f}{\partial z'} & \dfrac{\partial y}{\partial z'} & \dfrac{\partial z}{\partial z'} \end{vmatrix}}{|\underline{J}|} \qquad \frac{\partial f}{\partial y} = \frac{\begin{vmatrix} \dfrac{\partial x}{\partial s} & \dfrac{\partial f}{\partial s} & \dfrac{\partial z}{\partial s} \\[2ex] \dfrac{\partial x}{\partial t} & \dfrac{\partial f}{\partial t} & \dfrac{\partial z}{\partial t} \\[2ex] \dfrac{\partial x}{\partial z'} & \dfrac{\partial f}{\partial z'} & \dfrac{\partial z}{\partial z'} \end{vmatrix}}{|\underline{J}|}$$

$$\frac{\partial f}{\partial z} = \frac{\begin{vmatrix} \dfrac{\partial x}{\partial s} & \dfrac{\partial y}{\partial s} & \dfrac{\partial f}{\partial s} \\[2ex] \dfrac{\partial x}{\partial t} & \dfrac{\partial y}{\partial t} & \dfrac{\partial f}{\partial t} \\[2ex] \dfrac{\partial x}{\partial z'} & \dfrac{\partial y}{\partial z'} & \dfrac{\partial f}{\partial z'} \end{vmatrix}}{|\underline{J}|} \tag{11.3.8}$$

Using Eqs. (11.3.8) by substituting u, v, and then w for f and using the definitions of the strains, we can express the strains in terms of natural coordinates (s, t, z') to obtain an equation similar to Eq. (10.3.14). In compact form, we can again express the strains in terms of the shape functions and global nodal coordinates similar to Eq. (10.3.15). The matrix $\underline{B}$, given by a form similar to Eq. (10.3.17), is now a function of s, t, and z' and is of order 6×24.

The 24×24 stiffness matrix is now given by

$$[k] = \int_{-1}^{1}\int_{-1}^{1}\int_{-1}^{1} [B]^T[D][B]|\underline{J}|\, ds\, dt\, dz' \tag{11.3.9}$$

Again, it is best to evaluate $[k]$ by numerical integration (also see Section 10.4); that is, we evaluate (integrate) the eight-node hexahedral element stiffness matrix using a $2 \times 2 \times 2$ rule (or two-point rule). Actually, eight points defined in Table 11–1 are used to evaluate $\underline{k}$ as

$$\underline{k} = \sum_{i=1}^{8} \underline{B}^T(s_i, t_i, z_i')\underline{D}\underline{B}(s_i, t_i, z_i')|\underline{J}(s_i, t_i, z_i')|\, W_i W_j W_k$$

where $W_i = W_j = W_k$ for the two-point rule.

As is true with the bilinear quadrilateral element described in Section 10.3, the eight-noded linear hexahedral element cannot model beam-bending action well because the element sides remain straight during the element deformation. During the bending process, the elements will be stretched and can shear lock. This concept of shear locking is described in more detail in [12] along with ways to remedy it. However, the quadratic hexahedral element described subsequently remedies the shear locking problem.

Table 11–1 Table of Gauss points for linear hexahedral element with associated weights[a]

Points, i	s_i	t_i	z'_i	Weight, W_i
1	$-1/\sqrt{3}$	$-1/\sqrt{3}$	$1/\sqrt{3}$	1
2	$1/\sqrt{3}$	$-1/\sqrt{3}$	$1/\sqrt{3}$	1
3	$1/\sqrt{3}$	$1/\sqrt{3}$	$1/\sqrt{3}$	1
4	$-1/\sqrt{3}$	$1/\sqrt{3}$	$1/\sqrt{3}$	1
5	$-1/\sqrt{3}$	$-1/\sqrt{3}$	$-1/\sqrt{3}$	1
6	$1/\sqrt{3}$	$-1/\sqrt{3}$	$-1/\sqrt{3}$	1
7	$1/\sqrt{3}$	$1/\sqrt{3}$	$-1/\sqrt{3}$	1
8	$-1/\sqrt{3}$	$1/\sqrt{3}$	$-1/\sqrt{3}$	1

[a] $1/\sqrt{3} = 0.57735$.

Quadratic Hexahedral Element

For the quadratic hexahedral element shown in Figure 11–6, we have a total of 20 nodes with the inclusion of a total of 12 midside nodes.

The function describing the element geometry for x in terms of the 20 a_i's is

$$\begin{aligned} x = a_1 + a_2 s + a_3 t + a_4 z' + a_5 st + a_6 tz' + a_7 z's + a_8 s^2 + a_9 t^2 \\ + a_{10} z'^2 + a_{11} s^2 t + a_{12} st^2 + a_{13} t^2 z' + a_{14} tz'^2 + a_{15} z'^2 s \\ + a_{16} z' s^2 + a_{17} stz' + a_{18} s^2 tz' + a_{19} st^2 z' + a_{20} stz'^2 \end{aligned} \tag{11.3.10}$$

Similar expressions describe the y and z coordinates.

The development of the stiffness matrix follows the same steps we outlined before for the linear hexahedral element, where the shape functions now take on new forms. Again, letting $s_i, t_i, z'_i = \pm 1$, we have for the corner nodes ($i = 1, 2, \ldots, 8$),

$$N_i = \frac{(1 + ss_i)(1 + tt_i)(1 + z'z'_i)}{8}(ss_i + tt_i + z'z'_i - 2) \tag{11.3.11}$$

For the midside nodes at $s_i = 0$, $t_i = \pm 1$, $z'_i = \pm 1$ ($i = 17, 18, 19, 20$), we have

$$N_i = \frac{(1 - s^2)(1 + tt_i)(1 + z'z'_i)}{4} \tag{11.3.12}$$

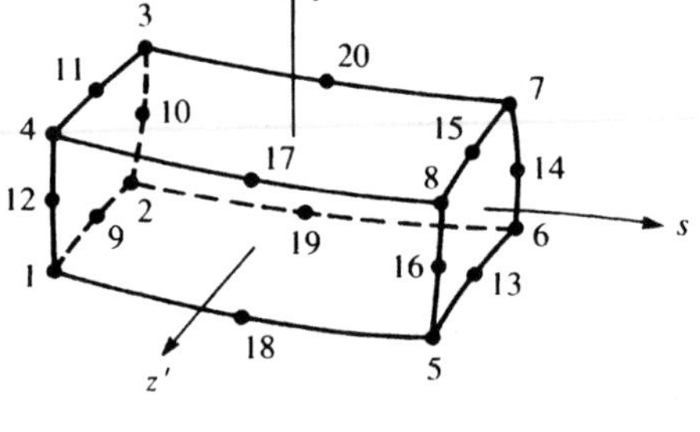

Figure 11–6 Quadratic hexahedral isoparametric element

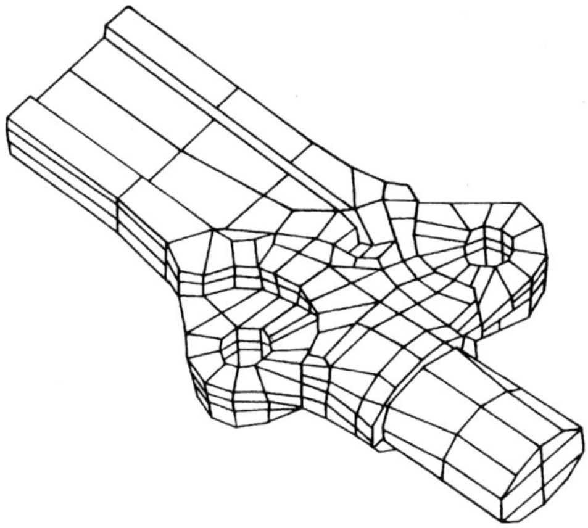

Figure 11–7 Finite element model of a forging using linear and quadratic solid elements

For the midside nodes at $s_i = \pm 1$, $t_i = 0$, $z_i' = \pm 1$ $(i = 10, 12, 14, 16)$, we have

$$N_i = \frac{(1 + ss_i)(1 - t^2)(1 + z'z_i')}{4} \tag{11.3.13}$$

Finally, for the midside nodes at $s_i = \pm 1$, $t_i = \pm 1$, $z_i' = 0$ $(i = 9, 11, 13, 15)$, we have

$$N_i = \frac{(1 + ss_i)(1 + tt_i)(1 - z'^2)}{4} \tag{11.3.14}$$

Note that the stiffness matrix for the quadratic solid element is of order 60×60 because it has 20 nodes and 3 degrees of freedom per node.

The stiffness matrix for this 20-node quadratic solid element can be evaluated using a $3 \times 3 \times 3$ rule (27 points). However, a special 14-point rule may be a better choice [9, 10].

As with the eight-noded plane element of Section 10.6 (Figure 10–16), the 20-node solid element is also called a serendipity element.

Figures 1–7 and 11–7 show applications of the use of linear and quadratic (curved sides) solid elements to model three-dimensional solids.

Finally, commercial computer programs, such as [11] (also see references [46–56] of Chapter 1), are available to solve three-dimensional problems. Figures 11–8, 11–9, and 11–10 show a steel foot pedal, a hollow cast-iron member, and an alternator bracket solved using a computer program [11]. We emphasize that these problems have been solved using the three-dimensional element as opposed to using a two-dimensional element, such as described in Chapters 6 and 8, as these problems have a three-dimensional stress state occurring in them. That is, the three normal and three shear stresses are of similar order of magnitude in some parts of the foot pedal, the cast-iron member, and the alternator bracket. The most accurate results will then occur when modeling these problems using the three-dimensional brick or tetrahedral elements (or a combination of both).

For the foot pedal, the largest principal stress was 4111 psi and the largest von Mises stress was 4023 psi, both located at the interior corner of the elbow.

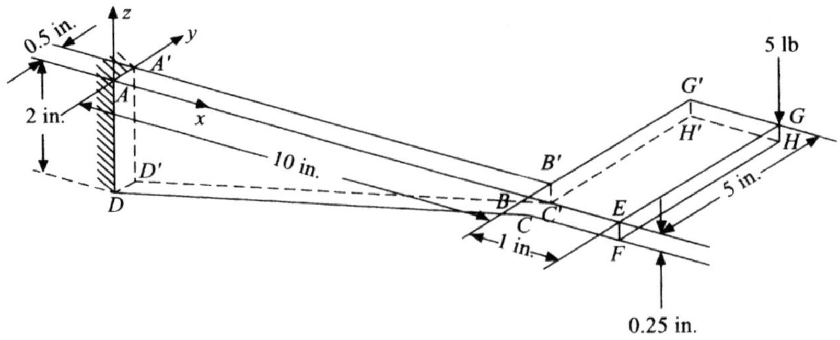

Figure 11–8 Three-dimensional steel foot pedal

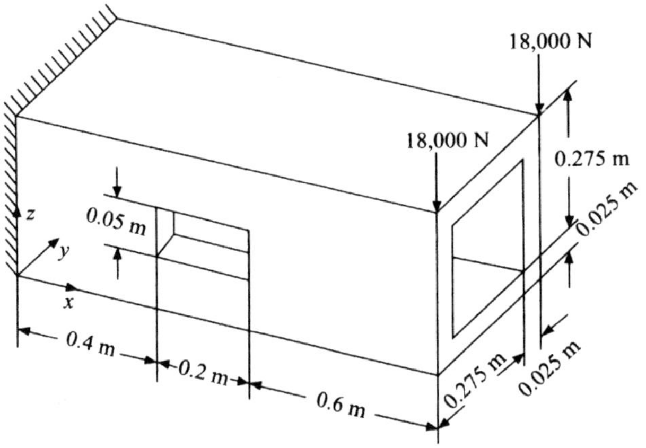

Figure 11–9 Cast iron hollow member, $E = 165$ GPa with opening on frontside

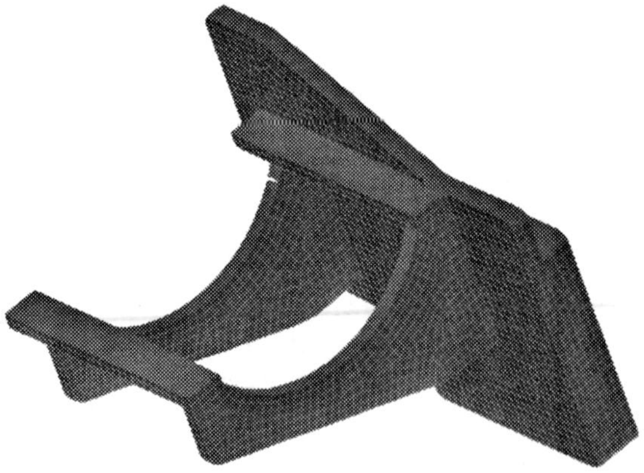

Figure 11–10 Meshed model of an alternator bracket (Courtesy of Andrew Heckman, Design Engineer, Seagrave Fire Apparatus, LLC)

The maximum displacement was 0.01234 in. down at the free end point H in Figure 11–8. The model had 889 nodes and 648 brick elements.

For the cast-iron member, the maximum principal stress was 19 MPa and the maximum von Mises stress was 23.2 MPa. Both results occurred at the top near the fixed wall support. The free end vertical displacement was −0.300 mm.

Table 11–2 Table comparing results for cantilever beam modeled using 4-noded-tetrahedral, 10-noded tetrahedral, 8-noded brick, and 20-noded brick element

Solid Element Used	Number of Nodes	Number of Degrees of Freedom	Number of Elements	Free End Displ., in.	Principal Stress, psi
4-noded tet	30	90	61	0.0053	562
4-noded tet	415	1245	1549	0.0282	2357
4-noded tet	896	2688	3729	0.0420	3284
4-noded tet	1658	4974	7268	0.0548	4056
10-noded tet	144	432	61	0.1172	6601
10-noded tet	2584	7752	1549	0.1277	7970
8-noded brick	64	192	27	0.1190	5893
8-noded brick	343	1029	216	0.1253	6507
8-noded brick	1331	3993	1000	0.1277	6836
20-noded brick	208	624	27	0.1250	7899
20-noded brick	1225	3675	216	0.1285	8350
20-noded brick	4961	14,883	1000	0.1297	8323
Classical solution				0.1286	6940

(Mr. William Gobeli for creating the results for Table 11–2)

For the alternator bracket made of ASTM-A36 hot-rolled steel, the model consisted of 13,298 solid brick elements and 10,425 nodes. A total load of 1000 lb was applied downward to the flat front face piece. The bracket back side was constrained against displacement. The largest von Mises stress was 11,538 psi located at the top surface near the center (narrowest) section of the bracket. The largest vertical deflection was 0.01623 in. at the front tip of the outer edge of the alternator bracket.

It has been shown [3] that use of the simple eight-noded hexahedral element yields better results than use of the constant-strain tetrahedral discussed in Section 11.1. Table 11.2 also illustrates the comparison between the corner-noded (constant-strain) tetrahedral, the linear-strain tetrahedral (mid-edge nodes added), the 8-noded brick, and the 20-noded brick models for a three-dimensional cantilever beam of length 100 in., base 6 in., and height 12 in. The beam has an end load of 10,000 lb acting upward and is made of steel ($E = 30 \times 10^6$ psi). A typical 8-noded brick model with the principal stress plot is shown in Figure 11–11. The classical beam theory solution for the vertical displacement and bending stress is also included for comparison. We can observe that the constant-strain tetrahedral gives very poor results, whereas the linear

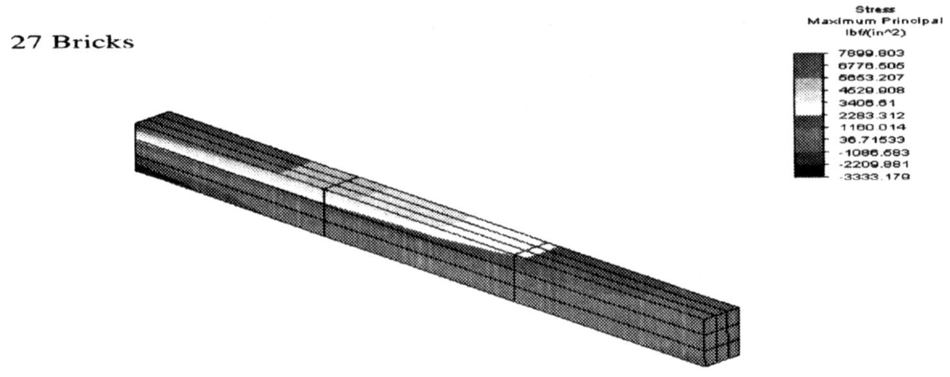

Figure 11–11 Eight-noded brick model (27 Bricks) showing principal stress plot

tetrahedral gives much better results. This is because the linear-strain model predicts the beam-bending behavior much better. The 8-noded and 20-noded brick models yield similar but accurate results compared to the classical beam theory results.

In summary, the use of the three-dimensional elements results in a large number of equations to be solved simultaneously. For instance, a model using a simple cube with, say, 20 by 20 by 20 nodes ($= 8000$ total nodes) for a region requires 8000 times 3 degrees of freedom per node ($= 24,000$) simultaneous equations.

References [4–7] report on early three-dimensional programs and analysis procedures using solid elements such as a family of subparametric curvilinear elements, linear tetrahedral elements, and 8-noded linear and 20-noded quadratic isoparametric elements.

References

[1] Martin, H. C., "Plane Elasticity Problems and the Direct Stiffness Method." *The Trend in Engineering*, Vol. 13, pp. 5–19, Jan. 1961.

[2] Gallagher, R. H., Padlog, J., and Bijlaard, P. P., "Stress Analysis of Heated Complex Shapes," *Journal of the American Rocket Society*, pp. 700–707, May 1962.

[3] Melosh, R. J., "Structural Analysis of Solids," *Journal of the Structural Division*, American Society of Civil Engineers, pp. 205–223, Aug. 1963.

[4] Chacour, S., "DANUTA, a Three-Dimensional Finite Element Program Used in the Analysis of Turbo-Machinery," Transactions of the American Society of Mechanical Engineers, *Journal of Basic Engineering*, March 1972.

[5] Rashid, Y. R., "Three-Dimensional Analysis of Elastic Solids-I: Analysis Procedure," *International Journal of Solids and Structures*, Vol. 5, pp. 1311–1331, 1969.

[6] Rashid, Y. R., "Three-Dimensional Analysis of Elastic Solids-II: The Computational Problem," *International Journal of Solids and Structures*, Vol. 6, pp. 195–207, 1970.

[7] *Three-Dimensional Continuum Computer Programs for Structural Analysis*, Cruse, T. A., and Griffin, D. S., eds., American Society of Mechanical Engineers, 1972.

[8] Zienkiewicz, O. C., *The Finite Element Method*, 3rd ed., McGraw-Hill, London, 1977.

[9] Irons, B. M., "Quadrature Rules for Brick Based Finite Elements," *International Journal for Numerical Methods in Engineering*, Vol. 3, No. 2, pp. 293–294, 1971.

[10] Hellen, T. K., "Effective Quadrature Rules for Quadratic Solid Isoparametric Finite Elements," *International Journal for Numerical Methods in Engineering*, Vol. 4, No. 4, pp. 597–599, 1972.

[11] Linear Stress and Dynamics Reference Division, Docutech On-line Documentation, Algor, Inc., Pittsburgh, PA.

[12] Cook, R. D., Malkus, D. S., Plesha, M. E., and Witt, R. J., *Concepts and Applications of Finite Element Analysis*, 4th ed., Wiley, New York, 2002.

Problems

11.1 Evaluate the matrix $\underline{B}$ for the tetrahedral solid element shown in Figure P11–1.

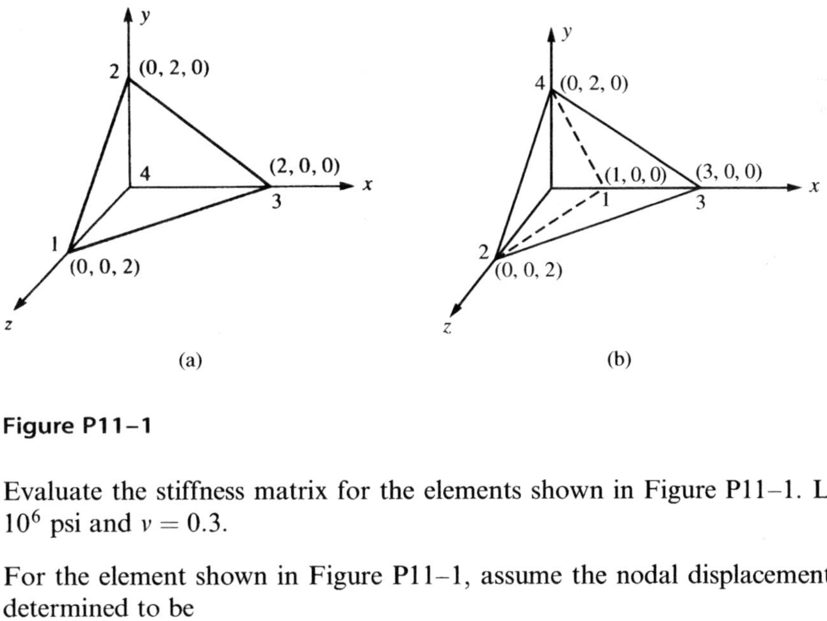

Figure P11–1

11.2 Evaluate the stiffness matrix for the elements shown in Figure P11–1. Let $E = 30 \times 10^6$ psi and $\nu = 0.3$.

11.3 For the element shown in Figure P11–1, assume the nodal displacements have been determined to be

$$\begin{array}{lll} u_1 = 0.005 \text{ in.} & v_1 = 0.0 & w_1 = 0.0 \\ u_2 = 0.001 \text{ in.} & v_2 = 0.0 & w_2 = 0.001 \text{ in.} \\ u_3 = 0.005 \text{ in.} & v_3 = 0.0 & w_3 = 0.0 \\ u_4 = -0.001 \text{ in.} & v_4 = 0.0 & w_4 = 0.005 \text{ in.} \end{array}$$

Determine the strains and then the stresses in the element. Let $E = 30 \times 10^6$ psi and $\nu = 0.3$.

11.4 What is special about the strains and stresses in the tetrahedral element?

11.5 Show that for constant body force Z_b acting on an element ($X_b = 0$ and $Y_b = 0$),

$$\{f_{bi}\} = \frac{V}{4}\begin{Bmatrix} 0 \\ 0 \\ Z_b \end{Bmatrix}$$

where $\{f_{bi}\}$ represents the body forces at node i of the element with volume V.

11.6 Evaluate the $\underline{B}$ matrix for the tetrahedral solid element shown in Figure P11–6. The coordinates are in units of millimeters.

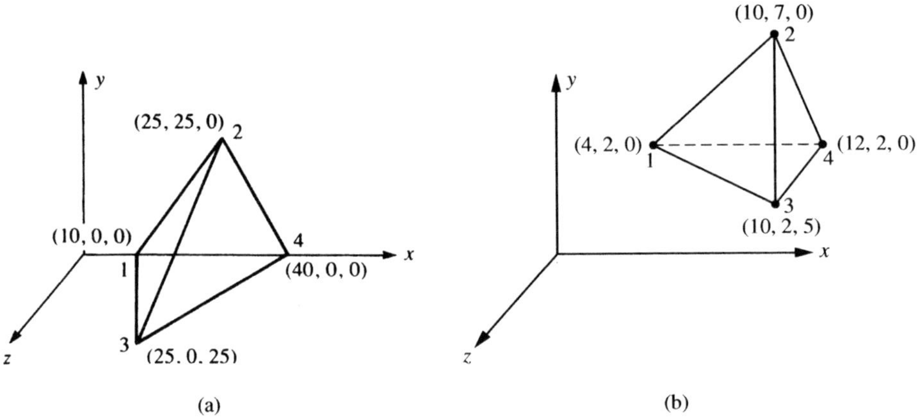

Figure P6–6

11.7 For the element shown in Figure P11–6, assume the nodal displacements have been determined to be

$u_1 = 0.0$	$v_1 = 0.0$	$w_1 = 0.0$
$u_2 = 0.01$ mm	$v_2 = 0.02$ mm	$w_2 = 0.01$ mm
$u_3 = 0.02$ mm	$v_3 = 0.01$ mm	$w_3 = 0.005$ mm
$u_4 = 0.0$	$v_4 = 0.01$ mm	$w_4 = 0.01$ mm

Determine the strains and then the stresses in the element. Let $E = 210$ GPa and $\nu = 0.3$.

11.8 For the linear strain tetrahedral element shown in Figure P11–8, (a) express the displacement fields u, v, and w in the x, y and z directions, respectively. Hint: There are

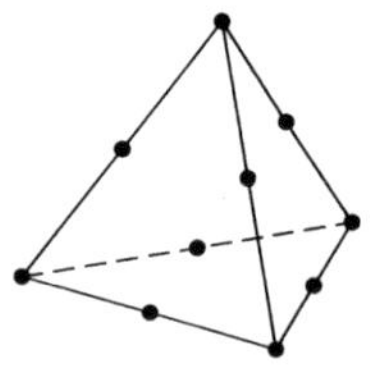

Figure P11–8

10 nodes each with three translational degrees of freedom, u_i, v_i, and w_i. Also look at the linear strain triangle given by Eq. (8.1.2) or the expansion of Eqs. (11.2.2).

11.9 Figure P11–9 shows how solid and plane elements may be connected. What restriction must be placed on the externally applied loads for this connection to be acceptable?

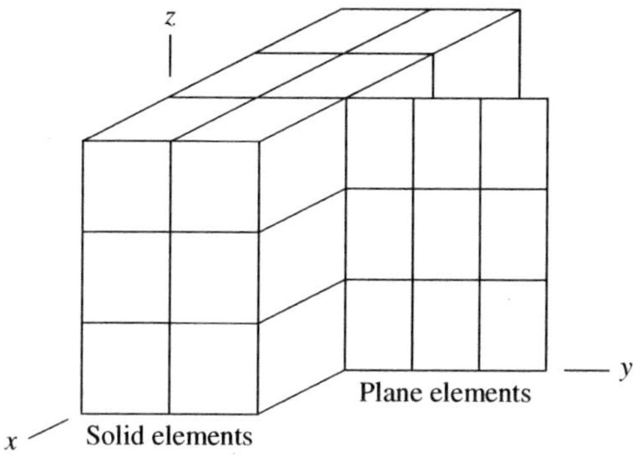

Figure P11–9

11.10 Express the explicit shape functions N_2 through N_8, similar to N_1 given by Eq. (11.3.4), for the linear hexahedral element shown in Figure 11–5 on page 501.

11.11 Express the explicit shape functions for the corner nodes of the quadratic hexahedral element shown in Figure 11–6 on page 504.

11.12 Write a computer program to evaluate $\underset{\sim}{k}$ of Eq. (11.3.9) using a $2 \times 2 \times 2$ Gaussian quadrature rule.

Solve the following problems using a computer program.

11.13 Determine the deflections at the four corners of the free end of the structural steel cantilever beam shown in Figure P11–13. Also determine the maximum principal stress.

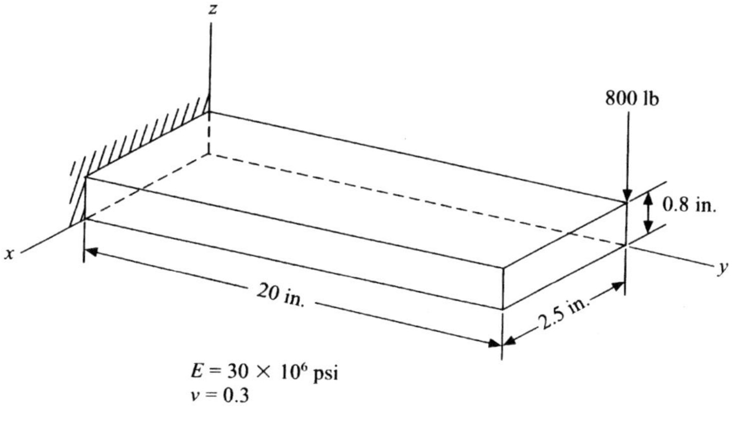

Figure P11–13

11.14 A portion of a structural steel brake pedal in a vehicle is modeled as shown in Figure P11–14. Determine the maximum deflection at the pedal under a line load of 5 lb/in. as shown.

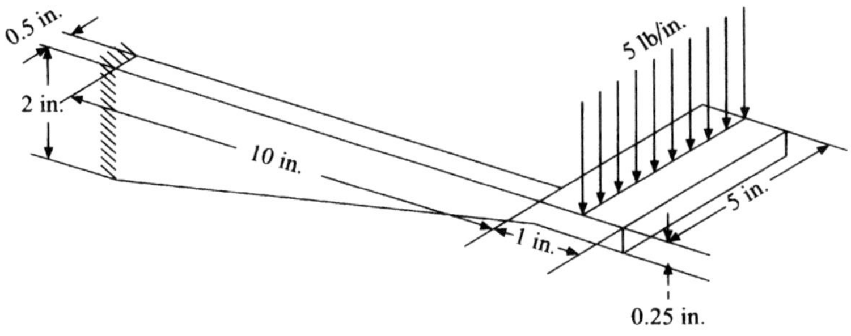

Figure P11–14

11.15 For the compressor flap valve shown in Figure P11–15, determine the maximum operating pressure such that the material yield stress is not exceeded with a factor of safety of two. The valve is made of hardened 1020 steel with a modulus of elasticity of 30 million psi and a yield strength of 62,000 psi. The valve thickness is a uniform 0.018 in. The value clip ears support the valve at opposite diameters. The pressure load is applied uniformly around the annular region.

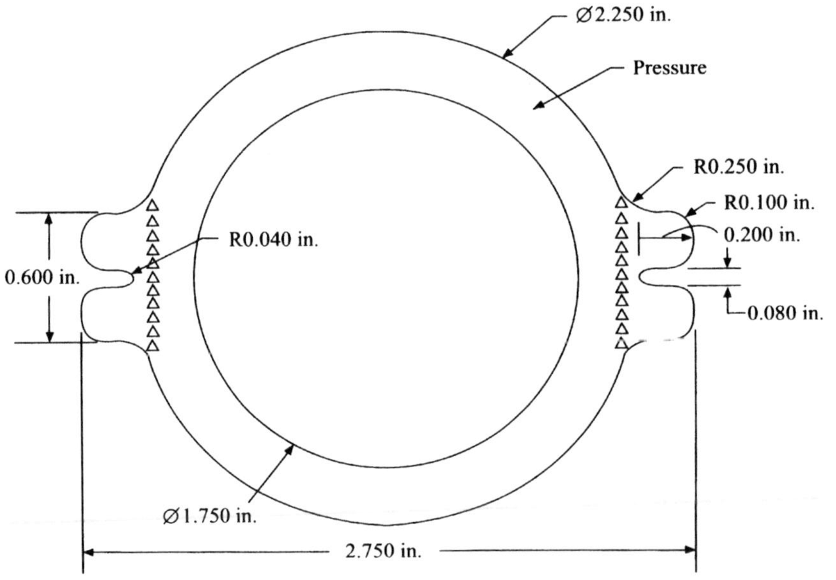

Figure P11–15

11.16 An S-shaped block used in force measurement as shown in Figure P11.16 is to be designed for a pressure of 1000 psi applied uniformly to the top surface. Determine the uniform thickness of the block needed such that the sensor is compressed no more than 0.05 in. Also make sure that the maximum stress from the maximum distortion energy failure theory is less than the yield strength of the material. Use a factor of safety of 1.5 on the stress only. The overall size of the block must fit in a 1.5-in.-high, 1-in.-wide, 1-in.-deep volume. The block should be made of steel.

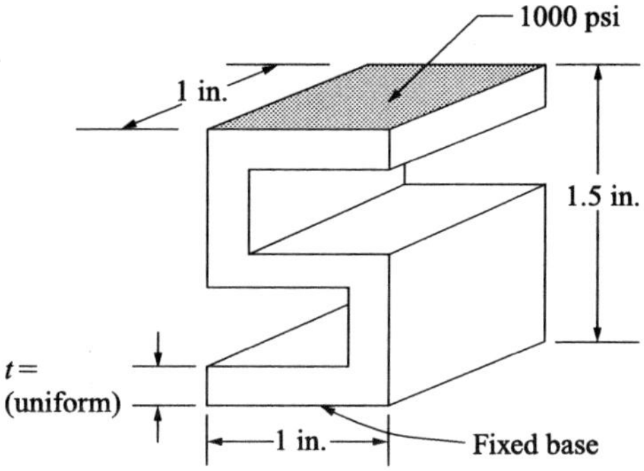

Figure P11–16 S-shaped block

11.17 A device is to be hydraulically loaded to resist an upward force $P = 6000$ lb as shown in Figure P11–17. Determine the thickness of the device such that the maximum deflection is 0.1 in. vertically and the maximum stress is less than the yield strength using a factor of safety of 2 (only on the stress). The device must fit in a space 7 in. high, 3 in. wide, and 2.3 in. deep. The top flange is bent vertically as shown, and the device is clamped to the floor. Use steel for the material.

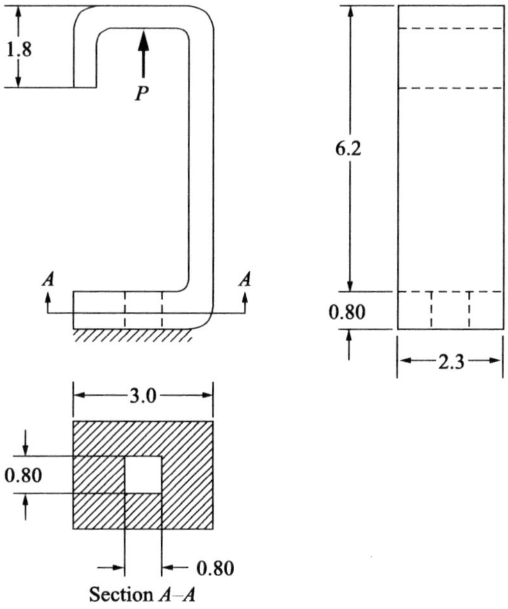

Figure P11–17 Hydraulically loaded device

Answers to Selected Problems

Chapter 2

2.1 **a.** $\underline{K} = \begin{bmatrix} k_1 & 0 & -k_1 & 0 \\ 0 & k_3 & 0 & -k_3 \\ -k_1 & 0 & k_1 + k_2 & -k_2 \\ 0 & -k_3 & -k_2 & k_2 + k_3 \end{bmatrix}$

b. $d_{3x} = \dfrac{k_2 P}{k_1 k_2 + k_1 k_3 + k_2 k_3}, \quad d_{4x} = \dfrac{(k_1 + k_2)P}{k_1 k_2 + k_1 k_3 + k_2 k_3}$

c. $F_{1x} = \dfrac{-k_1 k_2 P}{k_1 k_2 + k_1 k_3 + k_2 k_3}, \quad F_{2x} = \dfrac{-k_3(k_1 + k_2)P}{k_1 k_2 + k_1 k_3 + k_2 k_3}$

2.2 $d_{2x} = 0.5$ in., $\quad F_{3x} = 250$ lb, $\quad \hat{f}_{1x}^{(1)} = -\hat{f}_{2x}^{(1)} = -250$ lb, $\quad \hat{f}_{2x}^{(2)} = -\hat{f}_{3x}^{(2)} = -250$ lb

2.3 **a.** $\underline{K} = \begin{bmatrix} k & -k & 0 & 0 & 0 \\ -k & 2k & -k & 0 & 0 \\ 0 & -k & 2k & -k & 0 \\ 0 & 0 & -k & 2k & -k \\ 0 & 0 & 0 & -k & k \end{bmatrix}$

b. $d_{2x} = \dfrac{P}{2k}, \quad d_{3x} = \dfrac{P}{k}, \quad d_{4x} = \dfrac{P}{2k}$ **c.** $F_{1x} = -\dfrac{P}{2}, \quad F_{5x} = -\dfrac{P}{2}$

2.4 **a.** $\underline{K}$ same as 2.3a. **b.** $d_{2x} = \dfrac{\delta}{4}, \quad d_{3x} = \dfrac{\delta}{2}, \quad d_{4x} = \dfrac{3\delta}{4}$ **c.** $F_{1x} = \dfrac{-k\delta}{4}, \quad F_{5x} = \dfrac{k\delta}{4}$

2.5 $\underline{K} = \begin{bmatrix} 1 & -1 & 0 & 0 \\ -1 & 10 & 0 & -9 \\ 0 & 0 & 5 & -5 \\ 0 & -9 & -5 & 14 \end{bmatrix}$

2.6 $d_{2x} = 0.4746$ in

2.7 $d_{2x} = 1$ in., $d_{3x} = 2$ in.

$\hat{f}_{1x}^{(1)} = -\hat{f}_{2x}^{(1)} = -500$ lb, $\hat{f}_{2x}^{(2)} = -\hat{f}_{3x}^{(2)} = -500$ lb, $F_{1x} = -500$ lb

2.8 $d_{1x} = 0$, $d_{2x} = 3$ in., $d_{3x} = 7$ in., $d_{4x} = 11$ in.

$\hat{f}_{1x}^{(1)} = -\hat{f}_{2x}^{(1)} = -3000$ lb, $\hat{f}_{2x}^{(2)} = -\hat{f}_{3x}^{(2)} = -4000$ lb

$\hat{f}_{3x}^{(3)} = -\hat{f}_{4x}^{(3)} = -4000$ lb, $F_{1x} = -3000$ lb

2.9 $d_{2x} = -2$ in.

$\hat{f}_{1x}^{(1)} = -\hat{f}_{2x}^{(1)} = 2000$ lb, $\hat{f}_{2x}^{(2)} = -\hat{f}_{3x}^{(2)} = -1000$ lb

$\hat{f}_{2x}^{(3)} = -\hat{f}_{4x}^{(3)} = -1000$ lb, $F_{1x} = 2000$ lb, $F_{3x} = F_{4x} = 1000$ lb

2.10 $d_{2x} = 0.01$ m, $\hat{f}_{1x}^{(1)} = -\hat{f}_{2x}^{(1)} = -20$ N

$\hat{f}_{2x}^{(2)} = -\hat{f}_{3x}^{(2)} = -20$ N, $F_{1x} = -20$ N

2.11 $d_{2x} = 0.027$ m, $d_{3x} = 0.018$ m

$\hat{f}_{1x}^{(1)} = -\hat{f}_{2x}^{(1)} = -270$ N, $\hat{f}_{2x}^{(2)} = -\hat{f}_{3x}^{(2)} = 180$ N

$\hat{f}_{3x}^{(3)} = -\hat{f}_{4x}^{(3)} = 180$ N, $F_{1x} = -270$ N, $F_{4x} = -180$ N

2.12 $d_{2x} = 0.125$ m, $d_{3x} = 0.25$ m, $d_{4x} = 0.125$ m

$\hat{f}_{1x}^{(1)} = -\hat{f}_{2x}^{(1)} = -2.5$ kN, $\hat{f}_{2x}^{(2)} = -\hat{f}_{3x}^{(2)} = -2.5$ kN

$\hat{f}_{3x}^{(3)} = -\hat{f}_{4x}^{(3)} = 2.5$ kN, $\hat{f}_{4x}^{(4)} = -\hat{f}_{5x}^{(4)} = 2.5$ kN

$F_{1x} = -2.5$ kN, $F_{5x} = -2.5$ kN

2.13 $d_{2x} = -0.25$ m, $d_{3x} = -0.75$ m

$\hat{f}_{1x}^{(1)} = -\hat{f}_{2x}^{(1)} = 100$ N, $\hat{f}_{2x}^{(2)} = -\hat{f}_{3x}^{(2)} = 200$ N

$F_{1x} = 100$ N

2.14 $d_{3x} = 0.001$ m, $\hat{f}_{1x}^{(1)} = -\hat{f}_{3x}^{(1)} - -0.5$ kN

$\hat{f}_{2x}^{(2)} = -\hat{f}_{3x}^{(2)} = -0.5$ kN, $\hat{f}_{3x}^{(3)} = -\hat{f}_{4x}^{(3)} = 1$ kN

$F_{1x} = -0.5$ kN, $F_{2x} = -0.5$ kN, $F_{4x} = -1$ kN

2.15 $d_{2x} = 1/3$ in., $d_{3x} = -1/3$ in.

2.16 **a.** $x = 0.5$ in. ↓, $\pi_{p_{\min}} = -125$ lb-in.

b. $x = 2.0$ in. ←, $\pi_{p_{\min}} = -1000$ lb-in.

c. $x = 1.962$ mm ↓, $\pi_{p_{\min}} = -3849$ N · mm

d. $x = 2.4525$ mm →, $\pi_{p_{\min}} = -1203$ N · mm

2.17 $x = 2.0$ in. ↑

2.18 $x = 0.707$ in. $\leftarrow$, $\pi_{p_{\min}} = -235.7$ in-lb

2.19 Same as 2.10

2.20 Same as 2.15

Chapter 3

3.1 **a.** $\underline{K} = \begin{bmatrix} \dfrac{A_1E_1}{L_1} & \dfrac{-A_1E_1}{L_1} & 0 & 0 \\ \dfrac{-A_1E_1}{L_1} & \dfrac{A_1E_1}{L_1}+\dfrac{A_2E_2}{L_2} & \dfrac{-A_2E_2}{L_2} & 0 \\ 0 & \dfrac{-A_2E_2}{L_2} & \dfrac{A_2E_2}{L^2}+\dfrac{A_3E_3}{L_3} & \dfrac{-A_3E_3}{L_3} \\ 0 & 0 & \dfrac{-A_3E_3}{L_3} & \dfrac{A_3E_3}{L_3} \end{bmatrix}$

b. $d_{2x} = \dfrac{PL}{3AE}$, $d_{3x} = \dfrac{2PL}{3AE}$

c. **i.** $d_{2x} = 3.33 \times 10^{-4}$ in., $d_{3x} = 6.67 \times 10^{-4}$ in.

ii. $F_{1x} = -333$ lb, $F_{4x} = -667$ lb

iii. $\sigma^{(1)} = 333$ psi (T), $\sigma^{(2)} = 333$ psi (T), $\sigma^{(3)} = -667$ psi (C)

3.2 $d_{2x} = -0.595 \times 10^{-4}$ m, $d_{3x} = -1.19 \times 10^{-4}$ m, $F_{1x} = 5$ kN

$\hat{f}_{1x}^{(1)} = -\hat{f}_{2x}^{(1)} = 5$ kN, $\hat{f}_{2x}^{(2)} = -\hat{f}_{3x}^{(2)} = 5$ kN

3.3 $d_{2x} = 1.91 \times 10^{-3}$ in., $F_{1x} = -5715$ lb, $F_{3x} = -2286$ lb

$\hat{f}_{1x}^{(1)} = -\hat{f}_{2x}^{(1)} = -5715$ lb, $\hat{f}_{2x}^{(2)} = -\hat{f}_{3x}^{(2)} = 2286$ lb

3.4 $d_{2x} = -1.66 \times 10^{-4}$ in., $d_{3x} = -1.33 \times 10^{-3}$ in.

$F_{1x} = 667$ lb, $F_{4x} = 5333$ lb

$\hat{f}_{1x}^{(1)} = -\hat{f}_{2x}^{(1)} = 667$ lb, $\hat{f}_{2x}^{(2)} = -\hat{f}_{3x}^{(2)} = 4667$ lb

$\hat{f}_{3x}^{(3)} = -\hat{f}_{4x}^{(3)} = -5333$ lb

3.5 $d_{2x} = 0.003$ in., $d_{3x} = 0.009$ in., $F_{1x} = -15000$ lb

$\hat{f}_{1x}^{(1)} = -\hat{f}_{2x}^{(1)} = \hat{f}_{2x}^{(2)} = -\hat{f}_{3x}^{(2)} = -15000$ lb

3.6 $d_{2x} = 3.16 \times 10^{-3}$ in., $F_{1x} = -3790$ lb, $F_{3x} = F_{4x} = -2105$ lb

$\hat{f}_{1x}^{(1)} = -\hat{f}_{2x}^{(1)} = -3790$ lb, $\hat{f}_{2x}^{(2)} = -\hat{f}_{3x}^{(2)} = \hat{f}_{2x}^{(3)} = -\hat{f}_{4x}^{(3)} = 2105$ lb

3.7 $d_{2x} = 2.21 \times 10^{-5}$ in., $d_{3x} = 6.65 \times 10^{-3}$ in.

$F_{1x} = -33.15$ lb, $F_{4x} = -9975$ lb

$\hat{f}_{1x}^{(1)} = -\hat{f}_{2x}^{(1)} = \hat{f}_{2x}^{(2)} = -\hat{f}_{3x}^{(2)} = -33.15$ lb, $\hat{f}_{3x}^{(3)} = -\hat{f}_{4x}^{(3)} = 9975$ lb

3.8 $d_{2x} = -0.250$ mm, $d_{3x} = -1.678$ mm, $F_{1x} = 20$ kN

3.9 $d_{2x} = 0.01238$ m, $F_{1x} = -520$ kN, $F_{3x} = 530$ kN

$\hat{f}_{1x}^{(1)} = -\hat{f}_{2x}^{(1)} = -520$ kN, $\hat{f}_{2x}^{(2)} = -\hat{f}_{3x}^{(2)} = -530$ kN

3.10 $d_{2x} = 0.935 \times 10^{-3}$ m, $d_{3x} = 0.727 \times 10^{-3}$ m

$F_{1x} = -6.546$ kN, $F_{4x} = -1.455$ kN

$\hat{f}_{1x}^{(1)} = -\hat{f}_{2x}^{(1)} = -6.546$ kN, $\hat{f}_{2x}^{(2)} = -\hat{f}_{3x}^{(2)} = 1.455$ kN,

$\hat{f}_{3x}^{(3)} = -\hat{f}_{4x}^{(3)} = 1.455$ kN

3.11 $d_{2x} = 3.572 \times 10^{-4}$ m, $F_{1x} = -7.50$ kN, $F_{3x} = F_{4x} = F_{5x} = -7.50$ kN

$\hat{f}_{1x}^{(1)} = -\hat{f}_{2x}^{(1)} = -7.50$ kN,

$\hat{f}_{2x}^{(2)} = -\hat{f}_{3x}^{(2)} = \hat{f}_{2x}^{(3)} = -\hat{f}_{4x}^{(3)} = \hat{f}_{2x}^{(4)} = -\hat{f}_{5x}^{(4)} = 7.50$ kN

3.12 two-element solution, $d_{1x} = -0.686 \times 10^{-3}$ in.

one-element solution, $d_{1x} = -0.667 \times 10^{-3}$ in.

3.13 $\boldsymbol{B} = \left[-\dfrac{1}{L}+\dfrac{4x}{L^2} \quad \dfrac{-8x}{L^2} \quad \dfrac{1}{L}+\dfrac{4x}{L^2}\right]$, $\underline{k} = A\displaystyle\int_{-L/2}^{L/2} \underline{B}^T E \underline{B}\, dx$

3.15 **a.** $\underline{k} = 2.25 \times 10^6 \begin{bmatrix} 1 & 1 & -1 & -1 \\ 1 & 1 & -1 & -1 \\ -1 & -1 & 1 & 1 \\ -1 & -1 & 1 & 1 \end{bmatrix}$ lb/in.

b. $\underline{k} = \dfrac{10^6}{4} \begin{bmatrix} 1 & -\sqrt{3} & -1 & \sqrt{3} \\ -\sqrt{3} & 3 & \sqrt{3} & -3 \\ -1 & \sqrt{3} & 1 & -\sqrt{3} \\ \sqrt{3} & -3 & -\sqrt{3} & 3 \end{bmatrix}$ lb/in.

c. $\underline{k} = 7000 \begin{bmatrix} 3 & -\sqrt{3} & -3 & \sqrt{3} \\ -\sqrt{3} & 1 & \sqrt{3} & -1 \\ -3 & \sqrt{3} & 3 & -\sqrt{3} \\ \sqrt{3} & -1 & -\sqrt{3} & 1 \end{bmatrix}$ kN/m

d. $\underline{k} = 1.4 \times 10^4 \begin{bmatrix} 0.883 & 0.321 & -0.883 & -0.321 \\ 0.321 & 0.117 & -0.321 & -0.117 \\ -0.883 & -0.321 & 0.883 & 0.321 \\ -0.321 & -0.117 & 0.321 & 0.117 \end{bmatrix}$ kN/m

3.16 **a.** $\hat{d}_{1x} = 0.433$ in., $\hat{d}_{2x} = 0.592$ in.

b. $\hat{d}_{1x} = 0.433$ in., $\hat{d}_{2x} = -0.1585$ in.

3.17 **a.** $\hat{d}_{1x} = 2.165$ mm, $\hat{d}_{1y} = -1.25$ mm,

$\hat{d}_{2x} = 0.098$ mm, $\hat{d}_{2y} = -5.83$ mm

b. $\hat{d}_{1x} = -1.25$ mm, $\hat{d}_{1y} = 2.165$ mm,

$\hat{d}_{2x} = 3.03$ mm, $\hat{d}_{2y} = 5.098$ mm

3.18 **a.** $\sigma = 10{,}600$ psi, **b.** 45.47 MPa

3.19 **a.** $\underline{K} = k\begin{bmatrix} 2 & 0 & -\frac{1}{2} & \frac{1}{2} & -1 & 0 & -\frac{1}{2} & -\frac{1}{2} \\ 0 & 1 & \frac{1}{2} & -\frac{1}{2} & 0 & 0 & -\frac{1}{2} & -\frac{1}{2} \\ -\frac{1}{2} & \frac{1}{2} & \frac{1}{2} & -\frac{1}{2} & 0 & 0 & 0 & 0 \\ \frac{1}{2} & -\frac{1}{2} & -\frac{1}{2} & \frac{1}{2} & 0 & 0 & 0 & 0 \\ -1 & 0 & 0 & 0 & 1 & 0 & 0 & 0 \\ 0 & 0 & 0 & 0 & 0 & 0 & 0 & 0 \\ -\frac{1}{2} & -\frac{1}{2} & 0 & 0 & 0 & 0 & \frac{1}{2} & \frac{1}{2} \\ -\frac{1}{2} & -\frac{1}{2} & 0 & 0 & 0 & 0 & \frac{1}{2} & \frac{1}{2} \end{bmatrix}$

b. $d_{1x} = 0, \quad d_{1y} = \dfrac{-10}{k}$

3.20 $d_{2x} = 0, \quad d_{2y} = 0.142 \text{ in.}, \quad \sigma^{(1)} = \sigma^{(2)} = 707 \text{ psi (T)}$

3.21 $d_{1x} = \dfrac{231L}{AE}, \quad d_{1y} = \dfrac{43.5L}{AE}$

3.22 $d_{1x} = \dfrac{422L}{AE}, \quad d_{1y} = \dfrac{1570L}{AE}$

$\sigma^{(1)} = \dfrac{574}{A}\text{(C)}, \quad \sigma^{(2)} = \dfrac{422}{A}\text{(T)}, \quad \sigma^{(3)} = \dfrac{996}{A}\text{(T)}$

3.23 $d_{1x} = 0.24 \text{ in.}, \quad d_{1y} = 0, \quad \sigma^{(1)} = 12000 \text{ psi}$

3.24 $d_{2x} = \dfrac{26{,}675}{AE}, \quad d_{2y} = \dfrac{105{,}021}{AE}, \quad d_{3x} = \dfrac{-26{,}675}{AE}, \quad d_{3y} = \dfrac{105{,}021}{AE}$

$\hat{f}_{1x}^{(1)} = -\hat{f}_{2x}^{(1)} = -1333 \text{ lb}, \quad \hat{f}_{1x}^{(2)} = -\hat{f}_{3x}^{(2)} = -1667 \text{ lb}$

$\hat{f}_{2x}^{(3)} = -\hat{f}_{4x}^{(3)} = 1667 \text{ lb}, \quad \hat{f}_{2x}^{(4)} = -\hat{f}_{3x}^{(4)} = 0$

$\hat{f}_{3x}^{(5)} = -\hat{f}_{4x}^{(5)} = 1333 \text{ lb}, \quad \hat{f}_{1x}^{(6)} = -\hat{f}_{4x}^{(6)} = 0$

3.25 $d_{2x} = 0, \quad d_{2y} = \dfrac{225{,}000}{AE}, \quad d_{3x} = \dfrac{-53{,}340}{AE}, \quad d_{3y} = \dfrac{210{,}000}{AE}$

$\hat{f}_{1x}^{(1)} = -\hat{f}_{2x}^{(1)} = 0, \quad \hat{f}_{1x}^{(2)} = -\hat{f}_{3x}^{(2)} = -3333 \text{ lb}$

$\hat{f}_{2x}^{(4)} = -\hat{f}_{3x}^{(4)} = 1000 \text{ lb}, \quad \hat{f}_{3x}^{(5)} = -\hat{f}_{4x}^{(5)} = 2667 \text{ lb}$

$\hat{f}_{1x}^{(6)} = -\hat{f}_{4x}^{(6)} = 0$

3.26 No, the truss is unstable, $|\underline{K}| = 0$.

3.27 $d_{3x} = 0.0463 \text{ in.}, \quad d_{3y} = -0.0176 \text{ in.}$

$\hat{f}_{1x}^{(1)} = -\hat{f}_{3x}^{(1)} = -2.055 \text{ kip}, \quad \hat{f}_{2x}^{(2)} = -\hat{f}_{3x}^{(2)} = 6.279 \text{ kip}$

$\hat{f}_{3x}^{(3)} = -\hat{f}_{4x}^{(3)} = -6.6 \text{ kip}$

3.28 $\underline{T}^T = \begin{bmatrix} C & -S & 0 & 0 \\ S & C & 0 & 0 \\ 0 & 0 & C & -S \\ 0 & 0 & S & C \end{bmatrix}$ and $\underline{T}\underline{T}^T = \begin{bmatrix} 1 & 0 & 0 & 0 \\ 0 & 1 & 0 & 0 \\ 0 & 0 & 1 & 0 \\ 0 & 0 & 0 & 1 \end{bmatrix}$

$\therefore \underline{T}^T = \underline{T}^{-1}$

3.29 $d_{1x} = -0.893 \times 10^{-4} \text{ m}, \quad d_{1y} = -4.46 \times 10^{-4} \text{ m}$

$\sigma^{(1)} = 31.2 \text{ MPa (T)}, \quad \sigma^{(2)} = 26.5 \text{ MPa (T)}, \quad \sigma^{(3)} = 6.25 \text{ MPa (T)}$

3.30 $d_{1x} = 1.71 \times 10^{-4}$ m, $d_{1y} = -7.55 \times 10^{-4}$ m
$\sigma^{(1)} = 79.28$ MPa (T), $\sigma^{(2)} = 11.97$ MPa (T), $\sigma^{(3)} = -23.87$ MPa (C)

3.31 $d_{1x} = 8.25 \times 10^{-4}$ m, $d_{1y} = -3.65 \times 10^{-3}$ m
$\sigma^{(2)} = 57.74$ MPa (T), $\sigma^{(3)} = -115.5$ MPa (C)

3.32 $d_{2x} = 0.135 \times 10^{-2}$ m, $d_{2y} = -0.850 \times 10^{-2}$ m,
$d_{3y} = -0.137 \times 10^{-1}$ m, $d_{4y} = -0.164 \times 10^{-1}$ m,
$\sigma^{(1)} = -198$ MPa (C), $\sigma^{(2)} = 0$, $\sigma^{(3)} = 44.6$ MPa (T)
$\sigma^{(4)} = -31.6$ MPa (C), $\sigma^{(5)} = -191$ MPa (C),
$\sigma^{(6)} = -63.1$ MPa (C)

3.33 **a.** $d_{1x} = -3.448 \times 10^{-3}$ m, $d_{1y} = -6.896 \times 10^{-3}$ m
$\sigma^{(1)} = 102.4$ MPa (T), $\sigma^{(2)} = -72.4$ MPa (C)

3.34 $d_{4x} = 9.93 \times 10^{-3}$ in., $d_{4y} = -2.46 \times 10^{-3}$ in.
$\sigma^{(1)} = 31.25$ ksi (T), $\sigma^{(2)} = 3.459$ ksi (T), $\sigma^{(3)} = -1.538$ ksi (C)
$\sigma^{(4)} = -3.103$ ksi (C), $\sigma^{(5)} = 0$

3.35 $d_{1y} = -0.5 \times 10^{-3}$ in., $\sigma^{(1)} = 250$ psi (T)

3.36 $\hat{d}_{1x} = 0.212$ in.

3.37 $\hat{d}_{1x} = 0.0397$ in.

3.38 $\hat{d}_{2x} = 16.98$ mm

3.39 $\hat{d}_{2x} = 1.71$ mm

3.40 $d_{1x} = -3.018 \times 10^{-5}$ m, $d_{1y} = -1.517 \times 10^{-5}$ m,
$d_{1z} = 2.684 \times 10^{-5}$ m, $\sigma^{(1)} = -338$ kN/m^2 (C),
$\sigma^{(2)} = -1690$ kN/m^2 (C), $\sigma^{(3)} = -7965$ kN/m^2 (C)
$\sigma^{(4)} = -2726$ kN/m^2 (C)

3.41 $d_{1x} = 1.383 \times 10^{-3}$ m, $d_{1y} = -5.119 \times 10^{-5}$ m
$d_{1x} = 6.015 \times 10^{-5}$ m, $\sigma^{(1)} = 20.51$ MPa (T),
$\sigma^{(2)} = 4.21$ MPa (T), $\sigma^{(3)} = -5.29$ MPa (C)

3.42 $d_{5x} = 0.0014$ in., $d_{5y} = 0$, $d_{5z} = -0.00042$ in.
$\sigma^{(1)} = \sigma^{(4)} = 180$ psi (T), $\sigma^{(2)} = \sigma^{(3)} = 140$ psi (C)

3.43 $d_{4x} = 0.00863$ in., $d_{4y} = 0$, $d_{4z} = -0.00683$ in.
$\sigma^{(1)} = -916$ psi (C)

3.46 $d_{2y} = -0.0192$ in., $d_{3y} = -0.0168$ in.
$\sigma^{(1)} = -1668$ psi (C), $\sigma^{(2)} = 1332$ psi (T), $\sigma^{(3)} = 1000$ psi (T)

3.47 $d_{1x} = \dfrac{-110P}{AE}$ in., $d_{1y} = 0$, $d_{2x} = 0$, $d_{2y} = \dfrac{-405P}{AE}$ in.,

$d_{3x} = 0$, $d_{3y} = \dfrac{-433P}{AE}$ in., $d_{4x} = \dfrac{50P}{AE}$ in., $d_{4y} = \dfrac{-208P}{AE}$ in.

$\sigma^{(1)} = -0.156\dfrac{P}{A}$, $\sigma^{(2)} = -0.208\dfrac{P}{A}$, $\sigma^{(3)} = -1.16\dfrac{P}{A}$

$\sigma^{(4)} = 0.260\dfrac{P}{A}$, $\sigma^{(5)} = -0.573\dfrac{P}{A}$, $\sigma^{(6)} = 0.458\dfrac{P}{A}$

3.48 $d_{2y} = -0.955 \times 10^{-2}$ m, $d_{4y} = -1.03 \times 10^{-2}$ m,

$\sigma^{(1)} = 67.1$ MPa (C), $\sigma^{(2)} = 60.0$ MPa (T), $\sigma^{(3)} = 22.4$ MPa (C)

$\sigma^{(4)} = 44.7$ MPa (C), $\sigma^{(5)} = 20.0$ MPa (T)

3.49 $d'_{1x} = 0$, $d_{2y} = -0.00283$ in., $F_{2x} = 2000$ lb
$\sigma^{(1)} = 0$, $\sigma^{(2)} = 1414$ psi (T), $\sigma^{(3)} = 0$

3.50 $d_{2y} = -0.00283$ in.

3.51 $d'_{2x} = 0.002$ in., $f'_{1x} = -2800$ lb., $f'_{2x} = -2000$ lb

$F'_{2y} = -2828$ lb

3.52 **a.** $d_{1x} = 0.010$ in. ↓, $\pi_{p_{\min}} = -100$ lb-in.

b. $d_{1x} = 0.00833$ in. →, $\pi_{p_{\min}} = -41.67$ lb-in.

3.53 $\underline{k} = \dfrac{3A_0E}{2L}\begin{bmatrix} 1 & -1 \\ -1 & 1 \end{bmatrix}$

3.54 two-element solution: $d_{2x} = 0.00825$ in., $d_{3x} = 0.012$ in., $\sigma^{(1)} = 8250$ psi (T),
$\sigma^{(2)} = 3750$ psi (T),

3.55 two-element solution: $d_{2x} = 6.75 \times 10^{-3}$ in., $d_{3x} = 0.009$ in.

$\sigma^{(1)} = 6750$ psi (T), $\sigma^{(2)} = 2250$ psi (T)

3.56 $d_{2x} = 0.75 \times 10^{-3}$ in., $\sigma^{(1)} = 750$ psi (T)

3.57 $d_{1x} = \gamma L^2/(2E)$, $d_{2x} = 3\gamma L^2/(8E)$, $\sigma^{(1)} = \gamma L/8$, $\sigma^{(2)} = 3\gamma L/8$

3.58 **a.** $f_{1x} = 583.3$ lb, $f_{2x} = 666.7$ lb

b. $f_{1x} = 26.7$ kN, $f_{2x} = 80$ kN

Chapter 4

4.3 $d_{2y} = \dfrac{-7PL^3}{768EI}$, $\phi_1 = \dfrac{-PL^2}{32EI}$, $\phi_2 = \dfrac{PL^2}{128EI}$

$F_{1y} = \dfrac{5P}{16}$, $M_1 = 0$, $F_{3y} = \dfrac{11P}{16}$, $M_3 = \dfrac{-3PL}{16}$

4.4 $d_{1y} = \dfrac{-PL^3}{3EI}$, $\phi_1 = \dfrac{PL^3}{2EI}$, $F_{2y} = P$, $M_2 = -PL$

4.5 $d_{1y} = -2.688$ in., $\phi_1 = 0.0144$ rad, $\phi_2 = 0.0048$ rad
$F_{2y} = 2.5$ kip, $F_{3y} = -1.5$ kip, $M_3 = 10.0$ k-ft

4.6 $d_{3y} = -3.94$ in.

4.7 $d_{2y} = -0.105$ in., $\phi_2 = -0.003$ rad, $d_{3y} = -0.345$ in., $\phi_3 = -0.0045$ rad

4.8 $d_{2y} = -1.34 \times 10^{-4}$ m, $\phi_2 = 8.93 \times 10^{-5}$ rad
$F_{1y} = 10$ kN, $M_1 = 12.5$ kN · m, $F_{3y} = 1.87$ N, $M_3 = -2.5$ kN · m

4.9 $d_{3y} = -7.619 \times 10^{-4}$ m, $\phi_2 = -3.809 \times 10^{-4}$ rad, $\phi_1 = 1.904 \times 10^{-4}$ rad
$F_{1y} = -0.889$ kN, $F_{2y} = 4.889$ kN

4.10 $d_{2y} = -0.886$ in., $\phi_2 = -0.00554$ rad
$F_{1y} = 1115$ lb, $M_1 = -267$ k-in.

4.11 $d_{2y} = -7.934 \times 10^{-3}$ m, $\phi_1 = -2.975 \times 10^{-3}$ rad
$F_{1y} = 5.208$ kN, $F_{3y} = 5.208$ kN
$F_{\text{spring}} = 1.587$ kN

4.12 $d_{2y} = d_{4y} = \dfrac{-1wL^4}{607.5EI}, \quad d_{3y} = \dfrac{-wL^4}{507EI}$
$\phi_2 = \dfrac{-1wL^3}{270EI}, \quad \phi_4 = -\phi_2$
$F_{1y} = \dfrac{wL}{2}, \quad M_1 = \dfrac{wL^2}{12}$

4.13 $d_{2y} = \dfrac{-wL^4}{384EI}, \quad F_{1y} = \dfrac{wL}{2}, \quad M_1 = \dfrac{wL^2}{12}$

4.14 $d_{2y} = \dfrac{-5wL^4}{384EI}, \quad \phi_1 = -\phi_3 = \dfrac{-wL^3}{24EI}, \quad F_{1y} = \dfrac{wL}{2}$

4.15 $d_{3y} = \dfrac{-wL^4}{4EI}, \quad \phi_2 = \dfrac{-wL^3}{8EI}, \quad \phi_3 = \dfrac{-7wL^3}{24EI}$
$F_{1y} = \dfrac{-3wL}{4}, \quad M_1 = \dfrac{-wL^2}{4}, \quad F_{2y} = \dfrac{7wL}{4}$

4.16 $\hat{f}_{1y} = \dfrac{-3wL}{20}, \quad \hat{m}_1 = \dfrac{-wL^2}{30}, \quad \hat{f}_{2y} = \dfrac{-7wL}{20}, \quad \hat{m}_2 = \dfrac{wL^2}{20}$

4.17 $F_{1y} = \dfrac{3wL}{20}, \quad M_1 = \dfrac{wL^2}{30}, \quad F_{3y} = \dfrac{7wL}{20}, \quad M_3 = \dfrac{-wL^2}{20}$

4.18 $\phi_2 = \dfrac{wL^3}{80EI}, \quad F_{1y} = \dfrac{9wL}{40}, \quad M_1 = \dfrac{7wL^2}{120}, \quad F_{2y} = \dfrac{11wL}{40}$

4.19 $d_{3y} = -0.0244$ m, $\phi_3 = -0.0071$ rad, $\phi_2 = -0.00305$ rad
$F_{1y} = -24$ kN, $M_1 = -32$ kN · m, $F_{2y} = 56$ kN
$\hat{f}_{1y}^{(1)} = -\hat{f}_{2y}^{(1)} = -24$ kN, $\hat{m}_1^{(1)} = -32$ kN · m, $\hat{m}_2^{(1)} = -64$ kN · m
$\hat{f}_{2y}^{(2)} = 32$ kN, $\hat{m}_2^{(2)} = 64$ kN · m, $\hat{f}_{3y}^{(2)} = 0$, $\hat{m}_3^{(2)} = 0$

4.20 $\phi_1 = -0.0032$ rad, $d_{2y} = -0.0115$ m, $\phi_3 = 0.0032$ rad

$F_{1y} = 29.94$ kN, $F_{2y} = 0.1152$ kN, $F_{3y} = 29.94$ kN

$\hat{f}_{1y}^{(1)} = 29.94$ kN, $\hat{m}_1^{(1)} = 0$, $\hat{f}_{2y}^{(1)} = 0.058$ kN, $\hat{m}_2^{(1)} = 59.65$ kN · m

4.21 $d_{2y} = -2.514$ in., $\phi_2 = -0.00698$ rad, $\phi_3 = 0.0279$ rad

$F_{1y} = 37.5$ kip, $M_1 = 225$ k-in., $F_{3y} = 22.5$ kip

4.22 $d_{3y} = -3.277$ in., $\phi_3 = -0.0323$ rad, $\phi_2 = -0.0130$ rad

$F_{1y} = -20.5$ kip, $M_1 = -71.67$ k-ft, $F_{2y} = 60.5$ kip

4.23 $d_{2y} = -2.34$ in., $F_{1y} = 5325$ lb $= F_{3y}$, $M_1 = 19{,}900$ lb-ft $= -M_3$

4.24 $\phi_1 = -3.596 \times 10^{-4}$ rad, $\phi_2 = 9.92 \times 10^{-5}$ rad, $\phi_3 = 1.091 \times 10^{-4}$ rad

$F_{1y} = 9875$ N, $F_{2y} = 28{,}406$ N, $F_{3y} = 6719$ N

4.25 $d_{max} = -0.000756$ m at midspan of AB and BC

$\sigma_{max} = 34.3$ MPa at midspan of AB and BC

$\sigma_{min} = -51.0$ MPa at B

4.26 $d_{max} = -0.1953$ m at midspan of BC

$\sigma_{min} = -469$ MPa

4.27 $d_{max} = -1.028$ in. under 7.5 kip load at B

$\sigma_{max} = 34000$ psi

$\sigma_{min} = -65800$ psi

4.28 $d_{max} = -0.0419$ m at C

$\sigma_{max} = 66.97$ MPa at fixed end A

$\sigma_{min} = -133.9$ MPa at B

4.29 $d_{max} = -0.495$ in. at C

$\sigma_{max} = 5625$ psi at A

$\sigma_{min} = -22500$ psi at B

4.30 $d_{max} = -0.087$ m at C

$\sigma_{max} = 257$ MPa at B

4.37 $d_{2y} = \dfrac{-PL^3}{192EI} - \dfrac{wL^4}{384EI}$, $F_{1y} = \dfrac{P + wL}{2}$, $M_1 = \dfrac{PL}{8} + \dfrac{wL^2}{12}$

4.38 $d_{2y} = \dfrac{-5PL^3}{648EI}$

4.39 $d_{2y} = \dfrac{-(25P + 22wL)L^3}{240EI}$, $\phi_2 = \dfrac{-(PL^2 + wL^3)}{8EI}$

$F_{1y} = P + \dfrac{wL}{2}$, $M_1 = \dfrac{PL}{2} + \dfrac{wL^2}{3}$

4.40 $d_{2y} = -1.57 \times 10^{-4}$ m, $\phi_2 = 1.19 \times 10^{-4}$ rad

4.41 $d_{2y} = -3.18 \times 10^{-4}$ m, $\phi_2 = 1.58 \times 10^{-4}$ rad, $\phi_3 = 1.58 \times 10^{-4}$ rad

4.42 $d_{3y} = -2.13 \times 10^{-5}$ m, $\phi_2 = -1.28 \times 10^{-5}$ rad, $\phi_3 = 2.69 \times 10^{-5}$ rad

4.44 $\underline{k} = \dfrac{GA_W}{L}\begin{bmatrix} 1 & -1 \\ -1 & 1 \end{bmatrix}$

4.47 $\hat{\underline{k}} = EI\displaystyle\int_0^L [B]^T[B]\,d\hat{x} + k_f \int_0^L [N]^T[N]\,d\hat{x}$

Chapter 5

5.1 $d_{2x} = 0.0278$ in., $d_{2y} = 0$, $\phi_2 = -0.555 \times 10^{-4}$ rad
$\hat{f}_{1x}^{(1)} = -\hat{f}_{2x}^{(1)} = -8300$ lb, $\hat{f}_{1y}^{(1)} = -\hat{f}_{2y}^{(1)} = 4.6$ lb
$\hat{m}_1^{(1)} = 2775$ lb-in., $\hat{m}_2^{(1)} = 0$

5.2 $d_{2x} = d_{3x} = 0.688$ in., $d_{2y} = -d_{3y} = 0.00171$ in.
$\phi_2 = -\phi_3 = -0.00173$ rad
$\hat{f}_{1x}^{(1)} = -\hat{f}_{2x}^{(1)} = -2140$ lb, $\hat{f}_{1y}^{(1)} = -\hat{f}_{2y}^{(1)} = -2503$ lb
$\hat{m}_1^{(1)} = 343{,}600$ lb-in., $\hat{m}_2^{(1)} = 257{,}000$ lb-in.
$\hat{f}_{2x}^{(2)} = -\hat{f}_{3x}^{(2)} = 2497$ lb, $\hat{f}_{2y}^{(2)} = -\hat{f}_{3y}^{(2)} = -2140$ lb
$\hat{m}_2^{(2)} = -257{,}000$ lb-in., $\hat{m}_3^{(2)} = -256{,}600$ lb-in.
$\hat{f}_{3x}^{(3)} = -\hat{f}_{4x}^{(3)} = 2140$ lb, $\hat{f}_{3y}^{(3)} = -\hat{f}_{4y}^{(3)} = 2497$ lb
$\hat{m}_3^{(3)} = 256{,}600$ lb-in., $\hat{m}_4^{(3)} = 342{,}700$ lb-in.
$F_{1x} = F_{4x} = -2503$ lb, $F_{1y} = -F_{4y} = -2140$ lb
$M_1 = 343{,}600$ lb-in., $M_4 = 342{,}700$ lb-in.

5.3 Channel section 6×8.2 based on $M_{\max} = 106{,}900$ lb-in.

5.4 $d_{4x} = 0.00445$ in., $d_{4y} = -0.0123$ in., $\phi_4 = -0.00290$ rad
$\hat{f}_{1x}^{(1)} = -\hat{f}_{4x}^{(1)} = 4.04$ kip, $\hat{f}_{1y}^{(1)} = -\hat{f}_{4y}^{(1)} = -1.43$ kip
$\hat{m}_1^{(1)} = -254$ k-in., $\hat{m}_4^{(1)} = -513$ k-in.
$\hat{f}_{2x}^{(2)} = -\hat{f}_{4x}^{(2)} = 5.82$ kip, $\hat{f}_{2y}^{(2)} = -\hat{f}_{4y}^{(2)} = -1.45$ kip
$\hat{m}_2^{(2)} = -260$ k-in., $\hat{m}_4^{(2)} = -519$ k-in.
$F_{1x} = 3.1$ kip, $F_{1y} = 2.96$ kip, $M_1 = -254$ k-in.
$F_{2x} = -1.31$ kip, $F_{2y} = 5.86$ kip, $M_2 = -260$ k-in.
$F_{3x} = -1.78$ kip, $F_{3y} = 11.17$ kip, $M_3 = -1736$ k-in.

5.5 $d_{2x} = 0.05618$ in., $d_{2y} = -0.1792$ in., $\phi_2 = -0.00965$ rad
$\hat{f}_{1x}^{(1)} = 90.07$ kip, $\hat{f}_{1y}^{(1)} = 3.83$ kip, $\hat{m}_1^{(1)} = 361$ k-in.
$\hat{f}_{2x}^{(1)} = -73.43$ kip, $\hat{f}_{2y}^{(1)} = 7.27$ kip, $\hat{m}_2^{(1)} = -1106$ k-in.
$\hat{f}_{2x}^{(2)} = -\hat{f}_{3x}^{(2)} = 46.8$ kip, $\hat{f}_{2y}^{(2)} = 17.05$ kip, $\hat{m}_2^{(2)} = 1107$ k-in.

$\hat{f}_{3y}^{(2)} = 22.95$ kip, $\hat{m}_3^{(2)} = -2171$ k-in.

$F_{1x} = F_{3x} = 46.8$ kip, $F_{1y} = 77.1$ kip, $M_1 = 361$ k-in.

$F_{3y} = 22.95$ kip, $M_3 = 2171$ k-in.

5.6 $d_{2x} = -0.000269$ in., $d_{2y} = -0.0363$ in., $\phi_2 = -0.00347$ rad

$\hat{f}_{1x}^{(1)} = 46.6$ kip, $\hat{f}_{1y}^{(1)} = 6.07$ kip, $\hat{m}_1^{(1)} = 491.3$ k-in.

$\hat{f}_{2x}^{(1)} = -32.4$ kip, $\hat{f}_{2y}^{(1)} = 8.07$ kip, $\hat{m}_2^{(1)} = -831.3$ k-in.

$\hat{f}_{2x}^{(2)} = -\hat{f}_{3x}^{(2)} = -0.28$ kip, $\hat{f}_{2y}^{(2)} = 58.31$ kip, $\hat{m}_2^{(2)} = 1123.9$ k-in.

$\hat{f}_{3y}^{(2)} = 21.69$ kip, $\hat{m}_3^{(2)} = -1611.8$ k-in.

$\hat{f}_{4x}^{(3)} = -\hat{f}_{2x}^{(3)} = 50.2$ kip, $\hat{f}_{4y}^{(3)} = -\hat{f}_{2y}^{(3)} = -1.49$ kip, $\hat{m}_4^{(3)} = -154.2$ k-in.

$\hat{m}_2^{(3)} = -293.2$ k-in.

$F_{1x} = 28.65$ kip, $F_{1y} = 37.24$ kip, $M_1 = 491.3$ k-in.

$F_{3x} = 0.28$ kip, $F_{3y} = 21.69$ kip, $M_3 = -1611.8$ k-in.

$F_{4x} = -28.93$ kip, $F_{4y} = 41.05$ kip, $M_4 = -154.2$ k-in.

5.7 $d_{2x} = 0.4308 \times 10^{-4}$ m, $d_{2y} = -0.9067 \times 10^{-4}$ m,

$\phi_2 = -0.1403 \times 10^{-2}$ rad

$\hat{f}_{1x}^{(1)} = -\hat{f}_{2x}^{(1)} = 23.8$ kN, $\hat{f}_{1y}^{(1)} = 17.26$ kN, $\hat{m}_1^{(1)} = 32.77$ kN · m

$\hat{f}_{2y}^{(1)} = 22.74$ kN, $\hat{m}_2^{(1)} = -54.64$ kN · m

$\hat{f}_{2x}^{(2)} = -\hat{f}_{3x}^{(2)} = 11.31$ kN, $\hat{f}_{2y}^{(2)} = 37.19$ kN, $\hat{m}_2^{(2)} = 65.09$ kN · m

$\hat{f}_{3y}^{(2)} = 42.81$ kN, $\hat{m}_3^{(2)} = -87.54$ kN · m

$\hat{f}_{2x}^{(3)} = -\hat{f}_{4x}^{(3)} = 17.55$ kN, $\hat{f}_{2y}^{(3)} = -\hat{f}_{4y}^{(3)} = 1.40$ kN

$\hat{m}_2^{(3)} = -10.51$ kN · m, $\hat{m}_4^{(3)} = -5.30$ kN · m

$F_{1x} = -17.26$ kN, $F_{1y} = 23.80$ kN, $M_1 = 32.77$ kN · m

$F_{3x} = -11.31$ kN, $F_{3y} = 42.81$ kN, $M_3 = -87.54$ kN · m

$F_{4x} = -11.42$ kN, $F_{4y} = 13.40$ kN, $M_4 = -5.30$ kN · m

5.9 $d_{2x} = -4.95 \times 10^{-5}$ m, $d_{2y} = -2.56 \times 10^{-5}$ m, $\phi_2 = 2.66 \times 10^{-3}$ rad

$\hat{f}_{1x}^{(1)} = -\hat{f}_{2x}^{(1)} = 26.9$ kN, $\hat{f}_{1y}^{(1)} = -\hat{f}_{2y}^{(1)} = -42.0$ kN

$\hat{m}_1^{(1)} = 55.9$ kN · m, $\hat{m}_2^{(1)} = 111.7$ kN · m

$\hat{f}_{2x}^{(2)} = -\hat{f}_{3x}^{(2)} = -42.0$ kN, $\hat{f}_{2y}^{(2)} = -\hat{f}_{3y}^{(2)} = 26.9$ kN

$M_1 = 55.9$ kN · m, $M_3 = 44.7$ kN · m

5.10 $d_{2y} = -0.1423 \times 10^{-2}$ m, $\phi_2 = -0.5917 \times 10^{-3}$ rad

$\hat{f}_{1x}^{(1)} = 0$, $\hat{f}_{1y}^{(1)} = 10$ kN, $\hat{m}_1^{(1)} = 23.3$ kN · m, $\hat{f}_{2x}^{(1)} = 0$,

$\hat{f}_{2y}^{(1)} = -10$ kN, $\hat{m}_2^{(1)} = 6.7$ kN · m

5.11 $d_{2y} = -3.712 \times 10^{-5}$ m, $F_{1x} = 5440$ N, $F_{1y} = 10000$ N, $M_1 = 112$ N · m

5.12 $d_{1x} = -0.2143$ m, $d_{1y} = -0.250$ m, $\phi_1 = 0.0893$ rad, $d_{2x} = -0.2143$ m, $d_{2y} = -0.357 \times 10^{-4}$ m, $\phi_2 = 0.0714$ m

5.13 $d_{2x} = 0.0559$ in., $d_{2y} = 0.00382$ in., $\phi_2 = -0.000150$ rad

$d_{3x} = 0.0558$ in., $d_{3y} = -0.000133$ in., $\phi_3 = 0.000149$ rad

$F_{1x} = -198$ lb, $F_{1y} = -4770$ lb, $M_1 = 27460$ lb · in.

$F_{4x} = -4802$ lb, $F_{4y} = 4770$ lb, $M_4 = 27430$ lb · in.

5.14 $d_{2x} = 0.0174$ in., $d_{2y} = -0.0481$ in., $\phi_2 = -0.00165$ rad
$\hat{f}_{1x}^{(1)} = 19160$ lb, $\hat{f}_{1y}^{(1)} = -1385$ lb, $\hat{m}_1^{(1)} = -59050$ lb · in.
$\hat{f}_{2x}^{(1)} = -19160$ lb, $\hat{f}_{2y}^{(1)} = 1385$ lb, $\hat{m}_2^{(1)} = -176{,}000$ lb · in.

5.15 $d_{2x} = -1.76 \times 10^{-2}$ m, $d_{2y} = -1.87 \times 10^{-5}$ m, $\phi_2 = 5.00 \times 10^{-3}$ rad

$d_{3x} = -1.76 \times 10^{-2}$ m, $\phi_3 = -2.49 \times 10^{-3}$ rad

$F_{1x} = 20.0$ kN, $F_{1y} = 13.1$ kN, $M_1 = -57.4$ kN · m, $F_{3y} = -13.1$ kN

5.16 $d_{3y} = -2.83 \times 10^{-5}$ m, $d_{4x} = 1.0 \times 10^{-5}$ m, $d_{4y} = -2.83 \times 10^{-5}$ m

5.17 $d_{3y} = -0.397$ in., $\phi_3 = 0$

5.18 $d_{2x} = d_{2y} = -0.01 \times 10^{-3}$ m, $\phi_2 = 1.766 \times 10^{-4}$ rad

5.19 $d_{1x} = 0.702$ in., $d_{1y} = 0.00797$ in., $\phi_1 = -0.00446$ rad
$\hat{f}_{3x}^{(1)} = -\hat{f}_{1x}^{(1)} = -19.93$ kip, $\hat{f}_{3y}^{(1)} = -\hat{f}_{1y}^{(1)} = 18.1$ kip, $\hat{m}_3^{(1)} = 1309$ k · in.
$\hat{m}_1^{(1)} = 863$ k · in.

5.20 $d_{3x} = 1.24$ in., $d_{3y} = 0.00203$ in., $\phi_3 = -0.000556$ rad
$\hat{f}_{1x}^{(1)} = -2.76$ kip, $\hat{f}_{1y}^{(1)} = 1.79$ kip, $\hat{m}_1^{(1)} = 0$, $\hat{f}_{2x}^{(1)} = 2.76$ kip, $\hat{f}_{2y}^{(1)} = -1.79$ kip,
$\hat{m}_2^{(1)} = 322$ k · in.

5.21 Use a W16 × 31 for all sections

5.22 $\sigma_{\text{bending max}} = 11924$ psi

5.23 $d_{5x} = 0.0204$ in., $d_{5y} = 0.00122$ in., $\phi_5 = 0.000207$ rad

5.24 $d_{5x} = 2.82$ in., $d_{5y} = 0.00266$ in., $\phi_5 = -0.00139$ rad

5.25 **a.** $d_{2y} = -2.12 \times 10^{-3}$ in. **b.** $d_{3y} = -6.07 \times 10^{-2}$ in.

5.26 $d_{2x} = 0.596 \times 10^{-5}$ in., $d_{2y} = -0.332 \times 10^{-2}$ in., $\phi_2 = -0.100 \times 10^{-3}$ rad

$F_{1x} = 130$ lb, $F_{1y} = 10360$ lb, $F_{4x} = -130$ lb, $F_{4y} = 10360$ lb

5.27 $d_{3y} = -0.0153$ in., $f_{1x}^{(1)} = 30$ kN, $f_{1y}^{(1)} = -6.67$ kN, $m_1^{(1)} = 0$

5.28 $d_{2x} = 5.70$ mm, $d_{2y} = -0.0244$ mm, $\phi_2 = 0.00523$ rad

5.29 $d_{3y} = -1.83$ in., $d_{4y} = -1.22$ in.

5.30 $d_{3y} = 6.67$ in., $d_{4y} = -6.67$ in., $\phi_3 = -\phi_4 = -3.20$ rad

$F_{1x} = 11.69$ kN, $F_{1y} = 30$ kN, $M_1 = -1810$ kN · m

$F_{6x} = -11.69$ kN, $F_{6y} = 30$ kN, $M_6 = 1810$ kN · m

5.31 $d_{2y} = -1.58 \times 10^{-2}$ in.

5.32 $d_{2x} = 4.30$ mm, $\phi_2 = -0.241 \times 10^{-3}$ rad

$F_{1x} = -8339$ N, $F_{1y} = -4995$ N, $M_1 = 26{,}700$ N · m,

$F_{4x} = -6661$ N, $F_{4y} = 4995$ N, $M_4 = 23{,}330$ N · m

5.33 $d_{7x} = 0.0264$ m, $d_{7y} = 0.463 \times 10^{-4}$ m, $\phi_7 = 0.171 \times 10^{-2}$ rad

$\hat{f}_{1x}^{(1)} = -21.1$ N, $\hat{f}_{1y}^{(1)} = 30.4$ N, $\hat{m}_1^{(1)} = 74.95$ N · m

$\hat{f}_{3x}^{(1)} = 21.1$ N, $\hat{f}_{3y}^{(1)} = -30.4$ N, $\hat{m}_3^{(1)} = 46.65$ N · m

5.35 $d_{9x} = 0.0174$ m, $\hat{f}_{1x}^{(1)} = -22.6$ kN, $\hat{f}_{1y}^{(1)} = 16.0$ kN, $\hat{m}_1^{(1)} = 53.6$ kN · m

$\hat{f}_{3x}^{(1)} = 22.6$ kN, $\hat{f}_{3y}^{(1)} = -16.0$ kN, $\hat{m}_3^{(1)} = 42.4$ kN · m

5.36 $d_{6y} = -2.80 \times 10^{-7}$ m, $d_{7y} = -4.87 \times 10^{-7}$ m

5.37 $d_{5y} = -1.29 \times 10^{-2}$ m

5.38 $d_{2x} = 1.43 \times 10^{-1}$ m

5.39 Truss: $d_{7x} = 0.0260$ m, $d_{7y} = 0.00566$ m,

Frame: $d_{7x} = 0.0180$ m, $d_{7y} = 0.00424$ m

Truss, element 1: $\hat{f}_{1x} = -49{,}730$ N, $\hat{f}_{1y} = 0$

Frame, element 1: $\hat{f}_{1x} = -43{,}060$ N, $\hat{f}_{1y} = 22670$ N

5.40 $d_{max} = -0.0105$ m at midspan

$M_{max} = 1.568 \times 10^6$ N-m at C

5.41 $d_{max} = 0.0524$ m

$M_{max} = 6.22 \times 10^4$ N-m

5.45 Tapered beam $n = 3$

one element: $d_{1y} = -0.222 \times 10^{-1}$ in.

two elements: $d_{1y} = -0.189 \times 10^{-1}$ in.

four elements: $d_{1y} = -0.181 \times 10^{-1}$ in.

eight elements: $d_{1y} = -0.179 \times 10^{-1}$ in.

5.46 $\underline{K} = 15\dfrac{GJ_0}{L}\begin{bmatrix} 1 & -1 \\ -1 & 1 \end{bmatrix}$

5.48 $d_{2y} = -0.214$ in.

5.49 $d_{2y} = -0.729$ in.

5.51 $d_{1y} = -0.690 \times 10^{-2}$ m

5.52 $d_{5y} = -0.1776$ in.

5.53 $d_{4y} = -1.026$ in.

5.55 $d_{3y} = -2.54 \times 10^{-3}$ m

5.57 $d_{5y} = -2.22 \times 10^{-2}$ m

5.58 $d_{2y} = 0.491$ in., $d_{3z} = 0.837$ in.

5.59 $d_{7z} = -0.251$ in.

Chapter 6

6.1 Use Eq. (6.2.10) in Eq. (6.2.18) to show $N_i + N_j + N_m = 1$.

6.3 **a.** $$\underline{k} = 4.0 \times 10^6 \begin{bmatrix} 2.5 & 1.25 & -2.0 & -1.5 & -0.5 & 0.25 \\ & 4.375 & -1.0 & -0.75 & -0.25 & -3.625 \\ & & 4.0 & 0 & -2.0 & 1.0 \\ & & & 1.5 & 1.5 & -0.75 \\ & & & & 2.5 & -1.25 \\ \text{Symmetry} & & & & & 4.375 \end{bmatrix} \text{lb/in.}$$

b. $$\underline{k} = 13.33 \times 10^6 \begin{bmatrix} 1.54 & 0.75 & -1.0 & -0.45 & -0.54 & -0.3 \\ & 1.815 & -0.3 & -0.375 & -0.45 & -1.44 \\ & & 1.0 & 0 & 0 & 0.3 \\ & & & 0.375 & 0.45 & 0 \\ & & & & 0.54 & 0 \\ \text{Symmetry} & & & & & 1.44 \end{bmatrix} \text{lb/in.}$$

6.4 **a.** $\sigma_x = 19.2$ ksi, $\sigma_y = 4.8$ ksi, $\tau_{xy} = -15.0$ ksi

$\sigma_1 = 28.6$ ksi, $\sigma_2 = -4.64$ ksi, $\theta_p = -32.2^\circ$

b. $\sigma_x = 32.0$ ksi, $\sigma_y = 8.0$ ksi, $\tau_{xy} = -25.0$ ksi

$\sigma_1 = 47.7$ ksi, $\sigma_2 = -7.73$ ksi, $\theta_p = -32.2^\circ$

6.6 **a.** $$\underline{k} = 2.074 \times 10^5 \begin{bmatrix} 8437.5 & 1687.5 & -7762.5 & -337.5 & -675 & -1350 \\ 1687.5 & 3937.5 & 337.5 & -2137.5 & -2025 & -1800 \\ -7762.5 & 337.5 & 8437.5 & -1687.5 & -675 & 1350 \\ -337.5 & -2137.5 & -1687.5 & 3937.5 & 2025 & -1800 \\ -675 & -2025 & -675 & 2025 & 1350 & 0 \\ -1350 & -1800 & 1350 & -1800 & 0 & 3600 \end{bmatrix} \text{N/m}$$

b. $$\underline{k} = 4.48 \times 10^7 \begin{bmatrix} 25.0 & 0 & -12.5 & 6.25 & -12.5 & -6.25 \\ & 9.375 & 9.375 & -4.6875 & -9.375 & -4.6875 \\ & & 15.625 & -7.8125 & -3.125 & -1.5625 \\ & & & 27.343 & 1.5625 & -3.125 \\ & & & & 15.625 & 7.8125 \\ \text{Symmetry} & & & & & 27.343 \end{bmatrix} \text{N/m}$$

6.7 **a.** $\sigma_x = -5.289$ GPa, $\sigma_y = -0.156$ GPa, $\tau_{xy} = 0.233$ GPa

$\sigma_1 = -0.1459$ GPa, $\sigma_2 = -5.30$ GPa, $\theta_p = -2.59^\circ$

b. $\sigma_x = 0$, $\sigma_y = 42.0$ MPa, $\tau_{xy} = 33.6$ MPa

$\sigma_1 = 60.6$ MPa, $\sigma_2 = -18.6$ MPa, $\theta_p = -29^\circ$

6.9 **a.** $\sigma_x = -15.0$ ksi, $\sigma_y = -45.0$ ksi, $\tau_{xy} = -18.0$ ksi

$\sigma_1 = -6.57$ ksi, $\sigma_2 = -53.4$ ksi, $\theta_p = -25.1°$

b. $\sigma_x = -15.0$ ksi, $\sigma_y = -45$ ksi, $\tau_{xy} = -21.0$ ksi

$\sigma_1 = -4.19$ ksi, $\sigma_2 = -55.8$ ksi, $\theta_p = -27.2°$

c. $\sigma_x = -30$ ksi, $\sigma_y = -90$ ksi, $\tau_{xy} = -21$ ksi

$\sigma_1 = -23.38$ ksi, $\sigma_2 = -96.6$ ksi, $\theta_p = -17.47°$

f. $\sigma_x = -22.5$ ksi, $\sigma_y = -67.5$ ksi, $\tau_{xy} = -21.0$ ksi

$\sigma_1 = -14.2$ ksi, $\sigma_2 = -75.8$ ksi, $\theta_p = -21.5°$

6.10 **a.** $\sigma_x = -52.5$ MPa, $\sigma_y = -32.8$ MPa, $\tau_{xy} = -5.38$ MPa

$\sigma_1 = -31.4$ MPa, $\sigma_2 = -53.9$ MPa, $\theta_p = -14.3°$

b. $\sigma_x = -31.4$ MPa, $\sigma_y = -13.5$ MPa, $\tau_{xy} = 5.38$ MPa

$\sigma_1 = -12.0$ MPa, $\sigma_2 = -32.9$ MPa, $\theta_p = -15.5°$

c. $\sigma_x = -27.6$ MPa, $\sigma_y = -19.5$ MPa, $\tau_{xy} = 4.04$ MPa

$\sigma_1 = -17.9$ MPa, $\sigma_2 = -29.3$ MPa, $\theta_p = -22.5°$

d. $\sigma_x = -31.6$ MPa, $\sigma_y = -28.9$ MPa, $\tau_{xy} = -6.73$ MPa

$\sigma_1 = -23.0$ MPa, $\sigma_2 = -38.0$ MPa, $\theta_p = 39°$

6.11 **a.** $f_{s1x} = 0$, $f_{s1y} = 0$, $f_{s2x} = p_0 Lt/6$, $f_{s2y} = 0$

$f_{s3x} = p_0 Lt/3$, $f_{s3y} = 0$

b. $f_{s1x} = 0$, $f_{s2x} = p_0 Lt/12$, $f_{s3x} = p_0 Lt/4$

6.12 **b.** $f_{s1y} = f_{s2y} = p_0 Lt/\pi$

6.13 $d_{3x} = 0.5 \times 10^{-3}$ in., $d_{3y} = -0.275 \times 10^{-2}$ in.

$d_{4x} = -0.609 \times 10^{-3}$ in., $d_{4y} = -0.293 \times 10^{-2}$ in.

$\sigma_x^{(1)} = 824$ psi, $\sigma_y^{(1)} = 247$ psi, $\tau_{xy}^{(1)} = -1587$ psi

$\sigma_1^{(1)} = 2149$ psi, $\sigma_2^{(1)} = -1077$ psi, $\theta_p^{(1)} = -40°$

$\sigma_x^{(2)} = -826$ psi, $\sigma_y^{(2)} = 292$ psi, $\tau_{xy}^{(2)} = -411$ psi

$\sigma_1^{(2)} = 426$ psi, $\sigma_2^{(2)} = -960$ psi, $\theta_p^{(2)} = 18.15°$

6.14 **a.** $d_{2x} = 0.281 \times 10^{-4}$ m, $d_{2y} = -0.330 \times 10^{-4}$ m

$d_{5x} = 0.115 \times 10^{-4}$ m, $d_{5y} = -0.103 \times 10^{-4}$ m

$\sigma_x^{(2)} = 16.4$ MPa, $\sigma_y^{(2)} = 15.2$ MPa

$\tau_{xy}^{(2)} = -6.99$ MPa, $\sigma_1^{(2)} = 22.8$ MPa

$\sigma_2^{(2)} = 8.80$ MPa, $\theta_p^{(2)} = -42.7°$

$\sigma_x^{(1)} = 10.6$ MPa, $\sigma_y^{(1)} = 3.18$ MPa

$\tau_{xy}^{(1)} = -3.34$ MPa, $\sigma_1^{(1)} = 11.9$ MPa

$\sigma_2^{(1)} = 1.90$ MPa, $\theta_p^{(1)} = -21.0°$

b. $d_{1x} = -d_{2x} = -0.165 \times 10^{-5}$ m, $d_{1y} = d_{2y} = -0.125 \times 10^{-4}$ m

$d_{5x} = 0.274 \times 10^{-12}$ m, $d_{5y} = -0.163 \times 10^{-4}$ m

$\sigma_x^{(1)} = 5.99 \times 10^5$ N/m^2, $\sigma_y^{(1)} = -3.78 \times 10^6$ N/m^2

$\tau_{xy}^{(1)} = 4.05 \times 10^{-1}$ N/m^2, $\sigma_1^{(1)} = 5.99 \times 10^5$ N/m^2

$\sigma_2^{(1)} = -3.78 \times 10^6$ N/m^2, $\theta_p^{(1)} = 0°$, $\sigma_x^{(3)} = 5.64 \times 10^6$ N/m^2

$\sigma_y^{(3)} = 1.88 \times 10^7$ N/m^2, $\tau_{xy}^{(3)} = -1.11 \times 10^{-1}$ N/m^2

$\sigma_1^{(3)} = 1.88 \times 10^7$ N/m^2, $\sigma_2^{(3)} = 5.64 \times 10^6$ N/m^2, $\theta_p^{(3)} = -90°$

6.15 All f_{bx}'s are equal to 0.

a. $f_{b1y} = f_{b2y} = f_{b3y} = f_{b4y} = -10.28$ N, $f_{b5y} = -20.56$ N

b. $f_{b1y} = f_{b2y} = f_{b3y} = f_{b4y} = -8.03$ N, $f_{b5y} = -16.06$ N

6.18 **b.** Yes, **c.** Yes **g.** No

6.20 **a.** $n_b = 8$, **b.** $n_b = 12$

Chapter 7

7.9 $d_{2x} = d_{3x} = 0.647 \times 10^{-3}$ in., $d_{2y} = 0.666 \times 10^{-4}$ in.

$d_{3y} = -0.666 \times 10^{-4}$ in., skew effect

7.10 Stress approaches 2.5 psi near edge of whole for model of 70 nodes, 54 elements.

7.11 At depth 4 in. equal to width, stress approaches uniform $\sigma_y = -1000$ psi.

7.12 $\sigma_1 = 8836$ psi at top and bottom of hole

7.13 $\sigma_1 = 372$ psi at fillet

7.14 For refined mesh at re-entrant corner, $\sigma_1 = 20160$ psi.

7.15 $\sigma_{VM} = 93.7$ psi at load

7.17 For the model with 12 in. $\times \frac{1}{2}$ in. size elements, finite element solution yields free-end deflection of -0.499 in.; exact solution is -1.15 in. (See Table 7–1 in text for other results.)

7.19 $\sigma_1 = 3$ kN/m^2 (round hole model)

$\sigma_1 = 3.51$ kN/m^2 (square hole with corner radius)

7.21 $\sigma_{VM} = 8.1$ MPa

7.22 **a.** $\sigma_1 = 58700$ psi

7.23 $\sigma_1 = 19$ MPa at hole

7.25 Largest von Mises stress 35–45 MPa at inside edge at junction of narrow to larger section of wrench

7.27 Largest principal stress $\sigma_1 = 1005$ MPa at narrowest width of member (70-element, 94-node model)

7.35 For a 1 cm thick wrench, $\sigma_{VM} = 502$ MPa

Chapter 8

8.2 $$\varepsilon_x = \frac{1}{3b}(-u_1 + u_2 + 4u_4 - 4u_5), \quad \varepsilon_y = \frac{1}{3h}(-v_1 + v_3 + 4v_4 - 4v_6)$$

$$\gamma_{xy} = \frac{1}{3h}(-u_1 + u_3 + 4u_4 - 4u_6) + \frac{1}{3b}(-v_1 + v_3 + 4v_4 - 4v_6)$$

$$\sigma_x = \frac{E}{1-\nu^2}(\varepsilon_x + \nu\varepsilon_y), \quad \sigma_y = \frac{E}{1-\nu^2}(\varepsilon_y + \nu\varepsilon_x), \quad \tau_{xy} = G\gamma_{xy}$$

8.3 $$f_{s1x} = f_{s3x} = \frac{-pth}{6}, \quad f_{s5x} = \frac{-2pth}{3}$$

8.4 $$f_{s1x} = 0, \quad f_{s3x} = \frac{-p_0 \mathrm{th}}{6}, \quad f_{s5x} = \frac{-p_0 th}{3}$$

8.5 **a.** $\varepsilon_x = -5 \times 10^{-5}y + 2.5 \times 10^{-4}, \quad \varepsilon_y = -1.67 \times 10^{-4}x + 3.33 \times 10^{-5},$

$\gamma_{xy} = -5 \times 10^{-5}x - 1.11 \times 10^{-4}y + 4.17 \times 10^{-4}$

$\sigma_x = 3290$ psi, $\sigma_y = -4850$ psi, $\tau_{xy} = 1540$ psi

b. $\varepsilon_x = -5 \times 10^{-5}y + 1.67 \times 10^{-4}, \quad \varepsilon_y = -1.67 \times 10^{-4}x + 5 \times 10^{-5}$

$\gamma_{xy} = -5 \times 10^{-5}x - 4.17 \times 10^{-5}y + 2.08 \times 10^{-4}$

$\sigma_x = 928$ psi, $\sigma_y = -8290$ psi, $\tau_{xy} = 632$ psi

8.6 $\varepsilon_x = 2.54 \times 10^{-3}$

$\varepsilon_y = -7.62 \times 10^{-3}$

$\gamma_{xy} = -7.04 \times 10^{-3}$

8.7 $$N_1 = 1 - \frac{x}{20} + \frac{x^2}{1800}, \quad N_2 = \frac{-x+y}{60} + \frac{x^2+y^2}{1800} - \frac{xy}{900}$$

$$N_3 = \frac{-y}{60} + \frac{y^2}{1800}, \quad N_4 = \frac{xy}{900} - \frac{y^2}{900}, \quad N_5 = \frac{y}{15} - \frac{xy}{900}, \text{ etc.}$$

Chapter 9

9.1 **a.** $$\underline{K} = 25.132 \times 10^6 \begin{bmatrix} 5 & 1 & 0 & -1 & 1 & 0 \\ 1 & 4 & -2 & -1 & -2 & -3 \\ 0 & -2 & 8 & 0 & 4 & 2 \\ -1 & -1 & 0 & 1 & 1 & 0 \\ 1 & -2 & 4 & 1 & 4 & 1 \\ 0 & -3 & 2 & 0 & 1 & 3 \end{bmatrix} \text{ lb/in.}$$

b. $\underline{K} = 50.265 \times 10^6 \begin{bmatrix} 2.75 & 0 & -2.25 & 0.5 & 0.25 & -0.5 \\ 0 & 1 & 1 & -1 & -1 & 0 \\ -2.25 & 1 & 5.75 & -2.5 & 0.25 & 1.5 \\ 0.5 & -1 & -2.5 & 4 & 0.5 & -3 \\ 0.25 & -1 & 0.25 & 0.5 & 1.75 & 0.5 \\ -0.5 & 0 & 1.5 & -3 & 0.5 & 3 \end{bmatrix}$ lb/in.

9.2 $f_{s2r} = \dfrac{2\pi b p_0 h}{6}, \quad f_{s3r} = \dfrac{2\pi b p_0 h}{3}$

9.3 $f_{b1r} = f_{b2r} = f_{b3r} = 0.382$ lb
$f_{b1z} = f_{b2z} = f_{b3z} = -6.32$ lb

9.4 **a.** $\sigma_r = 8000$ psi, $\sigma_z = 0$, $\sigma_\theta = 8000$ psi, $\tau_{rz} = 1200$ psi
b. $\sigma_r = 5830$ psi, $\sigma_z = -3770$ psi, $\sigma_\theta = 3090$ psi, $\tau_{rz} = 400$ psi

9.6 **a.** $\underline{k} = 7.037 \begin{bmatrix} 3125 & 625 & 0 & -625 & 625 & 0 \\ & 2500 & -1250 & -625 & -1250 & -1875 \\ & & 5000 & 0 & 2500 & 1250 \\ & & & 625 & 625 & 0 \\ & & & & 2500 & 625 \\ \text{Symmetry} & & & & & 1875 \end{bmatrix}$ kN/mm

b. $\underline{k} = 11.73 \begin{bmatrix} 2475 & 0 & -2025 & 450 & 225 & -450 \\ & 900 & 900 & -900 & -900 & 0 \\ & & 5175 & -2250 & 225 & 1350 \\ & & & 3600 & 450 & -2700 \\ & & & & 1575 & 450 \\ \text{Symmetry} & & & & & 2700 \end{bmatrix}$ kN/mm

9.7 **a.** $\sigma_r = -84$ MPa, $\sigma_z = -84$ MPa, $\sigma_\theta = 252$ MPa, $\tau_{rz} = -101$ MPa
b. $\sigma_r = -103$ MPa, $\sigma_z = -103$ MPa, $\sigma_\theta = 112$ MPa, $\tau_{rz} = -73$ MPa

9.14 Using 0.5 in. radii in corners, $\sigma_1 = 7590$ psi at inside corner

9.18 $\sigma_1 = 4621$ psi outer edge of hole, along axis of symmetry

9.19 $\sigma_\theta = 22{,}711$ psi, $\sigma_r = -4984$ psi, $u_r = 0.037$ in.

9.20 $\sigma_1 = 64.1$ MPa, $u = 0.0782$ m top and bottom center of plates

9.24 $\sigma_{VM} = 5221$ psi at fillet, $\sigma_{VM} = 1637.5$ psi at groove

Chapter 10

10.2 **a.** $s = -\frac{1}{5}$, **b.** $N_1 = 0.4$, $N_2 = 0.6$

10.3 **a.** $s = 0$, **b.** $N_1 = 0.5$, $N_2 = 0.5$

10.5 **a.** $s = -0.5$, **b.** $N_1 = 0.375$, $N_2 = -0.125$, $N_3 = 0.75$

10.8 $d_{2x} = 4.859 \times 10^{-4}$m (right end), $d_{3x} = 2.793 \times 10^{-4}$m (center)

10.10 $\varepsilon_x = 0.0009375$ in./in., $\varepsilon_y = -0.00125$ in./in., $\gamma_{xy} = -0.000625$ rad

$\sigma_x = 18.5$ ksi, $\sigma_y = -31.9$ ksi, $\tau_{xy} = -7.21$ ksi

10.15 **a.** $f_{s3t} = 500$ lb, $f_{s4t} = 500$ lb, **b.** $f_{s1t} = 83.33$ lb, $f_{s4t} = 41.67$ lb

10.16 **a.** 1.917, **b.** 0.667, **c.** 0.400, **d.** 2.87, **f.** 0

Chapter 11

11.1 **a.**

$$\underline{B} = \frac{1}{8}\begin{bmatrix} 0 & 0 & 0 & 0 & 0 & 0 & 4 & 0 & 0 & -4 & 0 & 0 \\ 0 & 0 & 0 & 0 & 4 & 0 & 0 & 0 & 0 & 0 & -4 & 0 \\ 0 & 0 & 4 & 0 & 0 & 0 & 0 & 0 & 0 & 0 & 0 & -4 \\ 0 & 0 & 0 & 4 & 0 & 0 & 0 & 4 & 0 & -4 & -4 & 0 \\ 0 & 4 & 0 & 0 & 0 & 4 & 0 & 0 & 0 & 0 & -4 & -4 \\ 4 & 0 & 0 & 0 & 0 & 0 & 0 & 0 & 4 & -4 & 0 & -4 \end{bmatrix}$$

11.3 $\sigma_x = 77.9$ ksi, $\sigma_y = 8.65$ ksi, $\sigma_z = -49.0$ ksi

$\tau_{xy} = 11.5$ ksi, $\tau_{yz} = -23.1$ ksi, $\tau_{zx} = 5.77$ ksi

11.6 **a.**

$$\underline{B} = \frac{1}{18750}$$

$$\times \begin{bmatrix} -625 & 0 & 0 & 0 & 0 & 0 & 0 & 0 & 0 & 625 & 0 & 0 \\ 0 & -375 & 0 & 0 & 750 & 0 & 0 & 0 & 0 & 0 & -375 & 0 \\ 0 & 0 & -375 & 0 & 0 & 0 & 0 & 0 & 750 & 0 & 0 & -375 \\ -375 & -625 & 0 & 750 & 0 & 0 & 0 & 0 & 0 & -375 & 625 & 0 \\ 0 & -375 & -375 & 0 & 0 & 750 & 0 & 750 & 0 & 0 & -375 & -375 \\ -375 & 0 & -625 & 0 & 0 & 0 & 750 & 0 & 0 & -375 & 0 & 625 \end{bmatrix}$$

11.7 $\sigma_x = 72.7$ MPa, $\sigma_y = 169.6$ MPa, $\sigma_z = 72.7$ MPa

$\tau_{xy} = 59.2$ MPa, $\tau_{yz} = 32.3$ MPa, $\tau_{zx} = 91.5$ MPa

11.10 $N_2 = \dfrac{(1-s)(1-t)(1-z')}{8}$, $N_3 = \dfrac{(1-s)(1+t)(1-z')}{8}$,

$N_4 = \dfrac{(1-s)(1+t)(1+z')}{8}$,

$N_5 = \dfrac{(1+s)(1-t)(1+z')}{8}$, $N_6 = \dfrac{(1+s)(1-t)(1-z')}{8}$,

$N_7 = \dfrac{(1+s)(1+t)(1-z')}{8}$, $N_8 = \dfrac{(1+s)(1+t)(1+z')}{8}$

11.11 $N_1 = \dfrac{(1-s)(1-t)(1+z')(-s-t+z'-2)}{8}$,

$N_2 = \dfrac{(1-s)(1-t)(1-z')(-s-t-z'-2)}{8}$

11.13 $d_{max} = -0.662$ in. under the load

Chapter 13

13.1 $t_2 = 166.7\,^\circ\text{C}, \quad t_3 = 233.3\,^\circ\text{C}$

13.2 $t_2 = 150\,^\circ\text{F}, \quad t_3 = 100\,^\circ\text{F}, \quad t_4 = 50\,^\circ\text{F}$

13.3 $t_2 = 875\,^\circ\text{F}, \quad t_3 = 1250\,^\circ\text{F}, \quad F_1 = -180$ Btu/h

13.4 $t_1 = 151\,^\circ\text{F}, \quad t_2 = 148\,^\circ\text{F}, \quad t_3 = 140\,^\circ\text{F}, \quad t_4 = 125\,^\circ\text{F}$

13.5 $t_2 = 183\,^\circ\text{F}, \quad t_3 = 267\,^\circ\text{F}, \quad t_4 = 350\,^\circ\text{F}, \quad t_5 = 433\,^\circ\text{F}$

13.6 $t_2 = 421\,^\circ\text{C}, \quad t_3 = 121\,^\circ\text{C}, \quad q^{(3)} = 3975\ \text{W/m}^2$

13.7 $t_2 = 418.2\,^\circ\text{C}, \quad t_3 = 527.3\,^\circ\text{C}$

13.8 $t_2 = 20\,^\circ\text{C}, \quad t_3 = 20\,^\circ\text{C}, \ \bar{q}_{\text{max}} = 0.0009\ \text{W}, \ \bar{q}_{\text{min}} = -0.0009\ \text{W}$

13.9 $6\,^\circ\text{C}$ at center of wall, $\bar{q}_{\text{max}} = 5.54$ W, $\bar{q}_{\text{min}} = -5.54$ W

13.11 185°C at right end, $\bar{q}_{\text{max}} = 439$ W

13.14 $t_0 = 92.25°\text{C}, \ t_1 = 88.575°\text{C}, \ t_2 = 84.9°\text{C}, \ t_3 = 80°\text{C}$

13.16 $\underset{\sim}{k} = \dfrac{AK_{xx}}{L}\begin{bmatrix} 1 & -1 \\ -1 & 1 \end{bmatrix}$

13.18 $\underset{\sim}{k} = \begin{bmatrix} 39.57 & 7.076 & -5.417 \\ & 35.82 & -1.667 \\ & & 7.083 \end{bmatrix}, \quad \underset{\sim}{f} = \begin{Bmatrix} 2936 \\ 2936 \\ 50 \end{Bmatrix}$ Btu/h

13.19 $\underset{\sim}{f} = \begin{Bmatrix} 1291 \\ 27.3 \\ 1254 \end{Bmatrix}$ W

13.22 $t_4 = 75\,^\circ\text{F}, \quad t_5 = 25\,^\circ\text{F}$

13.36 $12\,^\circ\text{C}$ at 2.5 cm from top, $25\,^\circ\text{C}$ 1.25 cm from top, $\bar{q}_{\text{max}} = 1416$ W, $\bar{q}_{\text{min}} = -1083$ W

13.41 $\bar{q}_{\text{max}} = 3457$ W, $\bar{q}_{\text{min}} = -3848$ W

Chapter 14

14.1 $p_2 = 4.545$ m, $\quad p_3 = 1.818$ m, $\quad v_x^{(1)} = 10.91$ m/s, $\quad Q_f^{(1)} = 21.82\ \text{m}^3/\text{s}$

14.2 $p_2 = -15$ m, $\quad p_3 = 40$ m, $\quad p_4 = -65$ m, $\quad v_x^{(1)} = 25$ m/s, $\quad Q_1 = 50\ \text{m}^3/\text{s}$

14.3 $p_2 = 8.182$ in., $\quad p_3 = 5.455$ in., $\quad v_x^{(1)} = 0.182$ in./s, $\quad v_x^{(2)} = 0.273$ in./s,
$v_x^{(3)} = 0.545$ in./s, $\quad Q_f^{(1)} = 1.091\ \text{in}^3/\text{s}$

14.4 $p_2 = -3$ cm, $\quad p_3 = -8$ cm, $\quad v_x^{(1)} = 1.2$ cm/s, $\quad v_x^{(2)} = 2$ cm/s,
$Q_1 = Q_2 = 6\ \text{cm}^3/\text{s}$

14.6 $v^{(1)} = 2.0$ in./s, $\quad v^{(2)} = 4.0$ in./s, $\quad Q^{(1)} = Q^{(2)} = 4\ \text{in}^3/\text{s}$

14.7 $\underset{\sim}{f}_Q = \begin{Bmatrix} 54.76 \\ 28.57 \\ 16.67 \end{Bmatrix}\ \text{m}^3/\text{s}$

14.8 $f_1 = f_3 = 5 \text{ in}^3/\text{s}, \quad f_2 = 0$

14.9 $p_2 = p_3 = 12 \text{ m}, \quad p_5 = 11 \text{ m}$

Chapter 15

15.1 $d_{2x} = 0.021 \text{ in.}, \quad d_{3x} = 0.042 \text{ in.}, \quad \sigma_x = 0$

15.2 $d_{2x} = 0, \quad \sigma_x = 50.4 \text{ MPa}$

15.3 $d_{1x} = d_{1y} = -0.0175 \text{ in.}, \quad \sigma^{(1)} = 4350 \text{ psi (T)}$

$\sigma^{(2)} = -6150 \text{ psi (C)}, \quad \sigma^{(3)} = 4350 \text{ psi (T)}$

15.4 $d_{1x} = -0.0291 \text{ in.}, \quad d_{1y} = -0.0095 \text{ in.}$

$\sigma^{(1)} = -1370 \text{ psi (C)}, \quad \sigma^{(2)} = 2375 \text{ psi (T)}, \quad \sigma^{(3)} = -1370 \text{ psi (C)}$

15.5 $d_{2x} = 1.44 \times 10^{-4} \text{ m}, \quad \sigma^{(1)} = -20.2 \text{ MPa (C)}, \quad \sigma^{(2)} = \sigma^{(3)} = -10.1 \text{ MPa (C)}$

15.6 $d_{1x} = 0, \quad d_{1y} = 6.0 \times 10^{-4} \text{ m}, \quad \sigma^{(1)} = \sigma^{(3)} = -10.5 \text{ MPa (C)}$

$\sigma^{(2)} = 18.2 \text{ MPa (T)}$

15.7 $d_{1x} = 0, \quad d_{1y} = -3.6 \times 10^{-4} \text{ m}, \quad \sigma^{(1)} = \sigma^{(2)} = 0$

15.8 $d_{2x} = 0.0173 \text{ in.}, \quad \sigma_{st} = 840 \text{ psi (T)}, \quad \sigma_{br} = 1680 \text{ psi (C)}$

15.12 **a.** -0.001907 in. **b.** $\sigma_{br} = -28,600 \text{ psi}, \quad \sigma_{mg} = -19,067 \text{ psi}$

15.13 $f_{T1x} = -4464 \text{ lb}, \quad f_{T1y} = -8929 \text{ lb}, \quad f_{T2x} = 4464 \text{ lb}$

$f_{T2y} = -8929 \text{ lb}, \quad f_{T3x} = 0, \quad f_{T3y} = 17{,}857 \text{ lb}$

15.14 $f_{T1x} = -43.125 \text{ kN}, \quad f_{T1y} = 0, \quad f_{T2x} = 43.125 \text{ kN}, \quad f_{T2y} = -86.250 \text{ kN}$
$f_{T3x} = 0, \quad f_{T3y} = 86.250 \text{ kN}$

15.15 $f_{T1x} = -60.0 \text{ kip}, \quad f_{T1y} = -90 \text{ kip}, \quad f_{T2x} = 60 \text{ kip}, \quad f_{T2y} = 0,$
$f_{T3x} = 0, \quad f_{T3y} = 90 \text{ kip}$

15.16 $f_{T1x} = 134 \text{ kN}, \quad f_{T1y} = 134 \text{ kN}, \quad f_{T2x} = -134 \text{ kN}, \quad f_{T2y} = 0$
$f_{T3x} = 0, \quad f_{T3y} = -134 \text{ kN}$

15.17 $\sigma_x = \sigma_y = -8929 \text{ psi (C)}, \quad \tau_{xy} = 0$

15.18 $\sigma_x = 67.2 \text{ MPa}, \quad \sigma_y = 67.2 \text{ MPa}, \quad \tau_{xy} = 0$

15.19 $\{f_T\} = \dfrac{AE\alpha_0}{6}\begin{Bmatrix} -4t_1 - 5t_2 \\ 4t_1 + 5t_2 \end{Bmatrix}$

15.20 $\dfrac{AE\alpha}{2}\begin{Bmatrix} -t_1 - t_2 \\ t_1 + t_2 \end{Bmatrix}$

15.21 $\{f_T\} = \dfrac{2\pi\bar{r}AE\alpha(\Delta T)[\bar{B}]^T}{1-2v}\begin{Bmatrix}1\\1\\1\\0\end{Bmatrix}$

15.22 $d_{2x} = 0.8 \times 10^{-3}$ in., $d_{3x} = 0$, $d_{3y} = 0.8 \times 10^{-3}$ in.
$d_{4x} = d_{4y} = 0.8 \times 10^{-3}$ in.; stresses are zero

15.23 $d_{2x} = 0.989 \times 10^{-3}$ in., $d_{3x} = -0.756 \times 10^{-3}$ in.,
$d_{3y} = 0.989 \times 10^{-3}$ in., $d_{4x} = 0.132 \times 10^{-2}$ in.,
$d_{4y} = 0.2045 \times 10^{-2}$ in., $\sigma_1^{(1)} = 17$ ksi, $\sigma_2^{(2)} = -17$ ksi

Chapter 16

16.1 $[M] = \dfrac{\rho AL}{6}\begin{bmatrix}2 & 1 & 0\\1 & 4 & 1\\0 & 1 & 2\end{bmatrix}$

16.2 **a.** $[M] = \dfrac{\rho AL}{2}\begin{bmatrix}1 & 0 & 0 & 0\\0 & 2 & 0 & 0\\0 & 0 & 2 & 0\\0 & 0 & 0 & 1\end{bmatrix}$

b. $[M] = \dfrac{\rho AL}{6}\begin{bmatrix}2 & 1 & 0 & 0\\1 & 4 & 1 & 0\\0 & 1 & 4 & 1\\0 & 0 & 1 & 2\end{bmatrix}$

16.3 $\omega_1 = 0.806\sqrt{u}$, $\omega_2 = 2.81\sqrt{\mu}$

16.4 $\omega_1 = 5.368 \times 10^3$ rad/s, $\omega_2 = 17.556 \times 10^3$ rad/s

16.5 **a.**

t (s)	d_i (ft)	$\dot{d}_i$ (ft/s)	$\ddot{d}_i$ (ft/s^2)
0	0	0	25
0.03	0.01125	0.71	22.09
0.06	0.04238	1.03	−0.715
0.09	0.07287	0.67	−22.87
0.12	0.08278	−0.35	−45.28
0.15	0.05194	−1.43	−26.94

16.6 **a.**

t (s)	d_i (ft)	$\dot{d}_i$ (ft/s)	$\ddot{d}_i$ (ft/s^2)
0	0	0	10.00
0.02	0.0020	0.168	6.80
0.04	0.00672	0.256	1.968
0.06	0.01223	0.242	−3.338
0.08	0.01640	0.130	−7.84
0.10	0.01743	−0.053	−10.46

b.

t (s)	d_i (ft)	$\dot{d}_i$ (ft/s)	$\ddot{d}_i$ (ft/s^2)	F(t) (lb)
0.00	0.00000	0.000	10.000	20.0
0.02	0.00179	0.169	6.923	16.0
0.04	0.00625	0.263	2.248	12.0
0.06	0.0115	0.254	−2.945	8.0
0.08	0.0157	0.150	−7.458	4.0
0.10	0.0169	−0.0147	−10.251	0.0

16.7

Node	t (s)	d_i (in.)	$\dot{d}_i$ (in./s)	$\ddot{d}_i$ (in./s^2)
2	0	0	0	0
	0.00025	2.6E-6	0.031	249.6
	0.00050	3.4E-5	0.284	1768.9
	0.00075	1.9E-4	1.085	4641.9
	0.0010	6.36E-4	2.605	7519.3
3	0	0	0	0
	0.00025	6.59E-5	0.791	6328.8
	0.00050	4.99E-4	2.817	9881.2
	0.00075	1.51E-3	5.265	9701.7
	0.0010	3.10E-3	7.369	7128.3

16.8 Using Newmark's method with $\gamma = \frac{1}{2}$, $\beta = \frac{1}{6}$

Node	t (s)	d_i (in.)	$\dot{d}_i$ (in./s)	$\ddot{d}_i$ (in./s^2)	F(t) (lb)
2	0	0	0	0	0
	0.05	0.00172	0.103	4.131	0
	0.10	0.01544	0.513	12.27	0
3	0	0	0	40.0	2000
	0.05	0.0448	1.685	27.39	1800
	0.10	0.1536	2.479	4.37	1600

16.11 **a.** $\omega_1 = \dfrac{3.15}{L^2}\left(\dfrac{EI}{\rho A}\right)^{1/2}$, $\omega_2 = \dfrac{16.24}{L^2}\left(\dfrac{EI}{\rho A}\right)^{1/2}$, **c.** $\omega_1 = \dfrac{9.8}{L^2}\left(\dfrac{EI}{\rho A}\right)^{1/2}$

d. $\omega = \dfrac{14.8}{L^2}\left(\dfrac{EI}{\rho A}\right)^{1/2}$

16.17

	Node:	1	2	3	4	5	6
i	t (s)	Temperature (°C)					
0	0	200	200	200	200	200	200
1	8	0	159.0095	191.4441	198.2110	199.6110	199.8444
2	16	0	135.5852	178.1491	193.6620	198.2112	199.1445
3	24	0	120.2309	165.7003	187.3485	195.5379	197.5152
4	32	0	109.1993	154.9587	180.4038	191.7446	194.8115
5	40	0	100.7600	145.7784	173.4129	187.1268	191.1242
6	48	0	94.00311	137.8529	166.6182	181.9599	186.6590
7	56	0	88.39929	130.9034	160.1012	176.4598	181.6395
8	64	0	83.61745	124.7101	153.8759	170.7856	176.2620
9	72	0	79.43935	119.1075	147.9316	165.0508	170.6822
10	80	0	75.71603	113.9733	142.2502	159.3352	165.0171

16.18

	Node		
Time (s)	1	2	3 (using consistent capacitance matrix)
	Temperature (°C)		
0	25	25	25
0.1	85	18.53611	26.36189
0.2	85	29.61303	21.63526
0.3	85	36.18435	22.42717
0.4	85	40.72491	25.30428
0.5	85	44.27834	28.85201
0.6	85	47.29072	32.49614
0.7	85	49.95809	36.01157
0.8	85	52.37152	39.31761
0.9	85	54.57756	42.39278

16.18

	Node		
Time (s)	1	2	3 (using consistent capacitance matrix)
	Temperature (°C)		
1	85	56.60353	45.23933
1.1	85	58.46814	47.86852
1.2	85	60.1859	50.29457
1.3	85	61.76908	52.53218
1.4	85	63.22852	54.59557
1.5	85	64.574	56.49814
1.6	85	65.81448	58.25235
1.7	85	66.95818	59.86974

16.18

Time (s)	Node 1	Node 2	Node 3
		Temperature (°C)	
0	25	25	25
1.8	85	68.01265	61.36096
1.9	85	68.98485	62.73586
2	85	69.88121	64.0035
2.1	85	70.70765	65.17226
2.2	85	71.46961	66.24984
2.3	85	72.17214	67.24336
2.4	85	72.81986	68.15938
2.5	85	73.41705	69.00393
2.6	85	73.96766	69.78261
2.7	85	74.47531	70.50053
2.8	85	74.94336	71.16246
2.9	85	75.3749	71.77274
3	85	75.77277	72.33542

Appendix A

A1. **a.** $\begin{bmatrix} 3 & 0 \\ -3 & 12 \end{bmatrix}$ **b.** Nonsense **c.** Nonsense

d. $\begin{Bmatrix} 11 \\ 9 \\ 11 \end{Bmatrix}$ **e.** Nonsense **f.** $\begin{bmatrix} 10 & 7 & 6 \\ 3 & -1 & 7 \end{bmatrix}$

A2. $\begin{bmatrix} 1 & 0 \\ \frac{1}{4} & \frac{1}{4} \end{bmatrix}$

A3. $\dfrac{1}{17}\begin{bmatrix} 12 & -3 & -8 \\ -3 & 5 & 2 \\ -8 & 2 & 11 \end{bmatrix}$

A4. Nonsense

A5. $\begin{bmatrix} \frac{1}{2} & 0 \\ \frac{1}{8} & \frac{1}{8} \end{bmatrix}$

A6. Same as A3

A8. $\begin{bmatrix} \cos\theta & -\sin\theta \\ \sin\theta & \cos\theta \end{bmatrix}$

Appendix B

B1. $x_1 = 3.15, \quad x_2 = 0.62$

B2. $x_1 = 3.15, \quad x_2 = 0.62$

B3. $x_1 = 2.5, \quad x_2 = -1, \quad x_3 = 0.5$

B4. $x_1 = 3, \quad x_2 = -1, \quad x_3 = -2$

B5. **a.** $\begin{Bmatrix} x_1 \\ x_2 \end{Bmatrix} = \begin{bmatrix} 2 & -1 \\ 1 & -1 \end{bmatrix} \begin{Bmatrix} y_1 \\ y_2 \end{Bmatrix}$ **b.** $\begin{Bmatrix} z_1 \\ z_2 \end{Bmatrix} = \begin{bmatrix} -3 & 2 \\ 5 & -3 \end{bmatrix} \begin{Bmatrix} y_1 \\ y_2 \end{Bmatrix}$

B6. $x_1 = 0, \quad x_2 = 1, \quad x_3 = 2, \quad x_4 = 2, \quad x_5 = 0$

B7. $x_1 = 3.15, \quad x_2 = 0.62$

B8. **a.** Unique **b.** Nonexistent **c.** Unique **d.** Nonunique

Appendix D

D1. **a.** $f_{1y} = f_{2y} = -5$ kip, $m_1 = -m_2 = -100$ k-ft
b. $f_{1y} = f_{2y} = -5$ kip, $m_1 = -m_2 = -18.75$ k-ft
c. $f_{1y} = f_{2y} = -15$ kip, $m_1 = -m_2 = -75$ k-ft
d. $f_{1y} = -18.75$ kip, $f_{2y} = -6.25$ kip, $m_1 = -58.3$ k-ft, $m_2 = 33.3$ k-ft
e. $f_{1y} = -6$ kip, $f_{2y} = -14$ kip, $m_1 = -26.67$ k-ft, $m_2 = 40$ k-ft
f. $f_{1y} = -0.99$ kN, $f_{2y} = -4.0$ kN, $m_1 = -2.04$ kN · m, $m_2 = 5.10$ kN · m
g. $f_{1y} = f_{2y} = -6$ kN, $m_1 = -m_2 = -7.5$ kN · m
h. $f_{1y} = f_{2y} = -10$ kN, $m_1 = -m_2 = -6.67$ kN · m

Index